Introduction to Forestry Data Analysis with R

Introduction to Forestry Data Analysis with R equips students and practitioners with the skills needed to move confidently between field measurements and modern analytical workflows. As forestry, ecology, and natural resource management become increasingly data-driven, professionals are expected not only to collect information but also to organize, analyze, visualize, and defend quantitative results. This book responds to that shift by integrating foundational forest inventory concepts with practical computing in R.

Distinct from both generic programming texts and traditional mensuration references, this volume teaches R through real forestry datasets and operational examples. The first half develops core programming skills—data wrangling, visualization, and reproducible workflows—while the second half applies these tools to forest inventory, monitoring, and estimation. Classical methods developed by forest biometricians are presented alongside transparent, step-by-step computational implementations, enabling readers to connect statistical theory with modern, repeatable analysis.

Key Features:

- Introduces R and the tidyverse using forestry-specific datasets and management questions
- Develops reproducible workflows for data import, cleaning, transformation, visualization, and reporting
- Presents forest inventory concepts including simple random, systematic, stratified, cluster, and multi-stage sampling
- Implements classical estimators using script-based computing
- Integrates spatial data handling and mapping for areal sampling frames and field-based applications
- Emphasizes practical problem-solving, code organization, and analytical habits that scale from single stands to large inventories

Introduction to Forestry Data Analysis with R is intended for undergraduate and graduate students in forestry, natural resources, and environmental science, as well as practitioners seeking to modernize and streamline their analytical workflows. Whether readers are learning R for the first time or adapting established inventory methods to contemporary datasets, it provides a clear, practical, and reproducible foundation for data-driven forest analysis.

Andrew O. Finley is a Professor in the Department of Forestry and the Department of Statistics and Probability at Michigan State University, USA.

Jeffrey W. Doser is an Assistant Professor in the Department of Forestry and Environmental Resources at North Carolina State University, USA.

I0894150

Chapman & Hall/CRC The R Series

Series Editors

Recently Published Titles

Tidy Finance with R
Christoph Scheuch, Stefan Voigt, and Patrick Weiss

Deep Learning and Scientific Computing with R torch
Sigrid Keydana

Model-Based Clustering, Classification, and Density Estimation Using mclust in R
Lucca Scrucca, Chris Fraley, T. Brendan Murphy, and Adrian E. Raftery

Spatial Data Science: With Applications in R
Edzer Pebesma and Roger Bivand

Modern Data Visualization with R
Robert Kabacoff

Learn R: As a Language, Second Edition
Pedro J. Aphalo

Spatial Analysis in Geology Using R
Pedro M. Nogueira

Analyzing Baseball Data with R, Third Edition
Jim Albert, Benjamin S. Baumer, and Max Marchi

Geocomputation with R, Second Edition
Robin Lovelace, Jakub Nowosad, Jannes Muenchow

Microeconometrics with R
Yves Croissant

Statistical Inference via Data Science
A ModernDive into R and the Tidyverse, Second Edition
Chester Ismay, Albert Y. Kim, and Arturo Valdivia

Copula Additive Distributional Regression Using R
Giampiero Marra and Rosalba Radice

Introduction to Political Analysis in R
H. Whitt Kilburn

Displaying Time Series, Spatial, and Space-Time Data with R
Oscar Perpiñán Lamigueiro

Interactively exploring high-dimensional data and models in R
Dianne Cook and Ursula Laa

Simulation and Power Analysis Using R
Brandon LeBeau

Introduction to Forestry Data Analysis with R
Andrew O. Finley and Jeffrey W. Doser

For more information about this series, please visit: https://www.crcpress.com/Chapman--HallCRC-The-R-Series/book-series/CRCTHERSER

Introduction to Forestry Data Analysis with R

Andrew O. Finley and Jeffrey W. Doser

CRC Press
Taylor & Francis Group
Boca Raton London New York

CRC Press is an imprint of the
Taylor & Francis Group, an **informa** business
A CHAPMAN & HALL BOOK

Designed cover image: Lauren Noel

First edition published 2027
by CRC Press
2385 NW Executive Center Drive, Suite 320, Boca Raton, FL 33431

and by CRC Press
4 Park Square, Milton Park, Abingdon, Oxon, OX14 4RN

CRC Press is an imprint of Taylor & Francis Group, LLC

ISBN: 978-1-032-12028-7 (hbk)
ISBN: 978-1-032-12027-0 (pbk)
ISBN: 978-1-003-22269-9 (ebk)

DOI: 10.1201/9781003222699

Typeset in Latin Modern font
by KnowledgeWorks Global Ltd.

Publisher's note: This book has been prepared from camera-ready copy provided by the authors.

Contents

List of figures

List of tables

Preface

The role of the forester and natural resource professional has expanded. Data, analysis, and interpretation are now more central to our work than ever before, and today's practitioners are expected to move confidently between field measurements and analytical workflows. With unprecedented volumes of information on forested systems produced through inventories, sensor networks, remote sensing platforms, and long-term monitoring programs, practitioners are increasingly called upon to act as data analysts—wrangling, visualizing, and defending quantitative results in support of sound management decisions.

This book is for readers on two paths: students beginning their analytical training and practitioners who want to modernize and streamline their workflows. Our goal is to help you build proficiency in R and the tidyverse ecosystem, and to connect familiar estimators and inventory principles to clear, repeatable, and reproducible data analysis pipelines. Whether you're new to these tools or looking for more efficient ways to implement classical methods, the emphasis is on transparent workflows you can adapt to common estimation settings.

Chapters 1 through 9 form an introduction to practical computing with R. We start from base R—objects, vectors, data frames, indexing, functions, and input/output—and build toward pipe-based workflows. Along the way, we develop data wrangling with `dplyr` (verbs and joins), reshaping with `tidyr`, iteration, small reusable functions, visualization with `ggplot2`, and reproducible project practices that scale from a single stand to large inventories. The aim is pragmatic: readable code, consistent structure, and routines that make forest data easier to organize, verify, summarize, and explain.

Chapters 10 through 18 turn to forest inventory and estimation. Here we build directly on the computing foundation established earlier and introduce and apply well-established statistical tools—simple random sampling, systematic designs, stratification, cluster and multistage sampling, and ratio and regression estimators. The mathematical foundations of these methods were developed by generations of forest biometricians and applied statisticians whose work continues to guide the profession. Our contribution is to bring these classical ideas into a modern workflow: clear illustrations, transparent calculation paths, and reproducible implementations that make it easier to move from field measurements to defensible estimates and uncertainty statements.

Companion datasets, scripts, and other book resources are available on the book's website at https://ifdar-book.org/.

This site provides the materials used throughout the chapters, along with updates, errata, and supplemental content as the book evolves. We encourage you to download these resources and follow along with the examples and exercises.

Our hope is that this book helps you develop an adaptable codebase and a reliable, efficient way of working—a workflow you can reuse, audit, and share. By the end, you should be able to take a raw inventory dataset, turn it into analysis-ready tables and figures, compute estimates and their associated uncertainty, and communicate results in a way that supports defensible decisions as a forester, analyst, and steward of forested landscapes.

Andrew O. Finley
East Lansing, Michigan

Jeffrey W. Doser
Raleigh, North Carolina

Acknowledgments

During the long process of writing this book, we benefited greatly from the generous reviews, technical comments, contributed material, and encouragement of Sudipto Banerjee, Anthony D'Amato, Ken Desmarais, Mike Eckley, Lutz Fehrmann, Anton Grafström, Ed Green, Juha Heikkinen, Kim Iles, Malcolm Itter, Annika Kangas, Braeden Klaty, Rayleigh Lei, David MacFarlane, John Paul McTague, Vincent Melfi, Arne Nothdurft, David Orwig, Jarred Saralecos, Elliot Shannon, Göran Ståhl, Tim Vredenburg, Aaron Weiskittel, and Grayson White. We are also grateful to the Cow Creek Band of Umpqua Tribe of Indians for sharing forest inventory data that form the basis of several exercises and examples in this book.

Dedication

A.O.F. For Sarah, Ava, Oliver, and Callum. I also dedicate this book to my father, James C. Finley, whose work in forestry and stewardship inspired my own.

J.W.D. For Gabriela and Lewis, and for my parents, James and Elizabeth.

About the authors

Andrew O. Finley is a Professor in the Department of Forestry and the Department of Statistics and Probability at Michigan State University, USA.

Jeffrey W. Doser is an Assistant Professor in the Department of Forestry and Environmental Resources at North Carolina State University, USA.

List of acronyms

AGB—Aboveground biomass
AGS—Acceptable growing stock
ALS—Aerial laser scanning
BA—Basal area
BAF—Basal area factor
CHM—Canopy height model
CMS—Compartment Management Study
CO_2—Carbon dioxide
CRAN—Comprehensive R Archive Network
CV—Coefficient of variation
DBH—Diameter at breast height
EAB—Emerald ash borer
EDA—Exploratory data analysis
FACE—Free-Air Carbon Dioxide Enrichment
FEF—Fernow Experimental Forest
FPC—Finite population correction
G-LiHT—Goddard's LiDAR, Hyperspectral, and Thermal Imager
GIS—Geographic Information Systems
GPS—Global Positioning System
GSV—Growing stock volume
HDM—Horizontal distance multiplier
HWA—Hemlock woolly adelgid
IDE—Integrated development environment
IFDAR—Introduction to Forestry Data Analysis with R
LiDAR—Light Detection and Ranging
LPM—Local pivotal method
MID—Management Intensity Demonstration
MRV—Monitoring, reporting, and verification
MU—Management unit
NFI—National Forest Inventory
O_3—Ozone
PEF—Penobscot Experimental Forest
PLS—Personal laser scanning
PSP—Permanent sample plot
PSU—Primary sampling unit
REDD—Reducing Emissions from Deforestation and Forest Degradation
SSU—Secondary sampling unit

TF—Tree factor
UNFCCC—United Nations Framework Convention on Climate Change
USFS—United States Forest Service
XML—Extensible Markup Language

1

Overview and motivating data

1.1 Aim and scope

We wrote this book with two aims in mind. The first is to introduce programming to those working with forestry and ecology data. The second is to guide and illustrate the implementation of fundamental forestry data analysis techniques using a contemporary and powerful programming language.

We recognize that understanding and responding to current environmental and management challenges requires strong quantitative and analytical skills. Increasingly, forestry and ecology professionals assume the role of data analyst due to unprecedented environmental data availability and the questions these data are used to answer. Following a definition given by Bravo et al. (2016), a *data analyst* has the ability to make appropriate calculations, convert data to graphical representations, interpret the information presented in graphical or mathematical forms, and make judgments or draw conclusions based on quantitative data analysis. In addition to these skills, much of an analyst's time is devoted to basic *data wrangling*, which is the task of transforming data from one format into another to make it more suitable for analysis. Some estimate that upwards of 50% to 80% of an analyst's time is spent data wrangling (Lohr 2014).

The analyst's job is best tackled using a flexible and easily understood programming language with high-level functionality—that is, lots of ready-made functions for routine data wrangling, analysis, visualization, and reporting tasks. While there are several high-level language contenders meeting these general criteria, currently R (R Core Team 2024) and Python (Python Core Team 2015) dominate the arena. Both are *open source*, meaning the code is freely available and can be freely redistributed and modified. One language isn't inherently better than the other—the choice depends on your use case and the questions you're trying to answer. While Python is often touted as a general-purpose language with an easy-to-understand syntax[1], the same could be said for R. Relative to Python, R's origin and evolution favored interactive data analysis, which gives it an edge for the type of work pursued in this book. The focus here is to provide an introduction to R, which has become the more prevalent of these two languages in the environmental sciences.

There are of course numerous excellent books and online resources for learning R, several of which we recommend in subsequent chapters to supplement and extend the topics covered.

[1]Syntax is the set of rules that defines combinations of symbols to structure statements or expressions in a computer language.

DOI: 10.1201/9781003222699-1

In fact, many of these books would serve well as your introduction to R. Our personal experience, however, is that it's easier and more enjoyable to learn challenging and abstract topics when their relevance is tied to your field of interest using real-world examples. To this end, we motivate programming topics using datasets collected to answer specific forestry and environmental science questions.

After working through the fundamentals of programming using R in the first half of this book, our attention turns more fully to the analysis of data collected for forest inventory and monitoring efforts. Following a definition given by Kershaw et al. (2016), *forest inventory* is the procedure for obtaining current information on the quantity, quality, and condition of the forest resource, associated vegetation and composition, and many of the characteristics of the land area on which the forest is located. *Forest monitoring* quantifies change in the composition, structure, size, and health of the forest over an extended time period.

Forest mensuration describes the tools and procedures for measuring forest characteristics, along with the statistical methodologies used to generalize those measurements. There are several contemporary forest mensuration books that offer a comprehensive treatment of forest measurement techniques and associated statistical methods for estimating a wide range of forest characteristics. Several books in particular, Laar and Akça (2007), Kershaw et al. (2016), and Burkhart et al. (2018) provide an excellent foundation for applied forest mensuration with a focus squarely on providing quantitative information to support silviculture and management decisions. We mention these books because they provide an introductory-level background and motivation for the practical programming and applied data analysis covered in this book. Because our focus is on implementing quantitative methods, this book also pairs nicely with material that takes a more statistical perspective, for example, forest sampling (Kangas and Maltamo 2006; Köhl et al. 2006; Gregoire and Valentine 2008; Mandallaz 2007), growth and yield modeling (Weiskittel et al. 2011), and advanced environmental data analysis topics using R (Robinson and Hamann 2011; Mehtätalo and Lappi 2020; Green et al. 2020).

1.2 Motivating forestry datasets

We are in a data-rich era that provides extraordinary opportunities to understand complex ecological processes at local to global scales. The proliferation of environmental data is the result of sustained investments in data collection for regulatory, monitoring, and resource management objectives, as well as technological advances in spatially enabled networks and geospatial information storage, analysis, and distribution systems. The sources generating this unprecedented volume of data are increasingly diverse and specialized, including monitoring station instruments, remotely located sensors, and georeferenced field measurements.

By automating time-consuming, tedious, and repetitive data wrangling tasks, computing environments, such as R, allow the analyst to focus on answering the questions for which the data were collected. In forestry, data are collected to answer a wide variety of questions.

For example, field crews collect forest inventory data to answer questions about the amount and location of timber or non-timber resources. Monitoring data are collected to understand changes in forest characteristics. Highly detailed individual tree data are collected to understand *allometry*, which is the growth and size relationship between different parts of an organism. Experimental manipulations of trees, stands, or forests are used to better understand how environmental change and disturbance events impact individual growth rates and trends in population demographics.

The next several sections give a high-level tour of forestry datasets collected to answer a range of questions central to traditional forestry operations, silviculture, and management, as well as broader research questions related to climate and environmental change. When an analyst receives a dataset, the first task is to understand its structure and identify features that will matter for subsequent wrangling, modeling, and analysis.

We first consider datasets with fairly simple structure, a single *flat file*, meaning the data are organized in rows and columns, typically with rows corresponding to observational units (things we measure, e.g., trees) and columns corresponding to the various measurements and characteristics recorded for each unit. For the most part, in these examples we have left the datasets in their original form and have done minimal cleaning and wrangling. For example, you'll notice the datasets use different measurement systems and different approaches for organizing information. We explore these and more complex datasets throughout the remainder of the book to motivate and illustrate programming tools for data cleaning, wrangling, summarization, and analysis.

We provide the following datasets, as well as others used in this book, as supplementary material.

1.2.1 Fernow Experimental Forest biomass dataset

Initially, we might have in mind a modest-sized and uncomplicated dataset that serves a fairly specific purpose. For example, in forestry it's convenient to have a mathematical formula that relates an easy-to-measure tree or stand characteristic to a more difficult-to-measure characteristic. These mathematical formulas, referred to as *allometric equations*, are a common feature in forestry tasks. For the practicing forester, these formulas are often encountered when computing stem volume or biomass using diameter at breast height (DBH; 4.5 feet or 1.37 meters from the ground) and perhaps species—in this case, DBH is the easily obtained measurement and volume or biomass is the more difficult-to-measure characteristic.

Creating an allometric equation typically requires fitting a regression model that relates the difficult-to-measure characteristic to one or more easily measured characteristics. For example, one would likely need a dataset that contains measurements of both DBH and stem biomass on a collection of trees to fit a model using regression techniques and generate a general allometric equation for this relationship. Clearly, such a dataset would be quite expensive and time-consuming to collect because it would entail harvesting trees to directly measure their biomass. Many such valuable datasets exist in the forestry literature; see, e.g., Jenkins et al. (2002) and Radtke et al. (2015).

TABLE 1.1: Five of the 88 trees in the Fernow Experimental Forest biomass dataset.

Species	DBH (in)	All woody dry (kg)	Stem dry (kg)
Liriodendron tulipifera	7.8	129.5	80.5
Prunus serotina	2.4	6.3	4.9
Acer rubrum	6.0	105.1	54.7
Prunus serotina	6.7	118.3	72.4
Prunus serotina	3.4	20.1	14.0

Table 1.1 contains data for five of the 88 trees felled and measured as part of a United States Forest Service (USFS) study on the Fernow Experimental Forest (FEF), located in the central Appalachians of West Virginia (Wood et al. 2016). The data include DBH, tree height, and green and dry weight of the tree stem, top, small branches, large branches, and leaves. Table 1.1 provides a subset of these data, where each row holds the measurements for one tree. Oddly, these data include a mix of the English system, i.e., DBH in inches, and International System of Units commonly known as the metric system, i.e., weights in kilograms. These data, like many Forest Service datasets collected on experimental forests, are well documented and include detailed *metadata*.[2] Metadata are data that provide information about other data, i.e., data about data, and typically describe data collection protocol, measurement units, and file organization.

Figure 1.1 shows the relationship between DBH and dry weight of the total tree weight (All woody), stem, all large branches, and all small branches, separated by tree species. The points in this figure are the 88 tree measurements, and blue best fit lines are from a statistical model that characterizes the relationship between DBH and component weight.[3] Given DBH and species of a *new* tree, its biomass components are estimated using the mathematical formula represented by the blue lines. We'll encounter these data again in Chapter 9 when we learn how to make informative graphics for effectively communicating data and analysis results.

1.2.2 Elk County timber cruise dataset

A *timber cruise* is a forest inventory to locate and estimate the quantity of timber in a given area according to species, size, quality, potential products, and other characteristics. We'll see later that many decisions go into designing an inventory. These decisions include the number and selection of measurement units (i.e., locations in the forest where trees are measured), which trees to measure around a given location, and what type of measurements to take on the selected trees.

Figure 1.2 shows the resulting dataset from a timber cruise of a 271 acre forested property in Elk County, Pennsylvania. The cruise data comprise 54 points where a forester measured

[2]The FEF metadata are available at https://www.fs.usda.gov/rds/archive/Catalog/RDS-2016-0016.

[3]To be consistent with tree weight measurements, we changed the DBH measurement units from English to the metric system.

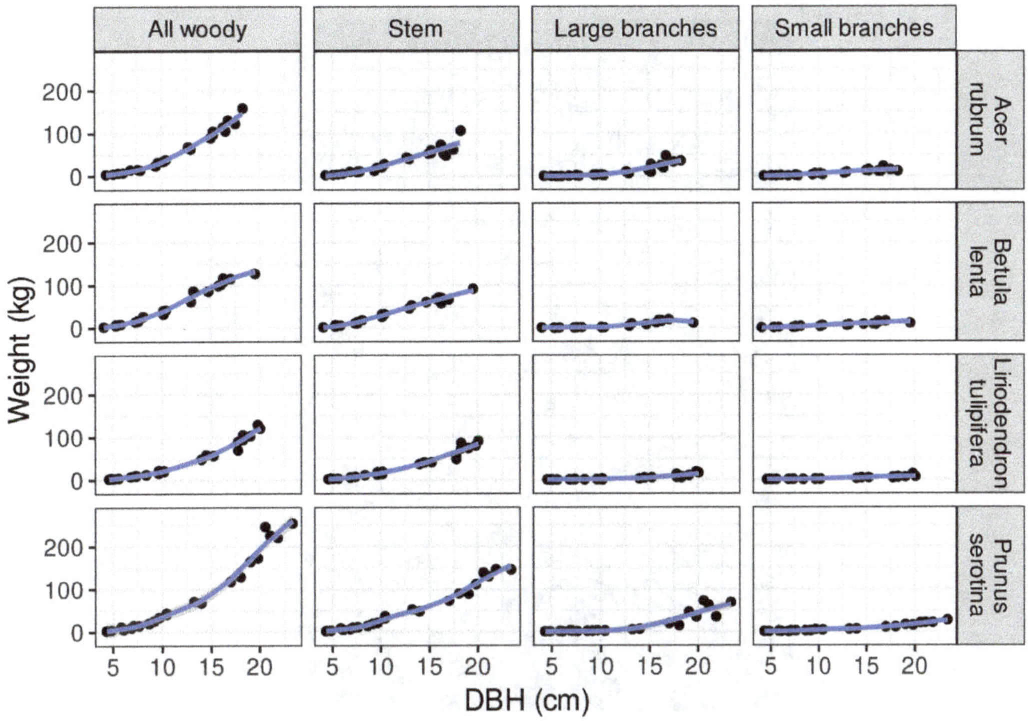

FIGURE 1.1: Relationship between tree DBH and component weight by species using the Fernow Experimental Forest biomass dataset. Points represent data observed on 88 trees, and blue lines are best fit lines through the observations.

characteristics such as species, DBH, height, and stock quality potential for trees around the point selected for measurement based on a certain rule (Chapter 12). Points were located on a systematic 7-by-7 chain (1 chain equals 66 feet) grid, resulting in about 1 point per 5 acres. We colored point locations based on the basal area (ft^2/ac) for trees recorded as acceptable growing stock (AGS). The dotted orange line shows the transect, or path, the forester walked between points.

Table 1.2 shows a subset of trees measured on the first and last point, where each row corresponds to a set of characteristics shared by one or more trees. Information in the first row of this table tells us inventory point 1 has one black cherry with a DBH of 12 (in) and

TABLE 1.2: A subset of trees from the first and last cruise point on the Elk County property.

Point ID	Trees	Species	DBH (in)	Height (logs)	Product
1	1	Black Cherry	12	1	AGS Sawtimber
1	1	Black Cherry	10	1.5	AGS Pulpwood
54	1	Beech	30	1	Snag
54	4	Beech	NA	NA	Regen

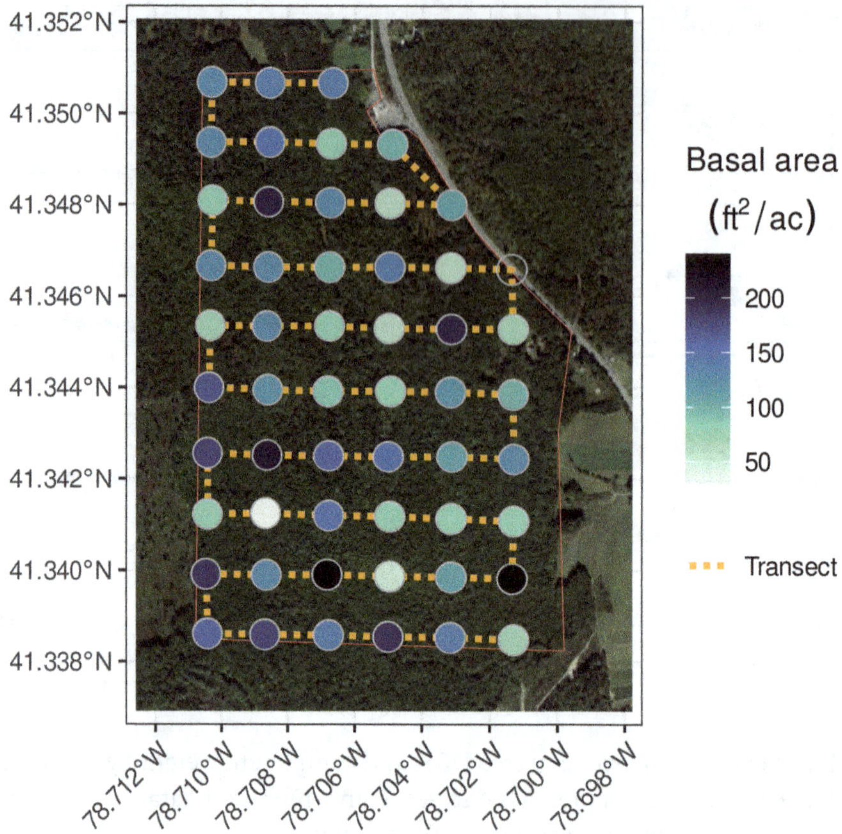

FIGURE 1.2: Location of timber cruise points on a 271 acre forested property in Elk County, Pennsylvania. Locations colored by acceptable growing stock basal area.

one 16 foot log that is designated as AGS sawtimber. Obviously, one would need to consult the cruise metadata to figure out the DBH and height units as well as the stem qualities needed for an AGS designation. The last row in Table 1.2 records four beech stems at point 54 that were considered regeneration. In such cruise data, regeneration typically refers to stems that are quite small with little or no DBH; hence, the NA values for DBH and height are expected. As we'll see later, NA values can indicate data missing for other reasons.

We'll revisit these data when we look at efficient ways to generate property-level tables and figures of key forest characteristics summarized by species and product type (Chapter 8). For consulting foresters, who collect and analyze these data routinely, canned programs that automate wrangling, summarizing, and reporting tasks are particularly valuable.

1.2.3 Free-Air Carbon Dioxide Enrichment experiment dataset

We often encounter data from highly structured and complex experiments. Such data typically present challenges in organization/storage, exploratory data analysis (EDA), statistical analysis, and interpretation of analysis results. An example dataset comes from the Aspen Free-Air Carbon Dioxide Enrichment (FACE) experiment conducted from 1997 to

TABLE 1.3: A subset of the Free-Air Carbon Dioxide Enrichment experiment dataset.

	Treat	Clone	Diam. 1997	Diam. 1998	Diam. 1999
1	Control	8L	0.4	1.34	2.52
2	Control	216	0.9	3.11	5.19
3	Control	8L	0.25	0.96	1.71
4	Control	216	0.78	3.03	5.03
5	Control	216	0.75	1.92	3.47
1017	Elevated O3	8L	0.6	2.01	3.12
1860	Elevated CO2+O3	259	0.68	1.22	1.54
679	Control	8L	NA	0.42	0.43
129	Control	42E	0.26	0.54	1.25
930	Elevated CO2	271	0.75	2.41	4.21

2009 on the Rhinelander Experimental Forest West Unit (US DOE 2020). The Aspen FACE Experiment was a multidisciplinary study to assess the effects of increasing tropospheric ozone (O_3) and carbon dioxide (CO_2) concentrations on the structure and functioning of northern forest ecosystems. The design allowed researchers to test the effects of these gases alone (and in combination) on many ecosystem attributes, including growth, leaf development, root characteristics, and soil carbon. The dataset considered here comprises tree diameter measurements at 10 (cm) above the ground from 1997 to 2008 for *Populus tremuloides* grown inside twelve 30 (m) diameter treatment tents with open tops and bottoms in which the concentrations of tropospheric O_3 and CO_2 were controlled (Kubiske 2013). Because there was no confinement, there was no significant change in the natural, ambient environment other than elevating these trace gas concentrations (Hendrey et al. 1999). Although individual tree measurements are similar to those in the FEF dataset in Section 1.2.1 (i.e., height and diameter), the study design specifies various tree genetics, varying gas treatments, and treatment replicates. These data are an example of longitudinal (time series) data, meaning researchers repeatedly measured individuals (trees in this case) over time. Table 1.3 provides a subset of measurements for the first five trees as well as five trees selected at random in the dataset. Here, a row identifies each tree's experimental treatment, genetic description (Clone), and diameter growth over time.

Notice the 1997 diameter measurements in Table 1.3 contain an NA, which indicates missing data—you'd see a lot more missing data if all the data were displayed. These missing observations resulted from tree mortality or data recording error. When analyzing these data to identify treatment effects, it's essential to make a decision on how to handle such missing data.

Figure 1.3 shows an EDA plot of the treatment effects compared with the control (i.e., no gases added). This figure suggests a possible positive growth response to elevated CO_2 and negative growth response to elevated O_3, compared with the control. This is consistent with other studies on atmospheric CO_2 fertilization, as well as studies on the damaging

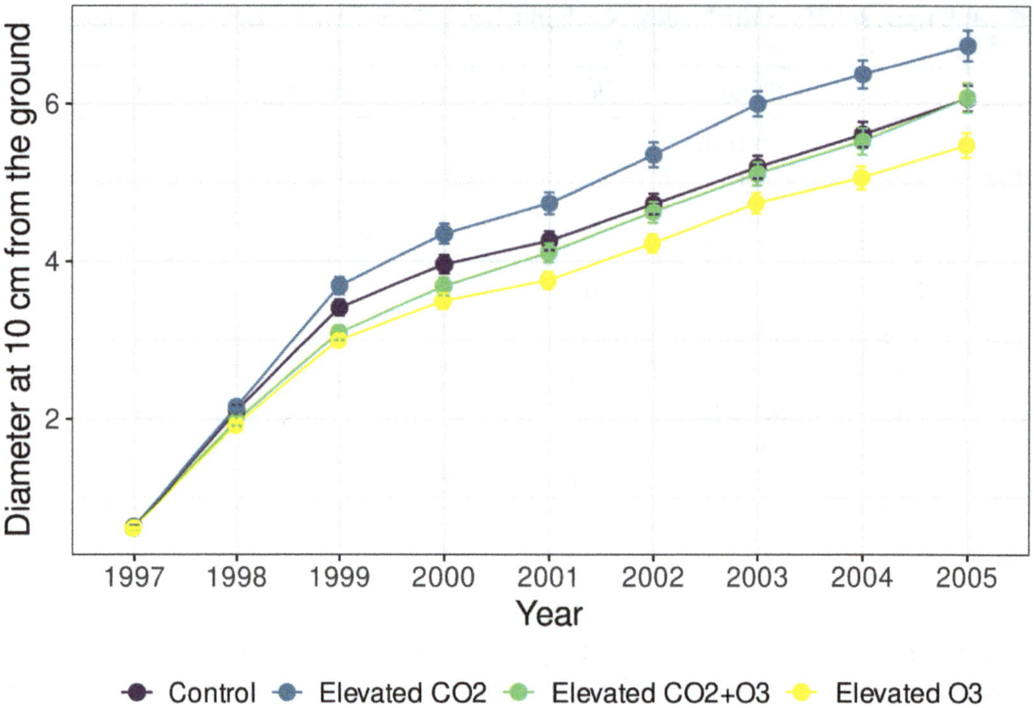

FIGURE 1.3: Free-Air Carbon Dioxide Enrichment experiment aspen diameter growth by treatment. Bars on each point indicate 95% confidence intervals.

effect of O_3 on leaf tissue. The figure also shows that experimental units that receive both elevated CO_2 and O_3 do not differ substantially from the control, which could suggest that any CO_2 fertilization is offset by O_3 damage.

Subsequent chapters focus on creating these and other EDA plots and summaries (Chapters 7 and 9), approaches to communicating uncertainty in analysis results (Chapter 11), and documenting data processing and analysis steps to promote open science, reproducibility, and transparency (Chapter 3; Powers and Hampton 2019).

1.2.4 Penobscot Experimental Forest inventory and LiDAR dataset

Coupling forest inventory with remotely sensed datasets using regression models offers an attractive approach to mapping forest characteristics at stand, regional, continental, and global scales. Light Detection and Ranging (LiDAR) data have shown great potential for use in estimating spatially explicit forest variables over a range of geographic scales (Næsset 2011; Babcock et al. 2013; Neigh et al. 2013; Finley et al. 2024; Nothdurft et al. 2025). Encouraging results from these and many other studies have spurred massive investment in LiDAR sensors as well as extensive campaigns to collect field-based calibration data.

Much of the interest in LiDAR-based forest mapping is to support carbon monitoring, reporting, and verification (MRV) initiatives, such as those defined by the United Nations

Programme on Reducing Emissions from Deforestation and Forest Degradation (UN-REDD 2008) and Paris Agreement (UNFCCC 2015). In these, and similar initiatives, aboveground biomass (AGB) is a primary forest characteristic of interest because it provides a nearly direct measure of forest carbon, i.e., carbon comprises ~47% of wood biomass (Doraisami et al. 2022). Most efforts to quantify and/or manage forest ecosystem services (e.g., carbon, biodiversity, water) seek high spatial resolution wall-to-wall data products such as gridded maps with associated measures of uncertainty. In fact the international initiatives noted above include language concerning the level of spatially explicit acceptable error in total forest carbon estimates.

A long-term monitoring effort on the Penobscot Experimental Forest[4] (PEF) in Bradley and Eddington, Maine, provides valuable data for assessing methods to quantify forest biomass change over time. The dataset comprises LiDAR-based canopy height estimates from the Goddard's LiDAR, Hyperspectral and Thermal Imager[5] (G-LiHT) collected in summer 2017, along with AGB measurements from a set of georeferenced forest inventory plots.

The G-LiHT LiDAR data were used to generate a canopy height model (CHM) at fine spatial resolution across the PEF. A CHM is a raster data layer that represents the height of the forest canopy above the ground surface at each location. We obtained the forest inventory data associated with each plot from the PEF's database of several ongoing, long-term silvicultural experiments described by Kenefic et al. (2025). These experiments are conducted within a series of management units (MUs) that partition the PEF.

Like the FACE data described in Section 1.2.3, the PEF data are longitudinal data that comprise repeated measurements on trees within permanent sample plots (PSPs). These periodic remeasurements are part of several PEF experiments designed to improve understanding of forest response to silvicultural treatments. Within each MU, plot remeasurements are indexed by consecutive inventory numbers. These inventory numbers are used instead of dates because sometimes multiple inventories occur within the same year and month. Table 1.4 shows 10 trees selected at random from the 316,837 tree measurements in the PEF dataset. Following the column labels, the first row describes a 0.7 (in) DBH tree that weighs 0.7 (lbs) measured on plot 41 in management unit 22 in inventory 7 conducted in 6/1988.

Figure 1.4 illustrates how plot-level AGB changed over time within a subset of MUs. As part of the experimental design, some MUs were not harvested (MU 10 and 32B), others were prescribed periodic light harvests (MU 12 and 20), and some received single heavy harvests (MU 15, 22, and 8). From an analyst's perspective, this figure is helpful for revealing patterns in growth, harvest, and when measurements were collected over time, e.g., it appears that plots were added in the mid-1990s to MU 32A and 32B.

Figure 1.5 shows the PEF MU boundaries and plot locations, colored by their most recent basal area measurements (ft^2/ac).

---

[4]https://research.fs.usda.gov/nrs/forestsandranges/locations/penobscot/

[5]https://gliht.gsfc.nasa.gov/

TABLE 1.4: A small selection of trees pulled from the Penobscot Experimental Forest inventory dataset.

MU	Plot	Inventory	Month	Year	Species code	DBH (in)	AGB (lbs)
22	41	7	6	1988	1	0.7	0.7
12	33	12	10	1994	1	2.4	15.5
28	42	10	7	1996	6	12.8	983.8
16	53	19	7	2002	6	6.4	176.2
27	63	10	8	1996	1	1.1	2.2
20	44	7	6	1977	15	0.9	2.3
90	12	22	6	2005	1	1.6	5.6
24	41	10	7	1995	6	1.4	4.1
29B	12	11	8	1997	4	0.7	1.1
20	22	9	6	1986	4	13.2	997.4

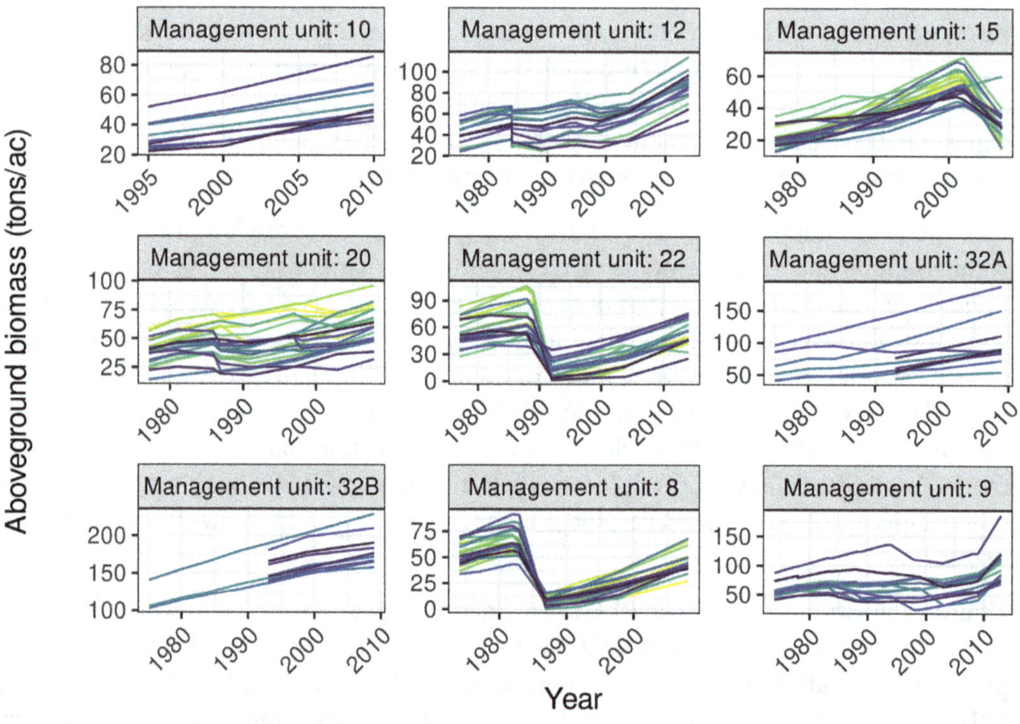

FIGURE 1.4: Change in plot-level aboveground biomass (AGB) within a subset of the PEF management units. Each line traces a permanent sample plot's AGB over time.

We'll revisit these data when practicing data aggregation approaches, e.g., how to move from tree-level measurements to plot-, MU-, and ultimately forest-level summaries (Chapter 12). We'll also exploit the longitudinal nature of these data (i.e., repeated measures over time) to explore monitoring methods designed to quantify change in forest characteristics. Finally, these data are particularly valuable for learning how remotely sensed data, such as LiDAR,

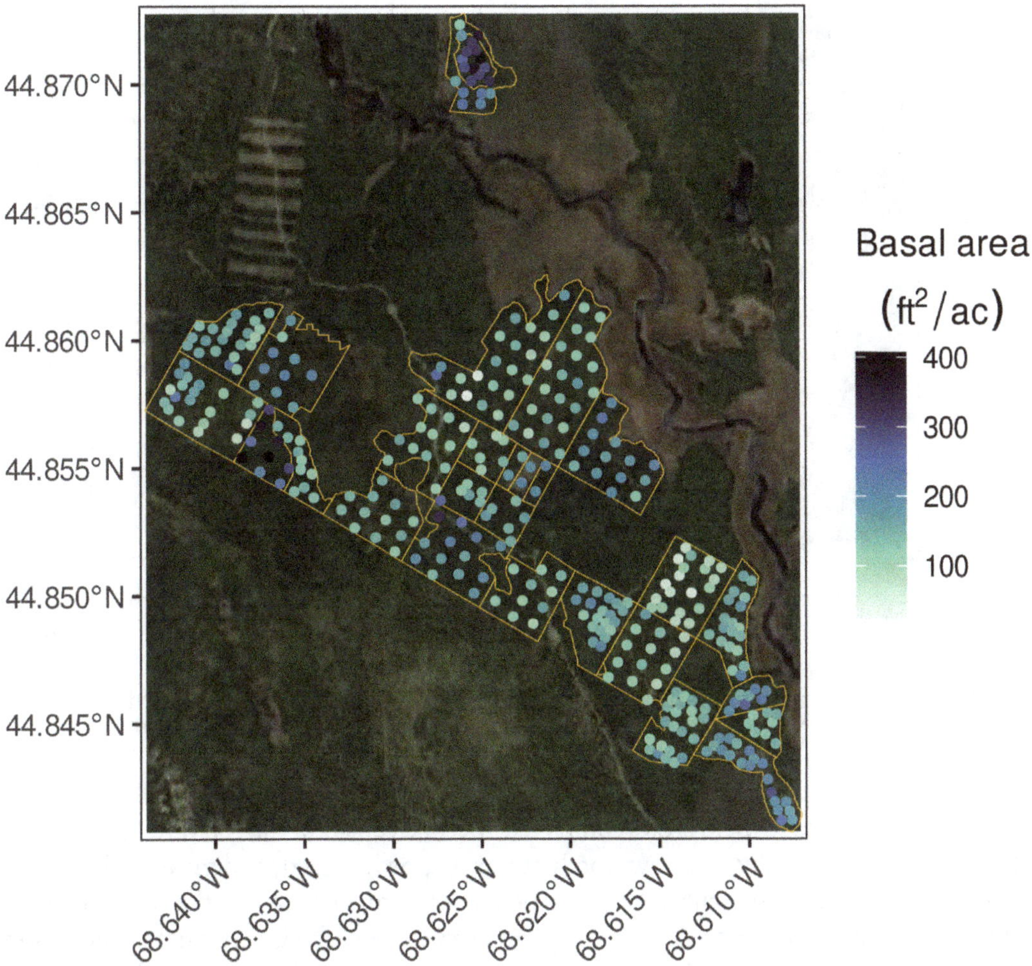

FIGURE 1.5: Penobscot Experimental Forest management unit boundaries and forest inventory plot locations colored by basal area.

can help inform forest inventory efforts and produce high-resolution, spatially-explicit dynamic maps.

1.2.5 Harvard Forest tree census dataset

In 2014, the Harvard Forest[6], located in Petersham, MA, in cooperation with the Smithsonian/CTFS Forest Global Earth Observatory[7] (ForestGEO), completed an initial mapping of a 35 ha (85 ac) forest dynamics monitoring plot. Every tree over 1 (cm) DBH in the plot was geolocated, tagged, and measured—83,801 trees in total. The Harvard Forest dataset is a *census*, or a complete enumeration of all units (e.g., trees) in a population. Such datasets are often referred to as *stem maps* because each tree is mapped using spatial coordinates.

[6]https://harvardforest.fas.harvard.edu/

[7]https://forestgeo.si.edu/

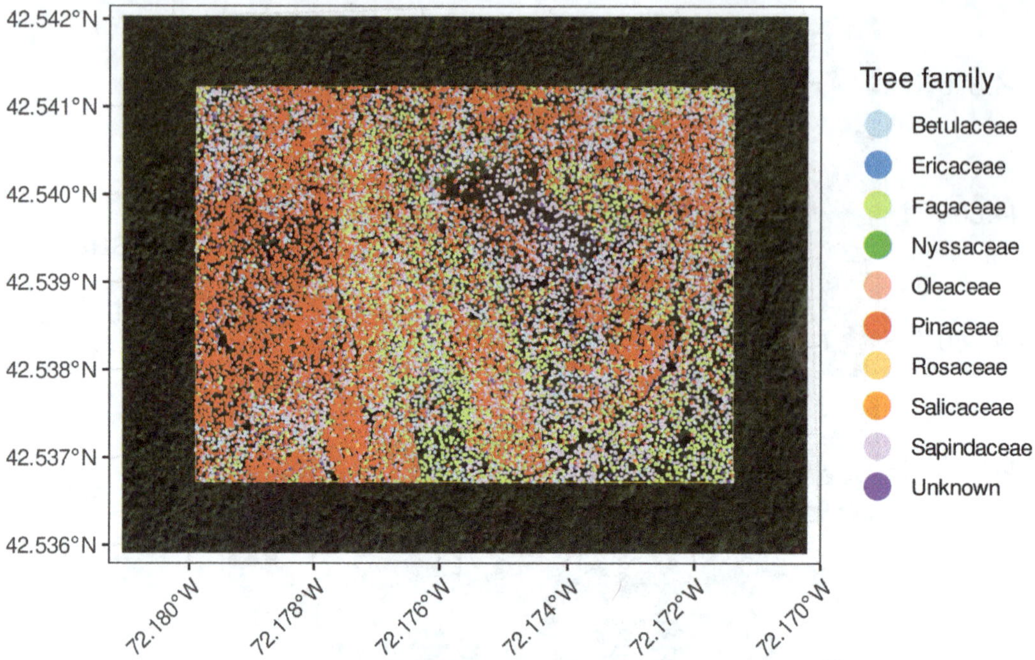

FIGURE 1.6: Stem map of all trees larger than 12.7 cm from the first Harvard Forest tree census.

Toward the end of 2020, field crews completed a second census of the Harvard Forest Global Earth Observatory, with 6,992 new woody stems mapped, tagged, and measured. Many of the longest-running research experiments on the Harvard Forest are in the ForestGEO plot footprint (Orwig et al. 2015). These closely monitored trees, along with numerous sensors the Harvard Forest has installed to track forest canopy gas exchange, provide invaluable information about tree- to forest-level physiology.

A census of trees within a large area is a particularly rare and valuable resource for learning about and testing different forest mensuration methods. We often want to use measurements from a small number of trees to learn about forest-wide characteristics. A census allows us to test such methods by comparing estimates based on a subset of the data to the true values for all trees in the census. Hence, we'll use the Harvard Forest tree census to explore the advantages and disadvantages of different sampling and statistical methods for forest inventory and monitoring.

Figure 1.6 shows the location of all trees greater than 12.7 (cm) measured in the first ForestGEO plot census with locations colored by tree family.

1.3 How to learn: The most important section in this book!

There are several ways to engage with the content in this book. One way is not to engage at all. Leave the book closed on your shelf and do something else with your time. That might or might not be a good strategy, depending on what else you do with your time, but you won't learn much from this book!

Another way is to read the book "passively" in a comfortable chair away from your computer. With this strategy you'll probably learn more than if you leave the book closed.

A third option is to read the book while you're at your computer, enter the R commands as you read about them, and work through the chapter exercises. You'll likely learn more this way.

A fourth strategy is even better. In addition to reading, entering the commands, and working through the practice exercises, you think about what you're doing, and ask yourself questions (which you then go on to answer). For example, after working through some R code that computes the logarithm of positive numbers, you might ask yourself, "What would R do if I asked it to calculate the logarithm of a negative number? What would R do if I asked it to calculate the logarithm of a really large number, such as one trillion?" You can explore these questions easily by just trying things out in the R console. As you read, you'll notice we occasionally use R functions with little or no explanation. In such cases, an "active" learning strategy is to use the help resources we suggest, read the function's manual page, and then test its behavior with your own example data.

If your goal is to maximize the time you have to binge-watch a series on your favorite streaming service, the first strategy might be optimal. But if your goal is to learn a lot about computing and analysis that will form a sound foundation upon which to build a new skill set or even career, the fourth strategy is probably going to be best.

2

Introduction to R and RStudio

As briefly discussed in Section 1.1, there are a variety of high-level programming languages commonly encountered in practice. Each has strengths and weaknesses, and it's not uncommon for two or more to be used in a single large project. This book focuses on R for several reasons.

1. R is free.
2. It's one of the most widely used software platforms in the environmental sciences.
3. R is under constant *open source* development by an expert core group. Open source means that anyone can inspect, modify, and enhance the code.[1]
4. It has an incredible variety of contributed packages.
5. A new user can (relatively) quickly gain enough skills to obtain, manage, and analyze data.

R has several enhanced interfaces, which are referred to as *integrated development environments* (IDEs). These interfaces simplify and facilitate software development. At minimum, an IDE consists of a source code editor and build automation tools. We'll use the RStudio IDE, which according to its developers "is a powerful productive user interface for R" (Posit team 2025). RStudio is well documented, under continuous development, and makes learning and using R more enjoyable.[2] Although we'll use RStudio, we can accomplish most of what we cover in this book in R (without an added interface) with few if any changes.

2.1 Installing R and RStudio

It's easy to install R on computers running Microsoft Windows, macOS, or Linux. R is available for download via the Comprehensive R Archive Network (CRAN) at https://cran.r-project.org/. You must install R prior to installing RStudio.

RStudio is also easy to install and is found at https://posit.co/downloads/ (here too, you'll find detailed and up-to-date instructions for downloading and installing R and RStudio).

[1]See more about the open source license at https://www.r-project.org/about.html.

[2]The company Posit developed and maintains RStudio. Posit was originally called RStudio, but the company changed its name to Posit in 2022 to reflect its development of data science products beyond RStudio. See the Posit website for more details (https://posit.co/).

DOI: 10.1201/9781003222699-2

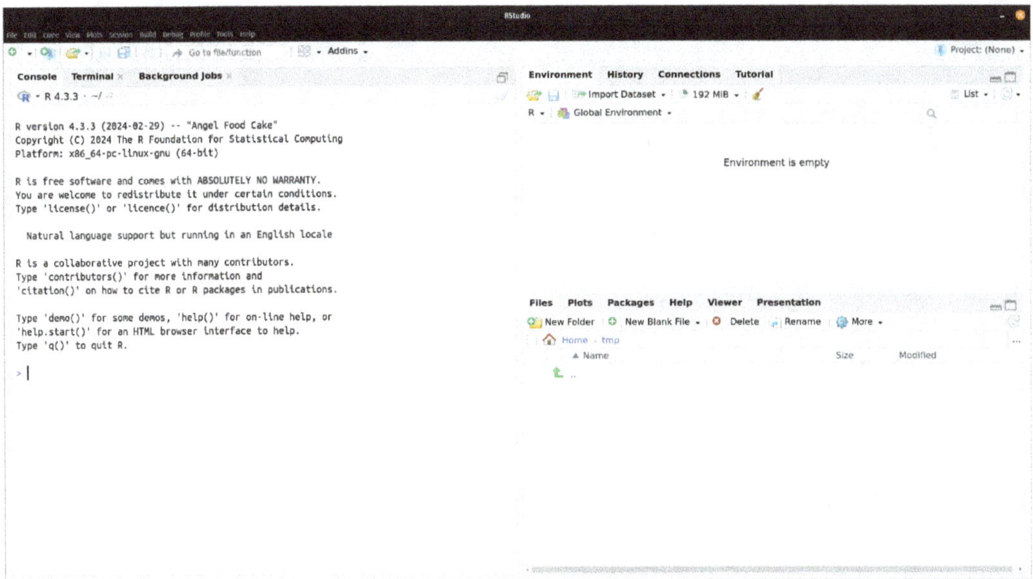

FIGURE 2.1: The RStudio IDE.

Start RStudio as you would any other program. For example, Microsoft Windows users will find it under the Start Menu or desktop shortcut (if the installation created a shortcut). Figure 2.1 displays a view of RStudio.

Initially the RStudio window contains three smaller windows. For now our main focus is on the large window on the left. This is the console window and it's where you type R commands. The next section provides a brief applied tour of RStudio using small non-complex data. Later in the book we'll work with larger and more complex datasets. To get more comfortable with RStudio, follow along with these sections at your computer, enter the commands, experiment with different commands, and explore.

2.1.1 First session

The command prompt is the greater than sign > and is located in the console window in Figure 2.1. Evaluate (or "run") code added after the command prompt by pressing the Enter key. Below, and throughout the book, code appears in a gray block and output (i.e., result) follows on subsequent lines and to the right of `#>`. The `#` appearing in the code blocks is the comment character, everything to the right of this character is ignored. Comments are used to add short explanations about code. Below you'll see a `[1]` printed before the output. We'll explain its relevance later in Section 4.2.3, but ignore this for now.

Below are a few commands followed by their output.

```
34 + 20 * sqrt(100)  # The +, -, *, / symbols have expected meanings.
```

```
#> [1] 234
```

```
exp(2)  # The exponential function.
```

```
#> [1] 7.3891
```

```
log(2)  # Base e logarithm function.
```

```
#> [1] 0.69315
```

```
2 ^ log(100, base = 2) # Base 2 logarithm function.
```

```
#> [1] 100
```

As illustrated above, functions perform different tasks. A function is called like this

```
function_name(argument_1 = value_1, argument_2 = value_2, ...)
```

where `value_1` is the value you provide to the first argument `argument_1`, and `value_2` is the value you provide to the second argument `argument_2`.[3] When you start typing the name of a function in the console, a pop-up shows possible completions and mousing over these options shows their function arguments and definitions. Pressing F1 on your keyboard opens the function's manual page in the Help tab in the lower right RStudio window. Functions can have any number of arguments: you'll see some with none and others with an indeterminate number. You'll also learn how to write your own functions in Chapter 5.

If a command is not completed, but the Enter key is pressed, the command prompt changes to a `+` sign, which is R's way of letting you know it's expecting additional code to complete the command. This often happens when you forget to close parentheses when typing a function or mathematical expression, i.e., typed a left parenthesis `(` but not the matching right parenthesis, or when you open a quote with the quotation symbol `"` but forget to close it with a second quotation symbol. To get back to the `>` command prompt, you can either type something to finish the command, e.g., right parentheses `)` or closing quotation `"`, or press the Esc key and retype the command.

Pressing the up arrow on your keyboard while in the console window cycles through the command history. This is useful when you want to rerun a command or tweak a command—press the up arrow until the command appears and then edit it before pressing Enter to run.

R has two widely used assignment operators: the left arrow, which consists of a less than sign followed immediately by a dash, `<-`, and the equals sign, `=`. R users continuously debate

[3]We use `code font` throughout the book to indicate R code and things to do with RStudio and file paths. Also, when referring to functions in the text, we'll append `()` to the function name, e.g., the square root function is `sqrt()`.

the merits of each assignment operator and their subtle differences. Many leading R style guides recommend the left arrow `<-` and we'll use this throughout the book. Because it's cumbersome to type `<-`, RStudio provides a keyboard shortcut `Alt` + `-` (hold the Alt key down and press the minus key). You can find this and many other shortcuts by going to the `Tools` dropdown menu in RStudio then `Keyboard Shortcuts Help`.

The line of code below assigns the value 5 to object `x`.

```
x <- 5
```

Notice we call `x` an object. You might think of `x` as a mathematical variable that's given the value 5. However, it's more appropriate to think of `x` as an object that can be described using a set of attributes, e.g., it's a numeric object and has a single value. These attributes define how a given object behaves in our code.

When you created `x`, you might have noticed it appeared in the top right `Global Environment` window, which keeps track of all objects you create in your RStudio session.

Typing an object's name in the console returns its value.

```
x
```

```
#> [1] 5
```

Careful, R is case sensitive. Try typing capital `X` in the console and you'll get an error because it was not defined.

```
X
```

```
#> Error: object 'X' not found
```

You can remove objects using the `rm()` function. For example, `rm(x)` removes `x` from the global environment.

It's easy to compute basic descriptive statistics and produce standard graphical representations of data. Let's consider the first 10 tree height and DBH observations in the FEF dataset introduced in Section 1.2.1. Begin by entering these data "by hand" using the `c()` function, which concatenates its indeterminate number of arguments into a vector.[4] For larger datasets we'll clearly want a more efficient way to enter data.

```
dbh <- c(6, 6.9, 6.4, 6.5, 7.2, 3.1, 2, 4.1, 2.4, 2.7)
ht <- c(48, 48, 48, 49, 51, 40, 30.5, 50, 28, 40.4)
```

[4] We'll provide a much more detailed introduction to vectors in Section 4.2. We briefly mention them here as a teaser and to motivate initial exploration.

When you create `dbh` and `ht`, you'll notice their names and a short description of their attributes appear in the `Global Environment` window. Again, typing `dbh` and `ht` in the console returns their values.

```
dbh
```

```
#>  [1] 6.0 6.9 6.4 6.5 7.2 3.1 2.0 4.1 2.4 2.7
```

```
ht
```

```
#>  [1] 48.0 48.0 48.0 49.0 51.0 40.0 30.5 50.0 28.0 40.4
```

Try out the `summary()` function on `dbh` or `ht`. This function computes some basic summary statistics for numeric objects. For example, from the output below we see `dbh` has a minimum of 2, maximum 7.2, and median of 5.05. These and other summary statistics are introduced in Chapter 11.

```
summary(dbh)
```

```
#>    Min. 1st Qu.  Median    Mean 3rd Qu.    Max.
#>    2.00    2.80    5.05    4.73    6.47    7.20
```

Here's a very basic scatter plot of `dbh` versus `ht`. We'll take a much more detailed tour of R's graphics capabilities in Chapter 6.

```
plot(dbh, ht)
```

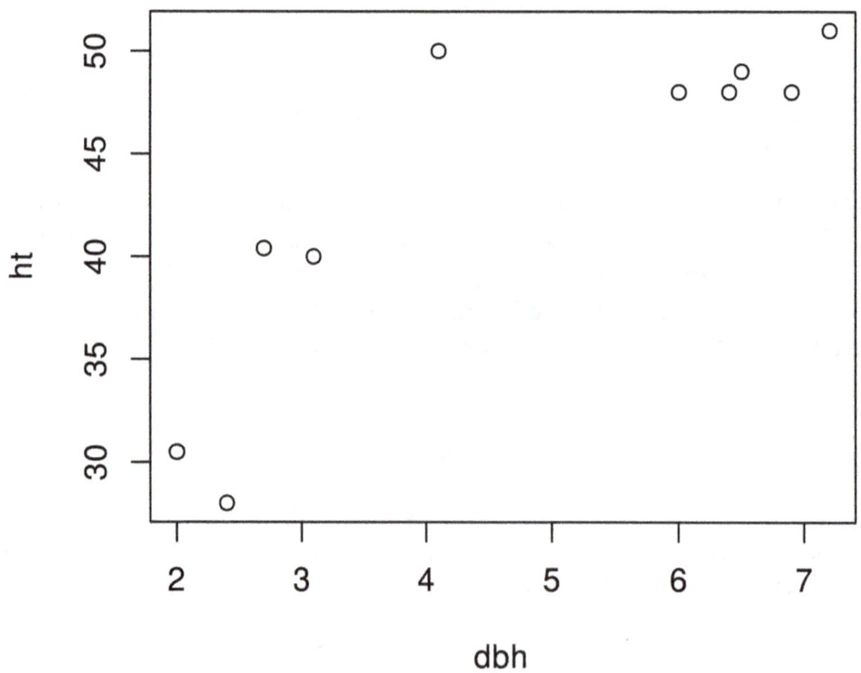

You'll notice the graphic output from `plot(dbh, ht)` pops up in the lower right window `Plots` tab. Unsurprisingly, the plot shows a positive relationship between tree DBH and height, i.e., as DBH increases so too does height, albeit in a nonlinear way (i.e., the points don't fall along a straight line).

2.2 Workspace and working directory

The *workspace* is your R session work environment and includes any objects you create. These objects are listed in the `Global Environment` window. The function `ls()`, which stands for list, also lists all objects in your workspace (note, this is the same list given in the `Global Environment` window).[5]

When you close RStudio, a dialog box will ask you if you want to save an image of the current workspace. If you choose to save your workspace, RStudio saves your session objects and information in a `.RData` file (the period makes it a hidden file) in your *working directory*. The next time you start R or RStudio, it looks for a `.RData` file in your working directory. If it finds one, it loads the file so your session picks up where you left off. If not, R starts fresh with an empty workspace. This brings us to the next question: what is a working directory?

Each R session is associated with a working directory. This is just a directory (i.e., folder) from which R reads and writes files, e.g., the `.RData` file, data files you want to analyze, or files you want to save. On macOS, when you start RStudio it sets the working directory to your home directory (e.g., `/Users/andy`). If you're on a different operating system, you can check where the default working directory is by typing `getwd()` in the console. You can change the default working directory under RStudio's `Global Options` dialog found under the `Tools` dropdown menu.

There are several ways to change the working directory once an R session is started in RStudio. One method is to click on the `Files` tab in the lower right window, navigate to the desired working directory, then select `Set As Working Directory` under the `More` dropdown menu. Alternatively, you can set the session's working directory using the `setwd()` function in the console. For example, on Microsoft Windows `setwd("C:/Users/andy/book/exercise1")` sets the working directory to `C:/Users/andy/book/exercise1`, assuming that file path and directory exist. Microsoft Windows file path uses a backslash, `\`, but in R the backslash is an escape character, hence specifying file paths in R on Windows uses the forward slash, i.e., `/`. Similarly on macOS you can use `setwd("/Users/andy/book/exercise1")`. Perhaps the simplest method is to use the `Session` dropdown menu in RStudio `Session > Set Working Directory > Choose Directory`, then navigate to and select the desired working directory. Later on, when we start reading and writing data, it'll be important that you can identify your current working directory relative to where data files are stored. We revisit these topics in Section 3.3.1 and 3.4.1.

[5]You might wonder about the empty parentheses after `ls`. Some functions don't require arguments; however, the parentheses still need to be added. Running a function name without parentheses returns the function's definition, which, at times, can be instructive.

2.3 Packages

As we've seen, functions are used to perform different tasks. These functions, along with their documentation and example datasets, are collected and distributed in packages. When you install and start R, a bundle of packages is automatically loaded and you can immediately use the functions they contain. These default packages, collectively referred to as base R, provide the language's basic functionality.

We'll eventually use packages written by R community members that contain functions tailored to specialized tasks and analyses. Many contributed packages are available on CRAN. As of October 2025, CRAN hosted 22,952 contributed packages! This staggering number is a testament to R's popularity and active community. Some of these packages are organized into CRAN task views[6] that provide guidance on which packages are relevant for various topics. There is, for example, an Analysis of Ecological and Environmental Data[7] task view that lists packages used for environmental data analysis.

To access functions stored in a package not included in base R, you need to do two things:

1. *Install the package on your computer*. You only need to install the package once. Use the `install.packages()` function to install a package. Alternatively, you can install the package via the `Install` button in the `Packages` tab located in RStudio's bottom right window. For example, in Chapter 6 we'll use the `dplyr` package installed using `install.packages("dplyr")`.
2. *Load the package*. You need to load the package every time you start a new session using the `library()` function. Loading the package gives you access to the package's functions. For example, `library(dplyr)` loads the `dplyr` package.

2.4 Getting help

A comprehensive, but overwhelming, RStudio cheatsheet is available under the `Help` drop-down menu `Help > Cheatsheets > RStudio IDE Cheat Sheet`. As you progress in learning R and RStudio, this cheatsheet will become more useful. For now, you might use the cheat-sheet to locate the various windows and functions identified in the coming chapters.

There are several help-related functions. For example, to learn about the `log()` function run `help(log)` or `?log` in the console to bring up its manual page. Also, as we saw in Section 2.1.1, typing the function name on the console and pressing F1 opens its manual page. The function `help.start()` opens a web page of documentation. To use this, run `help.start()` in the console window. The web page is opened by `help.start()` also provides links to several online manuals.

[6]https://cran.r-project.org/web/views/

[7]https://CRAN.R-project.org/view=Environmetrics

Many package authors provide *vignettes* as a user-friendly way to demonstrate package use through worked examples and more human readable details than found in manual pages. Run `browseVignettes(package="package_name")` in the console to access vignettes provided for the given `"package_name"`.[8]

For example, if you installed the `dplyr` package via `install.packages("dplyr")` as described in Section 2.3, then you can run `browseVignettes(package="dplyr")` to list all the vignettes available for the `dplyr` package. You can access a specific vignette by running `vignette("vignette_name", package = "package_name")`, e.g., run `vignette("base", package = "dplyr")` to read about how `dplyr` functions compare with equivalent operations in base R—topics we'll address later in this book. We'll direct you to relevant vignettes as packages are introduced in subsequent chapters.

Internet search engines provide another, sometimes more efficient and user-friendly, way to find answers to your questions. A Google search that includes "R" along with a short description of your question or issue is often fruitful. Such a search typically takes you to R help pages, tutorials, or general programming question/answer forums such as stackoverflow. If you're receiving a cryptic R warning or error message, a Google search often yields enough information to get you moving in the right direction.

In addition, R users have written many types of contributed documents. Some of this documentation is available at www.r-project.org/other-docs.html. Of course, there are also numerous books covering general and specialized R topics available for purchase.

2.5 Summary

We introduced the R programming language and the RStudio interface, which are our tools of choice for working with forestry and ecological datasets. We located the RStudio console window and practiced a few commands. We briefly introduced the workspace and working directory. We also discussed how functions are stored in packages, and how non-base R packages must be installed and loaded before their functions can be accessed.

Lastly, we covered the important topic of getting help when you're stuck. Knowing when and where to look for help, whether online or in the manual pages, is a skill developed over time. A significant portion of programming and data analysis involves finding and adapting code that others have written. In fact, some of the most efficient programmers and analysts are those who can recognize a problem, understand the general approach to solving it, and adapt existing code to accomplish the task at hand. We encourage you to explore the different options for getting help and to find those you consider most useful and informative.

[8]Many package developers only include vignettes on an associated website for the package, in which case the `browseVignettes()` function will not return any vignettes. The package website can be found by going to the URL section of the CRAN package page at `https://cran.r-project.org/web/packages/package_name/index.html`, where `package_name` is replaced with the package of interest.

The next chapter introduces tools and skills needed for efficient and reproducible analysis workflows.

2.6 Exercises

For Exercises 2.1-2.4, perform the following simple calculations in the R console.

Exercise 2.1. Calculate $4 + \sqrt{5}$.

Exercise 2.2. Calculate e^{10}, where e is the exponential function.

Exercise 2.3. Calculate $5^5 - 10^{10}$.

Exercise 2.4. Calculate $\frac{(5+5)}{(3\times4)}$. Make sure to include the parentheses in your code.

Exercise 2.5. Recall the plot we created of the `dbh` versus the `ht` values using `plot(dbh, ht)`. Run the code to recreate this plot in the console. Next, run the code `plot(dbh, ht, pch = 19)`. What does the `pch = 19` argument do? Further explore the `pch` argument by setting `pch` to values different from 19, e.g., 15, 16, 3, etc. Consult the manual page for `plot()`.

Exercise 2.6. Let's extend the plotting code a bit further. Run the code `plot(dbh, ht, pch = 19, col = 'blue')` in the console. What does the `col` argument do? Don't like blue? Choose a different color from the list of built-in colors displayed by typing `colors()` in the console.

Exercise 2.7. Run the code `plot(dbh, ht, pch = 19, col = 'blue', xlab = 'DBH')` in the console. What does the `xlab = 'DBH'` do? There is a similar argument called `ylab`. Try to figure out a way to change the label of the y-axis to `Height`.

3

Scripts and reproducible workflows

Data analysis, whether for a professional project or research, typically involves many steps and iterations. For example, creating a figure effective at communicating results can involve trying out several different graphical representations, and then tens if not hundreds of iterations to fine-tune the chosen representation. Each of these iterations might require several lines of code. Although we could accomplish this by typing and re-typing code lines in the console, it's easier and more efficient to write the code in a *script*—a plain text file that stores a sequence of commands—which we can submit to the console either a line at a time, several lines at a time, or all together.

In addition to making the workflow more efficient, scripts provide another great benefit. We often work on a project for a few hours, then move on to something else, and later return to it after a few days, months, or even years. In such cases, we may have forgotten how we created a particularly nice graphic or how to replicate an analysis, which can lead to several hours or even days spent rewriting the necessary code. If we save a script file, the ingredients are immediately available when we return to the project.

Next, consider a large scientific endeavor. Ideally, a scientific study is *reproducible*, meaning an independent group of researchers (or the original researchers) can duplicate the study. Reproducible research means anyone can replicate the steps in a study, from importing external datasets and wrangling them into analysis-ready formats, to creating tables and figures from the results. In principle we could detail this process in a step-by-step guide using words; however, in practice, it's typically difficult or impossible to reproduce a full data analysis based on a written explanation. It's much more effective to maintain code that accomplishes all steps in a single script file, with annotation included as comments.

In this chapter, we introduce scripts and their use to facilitate reproducible workflows. Because such workflows document the analysis of datasets, we cover common functions for reading and writing external data. Writing high-quality code and reproducible workflows means conforming to file naming, variable naming, code formatting, and file organization best practices. While many recommendations might initially seem tedious and unnecessary, we speak from experience: using scripts, staying organized, and working within a reproducible workflow will save tremendous amounts of time and headaches in the long run.

DOI: 10.1201/9781003222699-3

3.1 Scripts

Scripts facilitate efficient coding and provide a data management and analysis record. We illustrate these ideas by tracing steps taken to develop a figure using the FACE dataset introduced in Section 1.2.3. We'll walk through the analysis as a sequence of small steps, and then show how these same steps live together in a single script file. The illustration uses features that we've not yet discussed, so don't worry about the code details and focus instead on the script's role in the workflow.

First read the FACE data using `read.csv()`.

```
face_dat <- read.csv("datasets/FACE/FACE_aspen_core_growth.csv")
```

Let's create a scatter plot of 2008 DBH versus height and explore the data a little bit along the way. To do this, we create `dbh` and `ht` vectors using the 2008 DBH and height `face_dat` columns and print out the first ten values.

```
dbh <- face_dat$X2008_DBH
ht <- face_dat$X2008_Height
dbh[1:10]
```

```
#>  [1]   NA 9.55 2.00 9.00 3.11 6.35 4.60   NA   NA 1.42
```

```
ht[1:10]
```

```
#>  [1]   NA 1225  334 1079  370  859  818   NA   NA  268
```

The missing values (denoted by `NA`) suggest several trees were dead or not measured for some reason in 2008. At some point, it'd be good to investigate which trees have missing data and why, but we'll skip that for now. `plot()` omits missing values.

```
plot(dbh, ht)
```

Not surprisingly, the scatter plot shows DBH and height are positively correlated and their relationship is nonlinear. Now let's modify the basic scatter plot by adding a feature to identify trees belonging to the control and elevated CO_2 treatments. We do this by separating DBH and height into their respective treatment groups.

```
treat <- face_dat$Treat
dbh_treat_1 <- dbh[treat == 1] # Treatment 1 is the control.
```

```
ht_treat_1 <- ht[treat == 1]

dbh_treat_2 <- dbh[treat == 2] # Treatment 2 is the elevated CO2.
ht_treat_2 <- ht[treat == 2]
```

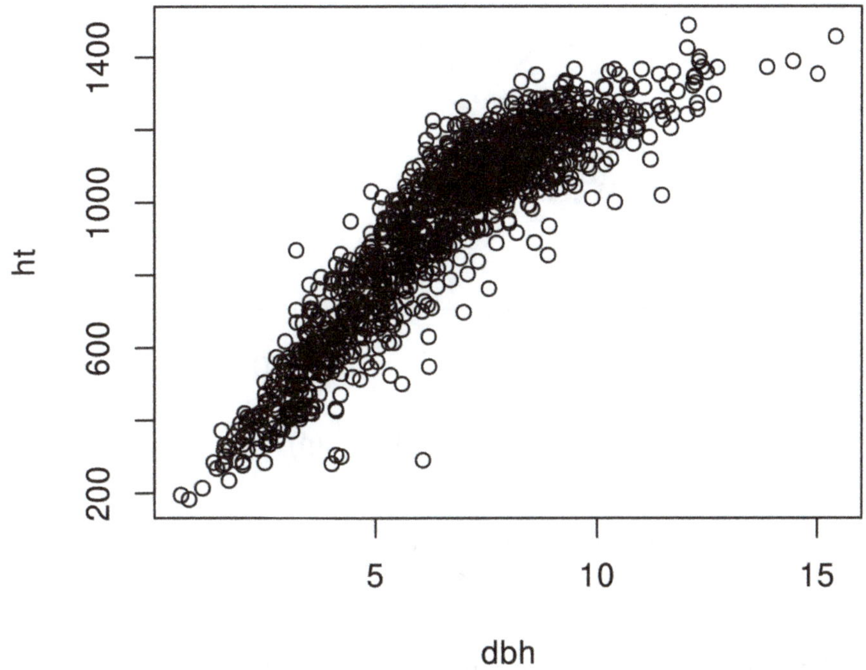

To make a more informative scatter plot we'll do two things. First, plot treatment 1 data, but ensure the plot region is large enough to show all the treatment 2 data. This is done by specifying sufficiently wide axis ranges via the `xlim` and `ylim` arguments in `plot()`. The `xlim` and `ylim` are set to the range of `dbh` and `ht` values, respectively, using the `range()` function.[1] Second add treatment 2 data via `points()`. There are several other arguments passed to `plot()`, but don't worry about these details for now.

We cover these and many other plot considerations in Chapter 9, where we'll learn about a package called `ggplot2` that provides another way to create such plots.

```
plot(dbh_treat_1, ht_treat_1, xlim = range(dbh, na.rm = TRUE),
     ylim = range(ht, na.rm = TRUE), pch = 19, col = "salmon3",
     cex = 0.5, xlab = "DBH (cm)", ylab = "Height (cm)")

points(dbh_treat_2, ht_treat_2, pch = 19, col = "blue", cex = 0.5)
```

[1]Try to figure out what the `na.rm` argument does in `range()`. Read about `range()` by typing `?range` in the console. Then create a short vector with a mix of numeric and `NA` values, and pass it to the `range()` function with `na.rm = TRUE`, then with `na.rm = FALSE`. As recommended in Section 1.3, it's these explorations that build your skills.

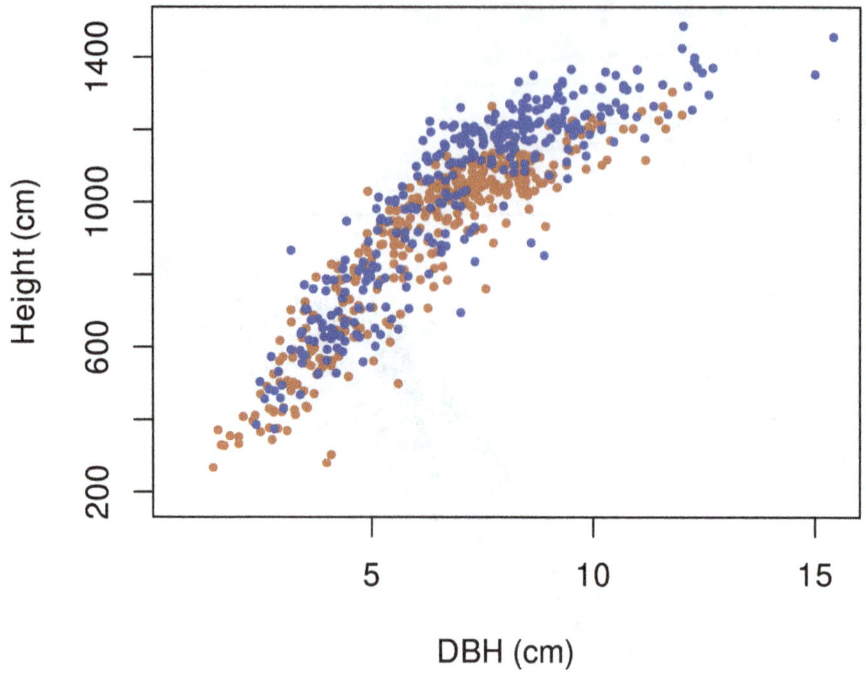

We should have a plot legend to tell the viewer which colors are associated with the treatments, as well as many other aesthetic refinements. For now, however, we'll resist such temptations.[2]

We only show some of the process leading to the completed plot above. We read in the data, created vectors of 2008 DBH and height measurements, took a look at a subset of measurement values, made an intermediate plot, partitioned the measurements by treatment, then made our final plot with points colored by treatment. However, we don't show a lot of the process. For example, we made several mistakes while getting the code and plot the way we wanted it for this book. We initially forgot to include `na.rm = TRUE` in `range()`, then spent some time fiddling with the treatment colors, point symbols, and axis labels.

Now imagine trying to recreate the plot a few days later. Possibly someone saw the plot and commented that it'd be interesting to see similar plots for each year in the study period. If we did all of this work only in the console—including all the false starts and refinements—it'd be hard to sort things out later. This would be especially true if a few months had passed, rather than just a few days. Creating the new plots would be much easier if we started from an existing script used to create our initial plot, especially if it included a few well-chosen comments to remind us what we were doing along the way.

Fortunately, it's easy to create and work with script files in RStudio. Just choose `File > New File > R script` and a script window will open in the upper left RStudio window. A script is

[2]As an aside, by only looking at the plotted data and thinking about basic plant physiology, can you guess which color is associated with the elevated CO_2 treatment?

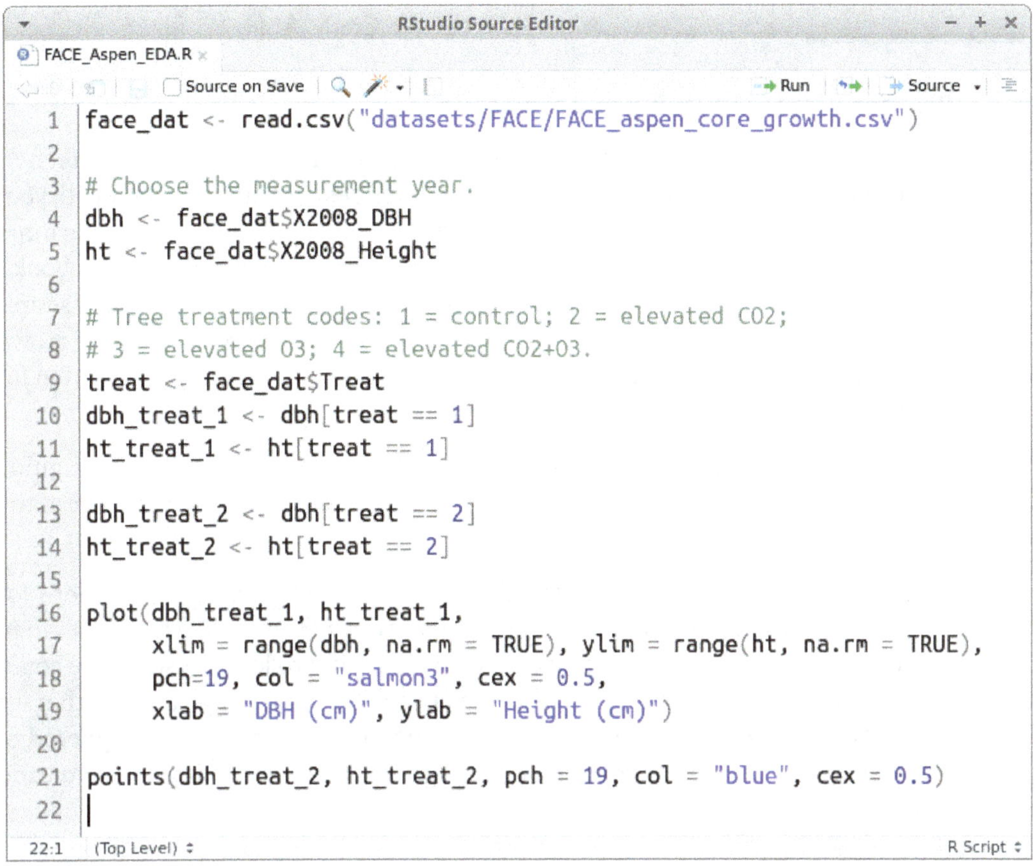

FIGURE 3.1: A script window in RStudio.

simply a text file that contains code and ends with the file suffix `.R`. Efficient use of scripts is a key step toward creating a reproducible and organized workflow.

Figure 3.1 shows the script window containing our FACE plot code. From the script window you can, among other things, save the script using either the disk icon near the top left of the window or via `File > Save` and run one or more lines of code using the `Run` icon in the window. By "run" we mean send code from your script to the console where it's executed. You run one line of code by placing your cursor anywhere on the line you want to run and pressing the `Run` icon. If you want to run several lines of code, highlight the desired lines and press the `Run` icon.[3] You can press the `Source` icon to run the entire script. In addition, there's a `Source on Save` checkbox. If this is checked, the entire script is run when the script file is saved (we rarely use this option). Generally, scripts are written incrementally—you write a few lines and make sure they run as expected, then write a few more lines and check them, and so forth until the script is complete.

[3]Alternatively, you can use a keyboard shortcut to run code and perform other tasks in RStudio. See `Help > Keyboard Shortcuts Help`.

3.2 Best practices for naming and formatting

Small choices about naming, formatting, and commenting have outsized effects on how usable your code will be over time. Following code style guides, staying consistent with your chosen file and variable naming, and using ample comments will make your code more accessible and reusable—both for you and for others. The scripts you write for this book's examples and exercises contribute to your codebase for future projects. A *codebase* is simply the collection of code used in analysis or software development. As your codebase grows, coding tasks become easier because you can reuse code snippets and functions written for previous tasks.

In subsequent sections, we detail a few key points from Hadley Wickham's R style guide available at https://style.tidyverse.org, along with some additional guidelines for maintaining high-quality code.

You can do your future self a favor by using clear and consistent naming conventions for files, variables, and functions. There have been more times than we like to admit when we've found ourselves digging through old directories trying to locate a script we could reuse for the task at hand. Or we spend too much time reading through a script that we or someone else wrote, trying to figure out what poorly named variables are supposed to represent. Following a simple and consistent naming rule is key to reducing confusion and saving time.

3.2.1 Script file naming

Script file names should be meaningful and end with a ".R" extension. They should also follow some basic naming rules. For example, say we're conducting an analysis on the emerald ash borer (*Agrilus planipennis*), referred to by its acronym EAB, distribution in Michigan and save the code in a script. Consider these possible script file names:

- GOOD: `EAB_MI_dist.R` (Naming rule called *delimiter naming* using an underscore. Here too we chose to capitalize acronyms.)
- GOOD: `emeraldAshBorerMIDist.R` (Naming rule called *camelCase*[4] where the first letter in each word is capitalized, except perhaps for the first word. It's called camelCase because the capital letters look like a camel's hump.)
- BAD: `insectDist.R` (camelCase but the file name is too ambiguous.)
- BAD: `insect.dist.R` (Too ambiguous and multiple periods can confuse the operating system's file type auto-detect.)
- BAD: `emeraldashborermidist.R` (Running the letters together without a delimiter or case change makes this too ambiguous and confusing.)

We'll generally use the underscore delimiter and capitalize acronyms for script and data file names.

[4] https://en.wikipedia.org/wiki/Camel_case

3.2.2 Variable naming

R has strict rules for naming variables. Syntactic names (i.e., valid names) consist of letters, digits, `.`, and `_` but can't begin with `_` or a digit. Also, you can't use any reserved words like `NA`, `NULL`, `TRUE`, `FALSE`, `if`, and `function` (see the complete list by running `?Reserved` in your console). A name that doesn't follow these rules is called a non-syntactic name (i.e., an invalid name) and will result in an error if you try to use it. The code below attempts to use a non-syntactic name.

```
1dbh <- 1
```

```
#> Error in parse(text = input): <text>:1:2: unexpected symbol
#> 1: 1dbh
#>      ^
```

Although we don't recommend it, you can override these rules by surrounding a non-syntactic name with backticks.

```
`1dbh` <- 1
```

In addition to adhering to syntactic name rules, there's value in following a naming convention for variables. Here are some examples of good and bad choices for a fictitious variable that holds EAB counts used in the previous section's EAB analysis:

- GOOD: `eab_count` (Delimiter naming using an underscore. Here we chose to keep all letters lowercase, unlike the file example of this naming convention in the previous section.)
- GOOD: `eab.count` (Delimiter naming using a period and all lowercase.)
- GOOD: `EABCount` (CamelCase naming again.)
- BAD: `eabcount` (Running the letters together without a delimiter or case change makes this too ambiguous and confusing.)

3.2.3 Code formatting

Here are some other coding best practices that help readability:

- Keep code lines under 80 characters long. RStudio includes a vertical line in the script window at the 80th character to make this suggestion easy to follow.
- Allow some whitespace in a consistent way between code elements within lines. Include one space on either side of mathematical and logical operators. Don't include spaces on either side of the `:`, `::`, and `:::` operators, which we will encounter later. Like the English language, always add a space after a comma and never before. In an expression, add a space before a left parenthesis and after the right parenthesis. Parentheses and square brackets can be directly adjacent to whatever they're enclosing (unless the last character they enclose is a comma). There should be no space between a function name and left parenthesis that opens the function's arguments. Notice how the use of whitespace

rules below makes reading the code easier on the eyes:

  - GOOD: `d <- a + (b * sqrt(5 ^ 2)) / e[1:5, ]`
  - BAD: `d<-a+(b*sqrt ( 5^2 ))/e[ 1 : 5 ,]`

- Avoid unnecessary parentheses; rather, rely on order of operations (yes, remember PEMDAS, see Section 10.2.2):
  - GOOD: `(a + 4) / sqrt(a * b)`
  - BAD: `((a) + (4)) / (sqrt((a) * (b)))`
- Indent code two spaces within functions, loops, and decision control structures (RStudio does this by default when you press the TAB key); we'll revisit this point in Chapter 5 where these topics are introduced.
- Consolidate calls to packages (using `library()`) at the top of the script rather than interspersing them throughout the script (Section 2.3).

3.2.4 Customize your coding environment

While this section has nothing to do with developing reproducible workflows, it has everything to do with being a happy coder. RStudio offers great flexibility to customize the interface such as changing themes and *syntax highlighting*. Syntax highlighting means different code features (e.g., function names, operators, variables, comments) are displayed using different colors and fonts within scripts. To change themes, go to `Tools > Global Options` and then click on `Appearance` and choose among Modern, Classic, and Sky. Here too, you can change the font and color schemes used for the script and other interface windows.

We generally find Editor themes with a dark background and mellow colored syntax highlighting easier on the eyes and more helpful to quickly identify coding errors. For example, an RStudio window using the Modern theme and Material Editor theme is shown in Figure 3.2. Using a dark palette is especially useful during long work sessions.[5]

3.3 File paths and external data

3.3.1 Relative and absolute paths

Recall from Section 2.2 that the working directory is the directory associated with each session from which R reads and writes files. You can change the working directory using methods discussed in Section 2.2, such as `setwd()`.

Your working directory is shown at the top of the console window or by running `getwd()` in the console. For example, the working directory for the session shown in Figure 3.2 is `~/projects/IFDAR_book`, where the preceding tilde, i.e., `~`, stands for the home directory. The home directory for Microsoft Windows is <drive>/Users/<username>, where <drive>

[5] Search "Customizing RStudio" and "Using RStudio Themes" on the Posit website if you're interested in creating custom themes and syntax highlighting.

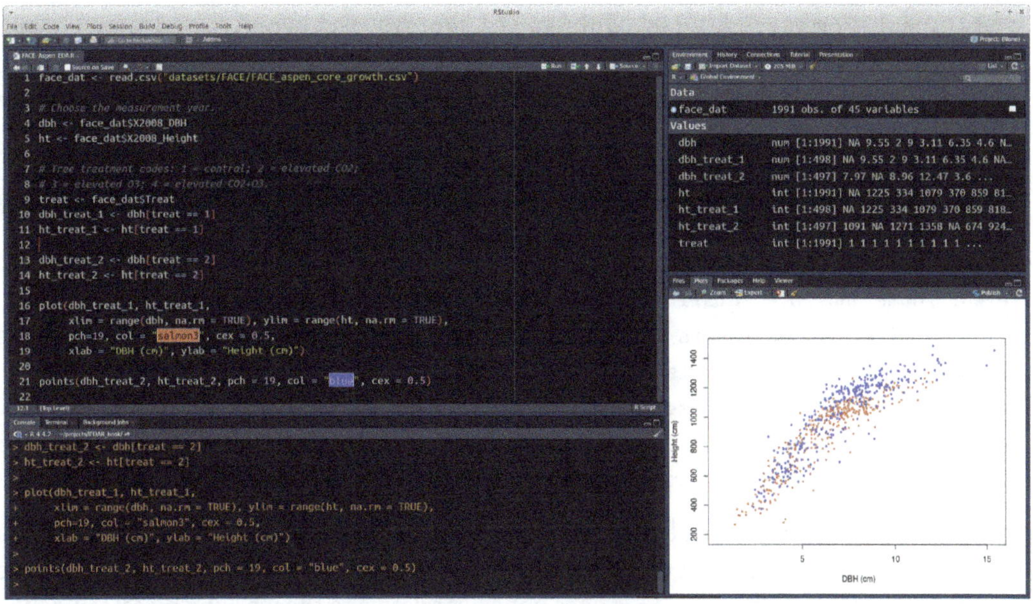

FIGURE 3.2: The RStudio IDE using the Modern theme and Material Editor theme.

is the name of the drive (e.g., `C:`) and `<username>` is your user name. For macOS, it's `/Users/<username>`, and for Linux it's `/home/<username>`. The root directory is `<drive>/` on Windows and the leading forward slash `/` on macOS and Linux.

Say we want to read the `FEF_trees.csv` data file into R. As we'll cover in Section 3.3.2, reading this file can be accomplished using `read.csv()`. The first argument in `read.csv()` is `file`, which is set equal to the quoted *path* to the file you want to read. There are two types of paths: *absolute* and *relative*. An absolute path is the directory hierarchy starting from the root directory and ending with the data file. A relative path is the directory hierarchy starting from the working directory and ending with the data file, i.e., the data file location *relative* to the working directory. The code below reads the `FEF_trees.csv` file on a Linux operating system first using the absolute path, then using the relative path.

```
# Reading data using the absolute path.
read.csv(file = "/home/andy/projects/IFDAR_book/datasets/FEF_trees.csv")

# Reading data using the relative path.
setwd("/home/andy/projects/IFDAR_book")
# Or equivalently.
setwd("~/projects/IFDAR_book")

read.csv(file = "datasets/FEF_trees.csv")
```

We **do not recommend** using absolute paths in your code. Using absolute paths makes fragile code, meaning it can break easily. Absolute paths in a script need to be changed every time the script is moved to a different directory or if someone else tries to run your script on their computer. Instead, use relative paths that read and write data relative to the working directory. As illustrated in the previous example, from the working directory your path can follow a directory hierarchy down into subdirectories (e.g., `datasets`). If you want to go up directory levels from the working directory, use two periods (`..`), which moves you up one level (to the *parent directory*).

Continuing with the example above and the working directory set to `setwd("/home/andy/projects/IFDAR_book")`, `read.csv(file = "../datasets/FEF_trees.csv")` would attempt to read `FEF_trees.csv` from `"/home/andy/projects/datasets/FEF_trees.csv"`, and `read.csv(file = "../../datasets/FEF_trees.csv")` would attempt to read `FEF_trees.csv` from `"/home/andy/datasets/FEF_trees.csv"`. Both attempts would result in an error because there is no directory called `datasets` in either the `projects` directory or its parent.

At some point, you'll try to read or write a file from a directory that doesn't exist. When you do, R will provide a somewhat cryptic warning and error message. Following the example above, `"../datasets/FEF_trees.csv"` does not exist relative to the working directory, so the code below returns an error. When you see this error, there's a very good chance the path you supplied to the read or write function isn't correct.

```
# File "../datasets/FEF_trees.csv" does not exist.
read.csv(file = "../datasets/FEF_trees.csv")
```

```
#> Warning in file(file, "rt"): cannot open file
#> '../datasets/FEF_trees.csv': No such file or directory

#> Error in file(file, "rt"): cannot open the connection
```

Setting and keeping track of a working directory might seem cumbersome. One way to make this a bit easier is to set the working directory at the top of your script using `setwd()`. While this isn't a terrible idea (and one we often use), a potentially more robust approach is to use an RStudio project, as we'll see in Section 3.4.

3.3.2 External data

Data come in a dizzying variety of formats. Sometimes data are stored in proprietary formats used by commercial software such as Excel, SPSS, or SAS. Sometimes they're structured using a relational model, for example, the USDA Forest Service Forest Inventory and Analysis database (Burrill et al. 2018). Sometimes we use a data-interchange format such as JSON (JavaScript Object Notation; Pezoa et al. (2016)), or a markup language format such as XML (Extensible Markup Language; Bray et al. (2008)), perhaps with specialized standards for describing ecological information such as EML (Ecological Metadata Language; Jones et al. (2019)). The good news is we've yet to encounter a data format unreadable by R. That's thanks to the large community of package developers and the many data formats supported within their respective disciplines. For example, the `foreign` (R Core Team 2025) and `haven`

(Wickham et al. 2025) packages provide functions for reading and writing data in common proprietary statistical software formats.

We cover a few popular base R functions for reading and writing plain-text rectangular files, often referred to as *flat files*. For most applications we consider, a flat file is organized in rows and columns, with rows corresponding to observational units (things we measure, e.g., trees) and columns corresponding to variables recorded for each unit.

There are many functions in base R and user-contributed packages that we don't cover (e.g., the `load()` and `save()` functions in base R for reading and writing R data files). Like everything else in this book, if the functions covered here don't meet your needs, then plenty of other options in R are only a Google search away.

Importing data

Fortunately, many datasets are (or can be) saved as text files. Most software can read and write such files, so our initial focus is on reading and writing text files.

`read.table()` and its offshoots, such as `read.csv()`, are used to read 2-dimensional data (i.e., flat files) from a text file. For example, as described in Section 1.2.1, the `FEF_trees.csv` file holds biomass information for trees measured on FEF plots within different watersheds across years. Below are the first 7 rows and columns of the file's 89 rows and 18 columns.

```
watershed,year,plot,species,dbh_in,height_ft,stem_green_kg
3,1991,29,Acer rubrum,6,48,92.2
3,1991,33,Acer rubrum,6.9,48,102.3
3,1991,35,Acer rubrum,6.4,48,124.4
3,1991,39,Acer rubrum,6.5,49,91.7
3,1991,44,Acer rubrum,7.2,51,186.2
3,1992,26,Acer rubrum,3.1,40,20.8
```

As is evident, the file's first row contains column names separated (delimited) by commas. Subsequent rows contain comma-delimited data on each tree's location (watershed code), measurement year, plot number, species, DBH, height, and weight of tree components (e.g., `stem_green_kg` is the green weight of the stem in kilograms).

`read.table()`, called in the code below, reads the `FEF_trees.csv` file and returns a data frame, which is assigned to `fef_trees`. A data frame is a 2-dimensional object that holds data in rows and columns and is formally introduced in Section 4.5—for now, picture a data frame as a spreadsheet or data table. We then use `nrow()` and `ncol()` to confirm R read in the correct number of data rows and columns.[6] Lastly, we call `head()` to print the first six rows and columns, again to reassure ourselves the data were read correctly.

```
fef_trees <- read.table(file = "datasets/FEF_trees.csv",
                        header = TRUE, sep = ",")
```

[6]You could instead use `dim()`, i.e., `dim(fef_trees)`, which returns a vector of length two with the first and second elements corresponding to number of rows and columns, respectively.

```
# Confirm the correct number of rows and columns were read.
nrow(fef_trees)
```

```
#> [1] 88
```

```
ncol(fef_trees)
```

```
#> [1] 18
```

```
# A look at the first six rows and columns.
head(fef_trees, n = c(6, 6))
```

```
#>   watershed year plot      species dbh_in height_ft
#> 1         3 1991   29 Acer rubrum     6.0        48
#> 2         3 1991   33 Acer rubrum     6.9        48
#> 3         3 1991   35 Acer rubrum     6.4        48
#> 4         3 1991   39 Acer rubrum     6.5        49
#> 5         3 1991   44 Acer rubrum     7.2        51
#> 6         3 1992   26 Acer rubrum     3.1        40
```

The `read.table()` arguments and associated values used above include:

- `file` identifies the file location. In this case, the quoted path `"datasets/FEF_trees.csv"` is the relative path from the working directory to `FEF_trees.csv` (Section 3.3.1).
- `header = TRUE` specifies the file's first row contains the column names, known as the "header."
- `sep = ","` defines the comma as the character delimiter between the file's columns.

The function `read.csv()` is the same as `read.table()` except the default delimiter is a comma, whereas the default delimiter for `read.table()` is whitespace.

The file `FEF_trees.tsv` contains the same data, but the columns are tab-delimited. The `sep` argument's value is the only change needed to read this file (and of course the `file` argument, because the data are stored in a different file).

```
fef_trees <- read.table(file = "datasets/FEF_trees.tsv",
                        header = TRUE, sep = "\t")
```

```
head(fef_trees, n = c(6, 6))
```

```
#>   watershed year plot      species dbh_in height_ft
#> 1         3 1991   29 Acer rubrum     6.0        48
#> 2         3 1991   33 Acer rubrum     6.9        48
```

```
#> 3          3 1991   35 Acer rubrum    6.4        48
#> 4          3 1991   39 Acer rubrum    6.5        49
#> 5          3 1991   44 Acer rubrum    7.2        51
#> 6          3 1992   26 Acer rubrum    3.1        40
```

File extensions, e.g., `.csv` or `.tsv`, are naming conventions and serve only to remind us how the columns are delimited, i.e., they don't influence how the read functions behave.

A third file, `FEF_trees.txt`, contains the same data, but with explanatory text added to the first several rows. The file also uses the whitespace delimiter rather than a comma or tab. Here are the first several rows of the file and seven columns of data.

```
This file includes felled tree biomass by tree component for hardwood
species sampled on the Fernow Experimental Forest (FEF), West Virginia.
A total of 88 trees were sampled from plots within two watersheds.

"watershed" "year" "plot" "species" "dbh_in" "height_ft" "stem_green_kg"
3 1991 29 "Acer rubrum" 6 48 92.2
3 1991 33 "Acer rubrum" 6.9 48 102.3
3 1991 35 "Acer rubrum" 6.4 48 124.4
3 1991 39 "Acer rubrum" 6.5 49 91.7
3 1991 44 "Acer rubrum" 7.2 51 186.2
3 1992 26 "Acer rubrum" 3.1 40 20.8
```

You'll notice that in this file, both the column names and the character column values are quoted. That's necessary because character strings could include whitespace, and otherwise the read function would treat that whitespace as a delimiter. For example, the space between *Acer* and *rubrum* would be mistaken for a column break if it weren't quoted.

`read.table()` needs to skip the first four rows above the header and know that whitespace is the delimiter. The `skip` and `sep` arguments control these behaviors.

```
fef_trees <- read.table(file = "datasets/FEF_trees.txt",
                        header = TRUE, skip = 4, sep = " ")

head(fef_trees, n = c(6, 6))
```

```
#>   watershed year plot     species dbh_in height_ft
#> 1         3 1991   29 Acer rubrum    6.0        48
#> 2         3 1991   33 Acer rubrum    6.9        48
#> 3         3 1991   35 Acer rubrum    6.4        48
#> 4         3 1991   39 Acer rubrum    6.5        49
#> 5         3 1991   44 Acer rubrum    7.2        51
#> 6         3 1992   26 Acer rubrum    3.1        40
```

For fun, see what happens when the `skip` argument is left out of `read.table()` in the code above.

Importing data with missing observations

Missing data are common in environmental data. Sometimes missing data are just that—missing—the place where an entry should be is blank. Other times we represent missing data with numbers (e.g., -9999) or symbols. `read.table()` has an argument `na.strings` used to specify how the file represents missing data. The file `FEF_trees_missing.csv` contains the same FEF data, except some measurements in the third row are missing. Here are the file's first seven rows and columns.

```
watershed,year,plot,species,dbh_in,height_ft,stem_green_kg
3,1991,29,Acer rubrum,6,48,92.2
3,1991,33,Acer rubrum,6.9,48,102.3
3,1991,35,Acer rubrum,,,
3,1991,39,Acer rubrum,6.5,49,91.7
3,1991,44,Acer rubrum,7.2,51,186.2
3,1992,26,Acer rubrum,3.1,40,20.8
```

Notice in the third row of data that measurements for `dbh_in`, `height_ft`, and `stem_green_kg` are missing. Rather than being marked with an explicit symbol (e.g., `NA`, which stands for not available), the missing measurements were simply left blank. As shown in the code below, setting the `na.strings` argument equal to `""` tells the read function that empty fields represent missing values. We again print a subset of the resulting data frame to make sure the missing values were replaced with `NA`.

```
fef_trees <- read.csv(file = "datasets/FEF_trees_missing.csv",
                      header = TRUE, na.strings = "")

head(fef_trees, n = c(6, 6))
```

```
#>   watershed year plot     species dbh_in height_ft
#> 1         3 1991   29 Acer rubrum    6.0        48
#> 2         3 1991   33 Acer rubrum    6.9        48
#> 3         3 1991   35 Acer rubrum     NA        NA
#> 4         3 1991   39 Acer rubrum    6.5        49
#> 5         3 1991   44 Acer rubrum    7.2        51
#> 6         3 1992   26 Acer rubrum    3.1        40
```

Take a look at the `read.table()` manual page to get a better sense of how flexible this and related functions are when reading file formats.

Importing data with non-syntactic column names

When read by `read.table()` and related functions, file column names that aren't syntactic (see Section 3.2.2) are modified to make them syntactic (see `?make.names` for the rules used to modify names).

Exporting data

Data files rarely come in an analysis-ready format. As further developed in Chapter 6, some level of *data cleaning* and *data wrangling* is typically needed prior to analysis. In addition to documenting such data preparation steps in a script, we often write the processed data to a file for further analysis, distribution, or archival in a data repository. `write.table()` writes data frames and data-frame-like objects to an external file.

To illustrate `write.table()`, the code below creates a new CSV file in our current working directory named `FEF_trees_new.csv` from the `fef_trees` data frame we previously created.

```
write.table(x = fef_trees, file = "FEF_trees_new.csv", sep = ",",
            row.names = FALSE)
```

Let's walk through the `write.table()` arguments and associated values used in the code above:

- `x = fef_trees` is the object we'll write to a file, in this case `fef_trees`.
- `file = "FEF_trees_new.csv"` is the file path and name to store the data specified in argument `x`. This is a relative path unless a root directory is included in the path. See Section 3.3.1 for more on relative and absolute paths.
- `sep = ","` defines the comma as the character delimiter between the file's columns.
- `row.names = FALSE` indicates row names won't be included as the first column in the file. This argument is `TRUE` by default. We prefer not including the row names.

`write.csv()` is a convenience version of `write.table()` for writing comma-delimited files.

3.3.3 Illustration

Let's review some key ideas from Sections 3.3.1 and 3.3.2 using an example. Say Sarah is analyzing some emerald ash borer data. Figure 3.3 illustrates her computer's directory structure, which is running the Microsoft Windows operating system. Her home directory `C:/Users/Sarah` contains three directories: `Documents`, `Research`, and `Archive`. To stay organized, she created a dedicated directory called `EAB` within `Research` where she keeps the analysis script and data. You'll also notice that, keeping with good file naming practices, she uses an underscore delimiter naming rule for her files; see, e.g., Section 3.2.1.

Code in the `EAB_analysis.R` script performs the analysis of data held in `EAB_data.csv`. We display the first two lines in this script below. The first line sets `EAB` as the working directory using its absolute path. Equivalently, she could've used `setwd("~/Research/EAB")` to set the working directory because `~` is shorthand for her home directory path `C:/Users/Sarah`. In most cases, it's much easier to use the tilde than to write out the full path from the root directory to the user's home directory. The next line of code reads the `EAB_data.csv` data file using the file's relative path from the `EAB` working directory. If the working directory wasn't set, or was set to some directory other than `EAB`, then `read.csv(file = "Data/EAB_data.csv")` would fail.

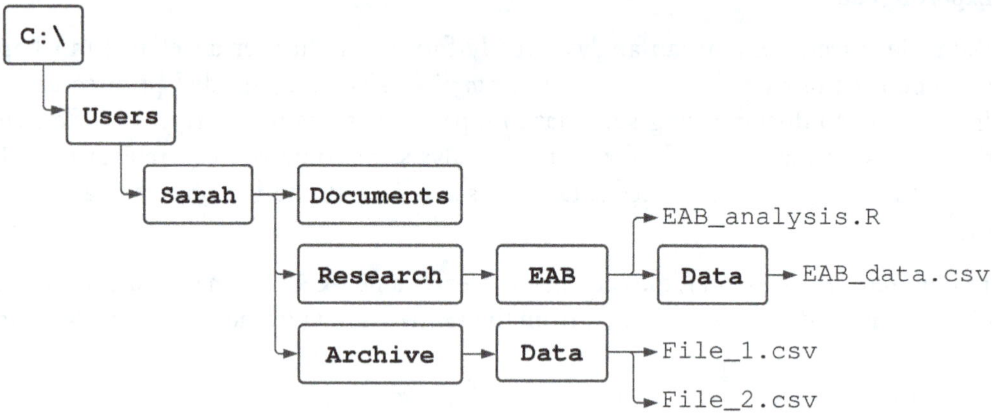

FIGURE 3.3: A directory structure used for illustration.

```
setwd("C:/Users/Sarah/Research/EAB")
# Or, equivalently, setwd("~/Research/EAB")

eab_dat <- read.csv(file = "Data/EAB_data.csv")
```

Say later in the script she needs to read `File_1.csv` from the `Archive/Data` directory. Given the `EAB` working directory, this file is read using the code below. Notice the use of `../..` to go up two directories to `Sarah` before going down into `Archive`.

```
data_1 <- read.csv(file = "../../Archive/Data/file_1.csv")
```

3.4 RStudio projects and keeping organized

When working on a project, it's generally best to keep associated files in an organized directory structure. Such a structure might comprise a main directory that contains subdirectories to hold data, scripts, metadata, and results; see, e.g., Section 3.4.1. If organized in this way, the main directory would likely serve well as the working directory. Each time you start an R session to work on the project, you'd need to set the working directory using `setwd()` or one of the other methods covered in Section 2.2. To eliminate the repetitive task of setting the working directory for each session, RStudio offers the *projects* feature.

We'll illustrate this feature by making an RStudio project you might use for working through this book's examples and exercises. In RStudio, go to `File > New Project`, then select `New Directory > New Project`. RStudio then asks for the directory name and its location. Let's call the project *IFDAR* (the acronym for the book's title) and choose a location that makes sense in your broader directory structure, then click `Create Project`. When these steps are complete, a new RStudio session will start, and you should see the project name tab (`IFDAR`) in the top right corner of the RStudio window. Clicking on the project name provides a menu with options to switch between projects, create a project, and control a project's default behavior.

By placing the book's `datasets` directory into your IFDAR project directory, code examples offered in the book should run without modifying the relative paths to the various illustrative datasets.

The primary benefit of an RStudio project is that we only need to specify the working directory once when setting up the project. In other words, when you open an RStudio project, the working directory is automatically set. You can confirm this by looking at the top of the console window after creating the `IFDAR` project, where you should see the working directory's absolute path. Each project is also associated with its own history, which can be useful if you ever forget to save some important code in a script. Working directory, history, and other project information are stored in a `.Rproj` file (e.g., `IFDAR.Rproj`), which can be opened from within RStudio via `File > Open Project` or used to open the project directly from your operating system's file browser. By automatically setting the working directory, RStudio projects allow for better organization, fewer file path errors, and improved reproducibility.

3.4.1 Directory structure

Whether you decide to use RStudio projects or not, establishing an organized directory structure around your working directory is key to efficiency and reproducibility. There is no single correct way to structure directories. As you learn R and become more familiar with data analysis, you'll develop preferences for subdirectory and file organization. To get you started, we'll outline an organization scheme that promotes reproducible research best practices suggested by Sandve et al. (2013) and Wilson et al. (2017). This scheme is illustrated in Figure 3.4 for an example project that aims to model emerald ash borer distribution in Michigan.

In the main project directory, we suggest having two files and four subdirectories. The first is the RStudio project file, e.g., `EAB_MI_dist.Rproj`. The second is a README file, typically a simple text file named `README.txt`, that provides a project description and a list of subdirectory contents with a brief description of each file's purpose. We recommend starting each project with a `README.txt` file and adding information to it as the project progresses. A README file not only helps you keep track of file organization, but is crucial for effective collaboration, reproducibility, and archiving.

The four subdirectories are called `data`, `results`, `doc`, and `src`. The `data` directory contains raw data and metadata used in the analysis. Depending on the amount and structure of the data, you might have multiple subdirectories within the `data` directory.

In Figure 3.4, the data directory holds two files used in the analysis, one containing EAB counts across Michigan (`EAB_MI_counts.csv`) and the other containing forest plot

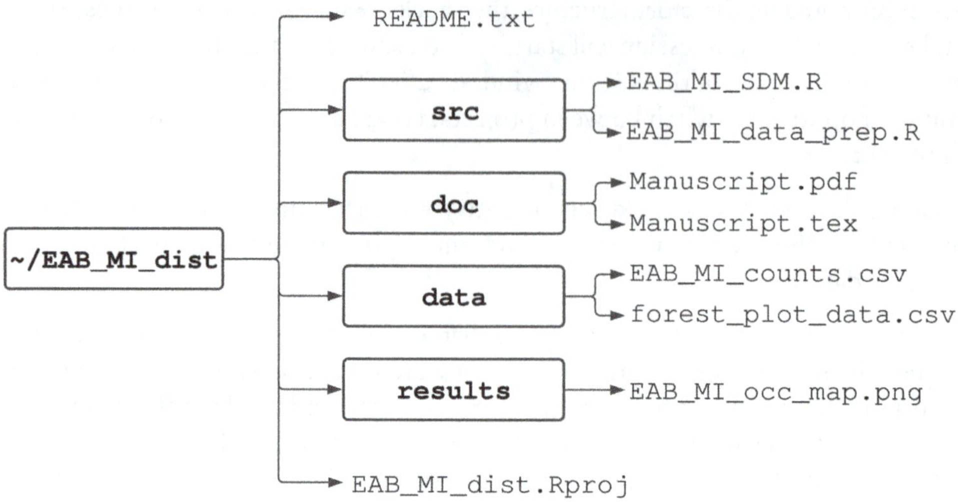

FIGURE 3.4: One suggested way to organize a project's directories and files.

characteristics (`forest_plot_data.csv`). The `results` directory should contain any files generated during the analysis, which might include figures (Chapter 9), tables, and archive-ready cleaned and tidy datasets (Chapter 8). For large projects, it's often useful to have a separate subdirectory inside the `results` directory called `figures` where final figures are stored. The `doc` directory contains documents associated with the project. Depending on project objectives, this can include files for manuscripts, notes, lectures, or professional reports. The `doc` directory contains files used to write a manuscript. Lastly, the `src` directory (short for "source code") contains all code written for the project, e.g., `EAB_MI_data_prep.R` to prepare the data for analysis conducted in `EAB_MI_SDM.R`.

While this suggested directory structure is quite specific, the important thing is that you follow a systematic and intuitive directory and file naming scheme that works best for you and allows others to reproduce your work with minimal effort.

3.5 Summary

In forestry, ecology, and the environmental sciences, there's a push for open science, reproducibility, and transparency (Ellison 2010), as evidenced by the increasing number of academic journals and government agencies requiring all data and code to be published along with final manuscripts and reports. Considering the scientific process more broadly, the goal of any scientific endeavor is to build upon previous knowledge. By using reproducible data analysis tools that make it easier for others (and your future self) to extend your work, you're doing your part to advance forestry and environmental sciences.

In this chapter, we introduced essential rules and tools to establish an organized and reproducible data analysis workflow. Maintaining organization throughout all aspects of an analysis can improve efficiency, reduce avoidable errors, and make troubleshooting problems much easier. An organized workflow is also essential for reproducible work.

Perhaps the easiest way to start documenting your workflow is to use scripts. Scripts are simply text files with a `.R` extension that hold code needed to reproduce steps in an analysis. The script window in RStudio allows you to write and run your code incrementally, save it, and easily return to it at some later point in time. Adding comments throughout your scripts is an efficient and convenient way to document your analysis and coding decisions.

The scripts you write to reproduce examples and complete exercises in this book can form the codebase from which you borrow code snippets for future work. The key to building a valuable codebase is to follow best practices when formatting code, naming variables and files, and keeping work in organized directory structures.

Maintaining an organized and intuitive directory structure, as well as using good coding practices and file naming conventions, is often overlooked as a crucial part of data analysis and scientific research (Wilson et al. 2017). Learning efficient coding, file organization, and data analysis techniques is analogous to learning proper field techniques for performing a timber inventory, regeneration survey, or any sort of ecological field work. Initially, it might seem tedious to pay attention to organizational and logistical details, but in the long run, using such techniques (or using efficient and accurate field techniques) leads to more reliable results and saves time and headaches.

We also covered some file reading and writing basics. These routine tasks, dependent on file and directory structure, are part of nearly every data analysis. We read and write external files using either their absolute or relative paths. The absolute path starts from the root directory. The relative path starts from the session's working directory, which is set at the beginning of each new session. RStudio projects remove the need to set a working directory for each new session and also provide some organization and session history niceties. A key skill moving forward is the ability to navigate your computer's directory structure relative to your chosen working directory or RStudio project.

In Chapter 4, we'll continue to introduce fundamental skills needed to work with data in R. With the basics of reproducible workflows established in this chapter, we'll move our discussion to tools in R used to organize, represent, access, and manipulate data.

3.6 Exercises

Exercise 3.1. Use Figure 3.3 to answer the following questions.

a. What is the root directory?

b. What is the home directory?

c. What is the absolute path to the working directory set using `setwd("~")`?

d. What is the absolute path to `File_2.csv`?

e. What is the relative path to `File_2.csv` if the working directory was set using `setwd("~/Research/EAB/Data")`?

f. Say `setwd("~/Research/EAB")` was used to set the working directory. Then, within RStudio, a new script file named `EAB_spatial_analysis.R` was created using `File > New File > R Script`. Where is this new script file saved by default?

Exercise 3.2. Consider your own computer when answering the following questions.

a. What operating system are you using, e.g., Microsoft Windows, macOS, Linux, Unix?

b. What is your root directory?

c. What is the absolute path to your home directory?

d. Where will you keep work associated with this book?

4

Data structures

A data structure is a type of object used to organize and store data so it can be accessed and manipulated in specific ways. Programming languages provide methods (or functions) designed to operate on different kinds of data structures.

In this chapter, we focus on data structures that serve as the essential building blocks for data analysis in R. To help with initial understanding, the data in this chapter are relatively modest in size and complexity. The ideas and methods, however, generalize to larger and more complex datasets.

The base data structures in R are vectors, matrices, arrays, data frames, and lists. The first three (vectors, matrices, and arrays) require all elements to be of the same type (i.e., numeric, character, or logical); hence, these structures are referred to as *homogeneous*. In R, such single-type objects are called *atomic*.

Data frames and lists allow elements to be of different types, e.g., some elements of a data frame may be numeric while other elements may be character. Perhaps not surprisingly, these structures are called *heterogeneous*. These base structures can also be organized by their dimensionality, i.e., 1-dimensional, 2-dimensional, or N-dimensional, as shown in Table 4.1.

R has no scalar types (i.e., individual values); instead, scalars are represented using a vector of length one.

More generally, we refer to data structures as objects. An efficient way to understand what comprises a given object is to use the `str()` function. `str()` is short for structure and prints a compact, human-readable description of any R data structure. For example, in the code below, we show that a scalar is actually a vector of length one.

TABLE 4.1: Dimensionality and element homogeneity of base R data structures.

Dimension	Homogeneous	Heterogeneous
1	Atomic vector	List
2	Matrix	Data frame
N	Array	

DOI: 10.1201/9781003222699-4

```r
a <- 1
str(a)
```

```
#>  num 1
```

```r
is.atomic(a)
```

```
#> [1] TRUE
```

```r
length(a)
```

```
#> [1] 1
```

Here we assigned `a` the scalar value 1. `str(a)` prints `num 1`, which tells us `a` is numeric of length one. We then use `is.atomic()` and `length()` to confirm that `a` is in fact an atomic vector of length one. There are similar logical tests for the other base data structures, e.g., `is.matrix()`, `is.array()`, `is.data.frame()`, and `is.list()`. These will all come in handy as we encounter different objects.

4.1 Types of variables

Before we dive into the different data structures in R, it's useful to define the different types of variables we may encounter in forestry and environmental datasets. Figure 4.1 displays the fundamental variable types we'll work with in this book, along with the R data type we'll use to represent each one. We define a *variable* as any characteristic that might vary across individual units (e.g., live trees) within a population of interest (e.g., all live trees in a forest).[1] Following Figure 4.1, a variable is initially classified as either *quantitative* or *qualitative*.

A quantitative variable has values that convey magnitude; that is, they provide a numerical measure. This numerical measure is either *continuous* or *discrete*. A quantitative continuous variable can take an infinite (uncountable) number of possible values, meaning it can vary in arbitrarily fine increments from one value to the next. Said differently, a continuous variable can have an arbitrary number of decimal places. Examples include height, weight, and volume. In contrast, a quantitative discrete variable can take only a finite (countable) set of possible values. These values are often (though not always) integers. Examples include age in whole years, number of trees, and number of fire events.

A qualitative variable (also referred to as a categorical variable) has values that represent different categories. Qualitative variables are either *nominal* or *ordinal*. A qualitative nominal variable takes values for which no ordering is possible or implied. For example, the

[1] We'll define the terms *unit* and *population* more formally in Chapter 10.

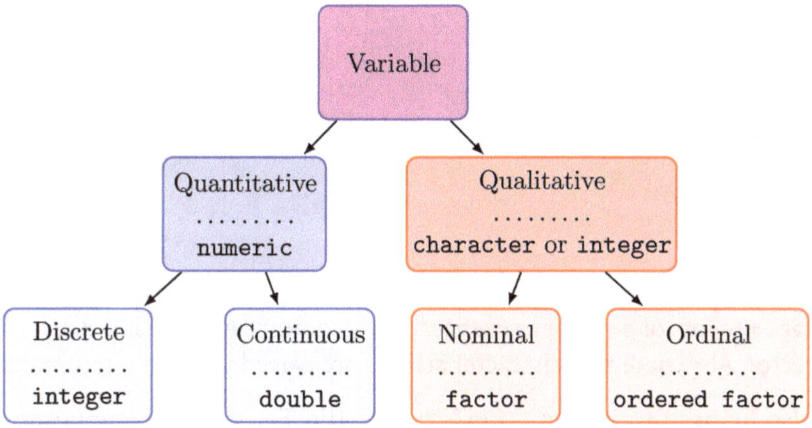

FIGURE 4.1: Variable types with corresponding R data types added below the dotted line.

variable species is nominal because there's no inherent or natural order among species names. Similarly, forest type coded as deciduous, coniferous, or mixed is nominal because there's no inherent ordering. In contrast, the categories of a qualitative ordinal variable have a natural ordering. For example, the variable tree canopy position can take values suppressed, intermediate, co-dominant, or dominant, where the categories imply an ordering. Further examples include disturbance severity coded as low, medium, and high, or forest or ecosystem succession stages. In these cases, there's a natural order to the categories.

4.2 Vectors

Think of a vector as a structure used to represent one variable in a dataset.[2] For example, a vector might hold the measured heights, in meters, of seven trees, while another vector might hold the tree species names. The `c()` function is useful for creating (small) vectors and for modifying existing vectors. Think of `c` as standing for 'combine.' Consider the following use of `c()` to create the `dbh` and `spp` vectors.

```
dbh <- c(20, 18, 13, 16, 10, 14)
dbh
```

```
#> [1] 20 18 13 16 10 14
```

[2]Technically, the objects described in this section are 'atomic' vectors (all elements of the same type). This distinction is important because lists, described in Section 4.6, are vectors that can have elements of different types. Unless specified otherwise, when we say vector, we're talking about an atomic vector.

```
spp <- c("Acer rubrum", "Acer rubrum", "Betula lenta",
         "Betula lenta", "Prunus serotina", "Prunus serotina")
spp
```

```
#> [1] "Acer rubrum"     "Acer rubrum"
#> [3] "Betula lenta"    "Betula lenta"
#> [5] "Prunus serotina" "Prunus serotina"
```

We see that elements of a vector are separated by commas when using the `c()` function to create a vector. Also, note that character strings are placed inside quotation marks.[3]

`c()` can also be used to add to an existing vector. For example, suppose we want to add a 13 inch DBH red maple to the dataset. We can modify our existing vectors as follows.

```
dbh <- c(dbh, 13)
spp <- c(spp, "Acer rubrum")
dbh
```

```
#> [1] 20 18 13 16 10 14 13
```

```
spp
```

```
#> [1] "Acer rubrum"     "Acer rubrum"
#> [3] "Betula lenta"    "Betula lenta"
#> [5] "Prunus serotina" "Prunus serotina"
#> [7] "Acer rubrum"
```

4.2.1 Types, conversion, and coercion

It's important to distinguish between different data types. For example, it makes sense to calculate the mean of DBH measurements stored in `dbh`, but it doesn't make sense to compute the mean of species stored in `spp`. Vectors can be of six different types: character, double, integer, logical, complex, and raw. We'll not encounter the complex and raw types in everyday data analysis, so we focus on the first four data types.

1. `character`: represents text. Our `spp` is a character vector.

   ```
   typeof(spp)
   ```

   ```
   #> [1] "character"
   ```

[3]Character strings can also be delimited using single quotations `'` rather than double quotations `"`. Double quotations are preferred for consistency with other languages where the two forms have different meanings.

We'll often use the phrase "character string" or simply "string" when talking about character types. A string is just an element comprising more than one character. For example, `"Acer rubrum"` is a character string.[4]

2. `double`: a numeric that can be an integer or non-integer value (e.g., 10, 4.2). Given the values in `dbh` are all integers, it might be surprising to see the output of `typeof(dbh)` below is `double`. By default, R creates a double type vector when numeric values are given via `c()`.

```
typeof(dbh)
```

```
#> [1] "double"
```

3. `integer`: a numeric that can only be an integer. We can create an integer vector by placing the letter `L` next to each number when forming the vector. For example, an integer vector of DBH values is created as follows.

```
dbh_int <- c(20L, 18L, 13L, 16L, 10L, 14L, 13L)
typeof(dbh_int)
```

```
#> [1] "integer"
```

4. `logical`: represents true or false values using `TRUE` or `FALSE`, respectively. To illustrate logical vectors, imagine the field technician who measured the trees in `dbh` and `spp` also indicated whether each tree was acceptable growing stock. Their determination was coded in `ags` as `TRUE` if the tree was acceptable and `FALSE` if it was not.

```
ags <- c(TRUE, TRUE, FALSE, TRUE, FALSE, FALSE, TRUE)
ags
```

```
#> [1]  TRUE  TRUE FALSE  TRUE FALSE FALSE  TRUE
```

```
typeof(ags)
```

```
#> [1] "logical"
```

When it makes sense, it's possible to convert vectors to different types. Consider the following examples.

```
dbh_int <- as.integer(dbh)
dbh_int
```

[4]Use `nchar()` to get the number of characters in a string, e.g., `nchar("Acer rubrum")`.

```
#> [1] 20 18 13 16 10 14 13
```

```
typeof(dbh_int)
```

```
#> [1] "integer"
```

```
dbh_char <- as.character(dbh)
dbh_char
```

```
#> [1] "20" "18" "13" "16" "10" "14" "13"
```

```
ags_double <- as.double(ags)
ags_double
```

```
#> [1] 1 1 0 1 0 0 1
```

```
spp_oops <- as.double(spp)
```

```
#> Warning: NAs introduced by coercion
```

```
spp_oops
```

```
#> [1] NA NA NA NA NA NA NA
```

```
sum(ags)
```

```
#> [1] 4
```

The integer version of `dbh` doesn't look any different, but it's stored differently, which can be important for computational efficiency and for interfacing with other software and programming languages. Converting `dbh` to character goes as expected—the quoted character representation of the numbers replaces the numbers themselves. Converting the logical vector `ags` to double is straightforward—`FALSE` is converted to 0 and `TRUE` is converted to 1. Now consider converting the character vector `spp` to a numeric double vector. It's not at all clear how to represent `"Acer rubrum"` as a number. In this case, R creates a double vector with each element set to `NA`, which is the representation of missing data, and throws a `warning` exception.[5] Finally, consider the code `sum(ags)`. `ags` is a logical vector, but when R sees that we're asking to sum this logical vector, it automatically converts it to a numeric

[5] In programming, *exception* is short for *exceptional event*. An exception occurs when the normal flow of a program is disrupted, typically by a syntax error or when the result might be unexpected, e.g., `as.double(spp)`. The phrase *throw an exception* implies our code might eventually *catch the exception* and deal with it appropriately, but that's a topic for later.

vector and then adds the 0's and 1's representing `FALSE` and `TRUE`, respectively (see more on this topic in the next section).

Similar to `typeof()`, there are functions that test whether a vector is of a particular type. Notice the test result is returned as a logical value.

```
is.double(dbh)
```

```
#> [1] TRUE
```

```
is.character(dbh)
```

```
#> [1] FALSE
```

```
is.integer(dbh_int)
```

```
#> [1] TRUE
```

```
is.logical(ags)
```

```
#> [1] TRUE
```

4.2.1.1 Coercion

An important concept in any programming language is the conversion of data types from one type to another. We've shown how to manually convert objects to different types using the `as.*` functions, where `*` represents the desired type. Automatic data type conversion also occurs, a process referred to as *coercion*. Consider the following examples.

```
xx <- c(1, 2, 3, TRUE)
xx
```

```
#> [1] 1 2 3 1
```

```
yy <- c(1, 2, 3, "aspen")
yy
```

```
#> [1] "1"     "2"     "3"     "aspen"
```

```
zz <- c(TRUE, FALSE, "aspen")
zz
```

```
#> [1] "TRUE"  "FALSE" "aspen"
```

```
dbh + ags
```

```
#> [1] 21 19 13 17 10 14 14
```

As noted previously, vectors can only contain elements of one type. If more than one type is included in a `c()` function, R silently *coerces* the vector to be of one type. The examples above illustrate R's coercion rules. If any element is a character, then the whole vector is character type. If there's a mix of numeric (either integer or double) and logical elements, then the whole vector is numeric. Note what happened when we added the numeric vector `dbh` to the logical vector `ags`. The logical vector was silently coerced to numeric, so that `FALSE` became 0 and `TRUE` became 1, and the two numeric vectors were then added *elementwise*, meaning each pair of corresponding elements in the two vectors was added together.

4.2.2 Patterned data

It's often convenient to generate patterned data for use in subsetting tasks, designing experiments, and identifying measurement units for some inventory designs. For example, we might need all the integers from 1 to 20 or a sequence of 100 values equally spaced between 0 and 1. The R functions `seq()` and `rep()` as well as the colon operator `:` are handy for generating such patterned data.

The colon operator generates a sequence of values with increments of 1.

```
1:10
```

```
#>  [1]  1  2  3  4  5  6  7  8  9 10
```

```
-5:3
```

```
#> [1] -5 -4 -3 -2 -1  0  1  2  3
```

```
10:4
```

```
#> [1] 10  9  8  7  6  5  4
```

```
pi:7
```

```
#> [1] 3.1416 4.1416 5.1416 6.1416
```

`seq()` generates either a sequence of pre-specified length or a sequence with pre-specified increments.

```
seq(from = 0, to = 1, length = 11)
```

```
#>  [1] 0.0 0.1 0.2 0.3 0.4 0.5 0.6 0.7 0.8 0.9 1.0
```

```
seq(from = 1, to = 5, by = 1/3)
```

```
#>  [1] 1.0000 1.3333 1.6667 2.0000 2.3333 2.6667 3.0000
#>  [8] 3.3333 3.6667 4.0000 4.3333 4.6667 5.0000
```

```
seq(from = 3, to = -1, length = 10)
```

```
#>  [1]  3.00000  2.55556  2.11111  1.66667  1.22222
#>  [6]  0.77778  0.33333 -0.11111 -0.55556 -1.00000
```

`rep()` replicates the values in a given vector.

```
rep(c(1, 2, 4), length = 5) # Recycle the vector to match length.
```

```
#> [1] 1 2 4 1 2
```

```
rep(c(1, 2, 4), times = 3) # Repeat the vector the number of times.
```

```
#> [1] 1 2 4 1 2 4 1 2 4
```

```
rep(c("a", "b", "c"), times = 3) # Same as above.
```

```
#> [1] "a" "b" "c" "a" "b" "c" "a" "b" "c"
```

```
rep(c("a", "b", "c"), each = 3) # Repeat each element "each" times.
```

```
#> [1] "a" "a" "a" "b" "b" "b" "c" "c" "c"
```

4.2.3 Indexing, accessing, and recycling elements

To access, and possibly change, a vector element's value, use the element's index (i.e., its position along the vector) in square brackets . For example, `dbh[4]` refers to the fourth element in the `dbh` vector, which has a value of 16. R starts numbering elements at 1, so the first element in the `dbh` vector is `dbh[1]`.

As a bit of an aside, when a vector is printed in the console window, R puts the index of the first value on a line in square brackets before the line of output. So the mysterious `[1]` before all output we've seen thus far in the book just means the index of the first value on the printed line is 1. If it's a long vector with more than one line of output, each new line is preceded by the index for the first value printed on the line. For example, if the output has 40 values, and 15 values appear on each line, then the first line will have `[1]` at the left, the second line will have `[16]` to the left, and the third line will have `[31]` to the left.

Here are a few indexing and replacement examples using the dbh vector.

```
dbh
```

```
#> [1] 20 18 13 16 10 14 13
```

```
dbh[5]
```

```
#> [1] 10
```

```
dbh[1:3]
```

```
#> [1] 20 18 13
```

```
dbh[length(dbh)]
```

```
#> [1] 13
```

```
dbh[]
```

```
#> [1] 20 18 13 16 10 14 13
```

```
dbh[3] <- 202 # Replace the third value with 202.
dbh
```

```
#> [1]  20  18 202  16  10  14  13
```

```
dbh[1:3] <- c(16, 8, 2) # Replace the first three values.
dbh
```

```
#> [1] 16  8  2 16 10 14 13
```

Notice in the code above that including nothing in the square brackets, i.e., `dbh[]`, results in the whole vector being returned. We can also replace vector values by accessing the elements where the new values will be assigned. For example, the value at `dbh[3]` is replaced with `202`, and then the vector's first three values `dbh[1:3]` are replaced with `16`, `8`, and `2`, respectively.

On your own, investigate the three replacement operations on the vector `x` in the code below.

```
x <- rep(0, times = 4) # Create a vector of four zeros.
x[1:3] <- c(1, 2)
x[1:4] <- c(3, 4)
x[1] <- c(5, 6)
```

Warnings thrown by two of the replacement operations above highlight a behavior called recycling. If R encounters two vectors of different lengths in an operation, it replicates (recycles) the smaller vector until its length equals the longer vector's length before performing the operation. This recycling happens in a replacement operation. If the number of elements to replace is different from the number of values supplied, then any necessary vector recycling occurs followed by truncation to match vector lengths, and a warning is thrown. Importantly, a warning is only thrown if truncation is necessary to match vector lengths. Therefore, if the number of elements to replace is a multiple of the number of elements supplied, then any necessary recycling occurs to match vector lengths, but no warning is thrown.

So, for the first replacement operation above, `c(1, 2)` is recycled to `c(1, 2, 1, 2)`, then the last `2` is truncated to make a vector of length three to replace `x` elements `1:3`, and a warning is thrown. The next replacement operation is similar to the first, but no warning is thrown because no truncation is necessary. For the last replacement operation, no recycling is necessary—only truncation of the value `6` to make the replacement vector length equal to 1. We'll see recycling again when working with comparison and mathematical operators in Section 4.7 and 10.2.3, respectively.

Negative indexes in the square brackets remove corresponding elements. A zero as an index returns nothing (more precisely, it returns a length zero vector of the appropriate type).

```
dbh[-3] # Print all values in dbh except element 3.
```

```
#> [1] 16  8 16 10 14 13
```

```
dbh[-length(dbh)] # Print all values in dbh except the last element.
```

```
#> [1] 16  8  2 16 10 14
```

```
fewer.dbh <- dbh[-c(1, 3 ,5)] # Remove elements 1, 3, and 5.
fewer.dbh
```

```
#> [1]  8 16 14 13
```

```
dbh[0]
```

```
#> numeric(0)
```

```
dbh[c(0, 2, 1)]
```

```
#> [1]  8 16
```

```
dbh[c(-1, 2)]
```

```
#> Error in dbh[c(-1, 2)]: only 0's may be mixed with negative subscripts
```

Notice mixing zero and other non-zero indexes is allowed, but mixing negative and positive indexes is not allowed.

What about the case where we don't know the positions of the elements we want to print or modify? For example, say we want the DBH of all acceptable growing stock trees in the `ags` vector. This is covered in Section 4.8, where we learn how to subset using logical values and conditional statements.

4.3 Factors

The `factor` data type represents categorical, or qualitative, variables (Figure 4.1). When represented using a `factor`, the different categories (or values) a categorical variable can take are called *levels*. For example, if the `spp` character vector created at the beginning of Section 4.2 was a `factor`, then species names would be the *levels*.

In many analyses, representing a categorical variable using a character vector (or perhaps an integer vector) is sufficient. However, for some analyses, representing the categorical variable using a `factor` has advantages.

Consider two categorical variables: one representing tree crown class with levels `S`, `I`, `C`, and `D` (i.e., *Suppressed*, *Intermediate*, *Codominant*, *Dominant*), and another representing sawlog quality grade of the first log with levels *Grade 1*, *Grade 2*, and *Grade 3*.

For the first variable, suppose that in a small dataset, all trees are either Dominant or Codominant crown class. If we represented the crown class variable using a character vector, there wouldn't be any way to know that the Suppressed and Intermediate categories exist, because they happen to not be present in the dataset.

For the log grade variable, a character vector representation doesn't explicitly indicate the natural ordering of the levels. That is, *Grade 1* is better than *Grade 2*, and *Grade 2* is better than *Grade 3*.

Both of these examples are illustrated in the code below.

```
# The nominal crown class variable.
crown_class <- c("D", "D", "C", "D", "D", "C", "C")
crown_class # Character.
```

```
#> [1] "D" "D" "C" "D" "D" "C" "C"
```

```
crown_class <- factor(crown_class, levels = c("S", "I", "C", "D"))
crown_class # Factor data type.
```

```
#> [1] D D C D D C C
#> Levels: S I C D
```

```
# The ordinal log grade variable.
grade <- c("Grade 1", "Grade 1", "Grade 3", "Grade 2",
           "Grade 3", "Grade 3", "Grade 2")
grade <- factor(grade,
                levels = c("Grade 3", "Grade 2", "Grade 1"),
                ordered = TRUE)
grade
```

```
#> [1] Grade 1 Grade 1 Grade 3 Grade 2 Grade 3 Grade 3
#> [7] Grade 2
#> Levels: Grade 3 < Grade 2 < Grade 1
```

In the `factor` version of crown class, levels are explicitly listed, so it's clear the two included levels aren't all possible levels. In the `factor` version of log grade, the ordering is explicit and preserved for use in subsequent analysis.

4.4 Missing data, infinity, and other special data types

Most real-world datasets have variables where some observations are missing. In longitudinal studies of tree growth (i.e., where trees are measured over time), it's common that trees die or can't be located for subsequent remeasurement. Statistical software should be able to represent missing data and analyze datasets in which some data are missing.

As we've already seen several times, `NA` is the missing data value in R. Because missing values may occur in numeric, character, and other data types, and because R requires vectors containing only elements of one type, there are different types of `NA` values. Usually, R determines the appropriate type of `NA` value automatically. It's worth noting that `NA` is not the same as the character string `"NA"`, as illustrated below with the aid of the `is.na()` function.

```
missing_species <- c("Quercus rubra", NA, "Acer saccharum")

is.na(missing_species)
```

```
#> [1] FALSE  TRUE FALSE
```

Now add a quoted NA to the end of missing_species and test again using is.na(). The is.na() result shows the quoted NA is not a missing value but rather a character string.

```
missing_species <- c(missing_species, "NA")
missing_species
```

```
#> [1] "Quercus rubra"  NA               "Acer saccharum"
#> [4] "NA"
```

```
is.na(missing_species)
```

```
#> [1] FALSE  TRUE FALSE FALSE
```

How should missing data be treated in computations, such as finding a vector's sum or mean? One possibility is to return an NA if missing values are encountered in the computation. Another is to remove any missing values prior to performing the computation.

The example below shows the mean() function's default behavior is to return NA if any missing values are present in the input. If you wish to remove missing values prior to computing the mean, then set the mean() function's na.rm argument to TRUE. While the na.rm argument is available in many mathematical functions, different functions have different default behaviors, so be sure to consult their manual pages for details.

```
mean(c(1 ,2, 3, NA, 5))
```

```
#> [1] NA
```

```
mean(c(1, 2, 3 ,NA, 5), na.rm = TRUE)
```

```
#> [1] 2.75
```

4.4.1 Infinity and NaN

What happens if code requests division by zero or results in a number that's too large to be represented? Knowing how R handles these situations comes in handy when your code produces an unexpected result. Here are some examples.

```
x <- 0:4
x
```

```
#> [1] 0 1 2 3 4
```

```
1/x
```

```
#> [1]     Inf 1.00000 0.50000 0.33333 0.25000
```

```
x/x
```

```
#> [1] NaN   1   1   1   1
```

```
y <- c(10, 1000, 10000)
2^y
```

```
#> [1]  1.0240e+03 1.0715e+301          Inf
```

`Inf` and `-Inf` represent infinity and negative infinity (and numbers that are too large in magnitude to be represented as `double`). `NaN` occurs when the result of a calculation is undefined, such as dividing zero by zero. R, like most programming languages, follows the ANSI/IEEE 754 Floating-Point Standard that prescribes behavior of mathematical operations involving `Inf`, `-Inf`, and `NaN` (IEEE 2008).

4.5 Data frames

Data are commonly rectangular in form, with variables as columns and measurement units as rows. In R, such data are held in a data frame, where columns (variables) are equal-length vectors that may be of different data types. A data frame constructed from a set of vectors is illustrated in Figure 4.2.

Let's continue using the species, DBH, and acceptable growing stock vectors defined at the beginning of Section 4.2. The code below follows Figure 4.2 to form a data frame using the

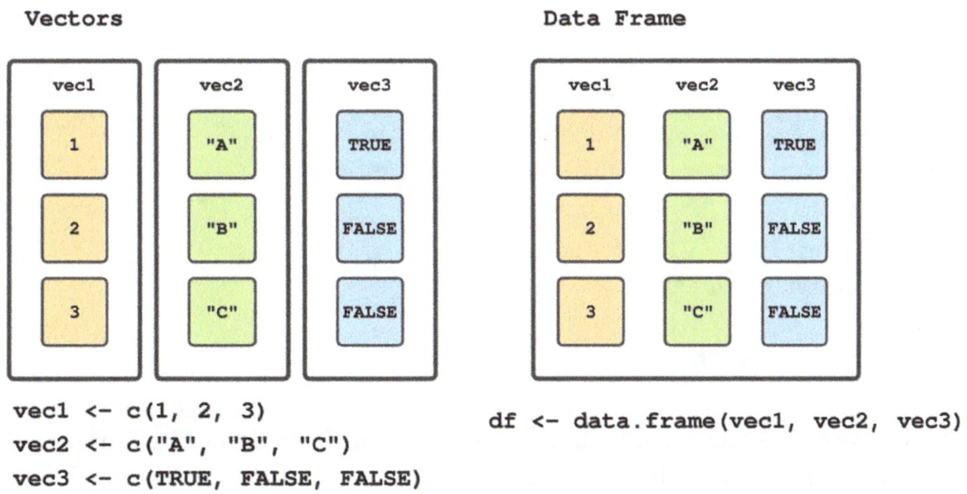

FIGURE 4.2: Relationship between vectors and data frames in R.

three vectors, with columns `spp`, `dbh`, and `ags`, and rows corresponding to measurements on each tree.[6]

```
trees <- data.frame(Species = spp, DBH = dbh, AGS = ags)
trees
```

```
#>             Species DBH   AGS
#> 1       Acer rubrum  16  TRUE
#> 2       Acer rubrum   8  TRUE
#> 3      Betula lenta   2 FALSE
#> 4      Betula lenta  16  TRUE
#> 5 Prunus serotina  10 FALSE
#> 6 Prunus serotina  14 FALSE
#> 7       Acer rubrum  13  TRUE
```

The column names and associated vectors are passed as arguments to `data.frame()`. Notice that while defining the `data.frame()` function arguments, we chose column names that differ from the original vector names.

Column names can be extracted and changed using either `names()` or `colnames()`. Both functions return a vector that can be modified, as we saw in Section 4.2.3. Below, we change the column names to `species`, `dbh`, and `acceptable_stock`, then change them back to `Species`, `DBH`, and `AGS`.

```
colnames(trees)
```

```
#> [1] "Species" "DBH"     "AGS"
```

```
is.vector(colnames(trees))
```

```
#> [1] TRUE
```

```
colnames(trees) <- c("species", "dbh", "acceptable_stock")
colnames(trees)
```

```
#> [1] "species"          "dbh"
#> [3] "acceptable_stock"
```

```
colnames(trees) <- c("Species", "DBH", "AGS")
```

Rows can also have names, although if row names are desired, it's often more useful to include a separate column in the data frame that contains those names. The default row

[6]While you can create data frames using the `data.frame()` function, remember that functions used to read external data (see Section 3.3.2) typically return data frames as well.

names are sequential numbers (as character strings) from `"1"` to the number of rows in the data frame. Like column names, row names can be extracted and changed if desired.

```
rownames(trees)
```

```
#> [1] "1" "2" "3" "4" "5" "6" "7"
```

```
rownames(trees) <- paste("Tree", 1:nrow(trees))
rownames(trees)
```

```
#> [1] "Tree 1" "Tree 2" "Tree 3" "Tree 4" "Tree 5"
#> [6] "Tree 6" "Tree 7"
```

Finally, let's take a look at the data frame's dimensions. `dim()` returns a vector holding the number of rows and the number of columns, respectively. Also try the functions `nrow()` and `ncol()` on the data frame and see what happens (this should help explain the `paste()` function used to name rows in the code above).

```
dim(trees)
```

```
#> [1] 7 3
```

Next, we'll look at a slightly larger dataset comprising loblolly pine growth measurements. The code below loads the dataset using `read.csv()`, introduced in Section 3.3.2, and assigns the resulting data frame to `loblolly`. For fun, we use `class()` to confirm that `loblolly` is, in fact, a data frame.

```
loblolly <- read.csv("datasets/loblolly_trees.csv")
class(loblolly)
```

```
#> [1] "data.frame"
```

```
head(loblolly)
```

```
#>   height age seed
#> 1   4.51   3  301
#> 2  10.89   5  301
#> 3  28.72  10  301
#> 4  41.74  15  301
#> 5  52.70  20  301
#> 6  60.92  25  301
```

Each row in the `loblolly` dataset is a tree, and columns record `height` (ft), `age` (years), and seed source code (`seed`). How many trees are in this dataset?

4.5.1 Accessing elements

To access specific elements in a data frame, we need to specify both row and column indexes. Inside the square brackets, row indexes come first, followed by a comma, then column indexes.

```
loblolly[1, 2] # Row 1 and column 2.
```

```
#> [1] 3
```

```
loblolly[1:3, 2] # Rows 1 through 3 and column 2.
```

```
#> [1]  3  5 10
```

```
loblolly[1:3, 2:3] # Rows 1 through 3 and columns 2 through 3.
```

```
#>   age seed
#> 1   3  301
#> 2   5  301
#> 3  10  301
```

```
loblolly[, 1] # All rows and column 1.
```

```
#>  [1]  4.51 10.89 28.72 41.74 52.70 60.92  4.55 10.92
#>  [9] 29.07 42.83 53.88 63.39  4.79 11.37 30.21 44.40
#> [17] 55.82 64.10  3.91  9.48 25.66 39.07 50.78 59.07
#> [25]  4.81 11.20 28.66 41.66 53.31 63.05  3.88  9.40
#> [33] 25.99 39.55 51.46 59.64  4.32 10.43 27.16 40.85
#> [41] 51.33 60.07  4.57 10.57 27.90 41.13 52.43 60.69
#> [49]  3.77  9.03 25.45 38.98 49.76 60.28  4.33 10.79
#> [57] 28.97 42.44 53.17 61.62  4.38 10.48 27.93 40.20
#> [65] 50.06 58.49  4.12  9.92 26.54 37.82 48.43 56.81
#> [73]  3.93  9.34 26.08 37.79 48.31 56.43  3.46  9.05
#> [81] 25.85 39.15 49.12 59.49
```

Note that `loblolly[, 1]` returns all elements in the first column. This mirrors vector behavior: leaving an index blank returns all values along that dimension. Similarly, `loblolly[1, ]` returns all columns for row 1.

We can also access columns (or rows) using their names.

```
loblolly[1:4, "height"]
```

```
#> [1]  4.51 10.89 28.72 41.74
```

```
loblolly[1:4, c("age", "seed")]
```

```
#>   age seed
#> 1   3  301
#> 2   5  301
#> 3  10  301
#> 4  15  301
```

There's another way to access columns in a data frame, using the `$` notation.

```
loblolly$height
```

```
#>  [1]  4.51 10.89 28.72 41.74 52.70 60.92  4.55 10.92
#>  [9] 29.07 42.83 53.88 63.39  4.79 11.37 30.21 44.40
#> [17] 55.82 64.10  3.91  9.48 25.66 39.07 50.78 59.07
#> [25]  4.81 11.20 28.66 41.66 53.31 63.05  3.88  9.40
#> [33] 25.99 39.55 51.46 59.64  4.32 10.43 27.16 40.85
#> [41] 51.33 60.07  4.57 10.57 27.90 41.13 52.43 60.69
#> [49]  3.77  9.03 25.45 38.98 49.76 60.28  4.33 10.79
#> [57] 28.97 42.44 53.17 61.62  4.38 10.48 27.93 40.20
#> [65] 50.06 58.49  4.12  9.92 26.54 37.82 48.43 56.81
#> [73]  3.93  9.34 26.08 37.79 48.31 56.43  3.46  9.05
#> [81] 25.85 39.15 49.12 59.49
```

```
loblolly$age
```

```
#>  [1]  3  5 10 15 20 25  3  5 10 15 20 25  3  5 10 15 20
#> [18] 25  3  5 10 15 20 25  3  5 10 15 20 25  3  5 10 15
#> [35] 20 25  3  5 10 15 20 25  3  5 10 15 20 25  3  5 10
#> [52] 15 20 25  3  5 10 15 20 25  3  5 10 15 20 25  3  5
#> [69] 10 15 20 25  3  5 10 15 20 25  3  5 10 15 20 25
```

```
height
```

```
#> Error: object 'height' not found
```

```
age
```

```
#> Error: object 'age' not found
```

Notice that typing the variable name alone (e.g., `height` or `age`) doesn't work. These are columns in the `loblolly` data frame, not standalone objects in the R environment, so they must be referenced with the data frame name.

When a subset includes a single column (e.g., `loblolly[, 1]` or `loblolly[, "height"]`), the result is no longer a data frame. Instead, R *simplifies* it to a vector. This can be frustrating if you expect a data frame and get a vector instead. To prevent this, add `drop = FALSE`, for example: `loblolly[, "height", drop = FALSE]`. Alternatively, see Section 6.2 for a data frame extension that doesn't simplify.

4.5.2 Adding columns

Suppose we want to add two new columns to `loblolly` that give height in meters and centimeters (the current `height` column is in feet). The `$` notation makes this straightforward.

```
loblolly$height_m <- 0.3048 * loblolly$height
loblolly$height_cm <- 100 * loblolly$height_m
```

4.5.3 Removing columns

Suppose we realize the columns added to `loblolly` aren't needed anymore, so we want to remove them. Here are three ways to remove the `height_m` and `height_cm` columns.

1. Subset the data frame to keep only the columns of interest and reassign.

   ```
   loblolly <- loblolly[, 1:3]
   ```

2. Subset by excluding specific columns and reassign.

   ```
   loblolly <- loblolly[, -c(4, 5)]
   ```

3. Assign the columns to `NULL`.

   ```
   loblolly[, 4:5] <- NULL
   # Or.
   loblolly[, c("height_m", "height_cm")] <- NULL
   # Or.
   loblolly$height_m <- NULL
   loblolly$height_cm <- NULL
   ```

4.5.4 Transforming columns

We can also transform existing columns easily. Suppose we want `age` to represent years since a base age of 10.

```
loblolly$age <- loblolly$age - 10 # Redefine age.
head(loblolly$age)
```

```
#> [1] -7 -5  0  5 10 15
```

```
# Change age back to the number of years old.
loblolly$age <- loblolly$age + 10
```

4.5.5 Rearranging columns

While rearranging columns is usually not all that important for actual data analysis purposes, we often want to rearrange the columns in a dataset for visualization purposes. Say we want to move the `seed` and `age` columns before the `height` column. This can be done with a column subset operation using either the column names or indexes.

```
loblolly <- loblolly[, c(3, 2, 1)]
# Or.
loblolly <- loblolly[, c("seed", "age", "height")]

names(loblolly)
```

```
#> [1] "seed"   "age"    "height"
```

4.6 Lists

The third data structure we'll work with is a list. Technically, a list is a vector but one in which elements can be of different types. For example, a list may have one element that's a vector, one that's a data frame, and another that's a function. To appreciate a list's usefulness, consider designing a function that fits a simple linear regression model (you don't need to know about regression analysis to follow this example). Evaluating a regression model's performance typically involves looking at several different outputs, some of which are listed below.

- The slope and intercept coefficients (a numeric vector with two elements).
- The residuals (a numeric vector with n elements, where n is the number of data points).
- Fitted values for the data (a numeric vector with n elements).
- The names of the dependent and independent variables (a character vector with two elements).

In fact, `lm()` fits a regression model and returns a list of these and other useful outputs (see Section 16.1 for details). The code below fits a regression model to the `height` and `age` variables in the `loblolly` data frame.

```
ht_age_mod <- lm(height ~ age, data = loblolly)
mode(ht_age_mod)
```

```
#> [1] "list"
```

```
names(ht_age_mod)
```

```
#>  [1] "coefficients"  "residuals"     "effects"
#>  [4] "rank"          "fitted.values" "assign"
#>  [7] "qr"            "df.residual"   "xlevels"
#> [10] "call"          "terms"         "model"
```

```
ht_age_mod$coefficients
```

```
#> (Intercept)         age
#>     -1.3124      2.5905
```

```
length(ht_age_mod$residuals)
```

```
#> [1] 84
```

The list returned by `lm()` above is assigned to `ht_age_mod`. `mode()` returns the type or storage mode of an object and is used above just to prove that `lm()` returns a list.[7] The code also illustrates that named elements of a list can be accessed using the dollar sign notation, similar to data frames.[8] Notice the `coefficients` list element is a length 2 vector of regression coefficients, while the `residuals` element is a length 84 vector of residuals.

Similar to `data.frame()`, you can create a list using `list()` as illustrated below using some data structures introduced earlier in the chapter. The first list element, named `first`, holds the `dbh` vector we created in Section 4.2. The second list element, named `second`, holds the `trees` data frame. The third list element, named `third`, holds a list with elements named `a` and `b` that hold a vector of values 1 through 10, and a copy of the `trees` data frame, respectively.

```
example_list <- list(first = dbh, second = trees,
                     third = list(a = 1:10, b = trees))
example_list
```

---

[7]One could also consult `lm()`'s manual page which describes the returned list.

[8]A data frame is actually a specialized list consisting of equal length vectors (columns) and unique row names. You can prove this to yourself by running `is.list(loblolly)` and `is.data.frame(loblolly)`.

```
#> $first
#> [1] 16  8  2 16 10 14 13
#>
#> $second
#>                 Species DBH   AGS
#> Tree 1     Acer rubrum  16  TRUE
#> Tree 2     Acer rubrum   8  TRUE
#> Tree 3    Betula lenta   2 FALSE
#> Tree 4    Betula lenta  16  TRUE
#> Tree 5 Prunus serotina  10 FALSE
#> Tree 6 Prunus serotina  14 FALSE
#> Tree 7     Acer rubrum  13  TRUE
#>
#> $third
#> $third$a
#>  [1]  1  2  3  4  5  6  7  8  9 10
#>
#> $third$b
#>                 Species DBH   AGS
#> Tree 1     Acer rubrum  16  TRUE
#> Tree 2     Acer rubrum   8  TRUE
#> Tree 3    Betula lenta   2 FALSE
#> Tree 4    Betula lenta  16  TRUE
#> Tree 5 Prunus serotina  10 FALSE
#> Tree 6 Prunus serotina  14 FALSE
#> Tree 7     Acer rubrum  13  TRUE
```

4.6.1 Accessing elements

We've already seen that list elements can be accessed using the dollar sign notation. In addition, square bracket notation can be used. However, there is a subtle wrinkle—using either single or double square brackets. Let's try accessing the first element in our example list using the dollar sign `$`, double square bracket `[[]]`, then single square bracket `[]`.

```
example_list$first
```

```
#> [1] 16  8  2 16 10 14 13
```

```
mode(example_list$first)
```

```
#> [1] "numeric"
```

```
example_list[[1]]
```

```
#> [1] 16  8  2 16 10 14 13
```

```
mode(example_list[[1]])
```

```
#> [1] "numeric"
```

```
example_list[1]
```

```
#> $first
#> [1] 16  8  2 16 10 14 13
```

```
mode(example_list[1])
```

```
#> [1] "list"
```

Notice the dollar sign and double bracket return a numeric vector, while the single bracket returns a list. Also notice the difference in results below.

```
example_list[c(1,2)]
```

```
#> $first
#> [1] 16  8  2 16 10 14 13
#>
#> $second
#>                 Species DBH   AGS
#> Tree 1      Acer rubrum  16  TRUE
#> Tree 2      Acer rubrum   8  TRUE
#> Tree 3     Betula lenta   2 FALSE
#> Tree 4     Betula lenta  16  TRUE
#> Tree 5 Prunus serotina  10 FALSE
#> Tree 6 Prunus serotina  14 FALSE
#> Tree 7      Acer rubrum  13  TRUE
```

```
example_list[[c(1,2)]]
```

```
#> [1] 8
```

The single bracket form returns the first and second elements of the list, while the double bracket form returns the second element in the first element of the list. Generally, don't put a vector of indexes or names in a double bracket, as you'll likely get unexpected results. See, for example, the results below.[9]

[9]Try this example using only single brackets. It'll return a list holding elements `first`, `second`, and `third`.

```
example_list[[c(1,2,3)]]
```

```
#> Error in example_list[[c(1, 2, 3)]]: recursive indexing failed at level 2
```

To recap, the single bracket `[]` returns a list of object(s) held at the given indexes or names placed in the bracket, whereas the double brackets `[[]]` return the object itself held at the index or name placed in the innermost bracket. Put differently, a single bracket is used to access a range of list elements and returns a list, and a double bracket should be used to access a single list element and returns the object held in the given element.

4.6.2 Adding and removing elements

As you've now already seen, there are often several ways to accomplish the same task using R. Here are three options for adding the `loblolly` data frame to `example_list`. Run any one of the options. Again, use `str()` for an informative look at the list's structure.

```
example_list$loblolly <- loblolly
# Or.
example_list[["loblolly"]] <- loblolly
# Or, recalling a list is a vector, you can use the c() function.
example_list <- c(example_list, list(loblolly = loblolly))
```

You can remove list elements by name or index. Again, run any one of the options below.

```
example_list$loblolly <- NULL
# Or.
example_list[["loblolly"]] <- NULL
# Or.
example_list <- example_list[1:3]
# Or.
example_list <- example_list[-4]
```

4.7 Comparison and logical operators

Comparison operators are binary operators that test a comparative condition between operands and return a logical value to indicate the test result. You might, or might not, recall from your grade school years that an *operand* is what operators are applied to, and an operator is *binary* if it has two operands. For example, the *greater than* binary operator in $1 > 2$ tests whether the left operand 1 is greater than the right operand 2, the result of which is FALSE.

Let's walk through the comparison operators available in R. We'll present the operator and its definition, followed by an example using the `dbh` and `spp` vectors created in Section 4.2. First let's recall the values held in these vectors.

```
dbh
```

```
#> [1] 16  8  2 16 10 14 13
```

```
spp
```

```
#> [1] "Acer rubrum"     "Acer rubrum"
#> [3] "Betula lenta"    "Betula lenta"
#> [5] "Prunus serotina" "Prunus serotina"
#> [7] "Acer rubrum"
```

1. `==` the equality operator: The "double equals sign" tests if operands are equal. Below, we perform a logical test to determine which `spp` vector elements equal `Acer rubrum`.

   ```
   spp == "Acer rubrum"
   ```

   ```
   #> [1]  TRUE  TRUE FALSE FALSE FALSE FALSE  TRUE
   ```
   Not surprisingly, the first two and last elements return `TRUE` and the other four elements return `FALSE`. Notice we're using the `==` sign, not the `=` sign. Mixing up the comparison operator `==` and assignment operator `=` is a common error.

2. `!=` the inequality operator: Tests if operands are not equal and is thus the inverse of `==`. We see this by testing which elements in `spp` do not equal `Acer rubrum`.

   ```
   spp != "Acer rubrum"
   ```

   ```
   #> [1] FALSE FALSE  TRUE  TRUE  TRUE  TRUE FALSE
   ```

3. `<`, `<=`, `>`, `>=` less than, less than or equal to, greater than, and greater than or equal to operators, respectively: Using the `dbh` vector, determine which elements are greater than 16 and then greater than or equal to 16.

   ```
   dbh > 16
   ```

   ```
   #> [1] FALSE FALSE FALSE FALSE FALSE FALSE FALSE
   ```

   ```
   dbh >= 16
   ```

   ```
   #> [1]  TRUE FALSE FALSE  TRUE FALSE FALSE FALSE
   ```

While not immediately apparent from the examples above, recycling is occurring before comparison operations. Recall from Section 4.2.3, the shorter vector is recycled (repeated) then, if needed, truncated to match the length of the longer vector. If truncation occurs, a warning is thrown. In all cases above, the right operand vector of length 1 is recycled to match the left operand vector length before the comparison operator is applied. For example, the length 1 vector `"Acer rubrum"` is recycled to match the length 7 `spp` vector then the equality operator is applied elementwise between the two, now equal length, vectors.[10]

Suppose we want to know which `dbh` vector elements are greater than 14 *and* less than 20. Answering this question requires use of two comparison operators, i.e., $<$ and $>$. In such cases, logical operators are used to combine multiple comparison operations into a single logical statement. We consider the following *logical operators*: "and", "or", "xor", and "negation." Importantly, in order of operation, comparison operators precede logical operators. The `Syntax` manual page (i.e., run `?Syntax` on the console) lists R operators' order of operation, where you'll notice the comparison operators are listed before the logical operators in the precedence groups under the Details section.

Let's walk through each of the logical operators.

1. `&` the "and" operator: A comparison using the `&` operator returns `TRUE` when both operands are `TRUE` and `FALSE` otherwise. The `&` operator works elementwise on the logical vector operands. Consider the following example.

```
dbh < 16
```

```
#> [1] FALSE  TRUE  TRUE FALSE  TRUE  TRUE  TRUE
```

```
dbh > 10
```

```
#> [1]  TRUE FALSE FALSE  TRUE FALSE  TRUE  TRUE
```

```
dbh < 16 & dbh > 10
```

```
#> [1] FALSE FALSE FALSE FALSE FALSE  TRUE  TRUE
```

First we show the results of `dbh < 16` and `dbh > 10` separately. When combining the comparison operations using the `&` operator, R first performs `dbh < 16` and `dbh > 10`, then applies `&` elementwise on the operands. The elementwise `&` returns `TRUE` when the element in the `dbh < 16` vector is `TRUE` **and** the corresponding element in the `dbh > 10` vector is `TRUE`, which in this case is for DBH values 14 and 13. The key point to remember is that `&` returns `TRUE` only if both operands are `TRUE`.

2. `|` the "or" operator: A comparison using the `|` operator returns `TRUE` if at least one operand

[10]Test your understanding of this very important concept using a few simple examples. For instance, write out the result of `spp == c("Acer rubrum", "Betula lenta")` and `c(16, 10, 1) >= dbh` by hand, then run the code to check your result.

is TRUE and FALSE otherwise. Like the & operator, the | operator works elementwise on the operands. Let's use the same example as before, but now we'll look for trees with a DBH less than 16 or a DBH greater than 10.

```
dbh < 16 | dbh > 10
```

```
#> [1] TRUE TRUE TRUE TRUE TRUE TRUE TRUE
```

Not surprisingly, this operation returns TRUE for all elements, because all elements in dbh are either less than 16 or greater than 10.

3. xor the "exclusive or" operator: A comparison using the xor operator returns TRUE if only one of the operands is TRUE and FALSE otherwise.

```
xor(dbh < 16, dbh > 10)
```

```
#> [1]  TRUE  TRUE  TRUE  TRUE  TRUE FALSE FALSE
```

Above, the first five elements are TRUE because only one operand is TRUE, and the last two elements are FALSE because both operands are TRUE.
While we can imagine cases where this operator would be handy, we've never found the occasion to use it in our own code.

4. ! the "negation" or "not" operator: The exclamation point ! (called "bang" in programmer's slang) reverses a logical value, i.e. !TRUE is FALSE and !FALSE is TRUE. The code below returns TRUE for DBH values not greater than 14 (while not required, the parentheses emphasize the order of operation). Also notice ! is a *unary operator*, i.e., it acts on a single operand to its right.

```
!(dbh > 14)
```

```
#> [1] FALSE  TRUE  TRUE FALSE  TRUE  TRUE  TRUE
```

There is a "&&" and "||" variant of "&" and "|", respectively. These "double" operators examine only the first element of operand vectors in a comparison rather than comparing element by element. There are a few cases where && and || are useful when writing conditional statements in functions (see, e.g., Chapter 5); however, we'll generally not make use of them in this book.

4.7.1 The %in% operator

Suppose we want to identify the dbh vector elements equal to 13, 10, or 14. We can do this using the equality operator == and the | operator as follows.

```
dbh
```

```
#> [1] 16  8  2 16 10 14 13
```

```
dbh == 13 | dbh == 10 | dbh == 14
```

```
#> [1] FALSE FALSE FALSE FALSE  TRUE  TRUE  TRUE
```

However, this is a little clunky, involves a lot of typing, and generally makes code hard to read. Lucky for us, R has the "in" operator, `%in%`, to accomplish this task in a more intuitive and easy-to-read manner.

```
dbh %in% c(13, 10, 14)
```

```
#> [1] FALSE FALSE FALSE FALSE  TRUE  TRUE  TRUE
```

In the spirit of coding techniques to promote efficient and readable code, we'll use the `%in%` operator throughout the book.

Comparison and logical operators are invaluable to identify subsets of data that meet specified conditions. The next section explores how conditional and logical operators facilitate subsetting vectors, data frames, and lists.

4.8 Subsetting with logical vectors

Consider the `loblolly` data frame. How can we access only those trees with heights more than 50 m? How can we access the age of those trees taller than 50 m? How can we compute the mean height of all trees from seed source 301? The dataset is small enough that it wouldn't be too onerous to extract the values by hand. However, for larger or more complex datasets, this task would be very difficult or impossible in a reasonable amount of time.

R has a powerful method for solving these sorts of problems based on a variant of the subsetting methods that we've learned. When given a logical vector in square brackets, R returns the values corresponding to `TRUE`.

To begin, focus on the `dbh` and `spp` vectors created in Section 4.2. As we saw in Section 4.7, `dbh > 15` returns `TRUE` for each value of `dbh` greater than 15 and `FALSE` for each value of `dbh` less than or equal to 15. Similarly, `spp == "Betula lenta"` returns `TRUE` or `FALSE` depending on whether the `spp` element equals `"Betula lenta"`.

Consider the next three lines of code used to illustrate subsetting with logical vectors.

```
spp[dbh > 15]
```

```
#> [1] "Acer rubrum"  "Betula lenta"
```

```
dbh[dbh > 15]
```

```
#> [1] 16 16
```

```
dbh[spp == "Betula lenta"]
```

```
#> [1]  2 16
```

The first line uses the logical vector from `dbh > 15` to return those elements of `spp` where `dbh > 15` is `TRUE`. Similarly, the second example uses the same logical vector to return all DBH values greater than 15. The third line uses the logical vector from `spp == "Betula lenta"` to return the DBH of all *Betula lenta*.

Subsetting with logical vectors is an important tool for data analysis, but it can be challenging at first. When presented with an expression like `spp[dbh > 15]`, it helps to read it "from the inside, out." In other words, first evaluate `dbh > 15` *inside* the square brackets and understand that it returns a logical vector. Then recognize the subset operation, i.e., using square brackets, returns those elements of `spp` that correspond to the `TRUE` values in the logical vector. We'll get more practice with this in this chapter's exercises.

4.8.1 Modifying and creating objects via subsetting

Subsetting on the left side of an assignment is a common way to modify an existing object. Consider the following examples that demonstrate modifying values of an existing object and assigning a subset to a new object.

```
x <- 1:10 # Make a vector with values 1 through 10.
x
```

```
#>  [1]  1  2  3  4  5  6  7  8  9 10
```

```
x[x < 5] <- 0 # Set all values less than 5 to 0.
x
```

```
#>  [1]  0  0  0  0  5  6  7  8  9 10
```

```
y <- -3:9 # Make another vector.
y
```

```
#>  [1] -3 -2 -1  0  1  2  3  4  5  6  7  8  9
```

```
y[y < 0] <- NA # Set all values less than 0 to NA.
y
```

```
#>  [1] NA NA NA  0  1  2  3  4  5  6  7  8  9
```

```
# Make a vector z that contains only positive values of y.
z <- y[y > 0]
z
```

```
#>  [1] NA NA NA  1  2  3  4  5  6  7  8  9
```

```
z <- z[!is.na(z)] # Keep only non-NA values.
z
```

```
#> [1] 1 2 3 4 5 6 7 8 9
```

Above, our goal was for `z` to contain only positive values of `y` (i.e., we specified `y > 0`). However, our resulting `z` contained `y`'s positive and `NA` values. This is because `NA` tested in a comparison is `NA` (e.g., `NA > 0` is `NA`) and `NA` used in square brackets returns an `NA`. As a result, it took an additional logical subset to arrive at the desired `z`, i.e., `z[!is.na(z)]`. These two steps could be combined into a single statement, i.e., `z <- y[!is.na(y) & y > 0]`. Special care is needed when using comparison, conditional, and mathematical operators on data with `NA` values.

4.8.2 Subsetting data frames

Subsetting a data frame's rows and columns using logical vectors follows in a straightforward way from Section 4.5.1. Let's again consider the `loblolly` data frame.

In the code below, because the `$` is referencing a single column, we're really just subsetting a vector, i.e., `age`. The subset is the age of trees from seed source 301.

```
loblolly$age[loblolly$seed == 301]
```

```
#> [1]  3  5 10 15 20 25
```

Next, consider two examples that subset the data frame using square brackets.

```
loblolly[loblolly$seed == 301, ]
```

```
#>   seed age height
#> 1  301   3   4.51
#> 2  301   5  10.89
#> 3  301  10  28.72
#> 4  301  15  41.74
#> 5  301  20  52.70
#> 6  301  25  60.92
```

```
loblolly[loblolly$age > 20, 2:3]
```

```
#>    age height
#> 6   25  60.92
#> 12  25  63.39
#> 18  25  64.10
#> 24  25  59.07
#> 30  25  63.05
#> 36  25  59.64
#> 42  25  60.07
#> 48  25  60.69
#> 54  25  60.28
#> 60  25  61.62
#> 66  25  58.49
#> 72  25  56.81
#> 78  25  56.43
#> 84  25  59.49
```

Above, notice desired rows are selected using a logical vector to the left of the comma within the square brackets. The first example selects rows (trees) with seed source equal to 301 and all columns. The second line selects trees with `age` greater than 20 and columns 2 and 3.

Next, let's use the `&` logical operator to find all trees with age greater than 20 and from seed source 301.

```
loblolly[loblolly$age > 20 & loblolly$seed == 301, ]
```

```
#>   seed age height
#> 6  301  25  60.92
```

Often the logical test is based on a dataset dependent value. For example, say we want the tallest tree's age and seed source.

```
loblolly[loblolly$height == max(loblolly$height), ]
```

```
#>    seed age height
#> 18  305  25   64.1
```

Next consider the much larger and more complex Free-Air Carbon Dioxide Enrichment (FACE) experiment dataset introduced in Section 1.2.3. Recall that each row in the FACE dataset represents a tree in the experiment, and the columns record the trees' clone type, treatment, and measurements taken over time. We first read the FACE data from its file and assign it to the `face` data frame. The function `dim()` tells us that `face` has 1991 trees (rows) and 45 columns.[11]

```
face <- read.csv("datasets/FACE_aspen_core_growth.csv")
dim(face)
```

```
#> [1] 1991   45
```

Let's take a subset of `face` that includes only aspen Clone 8L trees (which incidentally, was shown by Kubiske et al. (2007) to perform better in elevated O3 than other aspen clones in the FACE experiment). Consulting the metadata referenced in Section 1.2.3, aspen Clone 8L are identified by `Clone` column value `8L`. We call the resulting subset `face_8L` in the code below.

```
face_8L <- face[face$Clone == "8L", ]
dim(face_8L)
```

```
#> [1] 331  45
```

Using `dim` we see there are 331 Clone 8L trees.

Next, let's create a data frame that only contains data for Clone 8L and no missing height measurements in 2007 (the height values are in column `X2007_Height`). Recall, we saw `!is.na()` in Section 4.8.1 to identify all non-`NA` values as `TRUE` and `FALSE` otherwise.

```
face_8L_all2007 <- face_8L[!is.na(face_8L$X2007_Height), ]
dim(face_8L_all2007)
```

```
#> [1] 222  45
```

From `dim` we see there are 222 Clone 8L trees with non-missing height measurements in 2007.

Return attention to the original `face_8L` data frame. How can we extract only those trees with no missing data across all columns (i.e., not just `X2007_Height`)? Consider the following simple example.

```
df <- data.frame(V1 = c(1, 2, 3, 4, NA),
                 V2 = c(NA, 1, 4, 5, NA),
                 V3 = c(1, 2, 3, 5, 7))
df
```

```
#>   V1 V2 V3
#> 1  1 NA  1
#> 2  2  1  2
```

[11]If you're following along at your console, run `str(face)` for a more detailed description of the data frame.

```
#> 3  3  4  3
#> 4  4  5  5
#> 5 NA NA  7
```

```
is.na(df)
```

```
#>         V1    V2    V3
#> [1,] FALSE  TRUE FALSE
#> [2,] FALSE FALSE FALSE
#> [3,] FALSE FALSE FALSE
#> [4,] FALSE FALSE FALSE
#> [5,]  TRUE  TRUE FALSE
```

```
rowSums(is.na(df))
```

```
#> [1] 1 0 0 0 2
```

First notice that `is.na()` will test each element of a data frame for the presence of `NA`. Also recall that if R is asked to sum a logical vector, it'll first coerce the logical vector to numeric and then compute the sum, which effectively counts the number of `TRUE` elements in the logical vector. `rowSums()` computes the sum of each row. So `rowSums(is.na(df))` returns a vector with element values equal to the number of `NA`s in the corresponding data frame row, so zero elements correspond to rows in `df` with no `NA`s. The code below uses this approach to implement a simple method to identify trees with no missing data.

```
dim(face_8L)
```

```
#> [1] 331  45
```

```
face_8L_complete <- face_8L[rowSums(is.na(face_8L)) == 0, ]
dim(face_8L_complete)
```

```
#> [1] 186  45
```

Out of the 331 trees in the original data frame, only 186 have no missing data (a statistician's nightmare)!

4.8.3 Subsetting with `which()`

We've seen how to subset vectors and data frames by placing logical vectors within square brackets. An alternative approach to logical subsetting is to use `which()`. `which()` accepts as input a logical vector and returns the indexes of the vector that are `TRUE`. Consider the following examples

```
which(c(TRUE, FALSE, TRUE, FALSE)) # Which elements have a TRUE value?
```

```
#> [1] 1 3
```

```
dbh # Recall the dbh vector values.
```

```
#> [1] 16  8  2 16 10 14 13
```

```
which(dbh > 15) # Which elements have values greater than 15?
```

```
#> [1] 1 4
```

```
spp # Recall the spp vector values.
```

```
#> [1] "Acer rubrum"     "Acer rubrum"
#> [3] "Betula lenta"    "Betula lenta"
#> [5] "Prunus serotina" "Prunus serotina"
#> [7] "Acer rubrum"
```

```
which(spp == 'Betula lenta') # Which element equals Betula lenta?
```

```
#> [1] 3 4
```

The first example returns a vector with values 1 and 3, which are the indexes of those elements with `TRUE` values. In the second example, the result is a vector with values 1 and 4, which correspond to the `dbh` vector indexes with element values greater than 15. Similarly, the third example returns a vector with values 3 and 4, which are the `spp` vector indexes with element values equal to `Betula lenta`. As an alternative to logical subsetting, we can use the results from `which()` to subset a vector or data frame of interest.

```
dbh[which(spp == "Betula lenta")] # Which (index) subsetting.
```

```
#> [1]  2 16
```

```
dbh[spp == "Betula lenta"] # Logical subsetting.
```

```
#> [1]  2 16
```

Logical subsetting requires a bit less typing, so we generally prefer it over using `which()`. However, if you find `which()` more intuitive, we encourage you to use it.

Two extensions of `which()` are `which.max()` and `which.min()` that return the index of the element with the maximum and minimum value in a numeric vector, respectively. For example, below we use `which.max()` to determine the index of the maximum value in the `dbh` vector.

```
which.max(dbh)
```

```
#> [1] 1
```

Notice 16 is the maximum value in `dbh` (i.e., `max(dbh)`). Despite there being two values of 16 in `dbh`, `which.max()` returns the index for the first element with value equal to 16. You might have expected the function to return the indexes of all elements with values equal to 16; however, that's not the case. The moral of the story is to always read a function's manual page to understand its behavior (see `?which.max` and `?which.min`).

4.9 Summary

Vectors, matrices, arrays, data frames, and lists are data structures used to organize, store, and access data in R.[12] Homogeneous vectors, matrices, and arrays require all elements to be of the same data type. Heterogeneous data frames and lists allow elements to be of different types, e.g., some elements of a data frame may be numeric while others may be character. These data structures (or objects) are data analysis building blocks.

We introduced four data types: character, double, integer, and logical. We learned functions to convert between data types, as well as ways that R automatically converts between data types (called coercion). We covered the all-important topic of indexing vectors using square bracket notation `[]`. This is an essential skill for data analysis in R, and we'll make use of it throughout the book. We then briefly discussed working with factors, which comes into play when working with categorical variables (e.g., species classification and stand treatment).

We extended the concept of vectors to two dimensions in our discussion of data frames used to store rectangular data. We saw how the square bracket notation extends to accommodate data frame rows and columns, and we also introduced `$` to access specific data frame columns. Next we saw that lists are used to store elements of potentially different objects. We dedicated a good amount of text to subsetting lists, which can be done using single brackets `[]` (which returns a list consisting of the specified elements) and double brackets `[[]]` (which returns the actual objects at the specified elements of the list).

Comparison and logical operators were introduced and used to form logical vectors that, in turn, were used to subset vectors and data frames. The ability to select elements that meet specific criteria is useful through all steps in data analysis and reporting.

[12]While we didn't spend time describing matrices and arrays, you'll find they're variations and extensions of the other data structures covered here.

With this foundational understanding of data structures and subsetting, we next take a more in-depth look at the methods we can use to manipulate such data structures (i.e., functions), as well as other foundational programming concepts.

4.10 Exercises

In Exercises 4.1 through 4.9, we'll work with the small dataset shown in Table 4.2 that contains the heights and ages of six trees measured in three forest stands.

Exercise 4.1. Using data in Table 4.2, create three numeric vectors called `stand`, `height`, and `age` corresponding to table columns `stand`, `Height`, and `Age`, respectively. Then compute the mean value for the `Height` and `Age` variables. Omit missing values from the mean calculations using the appropriate `mean()` function arguments.

Exercise 4.2. Following from Section 4.2.3, use the square brackets `[]` to change the values in the first three elements of the `age` vector to 18, 60, and 10, respectively. After making the change, print the vector, then change the elements back to their original values given in Table 4.2.

Exercise 4.3. Use the square brackets `[]` to perform the following subsets on the `height` vector.

a. Print the value of element 1.
b. Print the values of elements 1, 3, and 6.
c. Use the vector `c(1,3,6)` to print values at elements 2, 4, and 5. Hint, use the - operator.
d. Print the last value in the vector without using the index value 6. Hint, use the `length()` function.

Exercise 4.4. Create a data frame called `tree_data` containing the `stand`, `height`, and `age` vectors and set their corresponding columns names to `Stand`, `Height`, and `Age`, respectively

TABLE 4.2: Practice dataset of tree height (m) and age (yr).

Height	Stand	Age
1	20.23	20
1	30.41	50
2	10.32	15
2	5.38	11
3	20.43	17
3	NA	37

(to underscore that R is case sensitive, we changed from lowercase vector names to uppercase column names). Display the structure of the data frame using the `str()` function. Use this data frame in Exercises 4.5 through 4.9.

Exercise 4.5. Following from Section 4.5.1, use the square brackets `[]` to perform the following subsets.

a. Print all values in row 1.
b. Print all values in rows 1, 3, and 6.
c. Use the vector `c(1,3,6)` to print values in rows 2, 4, and 5. Hint, use the `-` operator.
d. Print all values in the last row without using the index value 6. Hint, use the `nrow()` function.
e. Print all rows in column 1.
f. Print all rows in columns 1 and 3 using the column names, i.e., `Height` and `Age`.
g. Print the values in rows 5 and 6 and columns `Height` and `Age`.

Exercise 4.6. Use logical subsetting to print the height of trees in stand 1.

Exercise 4.7. Use logical subsetting to print the height and age of trees in stands 1 and 3.

Exercise 4.8. Use logical subsetting to print all rows with trees older than 30 but younger than 50 years.

Exercise 4.9. Use logical subsetting to print all rows corresponding to trees with age greater than or equal to 20 and at least 10 meters tall. Careful here, the `NA` in the `Height` column will need to be dealt with appropriately, e.g., you might use the negation operator `!` in combination with `is.na()` in your conditional statement as illustrated in Section 4.8.1.

Exercise 4.10. *Hardcoding* is the practice of embedding specific values or assumptions directly in your code rather than reading them from an external source or defining them in a flexible, changeable way. In other words, the values are *fixed in the code itself* and can only be changed by editing the code.

Hardcoding is generally poor programming practice because it reduces readability, makes code harder to maintain, and limits reuse. If a value changes (for example, a file path, a set of category codes, or a threshold used in an analysis), you must hunt through the code and modify it manually, increasing the chance of errors.

Consider the following vector containing a few plant codes used by the US Forest Service. Notice that the vector contains two missing values, denoted by `NA`.

```
spp_codes <- c("ABIES", "ABBA", NA, "ABCO", "ABFR", NA)
```

Say we want to remove these `NA` values. A hardcoding approach would explicitly use the indexes for the `NA` as illustrated in the code below.

```
spp_codes[-c(3,6)]
```

```
#> [1] "ABIES" "ABBA"  "ABCO"  "ABFR"
```

However, let's say this data cleaning step will need to be performed on other species code vectors of a potentially different length. Develop a non-hardcoding approach to remove `NA` values from vectors of any lengths. Apply your approach to the `spp_codes` vector.

Exercise 4.11. Collecting field data is often a messy process that can result in data entry errors. Consider the following vector representing the number of trees tallied in ten experimental forest plots.

```
n_trees <- c(150, 138, 289, 239, 12, 103, 310, -200, 218, 178)
```

Your background knowledge about the forest tells you that any element less than 50 is likely a data entry error and shouldn't be used in subsequent analyses. Also assume this is an ongoing data collection effort, so you want to avoid hardcoding. Write some code that sets all values less than 50 to `NA` then takes the mean of the resulting vector. The mean should be computed without the `NA` values (see the `mean()` function manual page for guidance).

Exercise 4.12. Read in the FEF file `"datasets/FEF_trees.csv"` and assign the resulting data frame to `fef_trees`. You might need to adjust the file path, depending on where your working directory is set relative to the `FEF_trees.csv` file. Use `fef_trees` to answer the following questions.

a. Confirm `fef_trees` has 88 rows and 18 columns. Create a subset of `fef_trees` called `fef_trees_sub` that includes all rows but only columns 1 through 6 (i.e., columns `watershed`, `year`, `plot`, `species`, `dbh_in`, and `height_ft`).

b. Recall each row in `fef_trees` corresponds to a tree measurement. Print the `fef_trees_sub` row that holds the tallest tree (i.e., maximum `height_ft` value). `max()` should come in handy in your logical statement.

c. Add new columns `dbh_cm` and `height_m` to `fef_trees_sub`. As the name suggests, these new columns should hold DBH in centimeters and height in meters.

d. Remove `dbh_in` and `height_ft` from `fef_trees_sub`.

e. Compute the mean DBH in cm for all *Acer rubrum* in watershed 3.

5

Functions and functional programming

In the previous chapter, we introduced data structures that serve as the building blocks for holding, manipulating, and analyzing data. *Functions*—such as `c()`, `data.frame()`, `str()`, `length()`, `sum()`, and `mean()`—are used to build data structures, describe characteristics of those structures, and summarize or analyze data held within those structures. Functions are fundamental to the R language. Gaining a better understanding of existing functions and the ability to write your own functions dramatically increases what you can do with R.

In this chapter, we introduce some key elements of functional programming used in R and found in nearly all programming and scripting languages. We begin by describing characteristics of functions and how you can define and use your own functions. We then introduce several control statements commonly used in functions to manipulate inputs and form desired output. We motivate this initial exploration of functional programming using forestry topics and examples. We cover only those topics needed in subsequent chapters, so this is a limited introduction. References for further reading are provided in the chapter summary.

5.1 Introduction to functions

A function has four key components:

1. Name: adheres to R's naming syntax (Section 3.2.2).
2. Arguments: the inputs (optional). Arguments are variables in the function's body.
3. Body: performs some task.
4. Return: the output (optional).

Let's map these components to function *pseudocode*.[1]

```
name_of_function <- function(arg_1, arg_2) {
  do something
  return(something)
}
```

[1] Pseudocode is notation resembling a simplified programming language and used to convey code features but doesn't actually conform to syntax requirements and hence doesn't run.

DOI: 10.1201/9781003222699-5

1. Name: we assign the function to `name_of_function`.
2. Arguments: are defined between `function()` parentheses. The function has arguments `arg_1` and `arg_2`.
3. Body: what the function will do is defined between the left and right curly brace `{}`. We wrote `do something` to indicate this is where the body of the function is written.
4. Return: before closing the function body, we define the function output within `return()`. While this component is optional—not all functions return something—most functions we write do return something.

Let's look at a syntactically correct function.

```
my_first_function <- function() {
  print("Ever stop to think, and forget to start again?")
}
```

Consider the four components:

1. Name: `my_first_function`.
2. Arguments: this function has no arguments, i.e., there is nothing defined between the `function()` parentheses.
3. Body: tells R to print a fun quote by Alan A. Milne.
4. Return: no objects are returned by this function.

Once defined, the function can be called. Below we call `my_first_function()`.

```
my_first_function()
```

```
#> [1] "Ever stop to think, and forget to start again?"
```

Here's another function.

```
pow <- function(x, v) {
  result <- x^v
  return(result)
}
```

Consider the four components:

1. Name: `pow`.
2. Arguments: this function has two arguments, `x` and `v`.
3. Body: uses the values held in `x` and `v` to compute `x^v` and assigns the result to `result`.
4. Return: returns the value held in `result`.

Let's try it out.

```
pow(x = 2, v = 5)
```

```
#> [1] 32
```

```
pow(x = 2, v = 0)
```

```
#> [1] 1
```

```
pow(x = c(2, NA, 5), v = c(2, 4))
```

```
#> Warning in x^v: longer object length is not a multiple
#> of shorter object length
```

```
#> [1]  4 NA 25
```

Notice in the last call to `pow()`, the return vector has the length of the longest input vector (in this case, `x`), and a warning is thrown. Given R's recycling rules, this behavior is expected. See Section 4.2.3 for more on vector operations and recycling. Here, to match `length(x)`, `v` is recycled to `c(2, 4, 2)` before `x^v` is computed.

5.1.1 Specifying arguments

When calling a function, arguments can be specified using their full name, partial name, or position.

We demonstrate each approach using the function `args_fn()` defined below.

```
args_fn <- function(first_arg, second_arg) {
  return(c(first_arg, second_arg))
}
```

`args_fn()` takes two arguments, `first_arg` and `second_arg`. The values of these arguments are combined into a single vector and returned. To reduce clutter, we define the vector to be returned within `return()`.

Let's first call the function using the full name of each argument. Below, we set the arguments to values `1` and `2`, respectively.

```
args_fn(first_arg = 1, second_arg = 2)
```

```
#> [1] 1 2
```

When using argument names, the order in which they are passed doesn't matter.

```r
args_fn(second_arg = 2, first_arg = 1)
```

```
#> [1] 1 2
```

Partial matching exists to save you from typing long argument names. With partial matching, you can specify arguments using the first portion of an argument's name, as long as it's unambiguous with the names of other arguments. In the example below, `s` partially matches `second_arg` and `f` partially matches `first_arg`.

```r
args_fn(s = 2, f = 1)
```

```
#> [1] 1 2
```

Because partial matching can lead to mistakes and is harder to read if you don't immediately recall what the partial name refers to, we rarely use it in practice.

You can also call a function without giving argument names. In that case, the values must follow the function's defined argument order. This is often done when we know the order and are feeling a bit lazy (i.e., we don't want to type the argument names).

```r
args_fn(1, 2)
```

```
#> [1] 1 2
```

5.1.2 Default values for arguments

Functions can be defined with default values for arguments. In such cases, the user can override a default value by specifying their own value for the argument. Consider the following example that defines a function `fn()` with arguments `a` and `b`. Argument `b` is defined with default value 1. After defining the function, it's called using the default value for `b` and then again with `b` set to 2.

```r
fn <- function(a, b = 1) {
  return(a + b)
}

fn(a = 1)
```

```
#> [1] 2
```

```r
fn(a = 1, b = 2)
```

```
#> [1] 3
```

Most functions introduced in previous chapters have default values for arguments. Some functions we'll introduce later in the book have dozens of arguments and, fortunately, authors of those functions have set reasonable default values.

5.1.3 Scope rules

A variable defined using a function argument or within the function body is a *local variable*, meaning the variable doesn't exist outside the function. Consider the example below, where `a` and `b` are defined in `fn()` and hence local to `fn()`. When `a` or `b` are called outside the function, R throws an object not found error.

```
fn <- function(a) {
  b <- 1 # Local a and b.
  return(a + b)
}

fn(a = 1)
```

```
#> [1] 2
```

```
a # Not defined.
```

```
#> Error: object 'a' not found
```

```
b # Not defined.
```

```
#> Error: object 'b' not found
```

The code above illustrates a *scope rule*. A language's scope rules determine where variables exist, their values, and how those values can be changed. In the code above, these rules dictate that `a` and `b` do not exist outside of `fn()`.

Consider the next example, where `b` is used in `fn()` but not defined in `fn()`. Here, `b` is referred to as a *free variable* and its value will need to be set outside the function. Here too, `b` is a *global variable* because it's available both outside and inside `fn()`.

```
fn <- function(a) {
  return(a + b) # Local a and free b.
}

fn(a = 1) # Free b not defined.
```

```
#> Error in fn(a = 1): object 'b' not found
```

```
b <- 1 # Global b.
```

```
fn(a = 1)
```

```
#> [1] 2
```

```
b <- 10 # Global b redefined.
```

```
fn(a = 1)
```

```
#> [1] 11
```

Above, when `fn()` is called, R first looks inside the function body to see if `b` is defined. When it finds that `b` has not been defined in the function body, it looks outside the function, which in this case is the environment from which `fn()` was called. In the first call to `fn(a = 1)`, the value for `b` has yet to be defined, so an object not found error is thrown. Prior to the second call to `fn(a = 1)`, a global `b` is defined and equals 1, so the free `b` is 1 and the output is 2. Prior to the third call to `fn(a = 1)`, the global `b` is redefined to equal 10, so the free `b` is 10 and the output is 11.

What's the rule when `b` is defined both inside and outside the function? Consider the example below.

```
fn <- function() {
  b <- 1 # Local b.
  return(b)
}
```

```
b <- 10 # Separate from fn()'s local b.
```

```
fn() # Uses the local b.
```

```
#> [1] 1
```

```
b # Uses the b defined outside fn().
```

```
#> [1] 10
```

Above, when `fn()` is called, it always uses its local `b`, i.e., the local `b` is separate from the `b` defined outside `fn()`. Here, there are two variables with the same name, but they exist separately because of their scope.

There are often times when the same variable name is used inside and outside of functions. In such cases, it's critical to understand the basic scope rules illustrated above.

Functions are objects and hence follow the same scope rules as variables. You can have local and global functions. For example, a function defined within another function has local scope. Consider the following example.

```
fn_1 <- function() {

  # Define local fn_2().
  fn_2 <- function() {
    print("I'm fn_2() in fn_1().")
  }

  fn_2() # Call local fn_2().
}

fn_1() # Will call its local fn_2().
```

```
#> [1] "I'm fn_2() in fn_1()."
```

```
fn_2() # Not defined outside fn_1().
```

```
#> Error in fn_2(): could not find function "fn_2"
```

Above, `fn_2()` is local to `fn_1()`. Each time `fn_1()` is called, `fn_2()` is defined and then called to print `"I'm fn_2() in fn_1()"`. Consistent with local scope rules, when `fn_2()` is then called outside `fn_1()`, an error message lets you know that `fn_2()` couldn't be found.

We can make `fn_2()` global to `fn_1()` by defining it prior to calling `fn_1()`, as demonstrated below.

```
# Define global fn_2().
fn_2 <- function() {
    print("I'm global fn_2().")
}

fn_1 <- function() {
  fn_2() # Call global fn_2().
}

fn_1() # Will call global fn_2().
```

```
#> [1] "I'm global fn_2()."
```

Talking about scope using *local* and *global* terms is useful for building some initial intuition about how functions you write will behave, but it leads to a limited view of what's going on behind the scenes. We encourage you to eventually seek a more formal understanding of R environments and scope because it'll make you a more effective programmer and analyst. Wickham (2019) is an excellent resource for building understanding about these and related topics.

5.1.4 Checking function inputs

It's easy to make mistakes while coding. One common mistake is providing the wrong data type or object as input to a function. A goal when writing functions is to catch such input errors before they create an unexpected function error or output.

Consider our earlier `pow()` function that computes x^v.

```
pow <- function(x, v) {
  result <- x^v
  return(result)
}
```

Clearly x and v should be numeric. What happens if we send in a character data type?

```
pow(x = "a", v = 2)
```

```
#> Error in x^v: non-numeric argument to binary operator
```

As you can see from the output above, when attempting `"a"^2`, R throws an error that stops the code. It's easy to see where the error originates in this simple example. However, for more complex functions, an error might be more difficult to diagnose or, worse yet, the function doesn't throw an error and returns erroneous output. So, when writing a function, it's good practice to include input checks and, if needed, stop the function early and provide informative error messages.

A simple approach to including input checks is to use base R's `stopifnot()` (stop if not). This function takes any number of logical expressions as inputs, and if any expression is false it calls `stop()` and produces an error message indicating the first false expression. Below we add `stopifnot()` to `pow()` to check that `x` and `v` are numeric, then test it out.

```
pow <- function(x, v) {
  stopifnot(is.numeric(x), is.numeric(v))

  result <- x^v
  return(result)
}

# Try it out.
```

```
pow(x = "a", v = 2)
```

```
#> Error in pow(x = "a", v = 2): is.numeric(x) is not TRUE
```

You can add a name to each `stopifnot()` expression, which is printed if the expression is false. As demonstrated below, this name can provide user-friendly guidance.

```
pow <- function(x, v) {
  stopifnot("The x argument value should be numeric." = is.numeric(x),
            "The v argument value should be numeric." = is.numeric(v))

  result <- x^v
  return(result)
}

# Try it out.
pow(x = "a", v = 2)
```

```
#> Error in pow(x = "a", v = 2): The x argument value should be numeric.
```

5.2 Control statements

5.2.1 `if` and `if else` statements

We often want to run different code conditional on characteristics of the data or object. This can be done using an `if` statement, the pseudocode for which is given below.

```
if (condition) {
  do something
}
```

The `if` statement's `condition` argument is a single logical value. If `condition` is `TRUE`, the code within the `if` statement's body is run. If `condition` is `FALSE`, the code within the `if` statement's body is skipped. Importantly, the `if` statement considers a single logical value for `condition`; you cannot use a logical vector longer than 1. The `condition` value often involves comparison and/or logical operators (see Section 4.7).

Consider the following example that tests if spp is `"Quercus rubrum"`, and if true, the common name is printed.

```
spp <- "Quercus rubrum"

if (spp == "Quercus rubrum") {
```

```
  print("The species is northern red oak.")
}
```

```
#> [1] "The species is northern red oak."
```

Often we want to run one bit of code if some logical condition is true and run a different bit of code if the condition is false. This is accomplished using an `if else` statement, which has the following form.

```
if (condition) {
  do something
} else {
  do something else
}
```

Consider the following example that extends the previous `if` example to print a message that lets us know the common name was not found.

```
spp <- "Acer saccharum"

if (spp == "Quercus rubrum") {
  print("The species is northern red oak.")
} else {
  print("The species' common name was not found.")
}
```

```
#> [1] "The species' common name was not found."
```

The `if else` statement can be extended with an `else if` to consider multiple conditions. The code within the first `TRUE else if (condition)` is run and evaluation of any subsequent `else if` statements is skipped.

Below, we use `else if` to match `spp` with different Latin names before printing "The species' common name was not found."

```
spp <- "Populus tremuloides"

if (spp == "Quercus rubrum") {
  print("The species is northern red oak.")
} else if (spp == "Acer saccharum") {
  print("The species is sugar maple.")
} else if (spp == "Populus tremuloides") {
  print("The species is quaking aspen.")
} else {
  print("The species' common name was not found.")
}
```

```
#> [1] "The species is quaking aspen."
```

Above, `spp == "Quercus rubrum"` is `FALSE` so the code next tests `spp == "Acer saccharum"` which is also `FALSE`, then it tests `spp == "Populus tremuloides"` which is `TRUE` so `"The species is quaking aspen."` is printed.

Next, let's embed the `if else` code above in a function that prints the common name of a given species. We'll call our function `get_common()`.

```
get_common <- function(spp) {

  if (spp == "Quercus rubrum") {
    print("The species is northern red oak.")
  } else if(spp == "Acer saccharum") {
    print("The species is sugar maple.")
  } else if(spp == "Populus tremuloides") {
    print("The species is quaking aspen.")
  } else {
    print("The species' common name was not found.")
  }

}

# Try it out.
get_common("Quercus rubrum")
```

```
#> [1] "The species is northern red oak."
```

```
get_common("Populus tremuloides")
```

```
#> [1] "The species is quaking aspen."
```

```
get_common("Pseudotsuga menziesii")
```

```
#> [1] "The species' common name was not found."
```

5.2.2 Vectorized `ifelse`

As noted previously, the `if` statement accepts a logical `condition` of length 1. So what happens if you want to condition on a longer vector of logical values? For these cases, base R offers `ifelse()`, which is a vectorized function of the following form.

```
ifelse(condition, if true do something, if false do something else)
```

Here's an example of `ifelse()` that determines if values in `x` are even or odd (`%%` is the modulus operator which computes the remainder of `x / 2`).

```
x <- 1:5
ifelse(x %% 2 == 0, "even", "odd")
```

```
#> [1] "odd"  "even" "odd"  "even" "odd"
```

We'll work with this and related vectorized control functions in Section 7.6.

5.2.3 Loops

Loops are an important component of most programming languages, and R is no exception. That said, in R there are often more efficient and less code-intensive approaches to performing the types of iterative operations that typically employ loops. We'll cover a few of these more efficient approaches in Chapter 7.

There are, however, several good reasons for learning about loops. They are: 1) convenient in some situations; 2) common in R and other languages; 3) instructive for understanding iterative operations and translating mathematical notation to code.

We provide a very brief introduction to `while` and `for` loops.

A `while` loop has the following form.

```
while (condition) {
  do something
}
```

As long as `condition` is `TRUE`, `do something` is iteratively evaluated. Once `condition` is `FALSE`, the loop stops. For example, the body of the `while` loop below prints the value of `i`, then increments `i` by 1, until `i` is greater than 3.

```
i <- 1

while(i <= 3) {
  print(i)
  i <- i + 1
}
```

```
#> [1] 1
#> [1] 2
#> [1] 3
```

A `for` loop has the following form.

```
for (variable in vector) {
  do something
}
```

Here's a `for` loop example.

```
for(i in 1:3) {
  print(i)
}
```

```
#> [1] 1
#> [1] 2
#> [1] 3
```

As shown using `print()`, for each successive iteration in the loop, the variable `i` takes the associated increment value in the vector `1:3`.

This increment change in the variable makes `for` loops particularly useful for stepping through elements in a vector. As an example, let's use a `for` loop to sum a numeric vector's values. Below, `x` is the vector to be summed, `total` will hold the sum, and the `for` loop variable `i` indexes each successive element in `x`. Notice how `length(x)` is used to get the correct number of elements in `x`

```
x <- c(0, 1, 2)

total <- 0 #Initialize to zero.

for (i in 1:length(x)) {
  total <- total + x[i]
}

total
```

```
#> [1] 3
```

Now let's embed the `for` loop above in a function called `my_sum()`.

```
my_sum <- function(x) {

  total <- 0
  for(i in 1:length(x)) {
    total <- total + x[i]
  }

  return(total)
}

# Try it out.
v <- c(0, 1, 2)
```

```
my_sum(v)
```

```
#> [1] 3
```

5.3 Illustrations

5.3.1 Basal area

Basal area is the cross-sectional area of a tree stem at breast height (4.5 ft or 1.37 m above ground). Conceptually, it's the area of the circular slice you'd see if the tree were cut cleanly at that height. When summed across trees and expressed on a per unit area basis (e.g., square feet per acre or square meters per hectare), basal area provides a compact and widely used summary of stand structure.

The geometry behind basal area is straightforward. The cross section of a tree at breast height is approximated as a circle, so its area follows directly from the familiar formula for the area of a circle,

$$\text{area} = \pi r^2, \tag{5.1}$$

where r is the radius. In forestry, we don't measure radius directly. Instead, we measure diameter at breast height (DBH), and since $r = \text{DBH}/2$, basal area is simply $\pi(\text{DBH}/2)^2$, with an additional unit conversion depending on how DBH is measured and the desired basal area units.

If DBH is measured in inches and basal area is desired in square feet, square inches must be converted to square feet. Because there are 144 square inches in a square foot, the conversion factor 1/144 appears in the basal area formula,

$$\frac{\pi}{144} \cdot \left(\frac{\text{DBH}}{2}\right)^2 = \frac{\pi}{144 \cdot 4} \cdot \text{DBH}^2 = 0.005454 \cdot \text{DBH}^2. \tag{5.2}$$

Likewise, when DBH is measured in centimeters and basal area is expressed in square meters, the conversion is from square centimeters to square meters. Since there are 10,000 square centimeters in a square meter, the corresponding factor is 1/10,000, giving

$$\frac{\pi}{10000} \cdot \left(\frac{\text{DBH}}{2}\right)^2 = \frac{\pi}{10000 \cdot 4} \cdot \text{DBH}^2 = 0.00007854 \cdot \text{DBH}^2. \tag{5.3}$$

Seen this way, basal area is nothing more than circle area with unit bookkeeping layered on top. Looking at (5.2) and (5.3), we can write a general expression for basal area as

$$\text{BA} = c \cdot \text{DBH}^2, \tag{5.4}$$

where c is a unit-specific constant. This constant is often referred to as the *foresters constant* and takes the value 0.005454 when DBH is in inches and basal area is in square feet, or 0.00007854 when DBH is in centimeters and basal area is in square meters.

With that structure in mind, writing a basal area function becomes straightforward. Below, we define `basal_area()` to take DBH values measured in inches or centimeters and return basal area in square feet or square meters. The default arguments assume DBH in inches and basal area in square feet. Within the function body, we include simple input checks that clarify acceptable values for the `dbh_units` and `ba_units` arguments. After those checks, an `if` statement selects the appropriate constant c, an `else if` statement computes the basal area `ba`, and the result is returned.

```
basal_area <- function(dbh, dbh_units = "in", ba_units = "sq_ft") {

  # Some input checks.
  stopifnot("dbh should be numeric." =
              is.numeric(dbh),
            "dbh_units should be either 'in' or 'cm'." =
              dbh_units %in% c("in", "cm"),
            "ba_units should be either 'sq_ft' or 'sq_m'." =
              ba_units %in% c("sq_ft", "sq_m")
            )

  # Define c.
  if(ba_units == "sq_ft") {
    c <- pi / (4 * 144)
  } else{ # ba_units == "sq_m"
    c <- pi / (4 * 10000)
  }

  # Compute ba and return.
  if (dbh_units == "in" & ba_units == "sq_ft") {
    ba <- c * dbh^2
  } else if (dbh_units == "in" & ba_units == "sq_m") {
    ba <- c * (2.54 * dbh)^2
  } else if (dbh_units == "cm" & ba_units == "sq_m") {
    ba <- c * dbh^2
  } else{ # dbh_units == "cm" & ba_units == "sq_ft"
    ba <- c * (dbh / 2.54)^2
  }

  return(ba)
}
```

Let's try our basal area function.

```
# Basal area in square feet from 10 inches DBH.
basal_area(dbh = 10)
```

```
#> [1] 0.54542
```

```
# Basal area in square meters from 10 inches DBH.
basal_area(dbh = 10, ba_units = "sq_m")
```

```
#> [1] 0.050671
```

```
# Basal area in square meters from 10 centimeters DBH.
basal_area(dbh = 10, dbh_units = "cm", ba_units = "sq_m")
```

```
#> [1] 0.007854
```

```
# Basal area in square feet from 10 centimeters DBH.
basal_area(dbh = 10, dbh_units = "cm")
```

```
#> [1] 0.08454
```

```
# Some initial error checking.
basal_area(dbh = 10, dbh_units = "ft")
```

```
#> Error in basal_area(dbh = 10, dbh_units = "ft"): dbh_units should
be either 'in' or 'cm'.
```

This looks pretty good. The first four calls to `basal_area()` try the various input combinations. The last call to `basal_area()` tests some error handling by wrongly setting `dbh_units = "ft"`, which stops the function and throws an error with an informative message.

Because `if` and `else if` in `basal_area()` involve only `dbh_units` and `ba_units`, our function can work on a vector of multiple DBH values as demonstrated by the code below.

```
# Vector of DBH values.
dbh <- c(5, 15, 10, NA, 5)

basal_area(dbh = dbh)
```

```
#> [1] 0.13635 1.22718 0.54542      NA 0.13635
```

5.3.2 Aboveground biomass allometric equation

Aboveground biomass (AGB) is directly related to carbon stored in woody material and hence is a key variable used in forest carbon monitoring programs. A tree's total AGB is related to its DBH and species. Jenkins et al. (2003) provide the following allometric equation to estimate AGB given a tree's species and DBH.

$$\text{AGB} = \exp\left(\beta_0 + \beta_1 \ln(\text{DBH})\right), \tag{5.5}$$

where

- AGB is total aboveground biomass (kg dry weight) for trees 2.5 (cm) DBH and larger,
- DBH is diameter at breast height in centimeters,
- exp is the exponential function,
- ln is the natural logarithm, i.e., inverse function of the exponential function,
- β_0 and β_1 are species-specific regression coefficients with values given in Jenkins et al. (2003) Table 4.

Let's implement (5.5) in R. Our function will return AGB given a tree's DBH and species.

To motivate this example, we consider some trees measured on one forest inventory plot in the Penobscot Experimental Forest (PEF). You can read more about the PEF in Section 1.2.4. We'll work with a set of 99 trees measured in PEF management unit 10 on plot 11 in 2010. Below, these tree data are read into the `pef_trees` data frame and the data frame's structure and first few column values are printed using `str()`.

```
pef_trees <- read.csv("datasets/PEF/PEF_mu_10_plot_11.csv")

str(pef_trees) # Take a look.
```

```
#> 'data.frame':    99 obs. of  2 variables:
#>  $ common_name: chr  "red maple" "red maple" "balsam fir" "balsam fir" ...
#>  $ DBH_in     : num  0.7 0.5 1.8 6.4 1.7 ...
```

To give us a better sense of these data, the next bit of code prints the number of trees by species. The `sort()` function sorts the `table()` output from least to most abundant species. This table shows there are nine species present on the inventory plot and their respective stem counts.

```
sort(table(pef_trees$common_name))
```

```
#> 
#>  red or black spruce        American beech 
#>                    1                     2 
#>          paper birch northern white-cedar 
#>                    2                     3 
#>      eastern hemlock             white ash 
#>                    5                     7 
#>         yellow birch             red maple 
#>                   11                    22 
#>           balsam fir 
#>                   46
```

There are some features we'll need to accommodate in our function. Specifically, the allometric equation is parameterized for:

TABLE 5.1: Regression coefficient values used in allometric equation 5.5 for tree species present in PEF example data.

Species	β_0	β_1
American beech	−2.0127	2.4342
Balsam fir	−2.5384	2.4814
Eastern hemlock	−2.5384	2.4814
Northern white-cedar	−2.0336	2.2592
Paper birch	−1.9123	2.3651
Red maple	−1.9123	2.3651
Red or black spruce	−2.0773	2.3323
White ash	−2.48	2.4835
Yellow birch	−1.923	2.3651

1. DBH in centimeters, so the PEF's DBH values need to be converted from inches to centimeters;
2. trees of DBH 2.5 (cm) and larger, we'll want to return `NA` for all trees less than 2.5 (cm);
3. species groups, so we'll need to pass in species and have the function apply the correct set of regression coefficients.

For this example, we tailor the function to those species in the current dataset. If the function was called for a species not in the current dataset, then we'd like an informative error to be thrown.

Table 5.1 provides regression coefficient values from Jenkins et al. (2003) for those species in `pef_trees`. These coefficient values will be used in our function to compute species-specific AGB.

Given the desired functionality, we define `agb()` below. The general form of `agb()` with all its `if` and `if else` statements is a common approach to writing code for allometric equations with coefficient values that differ among species or other characteristics.

```
agb <- function(dbh, species, dbh_units = "cm") {

  # Some input checks.
  stopifnot("dbh should be a numeric vector of length 1." =
              is.numeric(dbh) & length(dbh) == 1,
            "species should be a character of length 1." =
              is.character(species) & length(species) == 1,
            "dbh_units should be either 'in' or 'cm'." =
              dbh_units %in% c("in", "cm"))

  # Convert DBH to cm if needed.
```

```
  if(dbh_units == "in") {
    dbh <- dbh * 2.54
  }

  # Assign the species-specific regression coefficients.
  if (dbh < 2.5) {
      beta_0 <- NA
      beta_1 <- NA
  } else if (species == "American beech") {
      beta_0 <- -2.0127
      beta_1 <- 2.4342
  } else if (species %in% c("balsam fir",
                             "eastern hemlock")) {
      beta_0 <- -2.5384
      beta_1 <- 2.4814
  } else if (species == "northern white-cedar") {
      beta_0 <- -2.0336
      beta_1 <- 2.2592
  } else if (species %in%
             c('paper birch', 'red maple', 'yellow birch')) {
      beta_0 <- -1.9123
      beta_1 <- 2.3651
  } else if (species == "red or black spruce") {
      beta_0 <- -2.0773
      beta_1 <- 2.3323
  } else if (species == 'white ash') {
      beta_0 <- -2.48
      beta_1 <- 2.4835
  } else {
      stop(paste0("No allometric equation specified for ", species, "."))
  }

  # Apply the regression coefficients and return biomass.
  agb_km <- exp(beta_0 + beta_1 * log(dbh))

  return(agb_km)
}
```

Let's step through some features and coding decisions in `agb()`.

1. Inputs are `dbh`, `species`, and `dbh_units` with a default value of `"cm"`.
2. Given the way in which the function assigns values to `beta_0` and `beta_1` within the `if else` statement, it cannot handle more than one `dbh` and `species` value pair at a time (i.e., `dbh` and `species` must be vectors of length 1). This vector length constraint and other input requirements are checked in the `stopifnot()`.

3. If `dbh_units == "in"`, then `dbh` is transformed to centimeters using an `if` statement.
4. The `if else` statement used to assign species-specific coefficient values to `beta_0` and `beta_1` (using values in Table 5.1), also serves to assign `NA` to trees with DBH < 2.5 (cm) and throws the desired error and informative message if the species name is not found.

Okay, let's compute some AGB.

```
agb(5, "balsam fir")
```

```
#> [1] 4.2856
```

```
agb(5, "balsam fir", dbh_units = "in")
```

```
#> [1] 43.308
```

```
agb(1, "balsam fir") # DBH < 2.5 cm, should be NA.
```

```
#> [1] NA
```

```
agb(3, "quaking aspen") # This species is not defined in agb().
```

```
#> Error in agb(3, "quaking aspen"): No allometric equation specified
for quaking aspen.
```

In practice, we've seen code written to implement allometric equations, like `agb()`, span hundreds of lines, with seemingly endless control structures, many of which are nested (e.g., `if` statements within `if` statements). Such approaches have substantial disadvantages. For example, they:

1) require a lot of code and hence present more opportunities to make mistakes;
2) are difficult to maintain and fragile, meaning small changes can break something in the control statement sequence;
3) are challenging to read, as following the flow of multiple branching logical statements is painful;
4) need to be embedded in even more code when applied to data (see continuation of the example below);
5) are computationally inefficient compared with other preferred approaches (see, e.g., Section 7.14.3).

With all that said, let's continue on and use `agb()` to compute AGB for `pef_trees`. In the code below, we define `AGB_kg` as a vector of zeros the same length as the number of trees in `pef_trees`. `AGB_kg` will eventually hold AGB for each tree computed using `agb()`. Then, using a `for` loop, we iterate over the `pef_trees` trees, passing each tree's DBH and species into agb() and storing the resulting AGB in AGB_kg. Finally, we add an AGB_kg column to pef_trees.

```
AGB_kg <- rep(0, nrow(pef_trees)) # Empty vector to hold AGB.

# i indexes rows in pef_trees and elements in AGB_kg.
for(i in 1:nrow(pef_trees)) {
  AGB_kg[i] <- agb(pef_trees$DBH_in[i],
                   pef_trees$common_name[i],
                   dbh_units = "in")
}

# Assign AGB_kg as a new column in pef_trees.
pef_trees$AGB_kg <- AGB_kg

# Take a look.
str(pef_trees)
```

```
#> 'data.frame':    99 obs. of  3 variables:
#>  $ common_name: chr  "red maple" "red maple" "balsam fir" "balsam fir" ...
#>  $ DBH_in     : num  0.7 0.5 1.8 6.4 1.7 ...
#>  $ AGB_kg     : num  NA NA 3.43 79.91 3.8 ...
```

We'll return to allometric equations in Section 7.14.3, where we present an alternative and preferred approach to implementation.

5.4 Keeping track of functions

As introduced in Section 3.1, scripts facilitate efficient coding and provide a reproducible data management and analysis record. To this end, it's advantageous to also keep track of functions you've written. Functions you write for use in a script should be placed at the top of the script. Alternatively, if you have functions that you routinely use, they can be collected into a single .R file and then "sourced" via source().

For example, we've written lots of functions to perform various tasks, e.g., compute basal area given DBH or convert various measurements from English to metric units. We keep these functions in a file called "utils.R". As the file name suggests, these are utility functions used in near-daily scripting. Instead of copying and pasting these functions into each new script, we simply load all the utility functions at once by placing source("utils.R") at the top of the script. Like any file read into R, "utils.R" needs to be in the working directory, or you can provide a path to the file's location on your computer.

Keeping track of the functions you write over time in a "utils.R" file (or a script name of your choosing) is a great way to build efficiency.

5.5 Summary

Choices and loops are the two primary structures used to control flow in a program. Choices, implemented using `if` and `if else`, allow different code to be run conditional on the result of a logical test. These logical tests typically involve comparison and/or logical operators introduced in Section 4.7. While not covered in this chapter, `switch` is an additional control structure (see, e.g., `?switch`) that is occasionally useful. We'll return to control structures later in Section 7.6, where we consider vectorized versions of `if else` and `switch`.

Loops, like `while` and `for`, allow you to iterate over a piece of code until some condition is met or a desired number of iterations is reached. Loops can be very useful; however, there are many scenarios where the desired iterative operation can be done using other, more efficient, functions.

As noted when introducing this chapter, our focus was narrow and high-level, covering only topics needed in subsequent chapters. Our treatment of functions touched on the basics, which is enough for you to start writing your own simple functions and dissecting more complex functions written by others. For a deeper understanding of functional programming in R, we encourage you to study topics on scope and environments, which are covered well in *Advanced R* by Wickham (2019) and in other freely available resources online. These references will also guide you if your eventual aim is to write robust software for public consumption.

5.6 Exercises

Exercises 5.1 - 5.6 use tree measurements recorded in Table 5.2.

TABLE 5.2: Tree measurements used for Exercises 5.1.

Tree id	Species	DBH (in)	Logs (16 foot)
1	Red oak	38	3.5
2	Sugar maple	10	2
3	Sugar maple	17	1.5
4	Bitternut hickory	22	3
5	Tulip poplar	32	3
6	Red oak	15	3
7	White oak	31	3
8	Sugar maple	40	2.5

Exercise 5.1. Create a data frame called `plot_trees` that holds data given in Table 5.2. Your data frame should match the following when printed.

```
plot_trees
```

```
#>   tree_id               spp dbh_in logs_16ft
#> 1       1           red oak     38       3.5
#> 2       2       sugar maple     10       2.0
#> 3       3       sugar maple     17       1.5
#> 4       4 bitternut hickory     22       3.0
#> 5       5      tulip poplar     32       3.0
#> 6       6           red oak     15       3.0
#> 7       7         white oak     31       3.0
#> 8       8       sugar maple     40       2.5
```

Exercise 5.2. Write a function to compute basal area in square feet given DBH in inches. Apply your function and add your resulting basal area as a new column in `plot_trees`. We called our new basal area column `BA_sqft`. Your updated `plot_trees` should look similar to ours below.[2]

```
#>   tree_id               spp dbh_in logs_16ft BA_sqft
#> 1       1           red oak     38       3.5 7.87580
#> 2       2       sugar maple     10       2.0 0.54542
#> 3       3       sugar maple     17       1.5 1.57625
#> 4       4 bitternut hickory     22       3.0 2.63981
#> 5       5      tulip poplar     32       3.0 5.58505
#> 6       6           red oak     15       3.0 1.22718
#> 7       7         white oak     31       3.0 5.24144
#> 8       8       sugar maple     40       2.5 8.72665
```

The following information is used for Exercises 5.3 - 5.6. Foresters use mathematical formulas to estimate the amount of lumber in logs. These formulas are referred to as *log rules* and typically take into account average taper in the tree's stem and wood waste from saw kerf, i.e., wood lost due to the blade width. Log taper is often species-specific; hence, foresters use *form classes* to describe taper characteristics of species groups. In the eastern US, for example, form class 78 is often used to characterize hardwood species taper.[3]

International 1/4-inch (Int-1/4), Scribner, and Doyle are the names of common log rules used in the US. Wiant (1986) published the formulas below that estimate board foot volume given DBH (in) and number of 16 foot logs (L) for form class 78 under Int-1/4, Scribner and

---

[2]Our basal area function uses `pi/(4*144)`. If you wrote your function using the forester's constant 0.005454 then your basal area will differ slightly from ours beyond the second or third decimal place.

[3]The form class number, in this case 78, is a form quotient and is the ratio, expressed as a percent, of some upper-stem diameter to DBH; see, e.g., Burkhart et al. (2018). A larger form quotient means less stem taper, i.e., more stem volume.

Doyle log rules.[4] The subscript on each estimated volume variable (Vol) below indicates the log rule.

Int-1/4 log rule:

$$\begin{aligned}\text{Vol}_I =&(-13.35212 + 9.58615\text{L} + 1.52968\text{L}^2)+\\&(1.79620 - 2.59995\text{L} - 0.27465\text{L}^2)\text{DBH}+\\&(0.04482 + 0.45997\text{L} - 0.00961\text{L}^2)\text{DBH}^2.\end{aligned} \tag{5.6}$$

Scribner log rule:

$$\begin{aligned}\text{Vol}_S =&(-22.50365 + 17.53508\text{L} - 0.59242\text{L}^2)+\\&(3.02988 - 4.34381\text{L} - 0.02302\text{L}^2)\text{DBH}+\\&(-0.01969 + 0.51593\text{L} - 0.02035\text{L}^2)\text{DBH}^2.\end{aligned} \tag{5.7}$$

Doyle log rule:

$$\begin{aligned}\text{Vol}_D =&(-29.37337 + 41.51275\text{L} + 0.55743\text{L}^2)+\\&(2.78043 - 8.77272\text{L} - 0.04516\text{L}^2)\text{DBH}+\\&(0.04177 + 0.59042\text{L} - 0.01578\text{L}^2)\text{DBH}^2.\end{aligned} \tag{5.8}$$

Exercise 5.3. Write a function called `vol_I()` that implements (5.6). Use arguments `DBH_in` and `L_16ft` for DBH (in) and number of logs (16-ft), respectively. Consistent with (5.6), your function should return board foot volume. Your function should accept vectors for `DBH_in` and `L_16ft` and return a corresponding volume vector.

Below are some tests using our `vol_I()` that you can use to test your implementation.

```
vol_I(DBH_in = 12, L_16ft = 1)
```

```
#> [1] 56.129
```

```
vol_I(DBH_in = 24, L_16ft = 2.5)
```

```
#> [1] 519.67
```

```
vol_I(DBH_in = c(12, 24), L_16ft = c(1, 2.5))
```

```
#> [1]  56.129 519.665
```

Exercise 5.4. Using `vol_I()` you wrote in Exercise 5.3, add a column called `vol_I_bdft` to `plot_trees`. You should be able to pass `plot_trees$dbh_in` and `plot_trees$logs_16ft` into `vol_I()` to compute `vol_I_16ft` values. Your result should match ours below (we omitted the `spp` columns so our output fits nicely on the page).

[4]One board foot has dimension $1 \times 12 \times 12$ (in) with volume 144 (in^3).

```
#>   tree_id dbh_in logs_16ft BA_sqft vol_I_bdft
#> 1       1     38       3.5 7.87580   1852.968
#> 2       2     10       2.0 0.54542     59.548
#> 3       3     17       1.5 1.57625    164.301
#> 4       4     22       3.0 2.63981    490.421
#> 5       5     32       3.0 5.58505   1128.315
#> 6       6     15       3.0 1.22718    203.145
#> 7       7     31       3.0 5.24144   1052.482
#> 8       8     40       2.5 8.72665   1578.856
```

Exercise 5.5. Extend the function you wrote for Exercise 5.4 to allow the user to choose between the three log rules given in (5.6), (5.7), and (5.8). Call your new function `vol_bdft()`. In addition to `DBH_in` and `L_16ft` arguments, your new function should have an argument called `log_rule` that takes values `"Int-1/4"`, `"Scribner"`, or `"Doyle"` that are used to control which log rule is computed. Make the default value for `log_rule` be `"Int-1/4"`. Have your function throw an informative error if the `log_rule` value provided by the user doesn't match one of the accepted values.

Below are some tests using our `vol_bdft()` that you can use to test your implementation.

```
vol_bdft(DBH_in = c(12, 24), L_16ft = c(1, 2.5)) # Int-1/4 default.
```

```
#> [1]  56.129 519.665
```

```
vol_bdft(DBH_in = c(12, 24), L_16ft = c(1, 2.5), log_rule = "Scribner")
```

```
#> [1]  46.924 484.605
```

```
vol_bdft(DBH_in = c(12, 24), L_16ft = c(1, 2.5), log_rule = "Doyle")
```

```
#> [1]  29.01 428.94
```

```
# Check log_rule error handling.
vol_bdft(DBH_in = c(12, 24), L_16ft = c(1, 2.5), log_rule = "Int-1/8")
```

```
#> Error in vol_bdft(DBH_in = c(12, 24), L_16ft = c(1, 2.5), log_rule = "Int-
1/8"): log_rule should be either 'Int-1/4',
#>              'Scribner', or 'Doyle'.
```

Exercise 5.6. Using `vol_bdft()` from Exercise 5.5, add estimated board foot volume using the three log rules to `plot_trees`. Your result should match ours below (we included only `tree_id` and the volume columns so our output fits nicely on the page).

```
#>   tree_id vol_I_bdft vol_S_bdft vol_D_bdft
#> 1       1   1852.968   1777.412   1805.779
#> 2       2     59.548     45.776     22.374
#> 3       3    164.301    147.059    113.743
#> 4       4    490.421    451.134    382.179
#> 5       5   1128.315   1075.319   1045.083
#> 6       6    203.145    174.253    116.996
#> 7       7   1052.482   1000.796    963.753
#> 8       8   1578.856   1527.407   1571.260
```

6

Data summary and analysis with `tidyverse`

This chapter opens a suite of chapters that cover packages, collectively known as the `tidyverse`, designed explicitly for data science. In the book *R for Data Science* (Wickham, Çetinkaya-Rundel, et al. 2023), the lead author Hadley Wickham, a key developer of many `tidyverse` packages, promotes a common grammar and function structure to simplify and streamline data manipulation and analysis. Here, the focus is on automating time-consuming, tedious, and routine data wrangling and data summary tasks, as well as creating publication-quality plots, graphics, and tables that effectively summarize and communicate analysis results.

Some seasoned R users prefer to work almost exclusively with base R, while others live almost entirely in the `tidyverse`. With its ever-expanding functionality, documentation, and user community, we find ourselves increasingly happy to spend more time exploring the `tidyverse`. We don't view base R and the `tidyverse` as mutually exclusive, but rather as complementary tools. To this end, the next several chapters highlight `tidyverse` packages we use most frequently when analyzing forestry data. We strongly encourage you to learn both base R and the `tidyverse`, and to use whichever functions you find most intuitive and convenient for the task at hand. A more comprehensive overview of the `tidyverse` is given by Wickham, Çetinkaya-Rundel, et al. (2023), and additional resources can be found at www.tidyverse.org. We focus our tour of the `tidyverse` on the following five packages:

1. `tibble`: improving on the data frame (Chapter 6),
2. `readr`: reading and writing files (Chapter 6),
3. `dplyr`: manipulating and summarizing data (Chapter 7),
4. `tidyr`: cleaning and reshaping data (Chapter 8),
5. `ggplot2`: creating graphics (Chapter 9).

In this chapter, we introduce the `tibble` and `readr` packages, which provide new ways to store, read, and write data. The entire `tidyverse` can be installed and loaded at once using the code below (which may take a few minutes if you don't have the packages installed).

```
install.packages("tidyverse")
library(tidyverse)
```

Immediately after running `library(tidyverse)`, you might see some messages about masking printed to your console. *Masking* occurs when two or more packages have objects (e.g.,

DOI: 10.1201/9781003222699-6

functions) with the same name. The masking messages tell you which package takes precedence when you call an object name. For instance, the `stats` package is loaded automatically when you start R. This package includes a `filter()` and a `lag()` function. The `tidyverse` package `dplyr` also includes functions named `filter()` and `lag()`. Due to these common names, the `stats` package functions are masked, such that a call to `filter()` will use `dplyr`'s `filter()` function. Use the `::` operator to explicitly identify the package from which you want to call a function; for example, if you want `stats`'s `filter()`, call `stats::filter()`.

6.1 Minnesota tree growth dataset

We motivate methods presented in this and subsequent `tidyverse` chapters using a dendrochronological dataset described in Foster et al. (2014) and subsequently reanalyzed by Itter et al. (2017). The data, collected in northeastern Minnesota, comprise growth ring widths for 2,291 trees. We refer to a tree's growth ring measurements over time as its chronology. A *chronology* describes a tree's history of growth, suppression, and release. Chronologies can help us understand effects of age, natural disturbance, and silvicultural treatments on trees and stands; see, e.g., Itter et al. (2017). Chronologies in this dataset all end in 2007 and some start as far back as 1897. The growth ring width measurements were taken from increment cores extracted at DBH using an increment borer. Crossdating, i.e., assigning a year and tree age to each growth ring, was done using standard dendrochronological techniques (see, e.g., Bunn 2010 for crossdating methods and R tools). Trees were located in 105 plots distributed across 35 forest stands (3 plots per stand). Each stand represented an area with similar species composition and approximately homogeneous forest characteristics (e.g., tree density, size distribution, age distribution).

With a total of 131,386 ring width measurements over the 2,291 trees, the dataset is quite large; hence, we'll work with a subset that includes only the first five stands.[1] The `mn_trees_subset.csv` file, read into the `mn_trees` data frame below, contains each tree's species (`species`; species codes are defined in Table 6.1), year (`year`) coinciding with the tree's age (`age`), and increment core derived measurements of annual radial growth increment (`rad_inc`; annual growth ring width in mm) and diameter at breast height (`DBH`; end of growing season inside bark in cm). The data frame also includes an identification number for stand, plot, and tree, i.e., `stand_id`, `plot_id`, and `tree_id`, respectively. A tree's measurements are uniquely identified by the combination of `stand_id`, `plot_id`, and `tree_id` values (i.e., tree numbers are unique within plot and plot numbers are unique within stand).

In the spirit of `tidyverse`, we use the `dplyr` package's `glimpse()` function—a flexible alternative to `str()` introduced in Chapter 7—to preview the dataset's columns.

---

[1]While R has no trouble working with the full dataset on most computers, we use a subset to accommodate computers with limited resources. The full dataset used by Itter et al. (2017) is available in `"datasets/mn_trees.csv"`.

TABLE 6.1: Species codes used in the Minnesota tree growth dataset.

ABBA *Abies balsamea*	LALA *Larix laricina*	PIST *Pinus strobus*
ACRU *Acer rubrum*	PIGL *Picea glauca*	POGR *Populus grandidentata*
ACSA *Acer saccharum*	PIMA *Picea mariana*	POTR *Populus tremuloides*
BEPA *Betula papyrifera*	PIBA *Pinus banksiana*	QURU *Quercus rubra*
FRNI *Fraxinux nigra*	PIRE *Pinus resinosa*	THOC *Thuja occidentalis*

```
mn_trees <- read.csv("datasets/mn_trees_subset.csv")
glimpse(mn_trees)
```

```
#> Rows: 11,649
#> Columns: 8
#> $ stand_id <int> 1, 1, 1, 1, 1, 1, 1, 1, 1, 1, 1, 1, ~
#> $ plot_id  <int> 1, 1, 1, 1, 1, 1, 1, 1, 1, 1, 1, 1, ~
#> $ tree_id  <int> 1, 1, 1, 1, 1, 1, 1, 1, 1, 1, 1, 1, ~
#> $ year     <int> 1960, 1961, 1962, 1963, 1964, 1965, ~
#> $ species  <chr> "ABBA", "ABBA", "ABBA", "ABBA", "ABB~
#> $ age      <int> 1, 2, 3, 4, 5, 6, 7, 8, 9, 10, 11, 1~
#> $ rad_inc  <dbl> 0.930, 0.950, 0.985, 0.985, 0.715, 0~
#> $ DBH      <dbl> 2.1563, 2.3463, 2.5433, 2.7403, 2.88~
```

Notice, in addition to printing the data frame's dimensions (i.e., number of rows and columns), `glimpse()` prints each column's data type[2] and the first several values. The `$` preceding each column name is reminiscent of how data frame column vectors are accessed, e.g., `mn_trees$stand_id` (see Section 4.5.1).

To gain a better sense of the `mn_trees` data, Figure 6.1 plots tree age (left) and DBH growth (right) over time for trees measured in one stand. The left figure shows that several trees were well established at the start of the chronology in 1897, the oldest among them being a 16 year old *Betula papyrifera*. This figure also shows that most of the *Abies balsamea* entered the stand after 1950, with the youngest being established in 1977. The right figure shows tree chronologies have varying growth rates, which are functions of species, age, density, and other tree and stand factors.[3]

---

[2]Integer, double, character, logical, and factor data types introduced in Chapter 4 are indicated using `<int>`, `<dbl>`, `<chr>`, `<lgl>`, and `<fct>`, respectively.

[3]Code to reproduce this figure is given in Chapter 9.

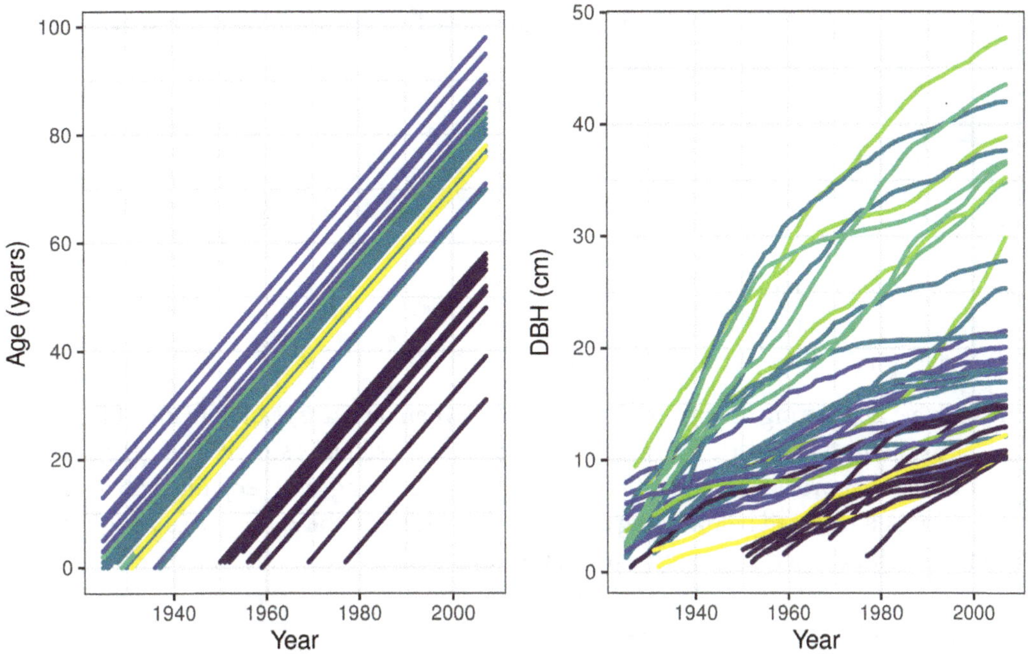

FIGURE 6.1: Tree increment core derived age and diameter at breast height (DBH) by year for trees in `mn_trees` Stand 5.

6.2 Improving data frames with `tibbles`

We begin our tour around the `tidyverse` with the `tibble` package, which provides an improved data frame called a tibble. Running `library(tidyverse)` automatically loads the `tibble` package (i.e., you don't need to run `library(tibble)`).

Like `data.frame()` defined in Section 4.5, given a set of vectors, `tibble()` returns a tibble. The code below creates a tibble called `trees` that we'll use to demonstrate `dplyr` functions in subsequent sections.[4]

```
trees <- tibble(id = as.integer(c(1, 1, 2, 2, 3)),
                year = as.integer(c(2020, 2021, 2020, 2021, 2021)),
                dbh = c(1.9, 2.1, 5.2, 5.5, 0.5))
```

Think of `trees` as a mini version of the Minnesota tree growth dataset `mn_trees` created back in Section 6.1. The `trees` dataset comprises measurements for three trees with columns holding unique tree identification number (`id`), measurement year (`year`), and DBH in

[4]While not critical for our use of `trees`, we coerce the numeric `id` and `year` vectors from double to the more appropriate integer data type (see Section 4.2.1).

inches (dbh). As you can see below, trees 1 and 2 have DBH measurements in years 2020 and 2021, whereas tree 3 was only measured in 2021.

```
glimpse(trees)
```

```
#> Rows: 5
#> Columns: 3
#> $ id   <int> 1, 1, 2, 2, 3
#> $ year <int> 2020, 2021, 2020, 2021, 2021
#> $ dbh  <dbl> 1.9, 2.1, 5.2, 5.5, 0.5
```

If you have an existing data frame (e.g., created when data are read in from a file, Section 3.3.2), as_tibble() converts it to a tibble. Importantly, a tibble is simply a wrapper around a data frame that provides some different printing, subsetting, and recycling behaviors. This last point is illustrated below using a few test functions on the mn_trees_tbl tibble created from the Minnesota tree growth data frame mn_trees.

```
mn_trees_tbl <- as_tibble(mn_trees) # Convert to a tibble.
is_tibble(mn_trees_tbl) # Confirm it's a tibble.
```

```
#> [1] TRUE
```

```
is.data.frame(mn_trees_tbl) # Note, it's also still a data frame.
```

```
#> [1] TRUE
```

A tibble has two key advantages over a data frame. First, when printing, its default behavior is to print the first ten rows and the columns that fit in the console window, as well as some additional information such as dimension (i.e., number of rows and columns), data types, and non-printed column names (e.g., because the DBH column did not fit in the console window below, DBH <dbl> is listed in the last line of output).

```
mn_trees_tbl # Implicitly calls print(mn_trees_tbl).
```

```
#> # A tibble: 11,649 x 8
#>    stand_id plot_id tree_id  year species   age rad_inc
#>       <int>   <int>   <int> <int> <chr>   <int>   <dbl>
#>  1        1       1       1  1960 ABBA        1   0.93
#>  2        1       1       1  1961 ABBA        2   0.95
#>  3        1       1       1  1962 ABBA        3   0.985
#>  4        1       1       1  1963 ABBA        4   0.985
#>  5        1       1       1  1964 ABBA        5   0.715
#>  6        1       1       1  1965 ABBA        6   0.84
#>  7        1       1       1  1966 ABBA        7   0.685
#>  8        1       1       1  1967 ABBA        8   0.94
```

```
#> 9            1         1         1  1968 ABBA         9   1.16
#> 10           1         1         1  1969 ABBA        10   0.775
#> # i 11,639 more rows
#> # i 1 more variable: DBH <dbl>
```

print(), with its default behavior, is invoked implicitly when the tibble object name is run on the console, as shown above. If you want a different print behavior, explicitly call print() with the arguments adjusted as desired. For example, the call to print() below includes the arguments n = 2 and width = Inf to print the first two rows of mn_trees_tbl and all columns, respectively.

```
print(mn_trees_tbl, n = 2, width = Inf)
```

```
#> # A tibble: 11,649 x 8
#>   stand_id plot_id tree_id  year species   age rad_inc
#>      <int>   <int>   <int> <int> <chr>   <int>   <dbl>
#> 1        1       1       1  1960 ABBA        1    0.93
#> 2        1       1       1  1961 ABBA        2    0.95
#>     DBH
#>   <dbl>
#> 1  2.16
#> 2  2.35
#> # i 11,647 more rows
```

If you want to print all rows but don't know how many rows there are, use nrow(), e.g., print(mn_trees_tbl, n = nrow(mn_trees_tbl)).

Second, recall from Section 4.5, a data frame subset operation that results in a single column is simplified to a vector. This might not seem like a big deal; however, it can be very frustrating and potentially break your code when you expect an object to behave like a data frame, but it doesn't because it's now a vector. A tibble doesn't have this behavior: a subset resulting in one column is still a tibble. The code below illustrates these different behaviors.

```
is.data.frame(mn_trees[, 1]) # No longer a data frame.
```

```
#> [1] FALSE
```

```
is_tibble(mn_trees_tbl[, 1]) # Check if it's in fact a tibble.
```

```
#> [1] TRUE
```

```
is.data.frame(mn_trees_tbl[, 1]) # Still a data frame and tibble.
```

```
#> [1] TRUE
```

As always, consult the package manual page to learn more about its functions (run `?tibble::tibble-package`). Also, run `browseVignettes(package="tibble")` in the console to access the vignettes in the `tibble` package. Specifically, take a look at `vignette(package="tibble")` to understand the different value recycling rules applied when constructing data frames and tibbles.

Functions in `tidyverse` packages are happy to work with either data frames or tibbles. Because we prefer their print and subset behavior, we'll generally work with tibbles moving forward.

6.3 Reading and writing files with `readr`

In Section 3.3.2, we introduced several base R functions for reading and writing plain-text flat files, e.g., `read.table()`, `write.table()`, `read.csv()`, `write.csv()`. As we've seen, these functions read external data into a data frame and write a data frame to an external file.

As with the `tibble` package introduced in Section 6.2, `tidyverse` packages aim to provide improved functionality and flexibility over base R. In this spirit, the `readr` package offers alternative functions for reading and writing plain-text flat files. The package's read and write functions provide a rich and flexible set of arguments to accommodate different column delimiters, data types, and file formats. Like equivalent functions in base R, `readr` provides a set of read and write functions for common column delimiters, including `read_table()` for white space, `read_csv()` for comma, and `read_tsv()` for tab. The `read_delim()` function reads files with any user defined delimiter. Like base R, `readr` read functions have corresponding write functions.

When consulting the `readr` manual page (run `?readr::read_delim`), you'll notice many function arguments are the same as those in equivalent base R functions—this means minimal changes are needed to migrate from base R to `readr` functions, e.g., swapping `read.csv()` with `read_csv()`. The package's vignette, accessed by running `vignette("readr")` in the console, offers an in-depth tour of the package's capabilities.

Here are a few good reasons to favor `readr` functions over equivalent base R functions.

1. Read functions return a `tibble` with all its added niceties; see, Section 6.2.
2. Messages and warnings are often helpful for diagnosing file formatting issues.
3. A progress bar provides file read and write speed.
4. Information about the file being read is printed to the console, e.g., number of rows and columns, delimiter, and column data types.
5. Depending on the dataset, read and write functions are up to 100x faster. This is particularly helpful when working with large files.
6. Read functions are often able to guess column data type, and it's easy to specify data type if guessed incorrectly.
7. Non-syntactic column names (see Section 3.2.2) are preserved by placing them within

backticks. Base R read functions modify non-syntactic names to make them syntactic (see Section 3.3.2 for details).

The code below uses `read_csv()` to read the Minnesota tree growth dataset into the `mn_trees` tibble.

```
library(readr) # Or loaded automatically with library(tidyverse).

mn_trees <- read_csv("datasets/mn_trees_subset.csv")
```

```
#> Rows: 11649 Columns: 8
#> -- Column specification ------------------------------------
#> Delimiter: ","
#> chr (1): species
#> dbl (7): stand_id, plot_id, tree_id, year, age, rad...
#>
#> i Use `spec()` to retrieve the full column specification for this data.
#> i Specify the column types or set `show_col_types = FALSE` to quiet this message.
```

Notice above, the data dimensions, delimiter, and column data types are printed after calling `read_csv()`. Two informational messages are also printed. The first says `spec()` provides a description of column data types (i.e., run `spec(mn_trees)`). The second says data types can be specified if needed, and that you can set the read function argument `show_col_types = FALSE` if you don't want this information printed while reading.

Although not strictly necessary for our subsequent use of `mn_trees`, let's go ahead and specify the integer columns. Consult the `readr` vignette and manual page to understand the `col_types` arguments used below.

```
mn_trees <- read_csv("datasets/mn_trees_subset.csv",
                     col_types = list(stand_id = col_integer(),
                                      plot_id = col_integer(),
                                      tree_id = col_integer(),
                                      year = col_integer(),
                                      age = col_integer()
                                      )
                     )
spec(mn_trees) # Confirm data types.
```

```
#> cols(
#>   stand_id = col_integer(),
#>   plot_id = col_integer(),
#>   tree_id = col_integer(),
#>   year = col_integer(),
#>   species = col_character(),
```

```
#>   age = col_integer(),
#>   rad_inc = col_double(),
#>   DBH = col_double()
#> )
```

```
glimpse(mn_trees) # Or confirm data types via glimpse().
```

```
#> Rows: 11,649
#> Columns: 8
#> $ stand_id <int> 1, 1, 1, 1, 1, 1, 1, 1, 1, 1, 1, 1, ~
#> $ plot_id  <int> 1, 1, 1, 1, 1, 1, 1, 1, 1, 1, 1, 1, ~
#> $ tree_id  <int> 1, 1, 1, 1, 1, 1, 1, 1, 1, 1, 1, 1, ~
#> $ year     <int> 1960, 1961, 1962, 1963, 1964, 1965, ~
#> $ species  <chr> "ABBA", "ABBA", "ABBA", "ABBA", "ABB~
#> $ age      <int> 1, 2, 3, 4, 5, 6, 7, 8, 9, 10, 11, 1~
#> $ rad_inc  <dbl> 0.930, 0.950, 0.985, 0.985, 0.715, 0~
#> $ DBH      <dbl> 2.1563, 2.3463, 2.5433, 2.7403, 2.88~
```

It's good practice to specify columns' data type—as illustrated above—because it forces you to think about how your data is represented and how it will enter into subsequent analysis. We'll use `readr` functions throughout the remainder of the book (but we'll often be lazy and not specify column data types).

6.4 Summary

This chapter opens a series of chapters that introduce a suite of packages collectively referred to as the `tidyverse`. As you'll see, these packages provide tools to efficiently manipulate, analyze, and graphically represent data. Most tasks completed using the `tidyverse` can also be accomplished using base R functions; however, the `tidyverse` provides (arguably) more intuitive and easier to apply solutions within a unified framework.

In this chapter, we introduced the `tibble` and `readr` `tidyverse` packages. A tibble is a wrapper[5] around a data frame that provides some different behaviors, particularly when printing and subsetting. The `tidyverse` packages covered in subsequent chapters work with either tibbles or data frames; however, the tibble's added niceties make them our preferred option. We briefly covered `readr` package functions for reading and writing flat files. Like the `tibble` package, `readr` provides alternatives to base R functions.

Our tour of the `tibble`, `readr`, and subsequent `tidyverse` packages is woefully incomplete. We're not able to cover all the important package details or behaviors you'll likely encounter in applications. Rather, to get you pointed in the right direction, we focus on an

[5] A *wrapper* is a function that encapsulates other objects and/or functions in a user-friendly and potentially more flexible interface.

introduction and some reasonably realistic examples. Given the common grammar and function structure throughout packages in the `tidyverse`, our introduction should make it easier to learn additional `tidyverse` packages for your specific data analysis needs. Manual pages, vignettes, Google searches that connect you to user forums, and excellent books by package authors such as Wickham, Çetinkaya-Rundel, et al. (2023) are critical learning resources on your journey through the `tidyverse`.

6.5 Exercises

Exercise 6.1. Use the data in Table 4.2 to create a data frame using `data.frame()` called `stands_df` and a tibble using `tibble()` called `stands_tbl`. While creating the data frame and tibble, be sure to set the `stand` and `age` columns to integers using `as.integer()` as illustrated in Section 6.2. Print both `stands_df` and `stands_tbl` to confirm the data were entered correctly. Also, use either base R's `str()` or `dplyr`'s `glimpse()` to confirm the columns have the desired data type.

Exercise 6.2. `class()` will show that `stands_tbl` is a tibble (i.e., run `class(stands_tbl)`). What are two other ways you can check that `stands_tbl` is a tibble?

Exercise 6.3. How do the objects returned by the following operations differ? Why might this difference cause unexpected behavior in subsequent operations?

```
stands_df[, "age"]
stands_tbl[, "age"]
```

Exercise 6.4. Use `write_csv()` to write `stands_tbl` to a file called `"stands_tbl.csv"`. Then, read `"stands_tbl.csv"` using `read_csv()` with the appropriate `col_types` arguments and assign the resulting tibble to `stands_tbl_2`. Use `base::all.equal()` to confirm the two tibbles are the same (i.e., `all.equal(stands_tbl, stands_tbl_2)` should be `TRUE`).

7

Manipulating and summarizing data with `dplyr`

The Minnesota tree growth dataset introduced in Section 6.1 and the datasets presented in Section 1.2 provide a look at different ways to organize data in a single table. In most cases, dataset rows correspond to observational units (e.g., plots, trees, or years), and columns correspond to variables. This is the data organization format envisioned when applying `dplyr` functions.

According to the package's author, Hadley Wickham, `dplyr` is designed to be "a grammar of data manipulation, providing a consistent set of verbs that help you solve the most common data manipulation challenges." The `dplyr` functions do, in fact, capture data analysis actions (i.e., verbs). These verbs are best organized into three categories based on the component of the dataset upon which they operate:

- Rows:
    - `filter()` subset rows based on column values.
    - `slice()` subset rows based on index.
    - `arrange()` change row order.
- Columns:
    - `select()` subset columns based on name or characteristics.
    - `rename()` change column names.
    - `mutate()` change column values or create new columns.
    - `relocate()` change column order.
- Groups of rows:
    - `summarize()` collapse multiple rows into a single row.

A detailed `dplyr` cheat sheet is available from within RStudio under `Help > Cheatsheets > Data Transformation with dplyr`, and more resources can be found by running `?dplyr::dplyr` in the console. In addition, Hadley Wickham wrote a vignette that compares `dplyr` functions to their base R equivalents. Run `vignette("base")` in the console to access this vignette. If you're new to `dplyr`, we suggest working through this chapter first, then returning to the vignette if you want a crosswalk between `dplyr` and base R operations.

Sections 7.1 through 7.8 introduce `dplyr` verb functions that operate on a single tibble. We then cover several support functions designed to extend the verb functions' usefulness and flexibility. Topics in each section are illustrated using a minimal example based on a

DOI: 10.1201/9781003222699-7

small toy tibble called `trees`, created in Section 6.2. The toy dataset example is followed by a more thorough exposition using the Minnesota tree growth dataset introduced in Section 6.1. Extensive `dplyr` workflow examples using the Minnesota tree growth dataset are provided in Section 7.13. Finally, Section 7.14 covers a set of `dplyr` verb functions for combining information from two or more tibbles.

7.1 Filtering rows

`filter()` creates tibble row subsets that satisfy a condition involving values in one or more columns. The first argument in `filter()` is the tibble from which you want to subset rows based on conditions defined in subsequent arguments. Conditional statements passed as separate arguments are combined with the `&` operator. Write conditional statements using the tibble column names and operators in Section 4.7.

Figure 7.1 stylistically represents the `trees` tibble (created in Section 6.2), application of `filter()` to retain only those rows with `dbh` greater than 2, and the resulting tibble.

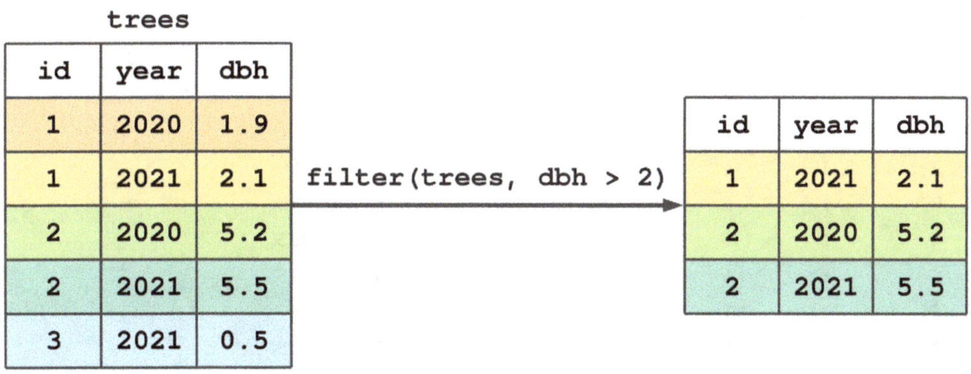

FIGURE 7.1: Using `filter()` to subset rows from `trees` with `dbh` greater than 2.

Next we filter `mn_trees` (created in Section 6.3) using several different conditions. For illustration, we simply print the filtered result, but you could assign it to a new tibble for subsequent manipulation. The call to `filter()` below retains only trees in stand 5.[1]

```
filter(mn_trees, stand_id == 5)
```

```
#> # A tibble: 3,020 x 8
```

[1]Notice that only the first 10 of 3,020 rows were printed. This is because `mn_trees` is a tibble, so the output of `filter()` is also a tibble and exhibits the nice print behavior described in Section 6.2.

```
#>    stand_id plot_id tree_id  year species   age rad_inc
#>       <int>   <int>   <int> <int> <chr>   <int>   <dbl>
#>  1        5       1       1  1969 ABBA        1    1.98
#>  2        5       1       1  1970 ABBA        2    2.19
#>  3        5       1       1  1971 ABBA        3    1.67
#>  4        5       1       1  1972 ABBA        4    2.34
#>  5        5       1       1  1973 ABBA        5    2.36
#>  6        5       1       1  1974 ABBA        6    2.29
#>  7        5       1       1  1975 ABBA        7    2.28
#>  8        5       1       1  1976 ABBA        8    2.46
#>  9        5       1       1  1977 ABBA        9    1.68
#> 10        5       1       1  1978 ABBA       10    2.54
#> # i 3,010 more rows
#> # i 1 more variable: DBH <dbl>
```

Below, we use `filter()` to retain only *Abies balsamea* and *Pinus strobus* species rows (refer to Table 6.1 for species codes).

```
filter(mn_trees, species %in% c("ABBA", "PIST"))
```

```
#> # A tibble: 5,936 x 8
#>    stand_id plot_id tree_id  year species   age rad_inc
#>       <int>   <int>   <int> <int> <chr>   <int>   <dbl>
#>  1        1       1       1  1960 ABBA        1   0.93
#>  2        1       1       1  1961 ABBA        2   0.95
#>  3        1       1       1  1962 ABBA        3   0.985
#>  4        1       1       1  1963 ABBA        4   0.985
#>  5        1       1       1  1964 ABBA        5   0.715
#>  6        1       1       1  1965 ABBA        6   0.84
#>  7        1       1       1  1966 ABBA        7   0.685
#>  8        1       1       1  1967 ABBA        8   0.94
#>  9        1       1       1  1968 ABBA        9   1.16
#> 10        1       1       1  1969 ABBA       10   0.775
#> # i 5,926 more rows
#> # i 1 more variable: DBH <dbl>
```

The next two calls to `filter()` illustrate equivalent conditions that retain rows with trees older than 250 years in 2007 (i.e., the most recent date in the chronology). The second call to `filter()` is equivalent to the first because an implicit `&` is placed between conditional statements passed as separate arguments.

```
filter(mn_trees, age > 250 & year == 2007)
```

```
#> # A tibble: 2 x 8
#>   stand_id plot_id tree_id  year species   age rad_inc
```

```
#>      <int>    <int>    <int> <int> <chr>    <int>   <dbl>
#> 1        2        2        2  2007 PIRE       269    0.37
#> 2        2        2        3  2007 PIRE       265    0.26
#> # i 1 more variable: DBH <dbl>
```

```
filter(mn_trees, age > 250,
       year == 2007) # Implicit & between filter arguments 2 and 3.
```

```
#> # A tibble: 2 x 8
#>   stand_id plot_id tree_id  year species   age rad_inc
#>      <int>   <int>   <int> <int> <chr>   <int>   <dbl>
#> 1        2       2       2  2007 PIRE      269    0.37
#> 2        2       2       3  2007 PIRE      265    0.26
#> # i 1 more variable: DBH <dbl>
```

7.2 Slicing rows

`slice()` creates tibble row subsets given row indexes. The first argument in `slice()` is the tibble from which you want to subset rows and the second argument is a vector holding the desired row indexes. Figure 7.2 illustrates how `slice()` is used to retain rows 1 and 4 from `trees`.

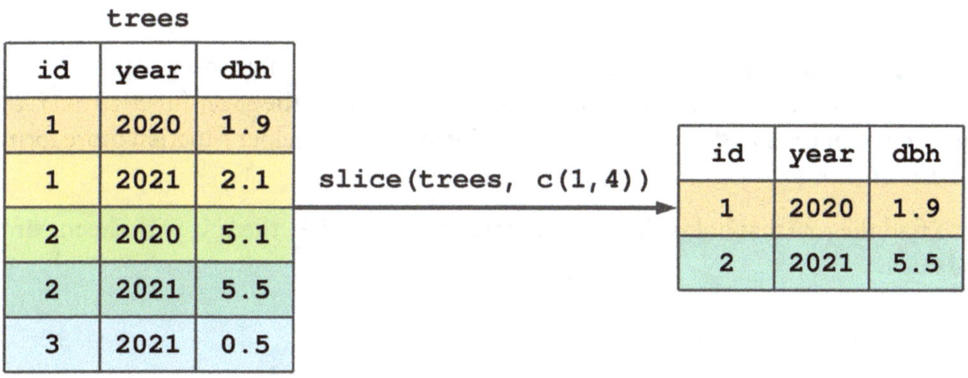

FIGURE 7.2: Using `slice()` to subset rows 1 and 4 of `trees`.

Putting a negation - operator in front of the second `slice()` argument produces the complement of the given row indexes. For example, following Figure 7.2, `slice(trees, -c(1,4))` would retain rows 2, 3, and 5.

`slice()` comes in different flavors, each designed to simplify a common row subsetting task. `slice_head()` and `slice_tail()` return the first and last row of the given tibble, respectively.

Similarly, `slice_min()` and `slice_max()` return the rows that hold the minimum and maximum value of a given column within the tibble. You can return a random subset of rows, with or without replacement and with specified selection probability, using `slice_sample()`. Change the `n` argument in these functions to return more than one row (i.e., the default argument is `n = 1`), or use the `prop` argument to return the desired proportion of rows. Let's take a look at a few examples using the Minnesota tree growth dataset.

Select the first row of `mn_trees`.

```
slice_head(mn_trees) # Or, equivalently, slice(mn_trees, 1)
```

```
#> # A tibble: 1 x 8
#>   stand_id plot_id tree_id  year species   age rad_inc
#>      <int>   <int>   <int> <int> <chr>   <int>   <dbl>
#> 1        1       1       1  1960 ABBA        1    0.93
#> # i 1 more variable: DBH <dbl>
```

Now, select the last two rows.

```
slice_tail(mn_trees, n = 2)
```

```
#> # A tibble: 2 x 8
#>   stand_id plot_id tree_id  year species   age rad_inc
#>      <int>   <int>   <int> <int> <chr>   <int>   <dbl>
#> 1        5       3      23  2006 THOC       75    1.04
#> 2        5       3      23  2007 THOC       76    0.95
#> # i 1 more variable: DBH <dbl>
```

A more convoluted, but instructive, alternative to return the last two rows is `slice(mn_trees, (n() - 1):n())` or `slice(mn_trees, c(n() - 1, n()))`, where the `dplyr` function `n()` returns the number of rows in the given tibble. We'll introduce this helper function more formally in Section 7.10.

Let's find the row that holds the oldest recorded tree age, i.e., the row with the maximum `age` value.

```
slice_max(mn_trees, age) # 269 years old!
```

```
#> # A tibble: 1 x 8
#>   stand_id plot_id tree_id  year species   age rad_inc
#>      <int>   <int>   <int> <int> <chr>   <int>   <dbl>
#> 1        2       2       2  2007 PIRE      269    0.37
#> # i 1 more variable: DBH <dbl>
```

You could also find this row using `filter(mn_trees, age == max(age))`. It'd be trickier to write a `filter()` statement to find, say, rows with the two greatest ages; however, such a query is easily tackled using `slice_max(mn_trees, age, n = 2)`.

7.3 Arranging rows by values

`arrange()` orders rows based on values in one or more columns. The first argument in `arrange()` is the tibble to be ordered, and subsequent arguments are the column names used to order the rows. If you specify more than one column, each additional column is used to break ties in the values of preceding columns. The default behavior is to sort column values in ascending order (i.e., smallest values at the top). If you want to sort column values in descending order, put the column name inside `desc()`.

`arrange()` doesn't change the number of rows, only their order. Row order doesn't typically matter when analyzing data; however, order is often important for understanding and communicating data structure.

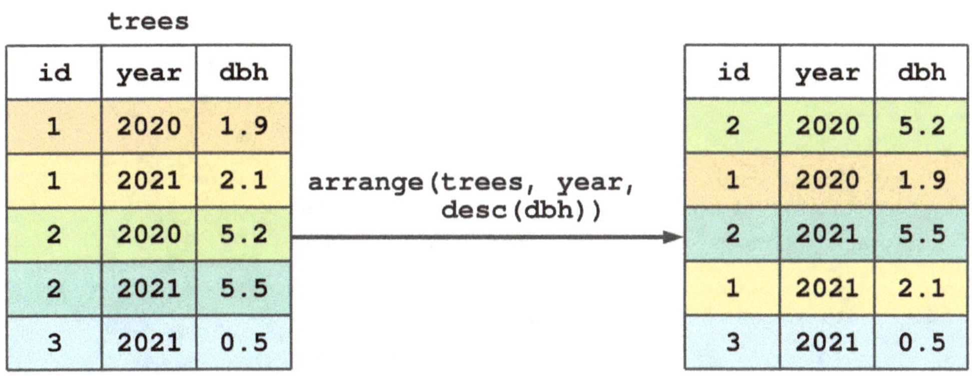

FIGURE 7.3: Using `arrange()` to change the row order of `trees`.

Figure 7.3 illustrates how to order `trees` by ascending `year`, then descending `dbh`. Notice that there are "ties" for 2020 and 2021; hence, the `desc(dbh)` argument causes rows within each year to be ordered by decreasing values of `dbh`.

The `"mn_trees_subset.csv"` file that we read into `mn_trees` is arranged in ascending order by `stand_id`, then `plot_id`, then `tree_id`, then `year`; thus, the code below doesn't change `mn_trees` row order.

```
mn_trees <- arrange(mn_trees, stand_id, plot_id, tree_id, year)
```

Let's use a subset of `mn_trees` to better explore `arrange()`. The code below filters trees in stand 1, plot 1, and year 2007. The resulting subset `s1_p1` is printed, then arranged in descending order by `tree_id`.

```
s1_p1 <- filter(mn_trees, stand_id == 1, plot_id == 1,
                year == 2007)
s1_p1
```

```
#> # A tibble: 7 x 8
#>   stand_id plot_id tree_id  year species   age rad_inc
#>      <int>   <int>   <int> <int> <chr>   <int>   <dbl>
#> 1        1       1       1  2007 ABBA       48   1.05
#> 2        1       1       2  2007 ABBA       63   0.49
#> 3        1       1       3  2007 BEPA       59   0.805
#> 4        1       1       4  2007 BEPA      119   0.69
#> 5        1       1       5  2007 BEPA      101   1.04
#> 6        1       1       6  2007 PIGL       54   1.7
#> 7        1       1       7  2007 PIGL       68   0.86
#> # i 1 more variable: DBH <dbl>
```

```
arrange(s1_p1, desc(tree_id))
```

```
#> # A tibble: 7 x 8
#>   stand_id plot_id tree_id  year species   age rad_inc
#>      <int>   <int>   <int> <int> <chr>   <int>   <dbl>
#> 1        1       1       7  2007 PIGL       68   0.86
#> 2        1       1       6  2007 PIGL       54   1.7
#> 3        1       1       5  2007 BEPA      101   1.04
#> 4        1       1       4  2007 BEPA      119   0.69
#> 5        1       1       3  2007 BEPA       59   0.805
#> 6        1       1       2  2007 ABBA       63   0.49
#> 7        1       1       1  2007 ABBA       48   1.05
#> # i 1 more variable: DBH <dbl>
```

Now let's arrange `s1_p1` by `desc(species)` and `age`.

```
arrange(s1_p1, desc(species), age)
```

```
#> # A tibble: 7 x 8
#>   stand_id plot_id tree_id  year species   age rad_inc
#>      <int>   <int>   <int> <int> <chr>   <int>   <dbl>
#> 1        1       1       6  2007 PIGL       54   1.7
#> 2        1       1       7  2007 PIGL       68   0.86
#> 3        1       1       3  2007 BEPA       59   0.805
#> 4        1       1       5  2007 BEPA      101   1.04
#> 5        1       1       4  2007 BEPA      119   0.69
#> 6        1       1       1  2007 ABBA       48   1.05
#> 7        1       1       2  2007 ABBA       63   0.49
```

```
#> # i 1 more variable: DBH <dbl>
```

Notice rows are ordered by the character column `species` in descending alphabetical order, then by ascending `age` within each species.

7.4 Selecting columns

Another common task is to select a subset of columns in the dataset. This is accomplished using `select()`. The first argument in `select()` is the tibble (or data frame) from which you want to subset columns, followed by additional arguments used to identify columns you want to retain.

Figure 7.4 offers a minimal `select()` example that lists column names to retain as separate arguments.

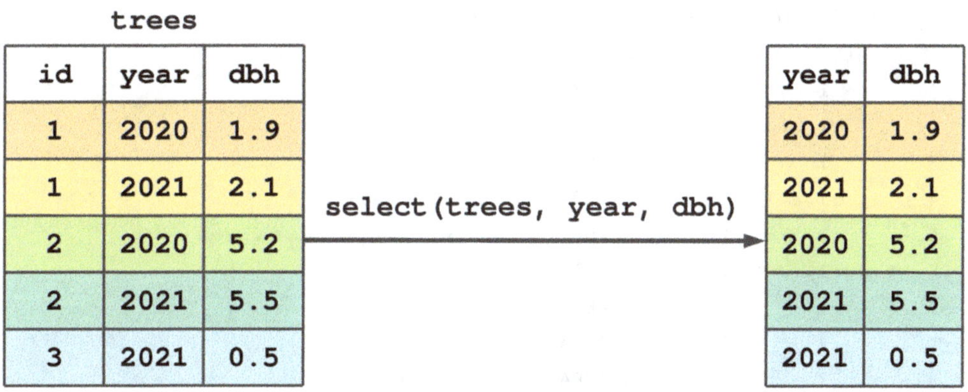

FIGURE 7.4: Using `select()` to subset columns from `trees`.

Consider how the next four calls to `select()` specify columns to retain. In particular, notice that:

1) columns can be selected by name or column index;
2) column names or indexes can be specified in separate arguments or passed all at once as a vector formed using `c()`;
3) the colon `:` operator can be used to specify consecutive column indexes;
4) the negation operator `!` returns the column set complement.

```
select(mn_trees, year, DBH) # c(year, DBH) gives the same result.
```

```
#> # A tibble: 11,649 x 2
```

```
#>     year   DBH
#>    <int> <dbl>
#>  1  1960  2.16
#>  2  1961  2.35
#>  3  1962  2.54
#>  4  1963  2.74
#>  5  1964  2.88
#>  6  1965  3.05
#>  7  1966  3.19
#>  8  1967  3.38
#>  9  1968  3.61
#> 10  1969  3.76
#> # i 11,639 more rows
```

```
select(mn_trees, !c(year, DBH)) # Negation operator.
```

```
#> # A tibble: 11,649 x 6
#>    stand_id plot_id tree_id species   age rad_inc
#>       <int>   <int>   <int> <chr>   <int>   <dbl>
#>  1        1       1       1 ABBA        1   0.93
#>  2        1       1       1 ABBA        2   0.95
#>  3        1       1       1 ABBA        3   0.985
#>  4        1       1       1 ABBA        4   0.985
#>  5        1       1       1 ABBA        5   0.715
#>  6        1       1       1 ABBA        6   0.84
#>  7        1       1       1 ABBA        7   0.685
#>  8        1       1       1 ABBA        8   0.94
#>  9        1       1       1 ABBA        9   1.16
#> 10        1       1       1 ABBA       10   0.775
#> # i 11,639 more rows
```

```
select(mn_trees, c(1:4, 6)) # Colon operator and vector of indexes.
```

```
#> # A tibble: 11,649 x 5
#>    stand_id plot_id tree_id  year   age
#>       <int>   <int>   <int> <int> <int>
#>  1        1       1       1  1960     1
#>  2        1       1       1  1961     2
#>  3        1       1       1  1962     3
#>  4        1       1       1  1963     4
#>  5        1       1       1  1964     5
#>  6        1       1       1  1965     6
#>  7        1       1       1  1966     7
#>  8        1       1       1  1967     8
```

```
#>  9          1         1         1  1968     9
#> 10          1         1         1  1969    10
#> # i 11,639 more rows
```

```
select(mn_trees, !c(1, 3, 4)) # Negation operator.
```

```
#> # A tibble: 11,649 x 5
#>    plot_id species   age rad_inc   DBH
#>      <int> <chr>   <int>   <dbl> <dbl>
#>  1       1 ABBA        1   0.93   2.16
#>  2       1 ABBA        2   0.95   2.35
#>  3       1 ABBA        3   0.985  2.54
#>  4       1 ABBA        4   0.985  2.74
#>  5       1 ABBA        5   0.715  2.88
#>  6       1 ABBA        6   0.84   3.05
#>  7       1 ABBA        7   0.685  3.19
#>  8       1 ABBA        8   0.94   3.38
#>  9       1 ABBA        9   1.16   3.61
#> 10       1 ABBA       10   0.775  3.76
#> # i 11,639 more rows
```

A set of *helper* functions aid in column selection. The next bit of code uses the `ends_with()` helper function to select all columns with names ending with "id".

```
select(mn_trees, ends_with("id"))
```

```
#> # A tibble: 11,649 x 3
#>    stand_id plot_id tree_id
#>       <int>   <int>   <int>
#>  1        1       1       1
#>  2        1       1       1
#>  3        1       1       1
#>  4        1       1       1
#>  5        1       1       1
#>  6        1       1       1
#>  7        1       1       1
#>  8        1       1       1
#>  9        1       1       1
#> 10        1       1       1
#> # i 11,639 more rows
```

Other handy helper functions you should investigate are `everything()`, `last_col()`, `starts_with()`, `contains()`, `matches()`, `num_range()`, `all_of()`, and `where()`. See `?dplyr::select` for details.

The & and | operators provide additional flexibility for specifying columns to retain or exclude. Use & to identify the intersection between two column sets (i.e., those columns that are in both sets). Use | to identify the union of two column sets (i.e., all columns in at least one set). For example, the code below returns the `tree_id` column because it's selected by both `ends_with("id")` **and** `contains("tree")`.

```
select(mn_trees, ends_with("id") & contains("tree"))
```

```
#> # A tibble: 11,649 x 1
#>    tree_id
#>      <int>
#>  1       1
#>  2       1
#>  3       1
#>  4       1
#>  5       1
#>  6       1
#>  7       1
#>  8       1
#>  9       1
#> 10       1
#> # i 11,639 more rows
```

`where()` is a bit different from other helper functions listed above. Use `where()` when you want to apply a function to all columns and select those for which the discerning function returns `TRUE`. The data type testing functions introduced in Section 4, e.g., `is.numeric()`, `is.integer()`, `is.double()`, `is.factor()`, and `is.character()` can be used with `where()`. For example, run `select(mn_trees, where(is.character))` to select only character data type columns, or `select(mn_trees, where(is.numeric))` to select all numeric columns.

How about selecting all columns of type integer that don't end in "id"? One approach is `select(mn_trees, where(is.integer) & !ends_with("id"))`.

As you'll see shortly, other `dplyr` functions use `select()` syntax to identify columns.

7.5 Renaming columns

It's easy to rename tibble columns using `rename()`. As illustrated in Figure 7.5, the first argument in `rename()` is the tibble containing the columns to rename, and subsequent arguments take the form `new_name = old_name`, where `old_name` is the existing column name and `new_name` is the name you'd like to change it to.

Let's change column names `rad_inc` and `DBH` in `mn_trees` to `rad_inc_mm` and `DBH_cm`, so we don't need to keep looking back at Section 6.1 to remember their units. Notice in the code below,

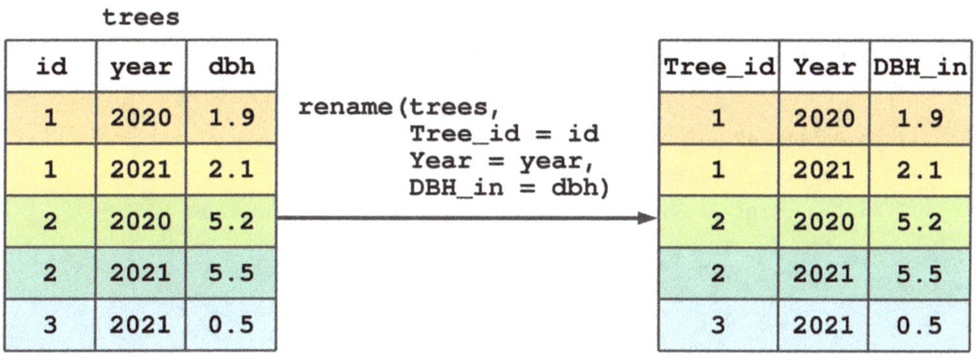

FIGURE 7.5: Using `rename()` to rename some columns in `trees`.

we assign the output of `rename()` back to `mn_trees` to make the name change permanent. Taking a `glimpse()` at `mn_trees` shows the column names are indeed modified.

```
mn_trees <- rename(mn_trees, rad_inc_mm = rad_inc,
                   DBH_cm = DBH)

glimpse(mn_trees)
```

```
#> Rows: 11,649
#> Columns: 8
#> $ stand_id   <int> 1, 1, 1, 1, 1, 1, 1, 1, 1, 1, 1, 1~
#> $ plot_id    <int> 1, 1, 1, 1, 1, 1, 1, 1, 1, 1, 1, 1~
#> $ tree_id    <int> 1, 1, 1, 1, 1, 1, 1, 1, 1, 1, 1, 1~
#> $ year       <int> 1960, 1961, 1962, 1963, 1964, 1965~
#> $ species    <chr> "ABBA", "ABBA", "ABBA", "ABBA", "A~
#> $ age        <int> 1, 2, 3, 4, 5, 6, 7, 8, 9, 10, 11,~
#> $ rad_inc_mm <dbl> 0.930, 0.950, 0.985, 0.985, 0.715,~
#> $ DBH_cm     <dbl> 2.1563, 2.3463, 2.5433, 2.7403, 2.~
```

`rename_with()` provides batch selection and function-based renaming. The first `rename_with()` argument is the name of the tibble containing columns to rename, the second argument is a function that takes a character string and returns a modified string (i.e., takes the old name and returns the new name), and the optional third argument specifies the columns to rename. The third argument uses `select()` syntax to select columns (see Section 7.4 and the example below). If the third argument is not used, the function in the second argument is applied to all columns.

Let's use the `tidyverse stringr` package's `str_to_title()` function to change only those columns that end with `id` to title case (i.e., capitalize the first letter). As the name suggests, the `stringr` package provides functions that work on character strings; see, e.g.,

?stringr::stringr.

```
rename_with(mn_trees, str_to_title, ends_with("id"))
```

```
#> # A tibble: 11,649 x 8
#>    Stand_id Plot_id Tree_id  year species   age
#>       <int>   <int>   <int> <int> <chr>   <int>
#>  1        1       1       1  1960 ABBA        1
#>  2        1       1       1  1961 ABBA        2
#>  3        1       1       1  1962 ABBA        3
#>  4        1       1       1  1963 ABBA        4
#>  5        1       1       1  1964 ABBA        5
#>  6        1       1       1  1965 ABBA        6
#>  7        1       1       1  1966 ABBA        7
#>  8        1       1       1  1967 ABBA        8
#>  9        1       1       1  1968 ABBA        9
#> 10        1       1       1  1969 ABBA       10
#> # i 11,639 more rows
#> # i 2 more variables: rad_inc_mm <dbl>, DBH_cm <dbl>
```

See `?dplyr::rename_with` for additional guidance and syntax to develop and apply more sophisticated renaming functions.

7.6 Creating new columns

`mutate()` creates new columns and adds them to the right side of an existing tibble. The first argument in `mutate()` is the tibble to be augmented and subsequent arguments define the new columns. This function is particularly useful because exiting columns in the tibble can be referenced when defining new columns. This feature is illustrated in Figure 7.6 using the existing `dbh` column to create a basal area (`ba`) column (see Section 5.3.1 for specifics on computing basal area).

Sometimes we want a tibble that includes only newly created columns or a combination of original and new columns. In this case, we use `transmute()`. For example, as illustrated in Figure 7.7, say we want to make `ba` using `dbh` in `trees`, but this time we want the new tibble to include only the columns `id`, `year`, and `ba` (i.e., we don't want to retain `dbh`). Also notice in Figure 7.7, we can carryover columns unchanged by just listing their names in transmute arguments.[2]

Let's use `mutate()` to add a basal area column to `mn_trees`. Given DBH in cm, we calculate BA $(\text{m}^2) = 0.00007854 \cdot \text{DBH}^2$, which is implemented in the code below to create `BA_m_sq`.

[2]You can also use `mutate()` and `transmute()` to rename columns, e.g., in Figure 7.7 you may replace `id` and `tree` with `Tree_id = id` and `Year = year`.

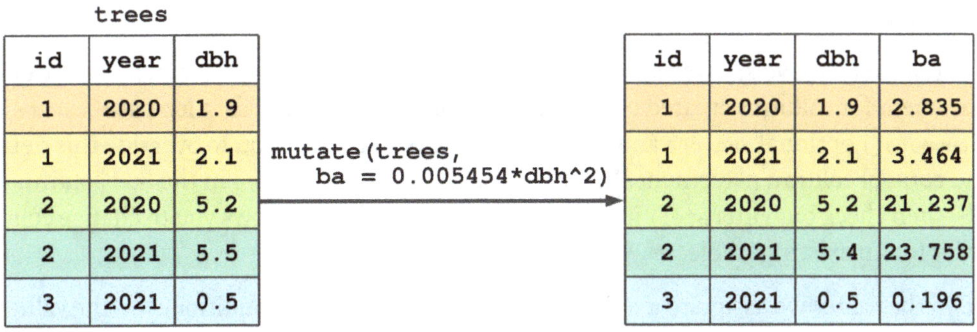

FIGURE 7.6: Using `mutate()` to add a basal area (`ba`) column to `trees`.

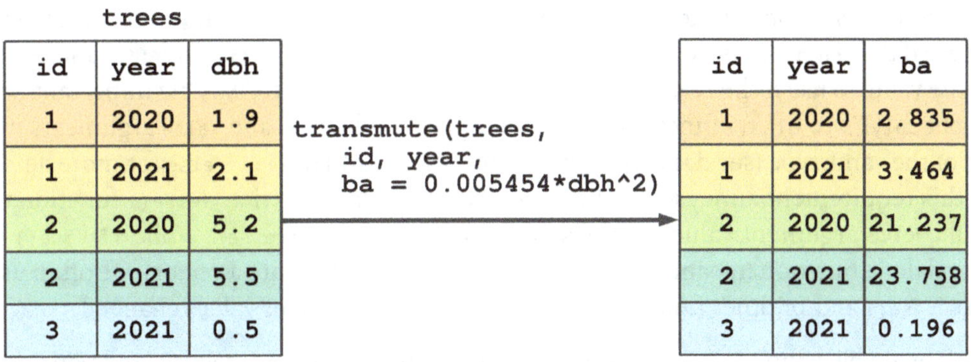

FIGURE 7.7: Using `transmute()` to create basal area (`ba`) and keep `id` and `year` columns.

```r
mn_trees <- mutate(mn_trees, BA_m_sq = 0.00007854 * DBH_cm^2)

glimpse(mn_trees)
```

```
#> Rows: 11,649
#> Columns: 9
#> $ stand_id   <int> 1, 1, 1, 1, 1, 1, 1, 1, 1, 1, 1, 1~
#> $ plot_id    <int> 1, 1, 1, 1, 1, 1, 1, 1, 1, 1, 1, 1~
#> $ tree_id    <int> 1, 1, 1, 1, 1, 1, 1, 1, 1, 1, 1, 1~
#> $ year       <int> 1960, 1961, 1962, 1963, 1964, 1965~
#> $ species    <chr> "ABBA", "ABBA", "ABBA", "ABBA", "A~
#> $ age        <int> 1, 2, 3, 4, 5, 6, 7, 8, 9, 10, 11,~
#> $ rad_inc_mm <dbl> 0.930, 0.950, 0.985, 0.985, 0.715,~
```

```
#> $ DBH_cm    <dbl> 2.1563, 2.3463, 2.5433, 2.7403, 2.~
#> $ BA_m_sq   <dbl> 0.00036518, 0.00043237, 0.00050802~
```

The `glimpse()` above shows `BA_m_sq` is now the last `mn_trees` column, which is the default placement for columns created using `mutate()` and `transmute()`. Take a look at `mutate()` and `transmute()` optional `.before` and `.after` arguments that allow you to override this default and control column placement. If you feel sad for `BA_m_sq`—being in the last column and all—then check out `relocate()` in the next section, which also allows you to change column order for an existing tibble.

Often the values we assign to a new column using `mutate()` are conditional on the values in one or more existing columns. For example, say we want to add a species column to `trees` and we know tree `id` 1 is *Larix laricina*, `id` 2 is *Picea glauca*, and `id` 3 is *Populus tremuloides*. How do we go about creating such a column?

One approach is to use the `dplyr`'s `if_else()` function, which is defined as `if_else(condition, true, false, missing = NULL)`, where `condition` is a logical vector created using one or more vectors and logical operators in Section 4.7, `true` is the value to use if the `condition` is `TRUE`, `false` is the value to use if the `condition` is `FALSE`, and `missing` is the value to use if the `condition` is neither `TRUE` nor `FALSE`, e.g., `NA`. One important point about `dplyr`'s `if_else()` function is the values given in the `true` and `false` arguments must be of the same type (see data types in Section 4.2.1 and `?dplyr::if_else` for more details). If this requirement annoys you, then consider base R's less strict `ifelse()` function that will coerce argument values to the same type (see `?base::ifelse`). While `if_else()` is a standalone function (meaning it can be used on its own to create a vector), we often use it in `mutate()` and/or other functions where vectors are created or values changed.

Let's begin by using `if_else()` within `mutate()` to create a `species` column and populate `id == 1` rows with "Larix laricina" and, for demonstration, all other rows with `NA`.[3]

```
mutate(trees, species = if_else(id == 1, "Larix laricina", NA))
```

```
#> # A tibble: 5 x 4
#>      id  year   dbh species
#>   <int> <int> <dbl> <chr>
#> 1     1  2020   1.9 Larix laricina
#> 2     1  2021   2.1 Larix laricina
#> 3     2  2020   5.2 <NA>
#> 4     2  2021   5.5 <NA>
#> 5     3  2021   0.5 <NA>
```

All rows in the `species` column can be filled in as follows.

[3]While briefly noted in Section 4.4, there are different `NA` types. The default `NA` type is logical, run `typeof(NA)`, the other types are `NA_integer_`, `NA_real_`, `NA_complex_`, and `NA_character_`, see `?base::NA`. Because `if_else()` is strict, if you want one of your arguments to be `NA` then it'll need to match the other argument's type, e.g., use `NA_character_` in example code above. Or use `ifelse()` which is happy to coerce to match types for you.

```
mutate(trees,
       species = if_else(id == 1, "Larix laricina", NA),
       species = if_else(id == 2, "Picea glauca", species),
       species = if_else(id == 3, "Populus tremuloides", species)
       )
```

```
#> # A tibble: 5 x 4
#>      id  year   dbh species
#>   <int> <int> <dbl> <chr>
#> 1     1  2020   1.9 Larix laricina
#> 2     1  2021   2.1 Larix laricina
#> 3     2  2020   5.2 Picea glauca
#> 4     2  2021   5.5 Picea glauca
#> 5     3  2021   0.5 Populus tremuloides
```

Notice above, `species` rows are incrementally filled, first for `id` 1, then 2, then 3. Alternatively, you can use nested conditional statements, where the `false` argument holds another `if_else()`.

```
mutate(trees,
       species = if_else(id == 1,
                         "Larix laricina",
                         if_else(id == 2,
                                 "Picea glauca",
                                 if_else(id == 3,
                                         "Populus tremuloides", NA)
                                 )
                         )
       )
```

```
#> # A tibble: 5 x 4
#>      id  year   dbh species
#>   <int> <int> <dbl> <chr>
#> 1     1  2020   1.9 Larix laricina
#> 2     1  2021   2.1 Larix laricina
#> 3     2  2020   5.2 Picea glauca
#> 4     2  2021   5.5 Picea glauca
#> 5     3  2021   0.5 Populus tremuloides
```

`if_else()` is a powerful tool in some situations, particularly when we need to test only one condition. However, as illustrated above, coding can be onerous when we need to test multiple conditions.

The alternative to writing multiple nested `if_else()` statements is `case_when()` (see `?dplyr::case_when`). Like `if_else()`, `case_when()` is a standalone function but is often used in `mutate()` to create or modify columns. `case_when()` takes one or more arguments, where each argument, or *case*, is a two-sided formula where the left side is separated by the right side by a tilde symbol `~`. The left side is a logical vector created using one or more vectors and logical operators in Section 4.7. The right side is the value to use when the left side is `TRUE`. If no cases are `TRUE` (i.e., none of the arguments' left sides are `TRUE`), then `NA` is returned. Like `if_else()`, `case_when()` requires values on the left side of cases to be the same type.

```
mutate(trees, species = case_when(
                                  id == 1 ~ "Larix laricina",
                                  id == 2 ~ "Picea glauca",
                                  id == 3 ~ "Populus tremuloides"
                                )
)
```

```
#> # A tibble: 5 x 4
#>      id  year   dbh species
#>   <int> <int> <dbl> <chr>
#> 1     1  2020   1.9 Larix laricina
#> 2     1  2021   2.1 Larix laricina
#> 3     2  2020   5.2 Picea glauca
#> 4     2  2021   5.5 Picea glauca
#> 5     3  2021   0.5 Populus tremuloides
```

As one final example, let's add a diameter class column to `mn_trees`. We'll use `transmute()` to return only `DBH_cm` and our new `DBH_class` columns. Diameter class criteria and associated naming are given on the left and right side, respectively, within `case_when()`.

```
transmute(mn_trees, DBH_cm,
          DBH_class = case_when(
                             DBH_cm < 2.5 ~ "seedling",
                             DBH_cm >= 2.5 & DBH_cm < 5 ~ "small sapling",
                             DBH_cm >= 5 & DBH_cm < 10 ~ "large sapling",
                             DBH_cm >= 10 & DBH_cm < 20 ~ "small pole",
                             DBH_cm >= 20 & DBH_cm < 30 ~ "large pole",
                             DBH_cm >= 30 & DBH_cm < 50 ~ "small sawlog",
                             DBH_cm >= 50                ~ "large sawlog"
                           )
)
```

```
#> # A tibble: 11,649 x 2
#>    DBH_cm DBH_class
#>     <dbl> <chr>
#>  1   2.16 seedling
#>  2   2.35 seedling
#>  3   2.54 small sapling
```

```
#>  4   2.74 small sapling
#>  5   2.88 small sapling
#>  6   3.05 small sapling
#>  7   3.19 small sapling
#>  8   3.38 small sapling
#>  9   3.61 small sapling
#> 10   3.76 small sapling
#> # i 11,639 more rows
```

7.7 Changing column order

Change column position within a tibble using `relocate()`. The first argument in `relocate()` is the tibble to be modified and subsequent arguments identify columns. Columns are identified using `select()` syntax, see Section 7.4. By default, selected columns are moved to the tibble's left side, as illustrated in Figure 7.8.

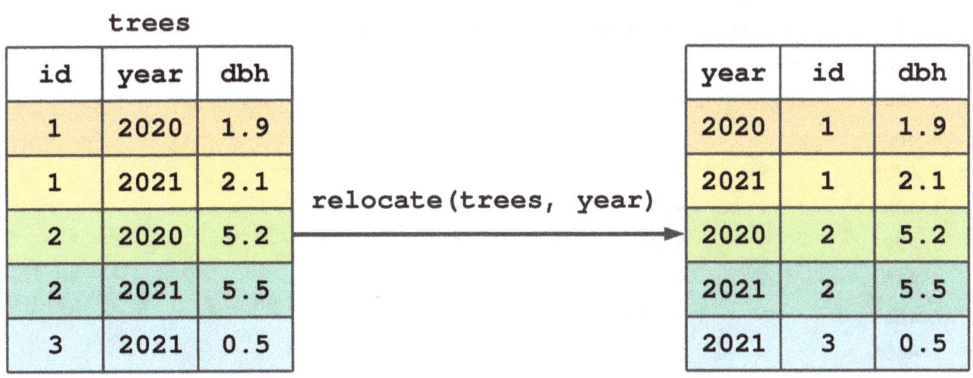

FIGURE 7.8: Using the default behavior of `relocate()` to move `year` to the left side of `trees`.

Use the `.before` or `.after` arguments to control column location, i.e., if you want the selected columns to be moved to "before" or "after" one or more specified columns. `select()` syntax is used when specifying columns for `.before` and `.after` arguments. You can also rename columns in the process of moving them as illustrated in Figure 7.9.

Let's now return to the `mn_trees` dataset to work a bit more with `relocate()`. For fun, we move `stand_id`, `plot_id`, `tree_id` and `year` from their original location in the first four columns to the right side of the last column. Notice how the `select()` helper functions `contains()`, `|`, and `last_col()` are used in `relocate()`.[4]

[4]The columns relocated to the tibble's right side are listed in the last few lines of the printed output because they don't fit on the page.

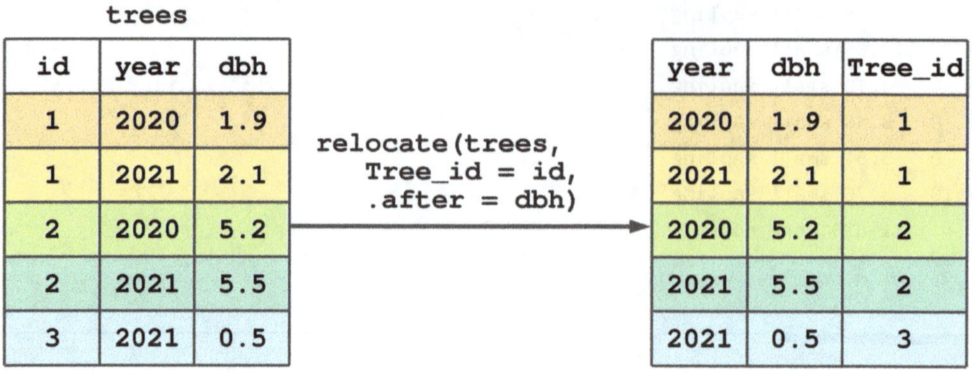

FIGURE 7.9: Using `relocate()` to rename and move `id` to after `dbh`.

```
relocate(mn_trees, contains("id") | year, .after = last_col())
```

```
#> # A tibble: 11,649 x 9
#>    species   age rad_inc_mm DBH_cm  BA_m_sq stand_id
#>    <chr>   <int>      <dbl>  <dbl>    <dbl>    <int>
#>  1 ABBA        1      0.93    2.16 0.000365        1
#>  2 ABBA        2      0.95    2.35 0.000432        1
#>  3 ABBA        3      0.985   2.54 0.000508        1
#>  4 ABBA        4      0.985   2.74 0.000590        1
#>  5 ABBA        5      0.715   2.88 0.000653        1
#>  6 ABBA        6      0.84    3.05 0.000731        1
#>  7 ABBA        7      0.685   3.19 0.000798        1
#>  8 ABBA        8      0.94    3.38 0.000895        1
#>  9 ABBA        9      1.16    3.61 0.00102         1
#> 10 ABBA       10      0.775   3.76 0.00111         1
#> # i 11,639 more rows
#> # i 3 more variables: plot_id <int>, tree_id <int>,
#> #   year <int>
```

Like row order, column order rarely matters for analysis; however, it's often important for recognizing and communicating data structure. We generally like to have those columns that describe observation nesting structure to be to the left of observation measurements. For the `mn_trees` dataset, this means `stand_id`, `plot_id`, `tree_id`, and `year` come before (i.e., to the left of) `species`, `age`, and tree-ring measurement variables—the way these data are organized in the "mn_trees_subset.csv" file.

7.8 Summarizing

`summarize()` collapses a tibble to a single row. We often use `summarize()` to compute summary statistics for one or more columns in a tibble. The first argument in `summarize()` is the tibble and subsequent arguments define how rows should be collapsed, i.e., summarized. For example, Figure 7.10 illustrates a call to `summarize()` that computes the dbh column's mean and standard deviation.

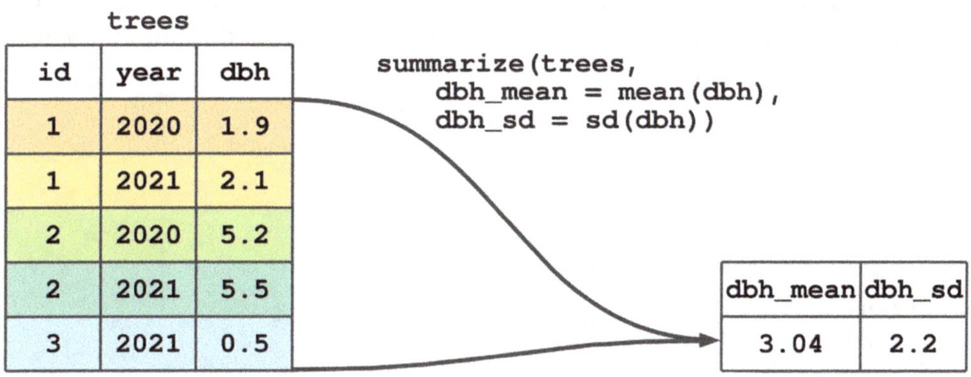

FIGURE 7.10: Using `summarize()` to compute the mean and standard deviation of `dbh`.

Returning for a moment to material in Chapter 4, the call to `summarize()` in Figure 7.10 is equivalent to the code below.

```
tibble(dbh_mean = mean(trees$dbh), dbh_sd = sd(trees$dbh))
```

```
#> # A tibble: 1 x 2
#>   dbh_mean dbh_sd
#>      <dbl>  <dbl>
#> 1     3.04   2.20
```

For our purposes, `summarize()` is most useful when applied to *grouped* data, which are introduced in Section 7.9.

7.9 Grouped data

Many datasets we encounter have a nested structure (e.g., measurements on trees within plots, plots within stands, stands within a forest) and/or populations of interest (e.g., species, geographic region, product type). When dataset rows correspond to observational units and columns correspond to variables, observation placement within nested levels or population membership is defined by values in one or more discrete, nominal, or ordinal variables (Figure 10.1). In this way, observations (i.e., rows) can be *grouped* by common nesting and/or population values. `group_by()` adds such row grouping information to a tibble and returns a *grouped* tibble.

The first argument in `group_by()` is the tibble to group, and subsequent arguments identify the columns or combination of columns that define row membership to groups. As we'll see, some `dplyr` verb functions behave differently when passed grouped data.

Let's return to the non-stylized `trees` data created in Section 6.2. The code below groups `trees` by `id`, assigns the result to `trees_by_id`, then prints `trees_by_id`. The second printed line, `Groups: id [3]`, tells us this is a grouped tibble, groups are defined by values in the `id` column, and there are 3 groups.[5]

```
trees_by_id <- group_by(trees, id)
trees_by_id
```

```
#> # A tibble: 5 x 3
#> # Groups:   id [3]
#>      id  year   dbh
#>   <int> <int> <dbl>
#> 1     1  2020   1.9
#> 2     1  2021   2.1
#> 3     2  2020   5.2
#> 4     2  2021   5.5
#> 5     3  2021   0.5
```

Multiple columns can be used to specify the grouping structure. The code below groups first by `id` and `year` with the result being 5 groups (i.e., 5 groups because every row in `trees` has a unique combination of `id` and `year` values).

```
trees_by_id_year <- group_by(trees, id, year)
trees_by_id_year
```

```
#> # A tibble: 5 x 3
#> # Groups:   id, year [5]
#>      id  year   dbh
#>   <int> <int> <dbl>
#> 1     1  2020   1.9
#> 2     1  2021   2.1
#> 3     2  2020   5.2
#> 4     2  2021   5.5
#> 5     3  2021   0.5
```

[5] This same group information is provided by `glimpse()`.

Additional grouping variables can be added to an already grouped tibble by setting the `group_by()` argument `.add` to `TRUE`. For example, `group_by(trees_by_id, year, .add = TRUE)` is equivalent to `group_by(trees, id, year)`.

If you want to know the grouping columns but don't want all the other information provided by `print()` or `glimpse()`, call `group_vars()` on the grouped tibble. See `?dplyr::group_data` for other functions used to extract group information.

We often want to remove or change a tibble's grouping structure. Use `ungroup()` to remove grouping information, i.e., make *grouped* data *ungrouped*. For example, remove all group information with `ungroup(trees_by_id_year)`, or remove individual grouping columns by passing them as subsequent arguments as illustrated below.

```
ungroup(trees_by_id_year, id) # Leave only year grouping.
```

```
#> # A tibble: 5 x 3
#> # Groups:   year [2]
#>      id  year   dbh
#>   <int> <int> <dbl>
#> 1     1  2020   1.9
#> 2     1  2021   2.1
#> 3     2  2020   5.2
#> 4     2  2021   5.5
#> 5     3  2021   0.5
```

So how do `dplyr` functions behave with a grouped tibble? We'll cover the various functions in the order of their initial presentation and start with functions that operate on tibble rows (i.e., `filter()`, `slice()`, and `arrange()`).

If `filter()` is passed a grouped tibble and the filtering condition uses a summary function or group characteristic (e.g., number of rows computed using `n()`, see Section 7.10), the result will be group specific. For example, the code below selects each tree's largest DBH measurement row. Here, `filter()` recognizes that `trees_by_id` is grouped by `id` and applies the summary function `max(dbh)` and condition `dbh == max(dbh)` separately within each tree's `dbh` values. Given there are three groups, the result comprises three rows, each holding a group-specific result.

```
filter(trees_by_id, dbh == max(dbh)) # Each tree's largest DBH.
```

```
#> # A tibble: 3 x 3
#> # Groups:   id [3]
#>      id  year   dbh
#>   <int> <int> <dbl>
#> 1     1  2021   2.1
#> 2     2  2021   5.5
#> 3     3  2021   0.5
```

Figure 7.11 is a stylized take on the code and output above, and useful for highlighting how `dbh == max(dbh)` is applied separately to each group of rows.

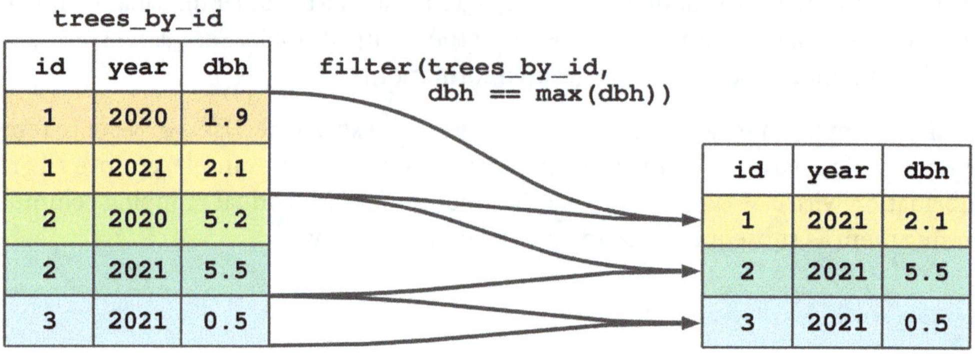

FIGURE 7.11: Using grouped `filter()` to subset rows with the maximum `dbh` by tree `id` from `trees`.

Next, let's consider a grouped `filter()` that uses a group characteristic in the filtering condition. Say we want all trees that have only one measurement. Recall, each row in `trees` represents a measurement, and the `n()` function returns the number of rows. So, `n()` applied to the grouped `trees_by_id` returns the number of measurements for each tree. This logic is used in the code below to return our desired result.

```
filter(trees_by_id, n() == 1) # Trees with only one measurement.
```

```
#> # A tibble: 1 x 3
#> # Groups:   id [1]
#>      id  year   dbh
#>   <int> <int> <dbl>
#> 1     3  2021   0.5
```

Similarly, `slice()` and its various flavors `slice_head()`, `slice_tail()`, `slice_min()`, `slice_max()`, and `slice_sample()` subset rows within each group. For example, the code below returns the row with the minimum year for each tree.

```
slice_min(trees_by_id, year) # Each tree's first measurement year.
```

```
#> # A tibble: 3 x 3
#> # Groups:   id [3]
#>      id  year   dbh
#>   <int> <int> <dbl>
#> 1     1  2020   1.9
#> 2     2  2020   5.2
#> 3     3  2021   0.5
```

Passing grouped and ungrouped data to `arrange()` gives the same result unless you set the function's optional argument `.by_group` to `TRUE`, in which case it will order first by the grouping.

Recall, the functions `select()`, `rename()`, `mutate()`, `transmute()`, and `relocate()` operate on tibble columns. Because `rename()` and `relocate()` only affect the column name and position, their behavior is the same given grouped or ungrouped data. Similarly, a grouped `select()` is the same as an ungrouped `select()`, except grouping columns are always included in the resulting column subset.

Similar to a grouped `filter()`, a grouped `mutate()` or grouped `transmute()` has different behavior if your new column definition uses a summary function or some other group specific characteristic. For example, the code below adds a new column to `trees` called `max_dbh` that holds the maximum DBH recorded for each tree.

```
mutate(trees_by_id, max_dbh = max(dbh))
```

```
#> # A tibble: 5 x 4
#> # Groups:   id [3]
#>      id  year   dbh max_dbh
#>   <int> <int> <dbl>   <dbl>
#> 1     1  2020   1.9     2.1
#> 2     1  2021   2.1     2.1
#> 3     2  2020   5.2     5.5
#> 4     2  2021   5.5     5.5
#> 5     3  2021   0.5     0.5
```

Notice the call to `mutate()` above uses the grouped `trees_by_id` tibble; hence, the summary function `max(dbh)` returns the maximum DBH found among each tree's measurements and recycles that value to match the number of rows in the given group. Breaking this down a bit more, for tree `id` 1 the maximum `dbh` is 2.1, i.e., `max(c(1.9, 2.1))`, so 2.1 is repeated for the first two rows of `max_dbh`. The `max_dbh` values for `id` 2 and 3 follow the same logic.

Let's consider a slightly more involved example. Say we want to add an annual DBH increment (`dbh_inc`) column to `trees`. Given `dbh` is the tree's diameter at the end of a growing season, `dbh_inc` is the difference between a tree's current and previous year DBH measurements, computed as follows.

```
mutate(trees_by_id, dbh_inc = dbh - lag(dbh))
```

```
#> # A tibble: 5 x 4
#> # Groups:   id [3]
#>      id  year   dbh dbh_inc
#>   <int> <int> <dbl>   <dbl>
#> 1     1  2020   1.9  NA
#> 2     1  2021   2.1   0.200
```

```
#> 3      2  2020   5.2  NA
#> 4      2  2021   5.5   0.300
#> 5      3  2021   0.5  NA
```

Because we pass the grouped `trees_by_id` to `mutate()` above, the `lag()` (see `?dplyr::lag`) function is applied to each tree's `dbh` vector. Specifically, for tree `id` 1, the first two values in the `dbh_inc` column equal `c(1.9, 2.1) - lag(c(1.9, 2.1))`. If it's not immediately apparent how `lag()` works or why `NA`s appear in the `dbh_inc` column, spend a little time reading the `lag()` manual page and work some simple examples, e.g., run `lag(c(1.9, 2.1))`.

As mentioned at the end of Section 7.8, `summarize()` is most useful when applied to grouped data. As illustrated in Figure 7.12, given a grouped tibble, column summaries defined in `summarize()` are applied separately to each group of rows. It's useful to compare Figures 7.12 and 7.10, and note the only difference is the code in Figure 7.12 is passed the grouped `trees_by_id` tibble. Again, it might be instructive to break down the steps leading to the output in Figure 7.12. For example, the mean and standard deviation values for tree `id` 1 are `mean(c(1.9, 2.1))` and `sd(c(1.9, 2.1))`, respectively.[6] Check your understanding about the grouped `summarize()` and computing the standard deviation by figuring out why the `NA` appears in the last row of `dbh_sd`.

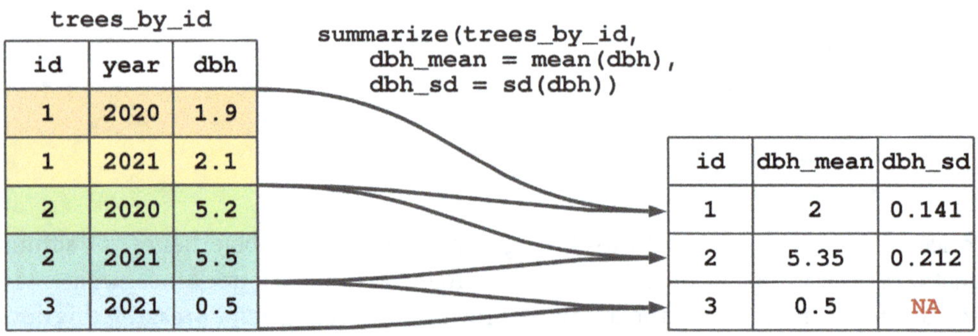

FIGURE 7.12: Using `summarize()` to compute tree specific mean and standard deviation of `dbh`.

Each time you apply `summarize()` to a grouped tibble, the default behavior is to remove the last grouping level (i.e., the right most group specified by `group_by()` and shown in `print()`, `glimpse()`, or `group_vars()`). The optional `summarize()` argument `.groups`, which is by default set to `"drop_last"`, controls if and how the output tibble is grouped. In most cases, this default behavior is sensible, because the output of `summarize()` is a summary of rows defined by the last grouping variable. However, when there are multiple grouping columns, you might want the output of `summarize()` to be ungrouped, which is accomplished by overriding the default behavior using `.groups = "drop"` or by passing the grouped `summarize()` output to the `ungroup()` function. We'll return to these very important points in Section 7.13 and Section 12.6 when working with tibbles grouped by multiple columns.

[6] Rounded to three digits.

A key dplyr strength is its ability to work with grouped data. For example, the grouped summarize() is so handy that it appears in nearly all subsequent estimation problems in Chapter 12 and beyond.

7.10 Counting

We often need to know the number of rows, perhaps by group, within a dataset. Rows are counted using the n() helper function, which is a *context dependent* function used within dplyr verb functions. It takes no arguments and returns the context specific group size. We see n() used most often in summarize() and mutate() to count rows in groups. Here are two examples that use the grouped trees_by_id tibble. The first uses n() in summarize() to count measurements for each tree id (i.e., rows of each id value).

```
summarize(trees_by_id,
          n_measurement = n()) # Measurements for each tree.
```

```
#> # A tibble: 3 x 2
#>      id n_measurement
#>   <int>         <int>
#> 1     1             2
#> 2     2             2
#> 3     3             1
```

Next, n() is used in mutate() to add a column containing the number of measurements for each tree id.

```
mutate(trees_by_id,
       n_measurement = n()) # Measurements for each tree.
```

```
#> # A tibble: 5 x 4
#> # Groups:   id [3]
#>      id  year   dbh n_measurement
#>   <int> <int> <dbl>         <int>
#> 1     1  2020   1.9             2
#> 2     1  2021   2.1             2
#> 3     2  2020   5.2             2
#> 4     2  2021   5.5             2
#> 5     3  2021   0.5             1
```

The number of distinct values in a discrete variable is also often of interest (e.g., number of distinct plot numbers or species). n_distinct() returns the number of unique values in one or more vectors, see ?dplyr::n_distinct for details. Like n(), when applied to a grouped tibble within a verb function, n_distinct() returns the number of group specific distinct values for the one or more columns provided as arguments. If more than one column is provided, then `n_distinct()` returns the number of distinct row combinations. For example, let's first count the number of unique measurement years within the ungrouped `trees` tibble.

```
summarize(trees, n_years = n_distinct(year))
```

```
#> # A tibble: 1 x 1
#>   n_years
#>     <int>
#> 1       2
```

The only two measurement years are 2020 and 2021; hence, `n_years` is 2 in the output above.

Now, how about the number of unique measurement years for each tree (i.e., using the grouped `trees_by_id` tibble)?

```
summarize(trees_by_id, n_years = n_distinct(year))
```

```
#> # A tibble: 3 x 2
#>      id n_years
#>   <int>   <int>
#> 1     1       2
#> 2     2       2
#> 3     3       1
```

Each row in the output above corresponds to an `id` and `n_years` holds the number of distinct years, i.e., 2020 and 2021 for `id` 1 and 2, and 2021 for `id` 3.

We'll get some more practice using `n()` and `n_distinct()` in Section 7.13.

7.11 Applying a function across columns

There's often a need to apply one or more functions to a set of columns in a tibble, which is accomplished using `dplyr::across()`. `across()` has two primary arguments. The first, `.cols`, uses `select()` syntax (see Section 7.4) to select columns to which a function(s) will be applied. The second, `.fns`, is a function or list of functions to apply to selected columns. While `across()` can be used in several `dplyr` functions, we use it most often in `mutate()` and `summarize()`.

To motivate some `across()` examples, let's add a tree height (ft) column to the `trees` tibble.

```
trees <- mutate(trees, height = c(23, 24.9, 47, 48.5, 8.8))
trees
```

```
#> # A tibble: 5 x 4
#>       id  year   dbh height
#>    <int> <int> <dbl>  <dbl>
#> 1      1  2020   1.9   23
#> 2      1  2021   2.1   24.9
#> 3      2  2020   5.2   47
#> 4      2  2021   5.5   48.5
#> 5      3  2021   0.5    8.8
```

Now, say we want to round `dbh` and `height` to whole numbers. One option is to apply the `round()` function to each column individually within `mutate()`, i.e., `mutate(trees, dbh = round(dbh), height = round(height))`. This works well, but it could get tedious if there are more than a few columns to round. The alternative is to use `across()` within `mutate()`, which is applied in the code below.[7]

```
mutate(trees, across(.cols = c(dbh, height), .fns = round))
```

```
#> # A tibble: 5 x 4
#>       id  year   dbh height
#>    <int> <int> <dbl>  <dbl>
#> 1      1  2020     2     23
#> 2      1  2021     2     25
#> 3      2  2020     5     47
#> 4      2  2021     6     48
#> 5      3  2021     0      9
```

Say you want to apply a function to all columns of a certain data type. In such cases, the `select()` syntax `where()` function (see Section 7.4) is useful, e.g., `mutate(trees, across(.cols = where(is.double), .fns = round))`.

If you want to specify non-default arguments in a function passed to `.fns`, or you'd like the columns defined in `.cols` to be passed to something other than the first argument in a `.fns` function, then you need to use an *anonymous function*. An anonymous function is just a function without a name. Continuing with the rounding example, say we want to round the `dbh` and `height` columns to one digit (the default of `round()` is `digits = 0`). This is accomplished using an anonymous function as follows.

```
mutate(trees, across(.cols = c(dbh, height),
                     .fns = function(x) round(x, digits = 1)))
```

---

[7]In the output of this code, you might be surprised to see the DBH of 0.5 for tree `id` 3 is rounded to zero, especially since the DBH of 5.5 for tree `id` 2 is rounded to 6. R's `round()` function, like other languages that follow the IEC 60559 standard (IEEE 2008), rounds values involving .5 to the next even number, e.g., try `round(1.5)` and `round(2.5)`, see `?base::round` for an explanation. To apply the more conventional rounding rule, see the `janitor` package's `round_half_up()` function or `round2()` function defined at https://stackoverflow.com/a/12688836.

```
#> # A tibble: 5 x 4
#>      id  year   dbh height
#>   <int> <int> <dbl>  <dbl>
#> 1     1  2020   1.9   23
#> 2     1  2021   2.1   24.9
#> 3     2  2020   5.2   47
#> 4     2  2021   5.5   48.5
#> 5     3  2021   0.5    8.8
```

The anonymous function `function(x) round(x, digits = 1)` calls `round(x, digits = 1)` with `x` set as a placeholder for `dbh` and `height`.

A less verbose alternative to anonymous function syntax is to use the `tidyverse` `purrr` package's lambda syntax.[8] Following the example above, we can replace the anonymous function `function(x) round(x, digits = 1)` with equivalent lambda syntax `~ round(.x, digits = 1)` where the tilde `~` says we're providing a function with `.x` set as a placeholder for columns given in `.cols`.

Let's use `across()` within `summarize()` to compute the mean for `dbh` and `height` using base R's `mean()` function, first using an anonymous function then using the equivalent lambda syntax.

```
summarize(trees, across(.cols = c(dbh, height),
                        .fns = function(x) mean(x, na.rm = FALSE)))
```

```
#> # A tibble: 1 x 2
#>     dbh height
#>   <dbl>  <dbl>
#> 1  3.04   30.4
```

```
summarize(trees, across(.cols = c(dbh, height),
                        .fns = ~ mean(.x, na.rm = FALSE)))
```

```
#> # A tibble: 1 x 2
#>     dbh height
#>   <dbl>  <dbl>
#> 1  3.04   30.4
```

Now, let's compute the mean and standard deviation for `dbh` and `height` by passing a list of lambda syntax calls to the `mean()` and `sd()` functions to `.fns`.

[8]The `purrr` package provides a complete and consistent set of tools for working with functions and vectors. While we don't cover `purrr` in this book, we strongly encourage you to check it out after working through the `tidyverse` chapters.

```
summarize(trees, across(.cols = c(dbh, height),
                        .fns = list(~ mean(.x, na.rm = FALSE),
                                    ~ sd(.x, na.rm = FALSE))))
```

```
#> # A tibble: 1 x 4
#>   dbh_1 dbh_2 height_1 height_2
#>   <dbl> <dbl>    <dbl>    <dbl>
#> 1  3.04  2.20     30.4     17.0
```

In the output above, the numbers appended to column names, i.e., `_1` and `_2`, correspond to function list element indexes. We can explicitly label output columns by passing a function list with element names to `.fns` as illustrated below.

```
summarize(trees, across(.cols = c(dbh, height),
                        .fns = list(the_mean = ~ mean(.x, na.rm = FALSE),
                                    the_sd = ~ sd(.x, na.rm = FALSE))),
          )
```

```
#> # A tibble: 1 x 4
#>   dbh_the_mean dbh_the_sd height_the_mean height_the_sd
#>          <dbl>      <dbl>           <dbl>         <dbl>
#> 1         3.04       2.20            30.4          17.0
```

Use the `.names` argument in `across()` for finer control over column names, see `?dplyr::across` for details.

See the end of Section 12.6 for a more extensive `across()` within `summarize()` example. Also, there's an `across()` vignette, accessed by running `vignette("colwise")` on the console, that provides extensive illustrations and details.

7.12 Piping

You might have noticed that `.data` is the first argument for all `dplyr` verb functions. This is an intentional design feature that, when combined with the `tidyverse` `magrittr` package pipe operator `%>%`, facilitates efficient and intuitive workflows. The pipe operator is available automatically when `dplyr` or other `tidyverse` packages are loaded.[9]

The pipe operator places the object on its left into a function on its right. By default, the operator places the object in the function's first argument. Here's an initial non-`dplyr` example that *pipes* the vector `x` into the `max()` function.

[9]In 2023, a new pipe operator `|>` was introduced in base R 4.1.0. The two pipe operators are very similar. We prefer the `%>%` operator and will use this throughout the remainder of the book. See https://www.tidyverse.org/blog/2023/04/base-vs-magrittr-pipe/ for a simple description of the differences between the two pipes.

```
x <- c(1, 5)
x %>% max() # Equivalent to max(x).
```

```
#> [1] 5
```

The pipe operator implicitly adds the object to the function's argument list (i.e., the object is added to the function's argument list, but you just don't see it). Additional function arguments follow the implicit argument. For example, let's redefine `x` to include an `NA` then add the optional `na.rm` argument to `max()`.

```
x <- c(1, 5, NA)
x %>% max(na.rm = TRUE) # Equivalent to max(x, na.rm =TRUE).
```

```
#> [1] 5
```

To appreciate the beauty of using pipes with `dplyr`, consider the following example. Returning to the `trees` data, say we want to keep only tree `id` 1 measurements, remove the `height` column that was recently added in Section 7.11, and sort rows by `year` in descending order. One possible approach to this three-step process is to create intermediate tibbles as illustrated below.

```
tbl_1 <- filter(trees, id == 1)
tbl_2 <- select(tbl_1, !height)
tbl_3 <- arrange(tbl_2, desc(year))
```

If you plan to use these intermediate tibbles for something else, and they can be meaningfully named, then perhaps this approach is a good one. However, if the intermediate tibbles are just a means to an end, then there are two good reasons this approach should be avoided. First, the environment and code becomes cluttered with unimportant objects, e.g., `tbl_1` and `tbl_2`. Second, if you're like us, you'll inevitably make an error incrementing tibble number suffixes or inputting the wrong tibble in a subsequent function.

You could do away with the intermediate tibbles by simply overwriting a tibble name.

```
tbl <- filter(trees, id == 1)
tbl <- select(tbl, !height)
tbl <- arrange(tbl, desc(year))
```

This too is a less-than-desirable approach. It's difficult to debug because you'll need to re-run all lines each time you make a change to the code. Also, it's hard to follow how `tbl` is modified by each line.

Another approach is to nest the calls to `filter()`, `select()`, and `arrange()`.

```
tbl <- arrange(
          select(
              filter(trees, id == 1),
            !height),
          desc(year)
        )
```

The downside of this approach is that it's challenging to write and even harder to read.

Using pipes is the preferred solution to move the result of one verb function to the next.

```
tbl <- trees %>%
  filter(id == 1) %>%
  select(!height) %>%
  arrange(desc(year))
```

Notice, when written using pipes, the code reads like an action-packed sentence. In your mind, replace the pipe operator `%>%` with "then" to say "take the `trees` tibble *then* filter *then* select *then* arrange."

Let's try another piped operation using `trees`. Recall back in Section 7.9 we selected each tree's first measurement year using `slice_min(trees_by_id, year)`, where `trees_by_id` is `trees` grouped by `id`. Here's that same query performed using pipes.

```
trees %>%
  group_by(id) %>%
  slice_min(year)
```

```
#> # A tibble: 3 x 4
#> # Groups:   id [3]
#>      id  year   dbh height
#>   <int> <int> <dbl>  <dbl>
#> 1     1  2020   1.9   23
#> 2     2  2020   5.2   47
#> 3     3  2021   0.5    8.8
```

Again, read it like a sentence, "take `trees` *then* group it by `id` *then* slice."

We'll make extensive use of pipes throughout this book as we find them to be intuitive and a simple way to make code more readable.

The pipe operator has several features beyond its default behavior. For example, as illustrated in the code below, use the dot `.` symbol to indicate a specific location in the function argument list to place the object (think of the dot as the object's placeholder).

```
TRUE %>% max(x, na.rm = .) # Equivalent to max(x, na.rm = TRUE).
```

```
#> [1] 5
```

Run `?magrittr::%>%` on the console to learn more about pipe functionality.

7.13 Building dplyr workflows

This section combines verb and helper functions in piped workflows to answer a series of questions about the Minnesota tree growth dataset, which was introduced in Section 6.1. In most cases, we pose a question about the data, then offer code to answer the question. Use this section to build understanding by methodically picking apart our code and referencing preceding sections, function manual pages, and other resources.

Before proceeding, it's important to fully understand how tree measurements are organized in `mn_trees`. Recall from Section 6.1, a tree's set of annual measurements are uniquely identified by the combination of their `stand_id`, `plot_id`, and `tree_id` values (i.e., tree numbers are unique within plot, and plot numbers are unique within stand). For example, in Figure 6.1, the annual `age` and `DBH_cm` measurements that define a given tree's line share common `stand_id`, `plot_id`, and `tree_id` values.

Say y is some variable (i.e., column) in `mn_trees`. For example, y may be `species`, `age`, `rad_inc_mm`, `DBH_cm`, or `BA_m_sq`.[10] Then, using subscript notation that we'll introduce formally in Section 10.2.1, a measurement for y (i.e., row value for column y) is indexed using y_{ijkl} for `stand_id` (i), `plot_id` (j), `tree_id` (k), and `year` (l). Given the data comprise three plots within each of five stands, index values are:

$$
\begin{aligned}
i &= (1, 2, \dots, 5),\\
j &= (1, 2, 3),\\
k &= (1, 2, \dots, n_{ij}), \text{ and}\\
l &= \left(\min(\text{years}_{ijk}), \min(\text{years}_{ijk}) + 1, \dots, \max(\text{years}_{ijk})\right),
\end{aligned}
$$

where there are n_{ij} trees in the j-th plot in the i-th stand, and years_{ijk} is the set of years measured for the k-th tree in the j-th plot in the i-th stand.

Take a scroll through the `mn_trees` rows to confirm this index notation makes sense (e.g., run `mn_trees %>% print(n = nrow(.))` to print all rows). It might also be instructive to see all combinations of the indexes using the code below (again, add `%>% print(n = nrow(.))` to the line below if you want to see all rows). `distinct()` returns distinct, i.e., unique, row value combinations given one or more columns.

[10]While the categorical `species` variable doesn't change over time like the quantitative variables, its values can still be represented using y and subsequent subscript notation.

```
mn_trees %>%
  distinct(stand_id, plot_id, tree_id)
```

```
#> # A tibble: 161 x 3
#>    stand_id plot_id tree_id
#>       <int>   <int>   <int>
#>  1        1       1       1
#>  2        1       1       2
#>  3        1       1       3
#>  4        1       1       4
#>  5        1       1       5
#>  6        1       1       6
#>  7        1       1       7
#>  8        1       2       1
#>  9        1       2       2
#> 10        1       2       3
#> # i 151 more rows
```

The output from `distinct()` tells us that for `stand_id` 1 ($i = 1$) and `plot_id` 1 ($j = 1$) there are $n_{ij} = 7$ trees measured.

Okay, let's start exploring these data.

How many total tree measurements were taken?

```
mn_trees %>%
  summarize(n_measurements = n())
```

```
#> # A tibble: 1 x 1
#>   n_measurements
#>            <int>
#> 1          11649
```

How many individual trees were measured?.

```
mn_trees %>%
  summarize(n_trees = n_distinct(stand_id, plot_id, tree_id))
```

```
#> # A tibble: 1 x 1
#>   n_trees
#>     <int>
#> 1     161
```

How many stands?

```r
mn_trees %>%
  summarize(n_trees = n_distinct(stand_id))
```

```
#> # A tibble: 1 x 1
#>   n_trees
#>     <int>
#> 1       5
```

How many plots within each stand?

```r
mn_trees %>%
  group_by(stand_id) %>%
  summarize(n_plots = n_distinct(plot_id))
```

```
#> # A tibble: 5 x 2
#>   stand_id n_plots
#>      <int>   <int>
#> 1        1       3
#> 2        2       3
#> 3        3       3
#> 4        4       3
#> 5        5       3
```

How many trees within each plot (note, this is the k index's n_{ij})?

```r
mn_trees %>%
  group_by(stand_id, plot_id) %>%
  summarize(n_plot_trees = n_distinct(tree_id))
```

```
#> `summarise()` has grouped output by 'stand_id'. You can
#> override using the `.groups` argument.
```

```
#> # A tibble: 15 x 3
#> # Groups:   stand_id [5]
#>    stand_id plot_id n_plot_trees
#>       <int>   <int>        <int>
#>  1        1       1            7
#>  2        1       2            5
#>  3        1       3            2
#>  4        2       1            8
#>  5        2       2            4
#>  6        2       3            6
#>  7        3       1           14
#>  8        3       2           20
#>  9        3       3            5
```

```
#> 10         4        1            17
#> 11         4        2            19
#> 12         4        3            13
#> 13         5        1             8
#> 14         5        2            10
#> 15         5        3            23
```

The query above requires grouping by both `stand_id` and `plot_id` to count the number of distinct `tree_id` values for each plot. Recall from Section 7.9, the output from a grouped `summarize()` is a tibble with the last grouping level removed. Hence, in this case, the output is grouped by only `stand_id`. This makes sense because each row in the output is the number of distinct `tree_id` values for each `plot_id`. The message printed directly following the call to `summarize()` and prior to the tibble values (i.e., "`summarize()` has grouped output by `stand_id`. You can override using the `.groups` argument.") is a friendly reminder the resulting tibble is still grouped.[11] Also, as noted in Section 7.9 and as suggested in the printed message, you can change this default grouping behavior using the `.groups` argument (see `?dplyr::summarize` for specifics). It's often useful to add the grouping argument to `summarize()` (even if it's the default argument value, i.e., `.groups = "drop_last"`) because it reinforces your awareness and suppresses the message printed when the output is grouped. We'll add the `.groups` argument to all subsequent calls to `summarize()`.

How many years were measured for each tree? Also, what was the minimum and maximum measurement year for each tree (note, these are the l index's $\min(\text{years}_{ijk})$ and $\max(\text{years}_{ijk})$)?

```
mn_trees %>%
  group_by(stand_id, plot_id, tree_id) %>%
  summarize(n_years = n_distinct(year),
            min_year = min(year),
            max_year = max(year),
            .groups = "drop")
```

```
#> # A tibble: 161 x 6
#>    stand_id plot_id tree_id n_years min_year max_year
#>       <int>   <int>   <int>   <int>    <int>    <int>
#>  1        1       1       1      48     1960     2007
#>  2        1       1       2      62     1946     2007
#>  3        1       1       3      59     1949     2007
#>  4        1       1       4     111     1897     2007
#>  5        1       1       5     101     1907     2007
#>  6        1       1       6      54     1954     2007
#>  7        1       1       7      68     1940     2007
#>  8        1       2       1      41     1967     2007
```

[11]It's important to know if and how a tibble is grouped because it can change the behavior of `dplyr` verbs as described in Section 7.9.

```
#>  9          1        2        2       38      1970      2007
#> 10          1        2        3      111      1897      2007
#> # i 151 more rows
```

As described toward the end of Section 7.9, setting `.groups = "drop"` in `summarize()` removes `stand_id` and `plot_id` grouping from the output tibble.[12] Alternatively, we can remove any unneeded or unwanted grouping by piping `summarize()` output to `ungroup()`.

How many trees are within each stand? Given trees have unique `tree_id` values within `plot_id`, and `plot_id` is unique within `stand_id`, some care is needed when crafting code to answer this question. Let's begin by making a mistake and group only by `stand_id` and count distinct `tree_id`.

```
# Incorrect solution.
mn_trees %>%
  group_by(stand_id) %>%
  summarize(n_stand_trees = n_distinct(tree_id))
```

```
#> # A tibble: 5 x 2
#>   stand_id n_stand_trees
#>      <int>         <int>
#> 1        1             7
#> 2        2             8
#> 3        3            20
#> 4        4            19
#> 5        5            23
```

The `n_stand_trees` above is wrong because there will be some `tree_id` values that are common among the three plots within each stand (e.g., `tree_id` 1 occurs on all three plots but it will only be counted once). We get the correct tree count within each stand using the code below, which first counts the trees on each plot, then sums these counts within each stand. Importantly, after the first `summarize()`, `mn_trees` is grouped by only `stand_id` and is ready for the second `summarize()`.[13]

```
mn_trees %>%
  group_by(stand_id, plot_id) %>%
  summarize(n_plot_trees = n_distinct(tree_id),
            .groups = "drop_last") %>%
  summarize(n_stand_trees = sum(n_plot_trees))
```

```
#> # A tibble: 5 x 2
```

[12] In this case we don't really care if the output tibble is grouped or ungrouped, we removed the grouping simply to illustration `.groups` use.

[13] Recall, each call to summarize removes the right most grouping variable specified in `group_by()`. Although it's redundant, you could add `group_by(stand_id)` after the first `summarize()`.

```
#>   stand_id n_stand_trees
#>      <int>         <int>
#> 1        1            14
#> 2        2            18
#> 3        3            39
#> 4        4            49
#> 5        5            41
```

By species, how many trees were measured and how many measurements were taken? Sort the result by the number of trees measured.

```
mn_trees %>%
  group_by(species) %>%
  summarize(n_trees = n_distinct(stand_id, plot_id, tree_id),
            n_measurements = n()) %>%
  arrange(n_measurements)
```

```
#> # A tibble: 8 x 3
#>   species n_trees n_measurements
#>   <chr>     <int>          <int>
#> 1 POTR          1             71
#> 2 THOC          2            153
#> 3 PIRE          7            642
#> 4 PIMA         10            797
#> 5 PIGL         16            926
#> 6 ABBA         40           1896
#> 7 BEPA         35           3124
#> 8 PIST         50           4040
```

The first line in the output above says there's only one *Populus tremuloides* (POTR) in `mn_trees`. Let's double check this is correct. Visual inspection of the raw data is often an easy way to check results. We'll limit the columns viewed to the first five.

```
mn_trees %>%
  filter(species == "POTR") %>%
  select(1:5)
```

```
#> # A tibble: 71 x 5
#>    stand_id plot_id tree_id  year species
#>       <int>   <int>   <int> <int> <chr>
#>  1        3       3       5  1937 POTR
#>  2        3       3       5  1938 POTR
#>  3        3       3       5  1939 POTR
#>  4        3       3       5  1940 POTR
#>  5        3       3       5  1941 POTR
```

```
#>  6        3        3        5  1942 POTR
#>  7        3        3        5  1943 POTR
#>  8        3        3        5  1944 POTR
#>  9        3        3        5  1945 POTR
#> 10        3        3        5  1946 POTR
#> # i 61 more rows
```

Given the output above has 71 rows and the same `stand_id`, `plot_id` and `tree_id` for each row (which you can see in its entirety by adding `%>% print(n = nrow(.))` after `select(1:5)`), we can be confident there is only one *Populus tremuloides* tree in the dataset. The code below is another approach to confirm there's only one *Populus tremuloides* recorded. Use this same approach to confirm there are 2 *Thuja occidentalis* (`THOC`) in `mn_trees`.

```
mn_trees %>%
  filter(species == "POTR") %>%
  distinct(stand_id, plot_id, tree_id)
```

```
#> # A tibble: 1 x 3
#>   stand_id plot_id tree_id
#>      <int>   <int>   <int>
#> 1        3       3       5
```

The next series of questions look at each individual tree's chronology (i.e., measurements on a given tree).

Which tree has the longest chronology and what's its age and species?

```
mn_trees %>%
  group_by(stand_id, plot_id, tree_id) %>%
  summarize(n_measurements = n(),
            age = max(age),
            species = first(species),
            .groups = "drop") %>%
  slice_max(n_measurements)
```

```
#> # A tibble: 3 x 6
#>   stand_id plot_id tree_id n_measurements   age species
#>      <int>   <int>   <int>          <int> <int> <chr>
#> 1        1       1       4            111   119 BEPA
#> 2        1       2       3            111   119 BEPA
#> 3        1       2       5            111   125 BEPA
```

The output above tells us three trees share the longest chronology of 111 years. Two trees are 119 years old and the third is 125 years old. All three trees are *Betula papyrifera*. Notice in the `summarize()` above we used the `first()` helper function, which returns the first value in a vector (there is also a `last()` and `nth()` helper function, see `?dplyr::first`). Because the species value is the same for each trees' measurements, it doesn't matter which element in species is selected, i.e., the first, last, or n-th value will be the same.

Try running the code above without first() (i.e., just set species = species in summarize()), you'll see a lot more rows are returned and a lot of repeated values in the first five columns. This is because summarize() returns all the species column values and duplicates the other columns' values to get the number of rows to match that of species (this is not what you want). Accidentally creating a vector in a summarize() calculation is a common error, which is often diagnosed when you end up with more output rows than anticipated.

In the code above, it's important to first ungroup the summarize() output (using either the .groups = "drop" argument or by piping through ungroup()); otherwise, slice_max() is applied to the grouped data (i.e., grouped by stand_id and plot_id), and you get the tree with the longest chronology within each plot—which, again, isn't what you want.

Which tree has the largest annual growth increment (rad_inc_mm) and what's its rad_inc_mm expressed as annual DBH increment (cm)?

```
mn_trees %>%
  group_by(stand_id, plot_id, tree_id) %>%
  summarize(tree_rad_inc_mm = max(rad_inc_mm),
            tree_dbh_inc_cm = 2 * tree_rad_inc_mm * 0.1,
            .groups = "drop") %>%
  slice_max(tree_rad_inc_mm) %>%
  glimpse() # So that all values can be seen in the output.
```

```
#> Rows: 1
#> Columns: 5
#> $ stand_id        <int> 3
#> $ plot_id         <int> 2
#> $ tree_id         <int> 3
#> $ tree_rad_inc_mm <dbl> 7.73
#> $ tree_dbh_inc_cm <dbl> 1.546
```

Given each tree's basal area over time (BA_m_sq), how do we compute an annual basal area increment? Recall from Section 6.1, rad_inc_mm is the annual radial growth increment (i.e., annual growth by the end of the growing season). Similarly, DBH_cm is the end of the growing season inside bark DBH. In Section 7.6, we used mutate() to create an end of the growing season inside bark basal area column (BA_m_sq). The code below computes the desired basal area increment column (BA_inc_m_sq) by subtracting current from previous year basal area.

```
mn_trees <- mn_trees %>%
  mutate(DBH_previous_cm = DBH_cm - 2 * rad_inc_mm * 0.1,
         BA_previous_m_sq = 0.00007854 * DBH_previous_cm^2,
         BA_inc_m_sq = BA_m_sq - BA_previous_m_sq) %>%
  select(-contains("previous"))
```

```
mn_trees %>% glimpse()
```

```
#> Rows: 11,649
#> Columns: 10
#> $ stand_id    <int> 1, 1, 1, 1, 1, 1, 1, 1, 1, 1, 1, ~
#> $ plot_id     <int> 1, 1, 1, 1, 1, 1, 1, 1, 1, 1, 1, ~
#> $ tree_id     <int> 1, 1, 1, 1, 1, 1, 1, 1, 1, 1, 1, ~
#> $ year        <int> 1960, 1961, 1962, 1963, 1964, 196~
#> $ species     <chr> "ABBA", "ABBA", "ABBA", "ABBA", "~
#> $ age         <int> 1, 2, 3, 4, 5, 6, 7, 8, 9, 10, 11~
#> $ rad_inc_mm  <dbl> 0.930, 0.950, 0.985, 0.985, 0.715~
#> $ DBH_cm      <dbl> 2.1563, 2.3463, 2.5433, 2.7403, 2~
#> $ BA_m_sq     <dbl> 0.00036518, 0.00043237, 0.0005080~
#> $ BA_inc_m_sq <dbl> 0.000060283, 0.000067190, 0.00007~
```

A few things to notice in the code above. First, there is no need to group because calculations only involve values within a given row. Second, creating temporary columns improves readability and facilitates debugging: end of the previous year's DBH (`DBH_previous_cm`) and its associated basal area (`BA_previous_m_sq`). Third, the desired column (`BA_inc_m_sq`) is the difference between the current and previous year basal area. Finally, we removed the temporary columns using `select()` prior to assigning the result back to `mn_trees`.

7.13.1 A cautionary and encouraging note

Tidyverse workflows, e.g., piped `dplyr` expressions seen above, are powerful data manipulation and analysis tools. When using such tools in our work, it's important to keep the following sage words in mind.

> With great power comes great responsibility.
>
> — Stan Lee (via Ben Parker)

To ensure our work is accurate and delivered efficiently, it's our responsibility to become proficient with the tools of our profession. As professionals, we conduct analyses for clients or publicly funded research projects and have an obligation to deliver high-quality analysis and results. A firm grasp of quantitative fundamentals and the software tools of our trade is critical, but what really ensures consistent, high-quality output is methodical and obsessive checking and rechecking of code and results.

As you've already seen, it's easy to generate output using R and tools like `dplyr`, and sometimes it's even easy to generate output that's not accompanied by warnings or errors. Especially for the latter case, it's critical there is a mind (your mind) checking the output is sensible and correct. Consider the following example.

Say the first step in your analysis is to compute the total tree-ring width for each `tree_id` value given in the `ex_trees` tibble below (i.e., each tree's inside bark radius).

```
ex_trees <- tibble(tree_id = c(1, 1, 2, 2),
                   ring_width = c(0.2, 0.1, NA, NA))
ex_trees
```

```
#> # A tibble: 4 x 2
#>   tree_id ring_width
#>     <dbl>      <dbl>
#> 1       1        0.2
#> 2       1        0.1
#> 3       2       NA
#> 4       2       NA
```

Seems simple enough. We can use a grouped `summarize()` and even take care of those pesky `NA` values using the `sum()` function's `na.rm` argument.

```
ex_trees %>%
  group_by(tree_id) %>%
  summarize(tree_rad = sum(ring_width, na.rm = TRUE))
```

```
#> # A tibble: 2 x 2
#>   tree_id tree_rad
#>     <dbl>    <dbl>
#> 1       1      0.3
#> 2       2      0
```

Looks good, no errors, and numbers came out. Next, we'll extend this code to compute the mean `tree_rad` over the two trees, using an additional `summarize()`.

```
ex_trees %>%
  group_by(tree_id) %>%
  summarize(tree_rad = sum(ring_width, na.rm = TRUE)) %>%
  summarize(mean_rad = mean(tree_rad))
```

```
#> # A tibble: 1 x 1
#>   mean_rad
#>      <dbl>
#> 1     0.15
```

Okay, mean tree radius (inside bark) for the two trees is 0.15! Unfortunately, despite the fact that the code is perfectly correct and there were clearly no warnings or errors, the answer is completely wrong.

Perhaps you spotted the impending issue after the first bit of code yielded, for `tree_id = 2`, a `tree_rad` of `0` and not the expected `NA`. After all, the sum of `NA` values is `NA` (try it: `sum(NA)`). However, again, the code is correct. So what happened? The sum of an empty set of numbers is zero (try it: `sum()`). When all values passed to `sum()` are `NA` and `na.rm = TRUE`, all values are removed, resulting in an empty set; hence, the result is `0` (i.e., `sum(NA, na.rm = TRUE)`).

This is subtle, but it's a good example of the type of errors to be on the lookout for. Not all errors are coding errors, and it's important to stay constantly mindful and test your intuition against what the software produces.

Okay, so what's a fix for this specific example? One approach is to write a new sum function that returns `NA` for an empty set, but that might cause other code to break. Perhaps the simplest fix is to avoid calling `sum()` for empty sets, which is done in the following revised code with the help of `if_else()` that returns `NA` if all `ring_width` values are `NA`, and the previously used summation code otherwise.

```
ex_trees %>%
  group_by(tree_id) %>%
  summarize(tree_rad =
              if_else(all(is.na(ring_width)),
                      NA,
                      sum(ring_width, na.rm = TRUE))
)
```

```
#> # A tibble: 2 x 2
#>   tree_id tree_rad
#>     <dbl>    <dbl>
#> 1       1      0.3
#> 2       2     NA
```

7.14 Combining multiple tibbles

Most analyses use variables and associated observations held in a single tibble. This tibble, however, is typically constructed from multiple data sources. Hence, a common pre-analysis task is to combine or join information from various sources into a single tibble. Fortunately, as part of its "grammar of data manipulation," `dplyr` provides functions to simplify this task. We'll focus initially on two suites of function to join tibbles.

- Mutating joins: combine the columns of tibbles `x` and `y`. A mutating join adds new columns to `x` from matching rows of `y`.

- Filtering joins: matches rows of tibbles `x` and `y` in the same way as mutating joins, but affects rows of `x`, not columns.

Notice how the term "filter" applies to manipulating rows and "mutate" applies to adding columns, which lines up with our previous experience with `dplyr`'s `filter()` and `mutate()` functions. In the next two sections, we introduce functions for mutating and filtering joins. In addition to each function's description, we provide an example using toy data provided in Figure 7.13. After covering mutating and filtering joins, we'll introduce set operations and a few other `dplyr` functions for combining information in tibbles.

Species codes like those in Table 6.1 are commonly used in the field when recording species occurrence.[14] When field measurements are encoded in data files and then analyzed, results are presented using either common and/or scientific names. This means, at some point, species codes need to be replaced with species names. Typically, the intended audience determines if common or scientific names are used in a report (e.g., peer reviewed publications use scientific names and inventory summaries given to landowners use common names). As a result, we often find ourselves joining species codes, common names, and scientific names, each of which might reside in a different tibble. This is the motivation behind the toy data in Figure 7.13.

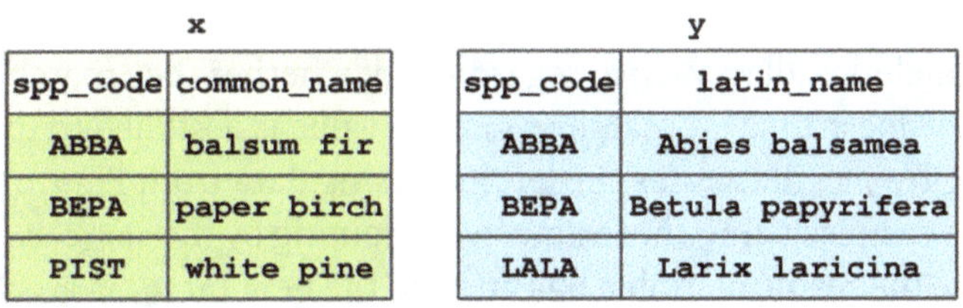

FIGURE 7.13: Stylized tibbles to illustrate `dplyr` join functions.

If you'd like to follow along with the examples, the code below creates the `x` and `y` tibbles used in Figure 7.13.

```
x <- tibble(spp_code = c("ABBA", "BEPA", "PIST"),
            common_name = c("balsam fir", "paper birch", "white pine"))

y <- tibble(spp_code = c("ABBA", "BEPA", "LALA"),
            latin_name = c("Abies balsamea", "Betula papyrifera",
                           "Larix laricina"))
```

For all join functions, the first two arguments, `x` and `y` respectively, define the tibbles to combine. All joins also return a new tibble with the same type as `x`. This consistent function input and output facilitates writing efficient piped workflows using all `dplyr` functions.

[14]With cold hands and water dripping off your nose, it's easier to scribble PSME than *Pseudotsuga menziesii* on your Rite in the Rain tally sheet.

In addition to the `x` and `y` arguments, join functions take a `by` argument that controls which columns are used to match row values across the two tibbles. The `by` argument can be specified either by giving column names in quotes (e.g., `"spp_code"`) or by using `join_by()` with unquoted column names. If the `by` argument is not supplied, then all columns with names common to `x` and `y` are used (this is referred to as a *natural join*). If you want to join using only a subset of common column names, specify them in `by`. For example, using the tibbles in Figure 7.13, to join `x` and `y` by `spp_code`, you can use either `by = "spp_code"` or `by = join_by(spp_code)`.[15]

To join using different column names in `x` and `y`, you can use a named character vector or a `join_by()` expression. For example, if `spp_code` in `y` were named `SPP_CODE`, then `by = c("spp_code" = "SPP_CODE")` or `by = join_by(spp_code == SPP_CODE)` will match rows in `x$spp_code` to `y$SPP_CODE`. After presenting the various mutating and filtering joins below, we'll work through some applied examples.

7.14.1 Mutating joins

Mutating joins add columns from `y` to `x` based on row matching. If a row in `x` matches multiple rows in `y`, the multiple matching rows from `y` will be included in the output.

`dplyr` offers four mutating joins.

1. `left_join()` includes all rows in `x`, regardless of whether they match values in `y`.[16] If a given row in `x` doesn't have a match in `y`, an `NA` value is assigned to that row in the new columns. Figure 7.14 illustrates the left join using data from Figure 7.13. Here, and in subsequent examples using these data, we're joining by the `spp_code` column. Because `PIST` isn't in `y`, the resulting tibble has an `NA` in the new `latin_name` column for the `PIST` row.

`left_join(x, y, by = "spp_code")`

spp_code	common_name	latin_name
ABBA	balsum fir	Abies balsamea
BEPA	paper birch	Betula papyrifera
PIST	white pine	NA

FIGURE 7.14: Left join by `spp_code` for tibbles in Figure 7.13.

2. `right_join()` includes all rows in `y`. It's equivalent to `left_join(y, x)`, but columns and rows will be ordered differently. Figure 7.15 illustrates the right join using data from

[15] Notice that a natural join of `x` and `y` in this example would also join only by `spp_code` because it's the only column name common to both tibbles.

[16] This join is used most commonly in practice because we're typically interested in keeping all rows in `x` and adding columns from `y`.

Figure 7.13. Because `LALA` is in `y` but not `x`, the resulting tibble has an `NA` in the new `common_name` column for the `LALA` row.

right_join(x, y, by = "spp_code")

spp_code	latin_name	common_name
ABBA	Abies balsamea	balsum fir
BEPA	Betula papyrifera	paper birch
LALA	Larix laricina	NA

FIGURE 7.15: Right join by `spp_code` for tibbles in Figure 7.13.

3. `inner_join()` includes only rows that match in both `x` and `y`. For the data in Figure 7.13, `ABBA` and `BEPA` are the only common `spp_code` values; hence, the inner join results shown in Figure 7.16.

inner_join(x, y, by = "spp_code")

spp_code	common_name	latin_name
ABBA	balsum fir	Abies balsamea
BEPA	paper birch	Betula papyrifera

FIGURE 7.16: Inner join by `spp_code` for tibbles in Figure 7.13.

4. `full_join()` includes all rows from `x` and `y`. Figure 7.17 illustrates the full join using data from Figure 7.13; notice that rows for all non-matching values receive an `NA` value.

full_join(x, y, by = "spp_code")

spp_code	common_name	latin_name
ABBA	balsum fir	Abies balsamea
BEPA	paper birch	Betula papyrifera
PIST	white pine	NA
LALA	NA	Larix laricina

FIGURE 7.17: Full join by `spp_code` for tibbles in Figure 7.13.

See ``?dplyr::`mutate-joins` `` for more details, optional arguments, and examples.

7.14.2 Filtering joins

Filtering joins filter rows from x based on the presence or absence of row matches in y. Filtering joins match rows in the same way as mutating joins, but the result is a row subset of x (i.e., same columns but possibly fewer rows of x).

Filtering joins are used most often to diagnose join mismatches that might cause unwanted or unexpected results when performing mutating joins.

There are two filtering joins:

1. `semi_join()` keeps all rows in x that have a match in y. For the data in Figure 7.13, ABBA and BEPA are the only common spp_code values in x and y; hence, the semi join results shown in Figure 7.18.

semi_join(x, y, by = "spp_code")

spp_code	common_name
ABBA	balsum fir
BEPA	paper birch

FIGURE 7.18: Semi-join by spp_code for tibbles in Figure 7.13.

2. `anti_join()` removes all rows in x that have a match in y. For the data in Figure 7.13, PIST is the only spp_code row in x that doesn't occur in y; hence, the anti join results shown in Figure 7.19.

anti_join(x, y, by = "spp_code")

spp_code	common_name
PIST	white pine

FIGURE 7.19: Anti-join by spp_code for tibbles in Figure 7.13.

See ?dplyr::`filter-joins` for more details, optional arguments, and examples.

7.14.3 Applying allometric equations using joins

Tree measurements for a typical timber cruise include species, DBH, number of logs, and log/stem quality. While height might be measured for a subset of trees, we rarely collect it for all measurement trees. Stem volume is typically an important variable that comes from a timber cruise; however, we don't measure stem volume directly on a tree. Stem volume and similar quantities such as biomass are computed using allometric equations (Chapter 5). Allometric equations are regression equations that relate measurements like species,

DBH, and perhaps height to more difficult and expensive to measure quantities such as stem volume or biomass.

Data to inform these allometric models is expensive to collect because trees must be harvested and meticulously partitioned to facilitate accurate dimension and weight measurements (see, e.g., Section 1.2.1). These efforts result in published allometric equations that take the inexpensive measurements as input (e.g., species and DBH) and return estimates of the expensive quantity (e.g., height, volume, biomass).

In this section, we'll work with species specific height models (e.g., given tree species and DBH, the model returns tree height) and volume models (e.g., given tree species, DBH, and height, the model returns tree volume). It's not necessary to understand model formulation and estimation for this section.

Let's begin by looking at a model used by the USDA Forest Service to estimate tree height (ft) given DBH (in) and species. The model, published in FVS (2021) Equation 4.1.2, is

$$\text{height} = 4.5 + \exp\left(\beta_1 + \frac{\beta_2}{(\text{DBH} + 1.0)}\right), \tag{7.1}$$

where regression coefficient values for β_1 and β_2 are provided in Table 4.1.2 of FVS (2021) for trees in the northeastern US. For convenience, we copied the coefficients from FVS (2021) to `"datasets/FVS_NE_coefficients.csv"`, which is read in below and named `ht_coeffs`.

```
ht_coeffs <- read_csv("datasets/FVS_NE_coefficients.csv")
```

```
#> Rows: 108 Columns: 3
#> -- Column specification ------------------------------------
#> Delimiter: ","
#> chr (1): USFS_FVS_code
#> dbl (2): beta_1, beta_2
#>
#> i Use `spec()` to retrieve the full column specification for this data.
#> i Specify the column types or set `show_col_types = FALSE` to quiet this message.
```

```
ht_coeffs %>%
  glimpse()
```

```
#> Rows: 108
#> Columns: 3
#> $ USFS_FVS_code <chr> "BF", "TA", "WS", "RS", "NS", "~
#> $ beta_1        <dbl> 4.5084, 4.5084, 4.5084, 4.5084,~
#> $ beta_2        <dbl> -6.0116, -6.0116, -6.0116, -6.0~
```

Each row in `ht_coeffs` holds species specific (`USFS_FVS_code`) regression coefficient values for β_1 (`beta_1`) and β_2 (`beta_2`) used to implement (7.1). For example, say you have a 10 (in) DBH balsam fir (*Abies balsamea*) and want to estimate its height (ft). The `USFS_FVS_code` for balsam fir is `BF` and associated coefficients are $\beta_1 = 4.5084$ and $\beta_2 = -6.0116$. The code below applies (7.1) using the given DBH, β_1, and β_2 values to produce the desired height estimate.

```
DBH_in <- 10
beta_1 <- 4.5084
beta_2 <- -6.0116
height_ft <- 4.5 + exp(beta_1 + beta_2/(DBH_in + 1))
height_ft
```

```
#> [1] 57.057
```

Now, say we want to use (7.1) to estimate the height for all trees measured in some cruise data where only DBH and species were collected. For illustration, we use a version of the two stand plot data introduced later in Section 12.3.6 and shown in Figure 12.7. As seen below, these data provide a stand and plot identification value and overstory tree DBH and species codes.

```
stands <- read_csv("datasets/two_stands_wo_ht_vol.csv")
```

```
#> Rows: 15 Columns: 4
#> -- Column specification ------------------------------------
#> Delimiter: ","
#> chr (1): scientific_name
#> dbl (3): stand_id, plot_id, DBH_in
#>
#> i Use `spec()` to retrieve the full column specification for this data.
#> i Specify the column types or set `show_col_types = FALSE` to quiet this message.
```

```
stands %>%
  glimpse()
```

```
#> Rows: 15
#> Columns: 4
#> $ stand_id        <dbl> 1, 1, 1, 1, 1, 2, 2, 2, 2, 2,~
#> $ plot_id         <dbl> 1, 1, 1, 3, 3, 1, 1, 1, 2, 2,~
#> $ scientific_name <chr> "Abies balsamea", "Pinus stro~
#> $ DBH_in          <dbl> 11.3, 9.8, 10.7, 15.4, 13.1, ~
```

We begin by joining coefficient columns in `ht_coeffs` to `stands` by species. Once that's complete, we'll use `mutate()` to add a new column `height_ft` with values defined by (7.1) and the columns `DBH_in`, `beta_1`, and `beta_2`.

There's one mild technical issue though—notice that `USFS_FVS_code` in `ht_coeffs` contains two-letter species codes, while `scientific_name` in `stands` contains scientific names.[17] Before

[17] We're adding this extra wrinkle because it's such a common issue and provides good practice with joins.

we can join `ht_coeffs` to `stands` by species, we need a common species identification column. Fortunately, this isn't the first time we've encountered this issue.

Columns in the `"datasets/USFS_species_codes.csv"` file, read in below, allow a crosswalk between species common name, scientific name, and three common USFS code types.[18]

```
spp_codes <- read_csv("datasets/USFS_species_codes.csv")
```

```
#> Rows: 108 Columns: 5
#> -- Column specification -----------------------------------
#> Delimiter: ","
#> chr (4): common_name, scientific_name, USFS_FVS_cod...
#> dbl (1): USFS_FIA_code
#>
#> i Use `spec()` to retrieve the full column specification for this data.
#> i Specify the column types or set `show_col_types = FALSE` to quiet this message.
```

```
spp_codes %>%
  glimpse()
```

```
#> Rows: 108
#> Columns: 5
#> $ common_name     <chr> "balsam fir", "tamarack", "wh~
#> $ scientific_name <chr> "Abies balsamea", "Larix lari~
#> $ USFS_FVS_code   <chr> "BF", "TA", "WS", "RS", "NS",~
#> $ USFS_FIA_code   <dbl> 12, 71, 94, 97, 91, 95, 90, 1~
#> $ PLANTS_code     <chr> "ABBA", "LALA", "PIGL", "PIRU~
```

The `USFS_FVS_code` column is common to both the `ht_coeffs` and `spp_codes` tibbles. The `scientific_name` column is common to both the `spp_codes` and `stands` tibbles. So, joining the `USFS_FVS_code` column from `spp_codes` to `stands` by `scientific_name` will let us subsequently join `ht_coeffs` to `stands` by `USFS_FVS_code`. The first step in this sequence of joins is given below.

```
stands <- left_join(stands, spp_codes, by = "scientific_name")

# Remove the columns we don't need.
stands <- stands %>% select(!c(common_name, USFS_FIA_code, PLANTS_code))
```

---

[18]Species codes held in the `USFS_FVS_code` and `USFS_FIA_code` columns are used by FVS (2021) and Burrill et al. (2018), respectively. Codes in the `PLANTS_code` column are from the PLANTS database (USDA 2021), which provides standardized information about the vascular plants, mosses, liverworts, hornworts, and lichens of the US and its territories.

```
# Check that all species have a scientific_name match.
any(is.na(stands$USFS_FVS_code)) # No NAs mean all matched (that's good)!
```

```
#> [1] FALSE
```

Let's step through the code above before looking at the resulting `stands`. `left_join()` adds columns from `spp_codes` to `stands` by matching `stands$scientific_name` to `spp_codes$scientific_name`. After the `left_join()`, all `spp_codes` columns are added to `stands`. The call to `select()` removes unneeded `spp_codes` columns (i.e., we only need `USFS_FVS_code`).

The final line of code checks that there are no `NA` values in `stands$USFS_FVS_code`. This is an important step because `NA` presence means not all species in `stands` were in `spp_codes`, which would present a problem as we move forward. Fortunately, no `NA` values were found, which means all rows in `stands` now have a valid `USFS_FVS_code` value.

An alternative, and perhaps better, approach to doing this check via `any(is.na(stands$USFS_FVS_code))` after the `left_join()` is to do it before the `left_join()` using an `anti_join()`. Recall, `anti_join()` removes all rows in `x` that have a match in `y`. So, if the `anti_join()` below returns a tibble with no rows, we know all species in `stands` have a match in `spp_codes`. We'll use anti joins for join diagnostics in subsequent code.

```
# Alternative way to check all species had a scientific_name match.
anti_join(stands, spp_codes, by = "scientific_name")
```

```
#> # A tibble: 0 x 5
#> # i 5 variables: stand_id <dbl>, plot_id <dbl>,
#> #   scientific_name <chr>, DBH_in <dbl>,
#> #   USFS_FVS_code <chr>
```

Okay, let's take a look at our work.

```
stands %>%
  glimpse()
```

```
#> Rows: 15
#> Columns: 5
#> $ stand_id        <dbl> 1, 1, 1, 1, 1, 2, 2, 2, 2, 2,~
#> $ plot_id         <dbl> 1, 1, 1, 3, 3, 1, 1, 1, 2, 2,~
#> $ scientific_name <chr> "Abies balsamea", "Pinus stro~
#> $ DBH_in          <dbl> 11.3, 9.8, 10.7, 15.4, 13.1, ~
#> $ USFS_FVS_code   <chr> "BF", "WP", "WP", "PB", "WP",~
```

The output from `glimpse()` above shows that `stands` now has the `USFS_FVS_code` column and we're ready to join it with `ht_coeffs`. Before proceeding with the left join of `stands` and `ht_coeffs` by `USFS_FVS_code`, we first use an anti join to check that all `stands$USFS_FVS_code` values are in `ht_coeffs$USFS_FVS_code`.

```
# Check that all USFS_FVS_code species in stands are in ht_coeffs.
anti_join(stands, ht_coeffs, by = "USFS_FVS_code")
```

```
#> # A tibble: 0 x 5
#> # i 5 variables: stand_id <dbl>, plot_id <dbl>,
#> #   scientific_name <chr>, DBH_in <dbl>,
#> #   USFS_FVS_code <chr>
```

The empty tibble returned by the anti join above tells us that all species in `stands` have a match in `ht_coeffs` and we can proceed with the left join to attach the regression coefficients in `ht_coeffs` to `stands`.

```
stands <- left_join(stands, ht_coeffs, by = "USFS_FVS_code")

stands %>%
  glimpse()
```

```
#> Rows: 15
#> Columns: 7
#> $ stand_id        <dbl> 1, 1, 1, 1, 1, 2, 2, 2, 2, 2,~
#> $ plot_id         <dbl> 1, 1, 1, 3, 3, 1, 1, 1, 2, 2,~
#> $ scientific_name <chr> "Abies balsamea", "Pinus stro~
#> $ DBH_in          <dbl> 11.3, 9.8, 10.7, 15.4, 13.1, ~
#> $ USFS_FVS_code   <chr> "BF", "WP", "WP", "PB", "WP",~
#> $ beta_1          <dbl> 4.5084, 4.6090, 4.6090, 4.438~
#> $ beta_2          <dbl> -6.0116, -6.1896, -6.1896, -4~
```

The output from `glimpse()` above shows that `beta_1` and `beta_2` are now added to `stands` and we can proceed with adding the new tree height column (`height_ft`) to `stands` using (7.1) within `mutate()`.

```
stands <- stands %>%
  mutate(height_ft = 4.5 + exp(beta_1 + beta_2/(DBH_in + 1)))

# Remove the now unnecessary coefficients and
# move USFS_FVS_code to after species.
stands <- stands %>%
  select(!c(beta_1, beta_2)) %>%
  relocate(USFS_FVS_code, .after = scientific_name)

stands %>%
  glimpse()
```

```
#> Rows: 15
#> Columns: 6
#> $ stand_id        <dbl> 1, 1, 1, 1, 1, 2, 2, 2, 2, 2,~
#> $ plot_id         <dbl> 1, 1, 1, 3, 3, 1, 1, 1, 2, 2,~
#> $ scientific_name <chr> "Abies balsamea", "Pinus stro~
#> $ USFS_FVS_code   <chr> "BF", "WP", "WP", "PB", "WP",~
#> $ DBH_in          <dbl> 11.3, 9.8, 10.7, 15.4, 13.1, ~
#> $ height_ft       <dbl> 60.182, 61.093, 63.644, 70.49~
```

There we go! We now have an estimate of tree height in our `stands` tibble.

Next, let's add stem volume estimates to our tree data. We'll follow the same process as adding height estimates, which starts with finding a suitable allometric model in the literature. Given we have species, DBH (in), and now height (ft), we can look for models that take these variables as input and return volume (m^3). A suitable model for this setting is given in Honer (1967) and defined as

$$\text{volume} = \frac{\text{DBH}^2}{\alpha_1 + \frac{\alpha_2}{\text{height}}}. \tag{7.2}$$

We copied species specific regression coefficient values for α_1 and α_2 from Honer (1967) to `"datasets/Honer_coefficients.csv"`, which is read in below and named `vol_coeffs`.

```
vol_coeffs <- read_csv("datasets/Honer_coefficients.csv")
```

```
#> Rows: 34 Columns: 3
#> -- Column specification ------------------------------------
#> Delimiter: ","
#> chr (1): USFS_FVS_code
#> dbl (2): alpha_1, alpha_2
#>
#> i Use `spec()` to retrieve the full column specification for this data.
#> i Specify the column types or set `show_col_types = FALSE` to quiet this message.
```

```
vol_coeffs %>%
  glimpse()
```

```
#> Rows: 34
#> Columns: 3
#> $ USFS_FVS_code <chr> "WP", "RP", "JP", "LP", "BS", "~
#> $ alpha_1       <dbl> 0.691, 0.710, 0.897, 0.694, 1.5~
#> $ alpha_2       <dbl> 363.68, 355.62, 348.53, 343.90,~
```

Columns `USFS_FVS_code`, `alpha_1`, and `alpha_2` in `vol_coeffs` hold values for species, α_1, and α_2, respectively.

Because `stands` and `vol_coeffs` use the same species codes (i.e., `USFS_FVS_code`), we can continue directly with checking that all species in `stands` are present in `vol_coeffs` via the anti join below.

```
anti_join(stands, vol_coeffs, by = "USFS_FVS_code")
```

```
#> # A tibble: 0 x 6
#> # i 6 variables: stand_id <dbl>, plot_id <dbl>,
#> #   scientific_name <chr>, USFS_FVS_code <chr>,
#> #   DBH_in <dbl>, height_ft <dbl>
```

The empty anti join tibble above confirms all species are present and we can proceed with the left join, compute tree volumes using (7.2) within `mutate()`, remove subsequently unneeded coefficient columns, and take a look at our new tree and species specific `volume_ft3` column.

```
stands <- left_join(stands, vol_coeffs, by = "USFS_FVS_code")

stands <- stands %>%
  mutate(volume_ft3 = DBH_in^2/(alpha_1 + alpha_2/height_ft))

# Remove the now unnecessary coefficients.
stands <- stands %>%
  select(!c(alpha_1, alpha_2))

stands %>%
  glimpse()
```

```
#> Rows: 15
#> Columns: 7
#> $ stand_id        <dbl> 1, 1, 1, 1, 1, 2, 2, 2, 2, 2,~
#> $ plot_id         <dbl> 1, 1, 1, 3, 3, 1, 1, 1, 2, 2,~
#> $ scientific_name <chr> "Abies balsamea", "Pinus stro~
#> $ USFS_FVS_code   <chr> "BF", "WP", "WP", "PB", "WP",~
#> $ DBH_in          <dbl> 11.3, 9.8, 10.7, 15.4, 13.1, ~
#> $ height_ft       <dbl> 60.182, 61.093, 63.644, 70.49~
#> $ volume_ft3      <dbl> 17.8147, 14.4556, 17.8745, 36~
```

As mentioned previously, applying allometric equations to estimate tree characteristics is a common task, and one greatly simplified using a `dplyr` join and verb function workflow.

7.14.4 Set operations

We often want to manipulate rows in two tibbles that have the same columns. For example, we might want to identify rows with identical values in two tibbles, or perhaps combine two tibbles but keep only one instance of duplicate rows. While such tasks could be accomplished using joins, `dplyr` provides functions that treat rows like sets (i.e., in the mathematical sense). Like joins, these functions accept two tibbles, `x` and `y`, and provide the following set operations.

- `intersect(x, y)`: return unique rows common to `x` and `y`.
- `setdiff(x, y)`: return unique rows in `x` that are not in `y`.
- `union(x, y)`: return unique rows in `x` or `y`.
- `union_all(x, y)`: return all rows (even duplicates) in `x` or `y`.

Let's try these functions out using the following example tibbles. Notice that rows one and two in `x` are duplicates (i.e., their tree `id` and `value` are the same). Also, the fourth row of `x` is the same as the first row of `y`.

```
x <- tibble(id = c(1, 1, 2, 3), value = c("a", "a", "b", "c"))
y <- tibble(id = 3:5, value = c("c", "d", "e"))
```

The result from `intersect()` identifies that both `x` and `y` have a row with `id = 3` and `value = "c"` in common.

```
intersect(x, y)
```

```
#> # A tibble: 1 x 2
#>      id value
#>   <dbl> <chr>
#> 1     3 c
```

The call to `setdiff()` shows the two unique rows in `x` that are not in `y`. Because set functions (with the exception of `union_all()`) return unique instances of rows, only one instance of `x`'s `id = 1` and `value = "a"` row is returned.

```
setdiff(x, y)
```

```
#> # A tibble: 2 x 2
#>      id value
#>   <dbl> <chr>
#> 1     1 a
#> 2     2 b
```

The result of `union()` is all rows in `x` and `y`, except duplicates of the rows with `id = 1, value = "a"` and `id = 3, value = "c"`.

```
union(x, y)
```

```
#> # A tibble: 5 x 2
#>      id value
#>   <dbl> <chr>
#> 1     1 a
#> 2     2 b
#> 3     3 c
```

```
#> 4       4 d
#> 5       5 e
```

Finally, `union_all()` is identical to `union()`, but its result includes the duplicates.

```
union_all(x, y)
```

```
#> # A tibble: 7 x 2
#>        id value
#>     <dbl> <chr>
#> 1       1 a
#> 2       1 a
#> 3       2 b
#> 4       3 c
#> 5       3 c
#> 6       4 d
#> 7       5 e
```

7.14.5 Row and column binds

Aside from joins and set operations that follow row matching rules, we frequently want simply to bind together two or more tibbles either by rows or columns. These operations are done using the `dplyr` `bind_rows()` and `bind_cols()` functions. Both functions take two or more tibbles to bind as separate arguments (or in a single list). The `bind_rows()` function has an optional argument `.id` that, if specified, will add a new column with values that link each row to the original tibbles. See `?dplyr::bind` for additional details about both functions.

We'll illustrate the bind functions using the `trees` tibble created in Section 6.2, which, recall, looks like the following.

```
trees
```

```
#> # A tibble: 5 x 4
#>        id  year   dbh height
#>     <int> <int> <dbl>  <dbl>
#> 1       1  2020   1.9   23
#> 2       1  2021   2.1   24.9
#> 3       2  2020   5.2   47
#> 4       2  2021   5.5   48.5
#> 5       3  2021   0.5    8.8
```

Say we collected core data for a few more trees and recorded it in the `more_trees` tibble below.

```
more_trees <- tibble(id = c(4, 4, 5),
                     year = c(2020, 2021, 2020),
                     dbh = c(10.2, 10.8, 14.0))
```

We can combine `trees` and `more_trees` into a single tibble as shown below.

```
bind_rows(trees, more_trees)
```

```
#> # A tibble: 8 x 4
#>      id  year   dbh height
#>   <dbl> <dbl> <dbl>  <dbl>
#> 1     1  2020   1.9   23
#> 2     1  2021   2.1   24.9
#> 3     2  2020   5.2   47
#> 4     2  2021   5.5   48.5
#> 5     3  2021   0.5    8.8
#> 6     4  2020  10.2   NA
#> 7     4  2021  10.8   NA
#> 8     5  2020  14     NA
```

The `.id` argument is added below, so we can keep track of the original tibbles.

```
bind_rows(trees, more_trees, .id = "coring_effort")
```

```
#> # A tibble: 8 x 5
#>   coring_effort    id  year   dbh height
#>   <chr>         <dbl> <dbl> <dbl>  <dbl>
#> 1 1                 1  2020   1.9   23
#> 2 1                 1  2021   2.1   24.9
#> 3 1                 2  2020   5.2   47
#> 4 1                 2  2021   5.5   48.5
#> 5 1                 3  2021   0.5    8.8
#> 6 2                 4  2020  10.2   NA
#> 7 2                 4  2021  10.8   NA
#> 8 2                 5  2020  14     NA
```

Say we also have species for `more_trees`. This information could be added when `more_trees` was formed, or afterwards using `bind_cols()` as follows.

```
species <- tibble(species = c("QURU", "QURU", "PIST"))

more_trees <- bind_cols(more_trees, species)

more_trees
```

```
#> # A tibble: 3 x 4
#>      id  year   dbh species
#>   <dbl> <dbl> <dbl> <chr>
#> 1     4  2020  10.2 QURU
#> 2     4  2021  10.8 QURU
#> 3     5  2020  14   PIST
```

If column names don't match, `bind_rows()` fills in missing values with `NA`. For example, the `species` column doesn't exist in `trees` but does in `more_trees`, hence the result below.

```
bind_rows(trees, more_trees, .id = "core_effort")
```

```
#> # A tibble: 8 x 6
#>   core_effort    id  year   dbh height species
#>   <chr>       <dbl> <dbl> <dbl>  <dbl> <chr>
#> 1 1               1  2020   1.9   23   <NA>
#> 2 1               1  2021   2.1   24.9 <NA>
#> 3 1               2  2020   5.2   47   <NA>
#> 4 1               2  2021   5.5   48.5 <NA>
#> 5 1               3  2021   0.5    8.8 <NA>
#> 6 2               4  2020  10.2   NA   QURU
#> 7 2               4  2021  10.8   NA   QURU
#> 8 2               5  2020  14     NA   PIST
```

7.15 Summary

Sections 7.1 through 7.8 covered core `dplyr` verb functions used to manipulate and summarize data held in a single tibble. To accommodate nested data structures, we introduced the *grouped* tibble in Section 7.9 along with functions to add, change, and remove groupings. Here too, we demonstrated how verb functions behave differently given a grouped tibble. In particular, we noted the grouped `summarize()` is a key tool used to solve many estimation problems we'll cover in subsequent chapters.

Section 7.10 highlighted two counting functions, `n()` and `n_distinct()`. When applied to grouped data, these counting functions are particularly useful for computing group-specific number of observations and enumerating unique value occurrences within one or more categorical variables (e.g., computing diversity metrics such as species richness).

`across()` was introduced in Section 7.11 as a tool to simplify code in cases when one or more functions are applied to multiple variables. We emphasized the pipe operator (`%>%`) in Section 7.12 for writing efficient and easy to read workflows.

Sections 7.13 served as a capstone—bringing together the preceding concepts and functions into workflows to explore the Minnesota tree growth datasets introduced in Sections 6.1.

Sections 7.14 turns to functions for combining two tibbles. Mutating joins, introduced in Section 7.14.1, use matching rows to add columns from the second tibble to the first tibble. Filtering joins, introduced in Section 7.14.2, subset rows in the first tibble given matching rows in the second tibble. Filtering joins are typically used to diagnose row mismatch issues prior to performing mutating joins. In Section 7.14.3, we stepped through a process to estimate tree height and volume using species-specific allometric equations. The process involved a series of filtering and mutating joins that created a single tibble that included all the columns needed to create the desired height and volume columns via calls to `mutate()`.

A suite of set operations were reviewed in Section 7.14.4. These functions operate on tibbles with the same columns and are used most often to identify unique or common rows between tibbles, or to combine tibbles while removing duplicate rows. Finally, Section 7.14.5 introduced the `bind_rows()` and `bind_cols()` functions used to augment tibble rows and columns, respectively.

7.16 Exercises

Figure 7.20 provides timber cruise data collected over three inventory plots and recorded in a field notebook. Rows in the notebook correspond to measurement trees and columns are defined as follows.

1. `Plot` plot number on which the tree was measured.
2. `Spp` species code.
3. `DBH` DBH in centimeters.
4. `Ht` total tree height in meters.
5. `Status` live (L) or dead (D).

Data provided in Figure 7.20 are used for Exercises 7.1 through 7.23. For most exercises, we provide the desired output and it's your job to write the `dplyr` code to reproduce the output. Use the pipe operator when possible.

Exercise 7.1. Create a tibble called `plots` that holds the cruise data provided in Figure 7.20. To do this, copy the data into a text file (you can separate columns using comma, space, or tab delimiters), or enter it into a spreadsheet and export it as a text file with appropriate column delimiters. Then read the file into R using a `readr` function (see Section 6.3). Once complete, your `plots` tibble should match the output below.

Hint, to get your column data types to match those shown below, use the `col_types` argument in your chosen `readr` read function.

```
plots
```

```
#> # A tibble: 7 x 5
```

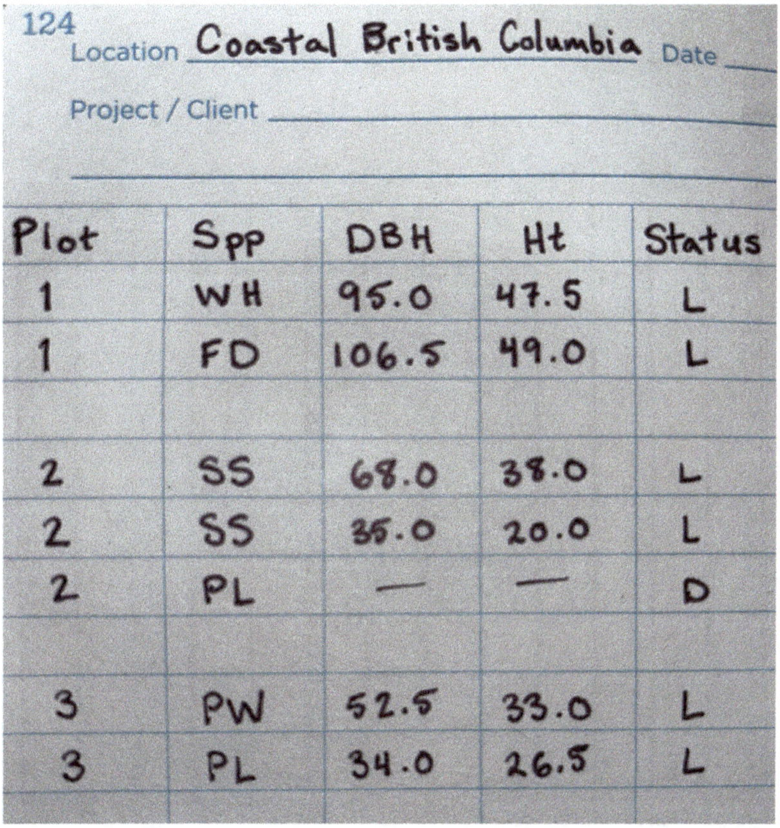

FIGURE 7.20: Example timber cruise data collected over three inventory plots and used for `dplyr` exercises.

```
#>     Plot Spp     DBH    Ht Status
#>    <int> <chr> <dbl> <dbl> <chr>
#> 1      1 WH     95    47.5 L
#> 2      1 FD    106.   49   L
#> 3      2 SS     68    38   L
#> 4      2 SS     35    20   L
#> 5      2 PL     NA    NA   D
#> 6      3 PW     52.5  33   L
#> 7      3 PL     34    26.5 L
```

Exercise 7.2. Given your output from Exercise 7.1, answer the following questions about `plots` (place your answer to each question behind a comment in your code).

a. How many rows and columns?
b. What are the column names?
c. What is the data type of each column?
d. Are there any `NA` values? If so, in which columns and which rows?

Exercise 7.3. Print rows with only dead trees.

```
#> # A tibble: 1 x 5
#>    Plot Spp     DBH    Ht Status
#>   <int> <chr> <dbl> <dbl> <chr>
#> 1     2 PL       NA    NA D
```

Exercise 7.4. Print rows with trees that do not have `NA` for DBH.

```
#> # A tibble: 6 x 5
#>    Plot Spp     DBH    Ht Status
#>   <int> <chr> <dbl> <dbl> <chr>
#> 1     1 WH     95    47.5 L
#> 2     1 FD    106.   49   L
#> 3     2 SS     68    38   L
#> 4     2 SS     35    20   L
#> 5     3 PW     52.5  33   L
#> 6     3 PL     34    26.5 L
```

Exercise 7.5. Print the row with the largest DBH.

```
#> # A tibble: 1 x 5
#>    Plot Spp     DBH    Ht Status
#>   <int> <chr> <dbl> <dbl> <chr>
#> 1     1 FD     106.    49 L
```

Exercise 7.6. Print rows with species code "SS" and DBH greater than or equal to 40 cm.

```
#> # A tibble: 1 x 5
#>    Plot Spp     DBH    Ht Status
#>   <int> <chr> <dbl> <dbl> <chr>
#> 1     2 SS       68    38 L
```

Exercise 7.7. Print rows that: 1) do not have species code "SS"; 2) are live; 3) have DBH greater than 50 cm.

```
#> # A tibble: 3 x 5
#>    Plot Spp     DBH    Ht Status
#>   <int> <chr> <dbl> <dbl> <chr>
#> 1     1 WH     95    47.5 L
#> 2     1 FD    106.   49   L
#> 3     3 PW     52.5  33   L
```

Exercise 7.8. Change column names `DBH` and `Ht` to `DBH_cm` and `Ht_m`. Be sure to assign your result back to `plots` to make the change permanent.

```
plots
```

```
#> # A tibble: 7 x 5
#>    Plot Spp   DBH_cm  Ht_m Status
#>   <int> <chr>  <dbl> <dbl> <chr>
#> 1     1 WH      95    47.5 L
#> 2     1 FD     106.   49   L
#> 3     2 SS      68    38   L
#> 4     2 SS      35    20   L
#> 5     2 PL      NA    NA   D
#> 6     3 PW      52.5  33   L
#> 7     3 PL      34    26.5 L
```

Exercise 7.9. Use `mutate()` to add a new column called `BA_m2` that holds each tree's basal area in square meters. Place this new column to the right of `DBH_cm`. Given DBH in centimeters, basal area in square meters is

$$\frac{\pi}{10000} \cdot \left(\frac{\text{DBH}}{2}\right)^2. \tag{7.3}$$

Your new `plots` tibble should look like the following.

```
plots
```

```
#> # A tibble: 7 x 6
#>    Plot Spp   DBH_cm   BA_m2  Ht_m Status
#>   <int> <chr>  <dbl>   <dbl> <dbl> <chr>
#> 1     1 WH      95    0.709   47.5 L
#> 2     1 FD     106.   0.891   49   L
#> 3     2 SS      68    0.363   38   L
#> 4     2 SS      35    0.0962  20   L
#> 5     2 PL      NA   NA       NA   D
#> 6     3 PW      52.5  0.216   33   L
#> 7     3 PL      34    0.0908  26.5 L
```

Exercise 7.10. Use the `mutate()` function to add a new column called `Live`. Values in `Live` should be `TRUE` when `Status` equals `L` and `FALSE` when `Status` equals `D`. Hint, consider using the `if_else()` or `case_when()` function within `mutate()` as described in Section 7.6.

```
plots
```

```
#> # A tibble: 7 x 7
#>    Plot Spp   DBH_cm   BA_m2  Ht_m Status Live
```

```
#>   <int> <chr> <dbl>   <dbl> <dbl> <chr>  <lgl>
#> 1     1 WH     95    0.709   47.5 L      TRUE
#> 2     1 FD    106.   0.891   49   L      TRUE
#> 3     2 SS     68    0.363   38   L      TRUE
#> 4     2 SS     35    0.0962  20   L      TRUE
#> 5     2 PL     NA   NA       NA   D      FALSE
#> 6     3 PW     52.5  0.216   33   L      TRUE
#> 7     3 PL     34    0.0908  26.5 L      TRUE
```

Exercise 7.11. Use the `select()` function to remove the `Status` column.

```
plots
```

```
#> # A tibble: 7 x 6
#>    Plot Spp   DBH_cm   BA_m2  Ht_m Live
#>   <int> <chr>  <dbl>   <dbl> <dbl> <lgl>
#> 1     1 WH      95    0.709   47.5 TRUE
#> 2     1 FD     106.   0.891   49   TRUE
#> 3     2 SS      68    0.363   38   TRUE
#> 4     2 SS      35    0.0962  20   TRUE
#> 5     2 PL      NA   NA       NA   FALSE
#> 6     3 PW      52.5  0.216   33   TRUE
#> 7     3 PL      34    0.0908  26.5 TRUE
```

Exercise 7.12. How many trees (i.e., rows) in each plot? Call your count variable `n_trees`. Hint, use `group_by()` in your piped `dplyr` expression.

```
#> # A tibble: 3 x 2
#>    Plot n_trees
#>   <int>   <int>
#> 1     1       2
#> 2     2       3
#> 3     3       2
```

Exercise 7.13. How many distinct species across all plots? Call your count variable `n_spp`.

```
#> # A tibble: 1 x 1
#>   n_spp
#>   <int>
#> 1     5
```

Exercise 7.14. How many distinct species within each plot? Call your count variable `n_spp`.

```
#> # A tibble: 3 x 2
#>    Plot n_spp
```

```
#>    <int> <int>
#> 1     1     2
#> 2     2     2
#> 3     3     2
```

Exercise 7.15. What are the distinct species codes across all plots?

```
#> # A tibble: 5 x 1
#>   Spp
#>   <chr>
#> 1 WH
#> 2 FD
#> 3 SS
#> 4 PL
#> 5 PW
```

Exercise 7.16. What are the distinct species codes within each plot?

```
#> # A tibble: 6 x 2
#> # Groups:   Plot [3]
#>    Plot Spp
#>   <int> <chr>
#> 1     1 WH
#> 2     1 FD
#> 3     2 SS
#> 4     2 PL
#> 5     3 PW
#> 6     3 PL
```

Exercise 7.17. Print rows with the largest DBH on each plot.

```
#> # A tibble: 3 x 6
#> # Groups:   Plot [3]
#>    Plot Spp   DBH_cm BA_m2  Ht_m Live
#>   <int> <chr>  <dbl> <dbl> <dbl> <lgl>
#> 1     1 FD      106. 0.891    49 TRUE
#> 2     2 SS       68  0.363    38 TRUE
#> 3     3 PW       52.5 0.216   33 TRUE
```

Exercise 7.18. What is the total BA on each plot? Call this total `total_BA_m2`. Hint, use `group_by()` in your piped `dplyr` expression that also involves calling the `sum()` function in `summarize()`. The sum should be done with `NA` values removed.

```
#> # A tibble: 3 x 2
#>    Plot total_BA_m2
#>   <int>       <dbl>
#> 1     1        1.60
```

```
#> 2     2          0.459
#> 3     3          0.307
```

Exercise 7.19. What is the total BA on each plot and by species? Call this total `total_BA_m2`. Hint, this should require modification of your 7.18 code's `group_by()` function and consultation with Section 7.13.1. The sum should be done with `NA` values removed.

```
#> # A tibble: 6 x 3
#> # Groups:   Plot [3]
#>    Plot Spp   total_BA_m2
#>   <int> <chr>       <dbl>
#> 1     1 FD         0.891
#> 2     1 WH         0.709
#> 3     2 PL         0
#> 4     2 SS         0.459
#> 5     3 PL         0.0908
#> 6     3 PW         0.216
```

Exercise 7.20. Next, let's add stem volume estimates using the tree measurements. These data were collected in coastal British Columbia, Canada. Given we have species, DBH (cm), and total height (m), we can look for models that take these variables as input and return volume (ft^3). A suitable model for this setting is defined as

$$\text{volume} = \exp(\beta_0) + \text{DBH}^{\beta_1} + \text{Ht}^{b_2}, \tag{7.4}$$

where β_0, β_1, and β_2 regression coefficient estimates for merchantable volume are provided in Table 6 of Nigh (2016). For convenience, we copied these coefficients from Nigh (2016) to `"datasets/Nigh_coefficients.csv"`.

Add a new column to `plots` called `vol_ft3` that holds merchantable volume for each tree and do some clean up when you're done. Write the code to replicate the following steps.

1) Make a coefficients tibble by reading in the species specific regression coefficients held in "Nigh_coefficients.csv". Hint, open "Nigh_coefficients.csv" in a text editor to see what it contains and how the columns are delimited.
2) Add these regression coefficients as new columns to `plots` via a join by `plots` column `Spp` and coefficients tibble `species_code`.
3) Specify (7.4) within `mutate()` to add the desired `vol_ft3` column. Place `vol_ft3` to the right of `Ht_m`.
4) Remove extraneous columns from `plots` (i.e., remove `common_name`, `scientific_name`, `species_code`, `beta_0`, `beta_1`, `beta_2`).

Your resulting `plots` should look like the following.

```
plots
```

```
#> # A tibble: 7 x 7
```

```
#>    Plot Spp   DBH_cm    BA_m2  Ht_m vol_ft3 Live
#>   <int> <chr>  <dbl>    <dbl> <dbl>   <dbl> <lgl>
#> 1     1 WH      95     0.709   47.5   4092. TRUE
#> 2     1 FD     106.    0.891   49     2959. TRUE
#> 3     2 SS      68     0.363   38     1996. TRUE
#> 4     2 SS      35     0.0962  20      617. TRUE
#> 5     2 PL      NA   NA        NA       NA  FALSE
#> 6     3 PW      52.5  0.216    33     1718. TRUE
#> 7     3 PL      34     0.0908  26.5    779. TRUE
```

Exercise 7.21. What is the total `vol_ft3` for each plot? Call your summarized variable `vol_ft3_per_plot`.

```
#> # A tibble: 3 x 2
#>    Plot vol_ft3_per_plot
#>   <int>            <dbl>
#> 1     1            7051.
#> 2     2            2613.
#> 3     3            2497.
```

Exercise 7.22. The area of each plot on which trees were measured is 0.04 ha. So your result from Exercise 7.21 represents tree volume per 0.04 ha on each plot. Multiply your volume result from Exercise 7.21 by 1/0.04 to represent tree volume per ha on each plot. Call your summarized variable `vol_ft3_per_ha`.

Hint, 1/0.04 is called an *expansion factor* and is used to convert values from one area basis to another. In this case, multiplying volumes per 0.04 ha by 1/0.04 converts them to volumes per ha, which is a more interpretable unit. We will formally introduce expansion factors and their use in Chapter 12. Sums should be computed with `NA` values removed.

```
#> # A tibble: 3 x 2
#>    Plot vol_ft3_per_ha
#>   <int>          <dbl>
#> 1     1        176267.
#> 2     2         65319.
#> 3     3         62416.
```

Exercise 7.23. What is the mean volume per ha across the three plots? That is, compute the mean of the `vol_ft3_per_ha` column you created in Exercise 7.22. Call your summarized variable `mean_vol_ft3_per_ha`.

Hint, pipe the result of your Exercise 7.22 into another `summarize()`. As we'll cover later in Chapter 12, this is our best estimate of mean volume per ha based on the cruise data from these three plots.

```
#> # A tibble: 1 x 1
#>   mean_vol_ft3_per_ha
#>                 <dbl>
#> 1             101334.
```

8

Tidying data with `tidyr`

Formats used to organize data are typically chosen to simplify recording, storage, analysis, or presentation. As we'll see, a format ideal for one purpose might not be ideal for other purposes. For example, a format useful for recording values in the field might not lend itself well to subsequent analysis or succinct data summaries.

The package `tidyr` provides functions to simplify moving between dataset formats and other common data wrangling and analysis tasks. Let's align our definitions for dataset components with those used in `tidyr`. In `tidyr`, a *dataset* is a collection of values, i.e., numbers, character strings, and other data types we covered. Every value is indexed by a variable and an observation. A variable contains all values that measure the same underlying characteristic (e.g., DBH, height, volume, or species) across observations. An observation (e.g., tree or plot) contains all values measured across variables.

Following Wickham, Çetinkaya-Rundel, et al. (2023), *tidy* formatting is a way to map a dataset's meaning to its structure. A dataset is messy or tidy depending on how rows and columns are matched with observations, variables, and values. Tidy data meet the following conditions.

1. Every column is a variable.
2. Every row is an observation.
3. Every row/column element is a single value.

Data are *messy* if they do not meet one or more of the tidy data conditions. Tidy data simplify many data summary and analysis tasks because they provide a standard way for structuring a dataset.

Consider the left dataset in Figure 8.1, called `dbh_wide`. Rows of `dbh_wide` correspond to three trees, columns are measurement years, and values are DBH measurements. As the dataset name suggests, these data are in *wide format*. This dataset is also messy because more than one column holds DBH measurements, and the column names themselves are values for a year variable.

As we'll explore further in Section 8.1, `tidyr`'s `pivot_longer()` function changes wide format to *long format*—a process illustrated by moving from `trees_wide` to `trees_long` in Figure 8.1. The `trees_long` dataset is tidy, as each column holds only one variable (`id`, `year`, and `dbh`), and each row corresponds to a single `id`, `year`, and `dbh` observation.

DOI: 10.1201/9781003222699-8

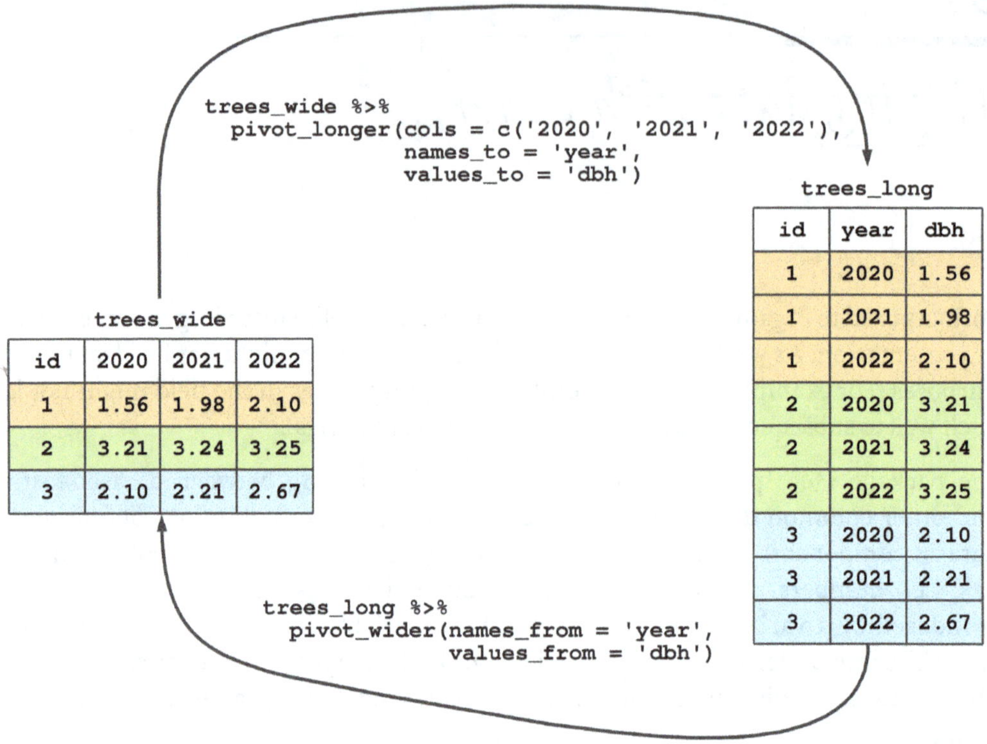

FIGURE 8.1: Changing longitudinal DBH measurements on three trees from wide to long format.

Most datasets we encounter are messy. Importantly, messy data are not bad or wrong; in fact, a messy table is often a very effective way to summarize data as we'll see in Section 8.3. However, from a data analysis standpoint, tidy data often simplify analysis and conversion to other formats. Receiving data in a messy format typically means you'll spend some time wrangling it into either a tidy format or another messy format to meet your needs. Subsequent sections cover some `tidyr` functions used to clean up messy data. Specifically, we'll cover how to remedy the following messy data features.

- Column headers are values, not variable names.
- Multiple variables are stored in one column.
- Variables are stored in both rows and columns.

The subsequent sections illustrate several `tidyr` functions using the messy FACE dataset introduced in Section 1.2.3. As a refresher, the Aspen FACE was designed to assess the effects of increasing tropospheric ozone and carbon dioxide concentrations on the structure and functioning of northern forest ecosystems. The code below loads packages we'll use in our tidying efforts and reads the FACE data into the `face` tibble.

```
library(readr)
library(dplyr)
library(tidyr)
face <- read_csv("datasets/FACE/FACE_aspen_core_growth.csv")
```

Run `face %>% glimpse()` to see column names and types. Also notice non-syntactic column names in the `FACE_aspen_core_growth.csv` file are placed within backticks in the `face` tibble.[1]

In the next few sections, the messy FACE dataset is used to introduce some `tidyr` functions. We focus on a subset of the FACE dataset that includes each tree's identifiers and height measurements for 2001 through 2005. The code below uses `dplyr` functions to select the desired columns and rename the `` `ID #` `` column name to `ID`.[2] In addition to unique tree `ID` and height measurements, we retain the experiment design variables replication (`Rep`), treatment (`Treat`), and tree clone identifier (`Clone`). Notice the use of the `contains()` helper function within `select()`.

```
face <- face %>%
  select(Rep, Treat, Clone, ID = `ID #`,
         contains(as.character(2001:2005)) & contains("Height"))
face
```

```
#> # A tibble: 1,991 x 9
#>      Rep Treat Clone    ID `2001_Height` `2002_Height`
#>    <dbl> <dbl> <dbl> <dbl>         <dbl>         <dbl>
#>  1     1     1     8    45            NA            NA
#>  2     1     1   216    44           547           622
#>  3     1     1     8    43           273           275
#>  4     1     1   216    42           526           619
#>  5     1     1   216    54           328           341
#>  6     1     1   271    55           543           590
#>  7     1     1   271    56           450           502
#>  8     1     1     8    57           217           227
#>  9     1     1   259    58           158           155
#> 10     1     1   271    59           230           241
#> # i 1,981 more rows
#> # i 3 more variables: `2003_Height` <dbl>,
#> #   `2004_Height` <dbl>, `2005_Height` <dbl>
```

---

[1]See Section 3.2.2 for syntactic name rules.

[2]As noted briefly in Section 6.3, `readr` read functions make non-syntactic column names syntactic by surrounding them by backticks. While this nice feature preserves the external file's column names, it can be cumbersome to include backticks when referencing columns in subsequent workflows.

8.1 Wide and long formats

Recall the FACE dataset is longitudinal, meaning measurements were taken repeatedly across time (in this case, years). Notice in the `face` tibble, height measurements are stored in successive columns, i.e., `` `2001_Height`...`2005_Height` ``. Like `trees_wide` in Figure 8.1, `face` is stored in wide format.

To tidy these messy data, our first task is to change `face` to a long format. We could imagine doing this in a spreadsheet software by copying and pasting, but this would be extremely tedious and would almost surely lead to errors. The `tidyr` package has two functions, `pivot_longer()` and `pivot_wider()`, to move between wide and long formats.

The call to `pivot_longer()` below moves `face` from wide to long format. In the `pivot_longer()` function, we specify the implicit `data = face`[3], columns to pivot into longer format (`cols = contains("Height")`), name to give the column created from the wide column names (`names_to = "Year_Type"`), and name we want to give to the column containing the measurements `values_to = 'Height_cm'`.

```
face_long <- face %>%
  pivot_longer(cols = contains("Height"),
               names_to = "Year_Type",
               values_to = "Height_cm")

face_long
```

```
#> # A tibble: 9,955 x 6
#>      Rep Treat Clone    ID Year_Type   Height_cm
#>    <dbl> <dbl> <dbl> <dbl> <chr>           <dbl>
#>  1     1     1     8    45 2001_Height        NA
#>  2     1     1     8    45 2002_Height        NA
#>  3     1     1     8    45 2003_Height        NA
#>  4     1     1     8    45 2004_Height        NA
#>  5     1     1     8    45 2005_Height        NA
#>  6     1     1   216    44 2001_Height       547
#>  7     1     1   216    44 2002_Height       622
#>  8     1     1   216    44 2003_Height       715
#>  9     1     1   216    44 2004_Height       716
#> 10     1     1   216    44 2005_Height       817
#> # i 9,945 more rows
```

Let's compare the dimensions of the wide format `face` with `face_long`.

[3]To facilitate piping, `data` is the first argument in `tidyr` functions.

```
dim(face)
```

```
#> [1] 1991    9
```

```
dim(face_long)
```

```
#> [1] 9955    6
```

There are more columns in the wide format `face` and more rows in the long format `face_long`—hence the reason these formats are referred to as wide and long.

`pivot_wider()` is used to widen a dataset in long format. For example, the code below moves `face_long` back to its original wide format.

```
face_wide <- face_long %>%
  pivot_wider(names_from = "Year_Type", values_from = "Height_cm")

# Check if face_wide and face are the same.
all.equal(face_wide, face)
```

```
#> [1] TRUE
```

In `pivot_wider()` above, we specify the implicit `data = face_long`, column from which to get wide column names (`names_from = "Year_Type"`), and column from which to get values (`values_from = "Height_cm"`). As expected and shown using `all.equal(face_wide, face)`, the call to `pivot_wider()` returns a tibble that's the same as `face`.

Looking at the `pivot_wider()` manual page, you'll see an argument `id_cols`. By default, `id_cols` includes all columns not specified in `names_from` or `values_from`. So, the call to `pivot_wider()` above used combinations of `Rep`, `Treat`, `Clone`, and `ID` values to identify rows for `face_wide`, which was our intention. There're many situations, however, when we want to override the `id_cols` default behavior and explicitly specify the columns used to identify rows; see, e.g., Section 8.3.

8.2 Separating and uniting columns

Let's take another look at the `Year_Type` column in the `face_long` dataset.

```
face_long %>%
  select(Year_Type)
```

```
#> # A tibble: 9,955 x 1
#>    Year_Type
```

```
#>    <chr>
#>  1 2001_Height
#>  2 2002_Height
#>  3 2003_Height
#>  4 2004_Height
#>  5 2005_Height
#>  6 2001_Height
#>  7 2002_Height
#>  8 2003_Height
#>  9 2004_Height
#> 10 2005_Height
#> # i 9,945 more rows
```

The `Year_Type` column contains two variables separated by an underscore—the measurement year and measurement type. To set `face` on the tidy data path, we need to separate `Year_Type` into its components. Below, we use `tidyr::separate_wider_delim()` to separate the `Year_Type` column into `Year` and `Type`. We want to retain the `Year` component, whereas the `Type` component is simply the value `"Height"` replicated for each row and can be discarded.

```
face_long <- face_long %>%
  separate_wider_delim(cols = "Year_Type",
                       delim = "_",
                       names = c("Year", NA))
face_long
```

```
#> # A tibble: 9,955 x 6
#>      Rep Treat Clone    ID Year  Height_cm
#>    <dbl> <dbl> <dbl> <dbl> <chr>     <dbl>
#>  1     1     1     8    45 2001         NA
#>  2     1     1     8    45 2002         NA
#>  3     1     1     8    45 2003         NA
#>  4     1     1     8    45 2004         NA
#>  5     1     1     8    45 2005         NA
#>  6     1     1   216    44 2001        547
#>  7     1     1   216    44 2002        622
#>  8     1     1   216    44 2003        715
#>  9     1     1   216    44 2004        716
#> 10     1     1   216    44 2005        817
#> # i 9,945 more rows
```

The call to `separate_wider_delim()` above specifies the implicit `data = face_long`, the column to separate (`cols = "Year_Type"`), the delimiter that separates the columns (`delim = "_"`), and names for the resulting columns (`names = c("Year", NA)`). The `NA` in the vector passed to the `names` argument means to remove the column that corresponds to the `Type` component of the `Year_Type` column. At this point, `face_long` is now tidy data!

Consult the manual page via ?tidyr::separate_wider_delim to learn more about this and related functions for separating delimited columns.

There are occasions when you might want to combine multiple columns into one. We can do this using unite(). For example, the code below unites the Rep, Treat, and Clone columns into a single column called Design, with each value separated by an underscore. See ?tidyr::unite for more details.

```
face_long %>%
  unite(col = "Design", Rep, Treat, Clone, sep = '_')
```

```
#> # A tibble: 9,955 x 4
#>    Design      ID Year  Height_cm
#>    <chr>    <dbl> <chr>     <dbl>
#>  1 1_1_8       45 2001         NA
#>  2 1_1_8       45 2002         NA
#>  3 1_1_8       45 2003         NA
#>  4 1_1_8       45 2004         NA
#>  5 1_1_8       45 2005         NA
#>  6 1_1_216     44 2001        547
#>  7 1_1_216     44 2002        622
#>  8 1_1_216     44 2003        715
#>  9 1_1_216     44 2004        716
#> 10 1_1_216     44 2005        817
#> # i 9,945 more rows
```

8.3 Application: Stand and stock tables

Data collected for forest research, inventory, and monitoring efforts are typically summarized using a variety of tables. For example, *stand* and *stock* tables are key data summary products that support a variety of forest management decisions, e.g., silvicultural prescription, economic valuation, and structural diversity assessment.

Recalling terminology introduced in Section 4.1, the stand table summarizes a *quantitative discrete* variable (e.g., stem count) grouped by one or more *categorical* (i.e., qualitative) variables (e.g., size class or species). Similarly, the stock table summarizes a *quantitative continuous* variable (e.g., volume, weight, or basal area) grouped by one or more *categorical* variables.

In forestry applications, stand tables often summarize the average number of trees per unit area (e.g., acre or hectare) within an area of interest (e.g., stand, management unit, or forest) by DBH class and species. Similarly, stock tables often summarize the average

TABLE 8.1: Estimates for number of trees and wood volume per acre by species and DBH used to illustrate building stand and stock tables.

Species	DBH (in)	Trees/ac	Volume (ft^3/ac)
Abies balsamea	11.3	8	142.8
Betula papyrifera	14.8	8	269.6
Betula papyrifera	15.4	8	293.7
Pinus strobus	9.8	8	116.3
Pinus strobus	10.7	8	143.6
Pinus strobus	13.1	8	231.9

wood volume or weight per unit area within an area of interest by DBH class and species. DBH classes are DBH intervals with widths chosen to communicate situation-specific information.

We motivate building stand and stock tables using forest stand estimates provided in Table 8.1. In this table, each row provides an estimated number of trees and volume on a per-acre basis for a given species and DBH. For example, the first row in Table 8.1 says a typical (or average) acre has 8 11.3-inch DBH *Abies balsamea* stems with a combined volume of 142.8 ft^3. We'll introduce the inventory data from which these estimates were derived later in Chapter 12. For now, however, these estimates are used to build the stand and stock tables given in Tables 8.2 and 8.3, respectively.

Before getting started, notice that stand and stock tables are not *tidy* according to the definition given at the opening of this chapter—they're wide and messy, and that's okay.

Let's begin with the stand table shown in Table 8.2. Stand table rows correspond to the distinct species in Table 8.1, columns are four-inch DBH classes, and values are the number of trees for the given row and column.

In Table 8.2, notice the parentheses and square bracket notation used to define DBH class intervals (i.e., column names). A *closed interval*, denoted by square brackets, means the endpoints are included in the interval. For example, $[6, 10]$ is a closed interval with 6- and 10-inch endpoints. Referencing Table 8.1, there is one *Pinus strobus* with a DBH that falls within the $[6, 10]$ inch DBH class interval, hence the corresponding 8 trees/ac value in Table 8.2. A closed interval can also be written using $\leq$ notation, e.g., $[6, 10] = \{6 \leq x \leq 10\}$ where x is DBH.

An *open interval*, denoted by parentheses, means the endpoints are not included in the interval. Following from open and closed notation, a *half-open* interval includes only one of its endpoints and is denoted by one parenthesis and one square bracket. The stand table has two half-open intervals, $(10, 14]$ and $(14, 18]$ (these can be written as $\{10 < x \leq 14\}$ and $\{14 < x \leq 18\}$). Table 8.1 shows one *Abies balsamea* and two *Pinus strobus* in the $(10, 14]$ DBH class, and two *Betula papyrifera* in the $(14, 18]$ DBH class. Stand table values are the

TABLE 8.2: Stand table for estimates given in Table 8.1. Values are number of trees per acre.

	DBH classes (in)			
Species	[6,10]	(10,14]	(14,18]	Totals
Abies balsamea	0	8	0	8
Betula papyrifera	0	0	16	16
Pinus strobus	8	16	0	24
Totals	8	24	16	48

TABLE 8.3: Stock table for estimates given in Table 8.1. Values are volume cubic feet per acre.

	DBH classes (in)			
Species	[6,10]	(10,14]	(14,18]	Totals
Abies balsamea	0.0	142.8	0.0	142.8
Betula papyrifera	0.0	0.0	563.3	563.3
Pinus strobus	116.3	375.5	0.0	491.8
Totals	116.3	518.3	563.3	1197.9

sum of trees for the given species and DBH class—hence there are 16 *Pinus strobus* in the $(10, 14]$ DBH class and 16 *Betula papyrifera* in the $(14, 18]$ DBH class.

The stock table is given in Table 8.3. This stock table is built the same way as the stand table, with the exception that its values are the sum of volume for each given species and DBH class. For example, the stand table's 8 *Pinus strobus* trees per acre in the $[6, 10]$ inch DBH class correspond to 116.3 ft^3 of volume per acre, and the 16 *Betula papyrifera* trees per acre in the $(14, 18]$ inch DBH class correspond to 563.3 ft^3 of volume per acre (i.e., $269.6 + 293.7$ in Table 8.1).

As we've seen, estimate values in Table 8.1 comprise the non-zero values in the resulting stand and stock tables. Zero values in stand or stock tables are just as important as non-zero ones—they indicate absence of specific species and DBH class combinations, which can influence management decisions and forest valuation. These *explicit* zeros in the stand and stock tables are *implicit* zeros in Table 8.1.

Lastly, it's useful to add row and column totals to the stand and stock table margins. For Tables 8.2 and 8.3, the row totals are the number of trees and volume per acre for each species, respectively. Similarly, column totals are the number of trees and volume per acre for each DBH class.

Drawing on the preceding sections in this and previous `tidyverse` chapters, we show how a piped series of functions can efficiently deliver stand, stock, and similar tables with a fraction of the pain incurred using hand calculations or traditional spreadsheets.

The following are a few considerations for building stand and stock tables from forest inventory data.

1. Apply the appropriate *estimator* to generate the desired values for the stand or stock table. These topics will be covered in Chapters 11, 12, and beyond.
2. Identify the quantitative discrete and continuous variables of interest as well as the categorical variables you'll use to define the table's rows and columns.
    - Often, we need to convert a continuous variable into a categorical variable to represent the table's rows or columns. For example, if DBH was collected as a continuous variable (which is often the case), then it'll need to be *discretized*, i.e., converted to a categorical variable, for use in a `dplyr` grouped summarize. How you discretize your continuous variable depends on what you want to communicate with the resulting table. For stand and stock tables, it's common to discretize continuous DBH measurements into 2 or 4 inch DBH classes (or equivalent widths in centimeters). The `ggplot2` package `cut_width()`, `cut_number()`, and `cut_interval()` functions provide convenient ways to discretize continuous variables.
3. Reshape your data to long format using `tidyr` functions, if needed, so they can be easily piped into a series of `dplyr` functions to form the desired stand or stock table.
    - In practice you'll encounter a variety of data collection, recording, and storage protocols that yield all sorts of data organization formats—these inevitably need wrangling into a format suitable for `dplyr`.

Our focus here is on the last two considerations. We'll work on the first consideration in subsequent chapters, specifically in Section 12.6.3. For now, however, we begin with the table of estimates provided in Table 8.1 and develop code to reproduce Tables 8.2 and 8.3.

We begin by reading tree estimates from Table 8.1 into the `ests` tibble.

```
ests <- read_csv("datasets/stand_stock_estimates.csv")
```

Next, in the code below, we use `cut_width()` to discretize `DBH_in` and add this new discretized column, called `DBH_4in`, back to `ests` via `mutate()`.[4] Also in the `mutate()` call, the `.after` argument positions the new `DBH_4in` column to the right of `DBH_in` to facilitate visual comparison.

`cut_width()` (and related functions `cut_number()` and `cut_interval()`) provide a lot of control over how intervals are formed and how values on their endpoints are handled.

```
ests <- ests %>%
  mutate(DBH_4in = cut_width(DBH_in, width = 4), .after = DBH_in)
ests
```

[4]Run `cut_width(DBH_in, width = 4)` in the console, try different `width` argument values, and consult its manual page via `?ggplot2::cut_width` to better understand the function's use and behavior.

```
#> # A tibble: 6 x 5
#>   Species           DBH_in DBH_4in trees_ac vol_cu_ft_ac
#>   <chr>              <dbl> <fct>      <dbl>        <dbl>
#> 1 Abies balsamea      11.3 (10,14]        8         143.
#> 2 Betula papyrife~    14.8 (14,18]        8         270.
#> 3 Betula papyrife~    15.4 (14,18]        8         294.
#> 4 Pinus strobus        9.8 [6,10]         8         116.
#> 5 Pinus strobus       10.7 (10,14]        8         144.
#> 6 Pinus strobus       13.1 (10,14]        8         232.
```

Take a look at the new `DBH_4in` column in the `ests` tibble above. Notice it's a factor data type and its values use parentheses and square bracket notation to define interval endpoints—the same notation used for column labels in the desired stand and stock tables. Also, confirm that values in `DBH_in` map to the correct `DBH_4in` interval.

Next, recall, explicit zeros in stand and stock tables are implicit zeros in Table 8.1. Let's add these implicit zeros to `ests` (i.e., make the implicit explicit by adding the species and DBH class combinations with zero trees and volume). This is done using `tidyr::complete()`. This function adds all missing value combinations for a discrete variable set. The first argument in `complete()` is the data, followed by column names used to form the combinations, then the `fill` argument is given a list of column names and their associated desired explicit values.

Our call to `complete()` below adds missing combinations of `Species` and `DBH_4in` and sets explicit values of `trees_ac` and `vol_cu_ft_ac` to zero. Notice, by not specifying an explicit value for `DBH_in` in the `fill` argument's list, the `NA` default value is used.[5]

```
ests <- ests %>%
  complete(Species, DBH_4in,
           fill = list(trees_ac = 0, vol_cu_ft_ac = 0))
ests
```

```
#> # A tibble: 11 x 5
#>    Species          DBH_4in DBH_in trees_ac vol_cu_ft_ac
#>    <chr>            <fct>    <dbl>    <dbl>        <dbl>
#>  1 Abies balsamea   [6,10]    NA          0            0
#>  2 Abies balsamea   (10,14]   11.3        8         143.
#>  3 Abies balsamea   (14,18]   NA          0            0
#>  4 Betula papyrif~ [6,10]    NA          0            0
#>  5 Betula papyrif~ (10,14]   NA          0            0
#>  6 Betula papyrif~ (14,18]   14.8        8         270.
#>  7 Betula papyrif~ (14,18]   15.4        8         294.
#>  8 Pinus strobus    [6,10]     9.8        8         116.
```

[5]The `NA` value for `DBH_in` is okay, because this variable is not needed to build subsequent stand and stock tables.

```
#>  9 Pinus strobus   (10,14]  10.7        8         144.
#> 10 Pinus strobus   (10,14]  13.1        8         232.
#> 11 Pinus strobus   (14,18]  NA          0           0
```

Next, sum values in the `trees_ac` and `vol_cu_ft_ac` columns that have the same species and DBH class.[6] We add the `.groups = "drop"` argument, as species grouping in the resulting tibble is not needed.

```
ests <- ests %>%
  group_by(Species, DBH_4in) %>%
  summarize(sum_trees_ac = sum(trees_ac),
            sum_vol_cu_ft_ac = sum(vol_cu_ft_ac),
            .groups = "drop")
ests
```

```
#> # A tibble: 9 x 4
#>   Species          DBH_4in sum_trees_ac sum_vol_cu_ft_ac
#>   <chr>            <fct>          <dbl>            <dbl>
#> 1 Abies balsamea   [6,10]             0                0
#> 2 Abies balsamea   (10,14]            8              143.
#> 3 Abies balsamea   (14,18]            0                0
#> 4 Betula papyrif~ [6,10]              0                0
#> 5 Betula papyrif~ (10,14]             0                0
#> 6 Betula papyrif~ (14,18]            16              563.
#> 7 Pinus strobus    [6,10]             8              116.
#> 8 Pinus strobus    (10,14]           16              376.
#> 9 Pinus strobus    (14,18]            0                0
```

Okay, now we're in business! The only thing left to do is reshape `ests` using `tidyr`'s `pivot_wider()` function to put `Species` along the rows and `DBH_4in` across the columns, and decide which variable to set as the resulting table's value. This is done in the code below to make the stand table.

```
stand <- ests %>%
  pivot_wider(id_cols = Species,
              names_from = DBH_4in,
              values_from = sum_trees_ac)

stand
```

```
#> # A tibble: 3 x 4
#>   Species            `[6,10]` `(10,14]` `(14,18]`
#>   <chr>                 <dbl>     <dbl>     <dbl>
```

[6]This step could be done before the call to `complete()` and yield the same result.

```
#> 1 Abies balsamea          0        8        0
#> 2 Betula papyrifera       0        0       16
#> 3 Pinus strobus           8       16        0
```

Notice in the code above, we specified `id_cols = Species`, which says we want the wide table rows to be defined by values in the `Species` column.[7]

The final touch is to add row and column totals—the code for which is provided below with minimal explanation. As you'll see, adding these totals is a bit convoluted; however, each code component has been covered previously, so see if you can work through it and understand the steps.

```
# Sum across rows.
stand <- stand %>%
  mutate(Totals = rowSums(across(where(is.numeric))))

# Sum down columns.
# This is a little tricky. Notice the . places the output from
# summarize() after stands in the bind_rows(stands, .), i.e., it
# binds the row of totals to the bottom of the stand tibble.
stand <- stand %>%
  summarize(Species = "Totals", across(where(is.numeric), sum)) %>%
  bind_rows(stand, .)

stand
```

```
#> # A tibble: 4 x 5
#>   Species           `[6,10]` `(10,14]` `(14,18]` Totals
#>   <chr>                <dbl>     <dbl>     <dbl>  <dbl>
#> 1 Abies balsamea           0         8         0      8
#> 2 Betula papyrifera        0         0        16     16
#> 3 Pinus strobus            8        16         0     24
#> 4 Totals                   8        24        16     48
```

The `stand` tibble above is the stand Table 8.2. Creating the stock table follows the exact same steps up to the call to `pivot_wider()`.

We took care to introduce the idea of implicit zeros in the forest inventory estimates and how `complete()` is used to make those implicit zeros explicit. In Chapter 12, making implicit zeros explicit via `complete()` will again be important for implementing methods that generate estimates from forest inventory data.

---

[7] If you remove `id_cols = Species`, you'll see the default behavior is to define rows using combinations of values from all columns other than those defined in `names_from` and `values_from`, i.e., `Species` and `sum_vol_cu_ft_ac`, which is not what we want.

When building stand and stock tables, specifying the optional `values_fill` argument in `pivot_wider()` allows you to skip the prior call to `complete()`. Like `complete()` and its `fill` argument, `pivot_wider()`'s `values_fill` argument lets you specify the explicit value for missing stand and stock table row and column combinations.

The code below rebuilds the stand table in a single workflow starting with a fresh copy of `ests` and taking advantage of `pivot_wider()`'s `values_fill` argument, which removes the need for `complete()` (we comment out the call to `complete()` and add `values_fill = 0` to `pivot_wider()`).

```
ests <- read_csv("datasets/stand_stock_estimates.csv")

stand <- ests %>%
  mutate(DBH_4in = cut_width(DBH_in, width = 4)) %>%
  # complete(Species, DBH_4in, fill = list(trees_ac = 0)) %>%
  group_by(Species, DBH_4in) %>%
  summarize(sum_trees_ac = sum(trees_ac), .groups = "drop") %>%
  pivot_wider(names_from = DBH_4in, values_from = sum_trees_ac,
              values_fill = 0) # Added values_fill argument.

# Add row and column totals.
stand <- stand %>%
  mutate(Totals = rowSums(across(where(is.numeric))))

# Sum down columns.
stand <- stand %>%
  summarize(Species = "Totals", across(where(is.numeric), sum)) %>%
  bind_rows(stand, .)

stand
```

```
#> # A tibble: 4 x 5
#>   Species            `(10,14]` `(14,18]` `[6,10]` Totals
#>   <chr>                  <dbl>     <dbl>    <dbl>  <dbl>
#> 1 Abies balsamea             8         0        0      8
#> 2 Betula papyrifera          0        16        0     16
#> 3 Pinus strobus             16         0        8     24
#> 4 Totals                    24        16        8     48
```

On your own, modify the code above to rebuild the stock table using `vol_cu_ft_ac` estimates. Your resulting stock table should match values in Table 8.3.

8.4 Summary

This chapter covered key functions in the `tidyr` package.[8] These functions simplify moving between dataset formats and other common data wrangling and analysis tasks.

The chapter opens with a definition of tidy data. Tidy data simplify analysis and conversion to other formats. As illustrated using the stand and stock tables in Section 8.3, messy data are not wrong or bad and are often effective formats for conveying information.

Section 8.1 introduced and illustrated `pivot_longer()` and `pivot_wider()` for moving between long and wide formats. Section 8.2 covered `separate_wider_delim()` and `unite()` used to split and combine character columns. Over these two sections, a piped workflow comprising `dplyr` and `tidyr` functions was developed to make the wide and messy FACE dataset tidy.

Forest stand and stock tables are fundamental to making informed forest management decisions. Generating such tables is a routine task and hence having an efficient and flexible software workflow to create these tables is important. Such a workflow was introduced in Section 8.3. An important workflow component was to make explicit the implicit zeros in the forest estimates via the `tidyr::complete()` function *or* by specifying `values_fill = 0` in `pivot_wider()`. The stand and stock table workflow will be extended in Section 12.6.3 with steps to generate estimates from forest inventory data.

8.5 Exercises

The exercises use the Elk County timber cruise dataset introduced in Section 1.2.2. These data were collected on a 271-acre forested property in Elk County, Pennsylvania, USA, to inform the owner's forest management plan.

In these exercises you'll develop stand and stock tables using per acre estimates for acceptable growing stock (AGS) sawtimber number of trees, basal area, and number of logs by species. Here, AGS is defined as a desirable species that contains at least one USFS Grade 3 log, now or in the future, and is likely to live another 15 years.[9]

For each sawtimber tree, the forester recorded the number of 16-foot logs to the nearest 1/2 log, to a minimum top diameter of 10 inches or to a point where the tree no longer meets USFS Grade 3 specifications due to forks, crook, excessive sweep, defect, etc. Tree diameter was measured at 4.5 feet above the ground (DBH) on the uphill side of the tree. All trees 1.0

[8]As with other programming topics introduced here, we provide only a partial tour and refer you to package manual pages and vignettes for more complete coverage, e.g., run `browseVignettes(package="tidyr")` on the console to see all `tidyr` vignettes.

[9]See Hanks (1976) for USFS log grade specifications.

inch DBH and larger were recorded by two-inch diameter classes using the mapping given in Table 8.4.

TABLE 8.4: DBH (in) range and associated class recorded in the Elk County cruise.

DBH range	DBH class
1.0–2.9	2
3.0–4.9	4
5.0–6.9	6
7.0–8.9	8
9.0–10.9	10
11.0–12.9	12
etc.	etc.

Estimates used to build the subsequent tables are provided in `"datasets/elk_spp_dbh_estimates.csv"`. Columns in this file are as follows:

1. `Species` tree species common name.
2. `DBH_2in` two-inch DBH class (see Table 8.4 for class range and endpoints).
3. `Trees_ac` estimated number of trees per acre.
4. `BA_ac` estimated square feet of basal area per acre.
5. `Logs_ac` estimated number of 16-foot logs per acre.

This data file is set up similarly to Table 8.1. Methods to generate these estimates from tree measurement data will be covered in Chapter 12.

For most exercises, we provide the desired output and it's your job to write the `tidyverse` code to reproduce it. Use piped `tidyverse` workflows to answer all questions.

Exercise 8.1. Create a tibble called `elk_ags` that holds the estimates described above and provided in `"datasets/elk_spp_dbh_estimates.csv"`. Once complete, your `elk_ags` tibble should match the output below. Use `elk_ags` for all subsequent questions.

```
elk_ags
```

```
#> # A tibble: 43 x 5
#>   Species     DBH_2in Trees_ac BA_ac Logs_ac
#>   <chr>         <dbl>    <dbl> <dbl>   <dbl>
#> 1 Beech            12   1.33   1.04    1.33
#> 2 Beech            14   0.585  0.625   0.682
#> 3 Beech            16   0.448  0.625   0.746
#> 4 Beech            20   0.0955 0.208   0.143
#> 5 Black Birch      12   1.06   0.833   1.06
#> 6 Black Birch      14   0.974  1.04    1.07
```

```
#>  7 Black Birch       16    0.895  1.25     1.42
#>  8 Black Birch       18    0.118  0.208    0.236
#>  9 Black Birch       20    0.0955 0.208    0.191
#> 10 Black Cherry      12    0.796  0.625    0.796
#> # i 33 more rows
```

Exercise 8.2. What are the distinct species? Your output might be formatted differently depending on the approach you choose to list the distinct species.

```
#> # A tibble: 9 x 1
#>   Species
#>   <chr>
#> 1 Beech
#> 2 Black Birch
#> 3 Black Cherry
#> 4 Cucumber Tree
#> 5 Northern Red Oak
#> 6 Red Maple
#> 7 Sugar Maple
#> 8 Yellow Birch
#> 9 Yellow Poplar
```

Exercise 8.3. By species, what is the estimate for: 1) trees per acre; 2) basal area per acre; 3) logs per acre? Call your summarized variables `sum_trees_ac`, `sum_ba_ac`, and `sum_logs_ac`.

Hint, this will involve a grouped `summarize()` that sums estimates by species.

```
#> # A tibble: 9 x 4
#>   Species          sum_trees_ac sum_ba_ac sum_logs_ac
#>   <chr>                   <dbl>     <dbl>       <dbl>
#> 1 Beech                  2.45       2.5         2.90
#> 2 Black Birch            3.14       3.54        3.98
#> 3 Black Cherry           4.52       6.25        7.63
#> 4 Cucumber Tree          0.821      1.46        2.09
#> 5 Northern Red Oak       0.864      2.08        1.82
#> 6 Red Maple             26.6       35.2        39.8
#> 7 Sugar Maple            2.54       3.12        3.43
#> 8 Yellow Birch           1.44       1.67        1.73
#> 9 Yellow Poplar          0.0565     0.208       0.170
```

Exercise 8.4. By the 2-inch DBH class (i.e., using `DBH_2in`), what is the estimated: 1) trees per acre; 2) basal area per acre; 3) logs per acre? Call your summarized variables `sum_trees_ac`, `sum_ba_ac`, and `sum_logs_ac`.

Hint, this will involve a grouped `summarize()` that sums estimates by DBH class.

```
#> # A tibble: 10 x 4
```

```
#>    DBH_2in sum_trees_ac sum_ba_ac sum_logs_ac
#>      <dbl>        <dbl>     <dbl>       <dbl>
#>  1      12       8.22       6.46       7.96
#>  2      14      16.4       17.5       20.8
#>  3      16       8.36      11.7       14.2
#>  4      18       4.72       8.33      10.3
#>  5      20       2.58       5.62       4.97
#>  6      22       1.58       4.17       3.75
#>  7      24       0.464      1.46       1.16
#>  8      26       0.0565     0.208      0.170
#>  9      28       0.0974     0.417      0.244
#> 10      30       0.0424     0.208      0.0849
```

Exercise 8.5. In preparation for building stand and stock tables, add a new column to `elk_ags` that explicitly defines each 2-inch DBH class's range and endpoints, as defined in Table 8.4. Call your new column `DBH_2in_range` and place it to the right of `DBH_2in`.

Hint, use `ggplot2::cut_width()` on `DBH_2in`. You'll likely need to consult the `cut_width()` manual page to define its `boundary` and `closed` arguments appropriately.

```
elk_ags
```

```
#> # A tibble: 43 x 6
#>    Species DBH_2in DBH_2in_range Trees_ac BA_ac Logs_ac
#>    <chr>     <dbl> <fct>            <dbl> <dbl>   <dbl>
#>  1 Beech        12 [11,13)         1.33   1.04    1.33
#>  2 Beech        14 [13,15)         0.585  0.625   0.682
#>  3 Beech        16 [15,17)         0.448  0.625   0.746
#>  4 Beech        20 [19,21)         0.0955 0.208   0.143
#>  5 Black ~      12 [11,13)         1.06   0.833   1.06
#>  6 Black ~      14 [13,15)         0.974  1.04    1.07
#>  7 Black ~      16 [15,17)         0.895  1.25    1.42
#>  8 Black ~      18 [17,19)         0.118  0.208   0.236
#>  9 Black ~      20 [19,21)         0.0955 0.208   0.191
#> 10 Black ~      12 [11,13)         0.796  0.625   0.796
#> # i 33 more rows
```

Exercise 8.6. Build a stand table with species along rows, DBH classes along columns (using `DBH_2in_range` created in Exercise 8.5), and trees per acre as values. Round the stand table's values to two digits.

In general, rounding should always be done after all calculations are complete. As covered in Section 7.11, an efficient way to round all numeric columns in your completed tibble is to pipe it into `mutate(across(where(is.numeric), ~ round(.x, digits = 2)))`. This call to `mutate()` applies a lambda syntax *anonymous function*—a function without a name—defined as `~`

`round(.x, digits = 2)` to each numeric column (see Section 7.11 for more details). Here, the anonymous function is `base::round()` with our desired number of digits set to 2. If you're happy with `round()`'s default `digits = 0`, then there's no need for the anonymous function and the call to `mutate()` is simply `mutate(across(where(is.numeric), round))`. Anonymous functions are only needed when you want to define your own function or specify non-default argument values in existing functions.

The stand table's first several columns are printed below, and the entire table can be found in `"datasets/tidyr_exr/tidyr_elk_ags_stand_table.tsv"`. Notice your row and column totals should equal the `sum_trees_ac` values found in Exercises 8.3 and 8.4, respectively.

```
#> # A tibble: 10 x 12
#>    Species       `[11,13)` `[13,15)` `[15,17)` `[19,21)`
#>    <chr>             <dbl>     <dbl>     <dbl>     <dbl>
#>  1 Beech              1.33      0.58      0.45      0.1
#>  2 Black Birch        1.06      0.97      0.9       0.1
#>  3 Black Cherry       0.8       1.17      1.19      0.38
#>  4 Cucumber Tr~       0         0.19      0.15      0.1
#>  5 Northern Re~       0         0.19      0         0
#>  6 Red Maple          4.51     10.9       5.07      1.91
#>  7 Sugar Maple        0.53      1.17      0.45      0
#>  8 Yellow Birch       0         1.17      0.15      0
#>  9 Yellow Popl~       0         0         0         0
#> 10 Totals             8.22     16.4       8.36      2.58
#> # i 7 more variables: `[17,19)` <dbl>,
#> #   `[21,23)` <dbl>, `[23,25)` <dbl>, `[27,29)` <dbl>,
#> #   `[29,31]` <dbl>, `[25,27)` <dbl>, Totals <dbl>
```

Exercise 8.7. Build a stock table with species along rows, DBH classes along columns (using `DBH_2in_range` created in Exercise 8.5), and basal area per acre as values. Round the stock table's values to two digits (see Exercise 8.6 for guidance on rounding column values).

Hint, it should take minimal changes to your Exercise 8.6 stand table workflow to produce the desired stock table.

The first several stock table columns are printed below, and the entire table can be found in `"datasets/tidyr_exr/tidyr_elk_ags_ba_stock_table.tsv"`. Notice your row and column totals should equal the `sum_ba_ac` values found in Exercises 8.3 and 8.4, respectively.

```
#> # A tibble: 10 x 12
#>    Species       `[11,13)` `[13,15)` `[15,17)` `[19,21)`
#>    <chr>             <dbl>     <dbl>     <dbl>     <dbl>
#>  1 Beech              1.04      0.62      0.62      0.21
#>  2 Black Birch        0.83      1.04      1.25      0.21
#>  3 Black Cherry       0.62      1.25      1.67      0.83
#>  4 Cucumber Tr~       0         0.21      0.21      0.21
#>  5 Northern Re~       0         0.21      0         0
```

```
#>   6 Red Maple         3.54      11.7        7.08       4.17
#>   7 Sugar Maple       0.42       1.25       0.62       0
#>   8 Yellow Birch      0          1.25       0.21       0
#>   9 Yellow Popl~      0          0          0          0
#> 10 Totals            6.46      17.5       11.7        5.62
#> # i 7 more variables: `[17,19)` <dbl>,
#> #   `[21,23)` <dbl>, `[23,25)` <dbl>, `[27,29)` <dbl>,
#> #   `[29,31]` <dbl>, `[25,27)` <dbl>, Totals <dbl>
```

Exercise 8.8. Build a stock table with species along rows, DBH classes along columns (using `DBH_2in_range` created in Exercise 8.5), and logs per acre as values. Round the stock table's values to two digits (see Exercise 8.6 for guidance on rounding column values).

The first several stock table columns are printed below, and the entire table can be found in `"datasets/tidyr_exr/tidyr_elk_ags_logs_stock_table.tsv"`. Notice your row and column totals should equal the `sum_logs_ac` values found in Exercises 8.3 and 8.4, respectively.

```
#> # A tibble: 10 x 12
#>    Species        `[11,13)` `[13,15)` `[15,17)` `[19,21)`
#>    <chr>              <dbl>     <dbl>     <dbl>     <dbl>
#>  1 Beech               1.33      0.68      0.75      0.14
#>  2 Black Birch         1.06      1.07      1.42      0.19
#>  3 Black Cherry        0.8       1.56      2.39      0.76
#>  4 Cucumber Tr~        0         0.29      0.37      0.29
#>  5 Northern Re~        0         0.19      0         0
#>  6 Red Maple           4.24     14.3       8.36      3.58
#>  7 Sugar Maple         0.53      1.36      0.75      0
#>  8 Yellow Birch        0         1.27      0.22      0
#>  9 Yellow Popl~        0         0         0         0
#> 10 Totals              7.96     20.8      14.2       4.97
#> # i 7 more variables: `[17,19)` <dbl>,
#> #   `[21,23)` <dbl>, `[23,25)` <dbl>, `[27,29)` <dbl>,
#> #   `[29,31]` <dbl>, `[25,27)` <dbl>, Totals <dbl>
```

Exercise 8.9. Take your wide messy stock table from Exercise 8.8, remove the total row and column, then reformat it to be long and tidy. The piped workflow will likely include a call to `filter()` or `slice()`, `select()`, and `pivot_longer()`. Set the `pivot_longer()` function's `names_to = "DBH_2in_range"` and `values_to = "Logs_ac"`.

Note, the resulting tibble will be like `elk_ags` but with `Logs_ac` summed within species and DBH class, and with explicit zero rows added for missing species and DBH class combinations. Given there are 9 species and 10 DBH classes, the resulting tibble should have 90 rows.

```
#> # A tibble: 90 x 3
#>    Species DBH_2in_range sum_logs_ac
#>    <chr>   <chr>               <dbl>
```

```
#>  1 Beech   [11,13)            1.33
#>  2 Beech   [13,15)            0.68
#>  3 Beech   [15,17)            0.75
#>  4 Beech   [19,21)            0.14
#>  5 Beech   [17,19)            0
#>  6 Beech   [21,23)            0
#>  7 Beech   [23,25)            0
#>  8 Beech   [27,29)            0
#>  9 Beech   [29,31]            0
#> 10 Beech   [25,27)            0
#> # i 80 more rows
```

As you'll see in the next chapter, this long and tidy format is used for most calls to `ggplot2` functions to make informative and beautiful graphics.

9

Creating graphics with `ggplot2`

Visualizing raw data and data summaries as graphics is key to EDA and to communicating analysis results. This chapter introduces the `ggplot2` package, which allows you to efficiently create beautiful and informative graphics. Unlike R's `base` graphics and other graphics packages that focus on providing functions for specific types of graphics (e.g., scatter plot, histogram, line graph), `ggplot2` takes a more flexible approach by defining and applying a *grammar* for constructing graphics.[1]

Like other `tidyverse` packages, Hadley Wickham is the lead author and developer of `ggplot2`. He based its grammar on earlier work in the book *Grammar of Graphics* (Wilkinson 2005), which lays out a grammar to describe the fundamental features underlying all statistical graphics. The `gg` in `ggplot2` stands for *Grammar of Graphics*. Wickham (2010) defines the grammar used by `ggplot2` and maps its components to those proposed in Wilkinson (2005).

The authoritative reference for learning `ggplot2` is the book *`ggplot2`: Elegant Graphics for Data Analysis* (Wickham, Navarro, et al. 2023) and its extensive online reference.[2] The aim of this chapter is to provide a brief introduction to the grammar components and how they work together to construct a graphic. This introduction will be done mostly through worked examples. We borrow heavily from the first chapter of Wickham, Navarro, et al. (2023) and refer you to this excellent book to further develop your understanding. Additional learning resources are provided in Section 9.5.

9.1 Grammar components

Grammar, according to Johnson (1969), is the "art of expressing relations of things in construction." In the `ggplot2` context, "things" are data or data summaries visualized using geometric objects (e.g., points, lines, bars). "Expressing relations" between data values is done using the geometric object's aesthetic attributes (e.g., location, color, shape, size). The geometric objects are set within a one- or two-dimensional space defined using some coordinate system. In this way, we are "constructing" a graphic to convey relationships within and across data or data summaries.

[1]In the current context, *grammar* is best defined as the "art of expressing relations of things in construction" (Johnson 1969) or as the "fundamental principles or rules of an art or science" (OED 2023).

[2]A free online version of this book is available at https://ggplot2-book.org and the reference is found at https://ggplot2.tidyverse.org.

DOI: 10.1201/9781003222699-9

The construction proceeds in layers, where each layer depicts one or more features of the *data* we wish to visualize. The grammar components include the following.

- *geometries* define the geometric objects seen in the graphic (e.g., points, lines, bars). Each geometry has a set of aesthetic attributes that map to data variable values. Different geometries can have different aesthetics.
- *mappings* connect the data's variables to aesthetic attributes of the geometric object (e.g., location, color, shape, size).
- *scales* translate a variable's values to visual representations (and vice versa). For example, scales assign a categorical variable's values to colors or shapes, or a continuous variable's values to locations in the graphic's one- or two-dimensional space.
- *statistical transformations* (*stats* for short) summarize the data's variables in some way. For example, if the desired graphic is a histogram of the number of trees by species, then stats summarize the input data to create the necessary tree count by species variables. Or a stat might discretize a continuous variable's values so they can be visualized as ordered categories (e.g., creating DBH classes like we saw in Section 8.3).
- *facets* display a common graphic across panels, each based on data subsets defined by one or more discrete variables. These are useful for visualizing how a variable's values differ across the discrete variable(s) used to define the facets.
- *coordinates* define the graphic's coordinate system into which the geometric objects are placed. Typically, this is a Cartesian coordinate system defined using a horizontal x-axis and a vertical y-axis, but other options exist (e.g., polar coordinates used for pie charts).
- *themes* control the non-data elements of the graphic, such as font size, text placement, legend orientation, and background color. While the defaults in `ggplot2` have been chosen with care, you may need to consult other references to create an attractive plot.

As we'll see in the next few sections, a graphic may comprise multiple layers, with each depicting different features of the data. Often, data are shared among layers, though that is not always the case. The next section provides some worked examples to help you get used to `ggplot2` syntax.

9.2 An initial look

We motivate this initial `ggplot2` exploration using the FEF dataset introduced in Section 1.2.1. Recall the FEF data were collected to create allometric equations that foresters could use to estimate tree component weight based on species and DBH. Here, we'll work with a simplified version of the FEF data provided in `"datasets/FEF_trees_simple.csv"`. In this file, each row corresponds to a set of tree measurements, and columns are as follows:

1. `Species` tree species scientific name.
2. `DBH_cm` DBH in centimeters.
3. `Height_m` total tree height in meters.
4. `Stem_kg` dry weight of the main stem in kilograms.
5. `Otherwoody_kg` dry weight of top and branch tree components in kilograms.

The code below loads the `readr` and `ggplot2` packages, reads the simplified FEF data into `fef`, and prints the first few rows and columns.

```
library(readr)
library(ggplot2)

fef <- read_csv("datasets/FEF_trees_simple.csv")
fef
```

```
#> # A tibble: 88 x 5
#>    Species      DBH_cm Height_m Stem_kg Otherwoody_kg
#>    <chr>         <dbl>    <dbl>   <dbl>         <dbl>
#>  1 Acer rubrum   15.2     14.6     54.7          50.4
#>  2 Acer rubrum   17.5     14.6     62.3          60.8
#>  3 Acer rubrum   16.3     14.6     73.3          41.2
#>  4 Acer rubrum   16.5     14.9     53.6          52.3
#>  5 Acer rubrum   18.3     15.5    106.           52.5
#>  6 Acer rubrum    7.87    12.2     11.7           2.5
#>  7 Acer rubrum    5.08     9.30     3.2           2  
#>  8 Acer rubrum   10.4     15.2     28.3          10.1
#>  9 Acer rubrum    6.10     8.53     5.5           2.5
#> 10 Acer rubrum    6.86    12.3     10.6           4.5
#> # i 78 more rows
```

The next several sections provide an applied look at each grammar component's role in constructing a basic graphic.

9.2.1 Geometries, mapping, and aesthetics

Scatter plots are a workhorse of data visualization and provide a good entry point to learning `ggplot2`. Let's begin by making a scatter plot of DBH versus stem weight using the `fef` data.

Code for all `ggplot2` graphics starts with `ggplot()`, followed by layers and modifying functions added with the + operator. Use of the + operator underscores the constructive nature of `ggplot2` graphics.

The code below includes a call to `ggplot()`, with the `data` argument set to `fef` and the `mapping` argument set to `aes()`, which is used to connect variables with the anticipated geometry's aesthetics.[3] Here, aesthetic `x` is mapped to `DBH_cm` and aesthetic `y` is mapped to `Stem_kg`. By itself, the call to `ggplot()` doesn't define the type of graphical display to draw; hence, the resulting graphic is empty except for the axes defined by the `x` and `y` aesthetics. The *scales* grammar component is at work here, applying defaults to set axis extents and value breaks, and determining many other visual features.

```
ggplot(data = fef, mapping = aes(x = DBH_cm, y = Stem_kg))
```

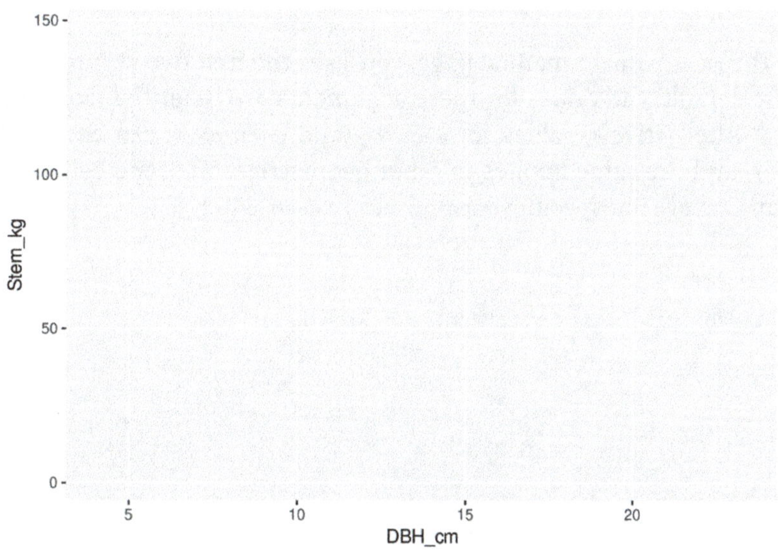

To draw objects in the graph plane, we need to specify a geometry that will do something with the supplied `x` and `y` aesthetics.[4] The desired scatter plot has points drawn at `x` and `y` value pairs. Point geometries are drawn using `geom_point()`. The code below adds (using the + operator) the point geometry.

```
ggplot(data = fef, mapping = aes(x = DBH_cm, y = Stem_kg)) +
  geom_point()
```

[3]See aesthetics manual page `?ggplot2::aes`.

[4]Take a look under the Geoms section at https://ggplot2.tidyverse.org/reference to get a sense of the variety of geometries that can be drawn.

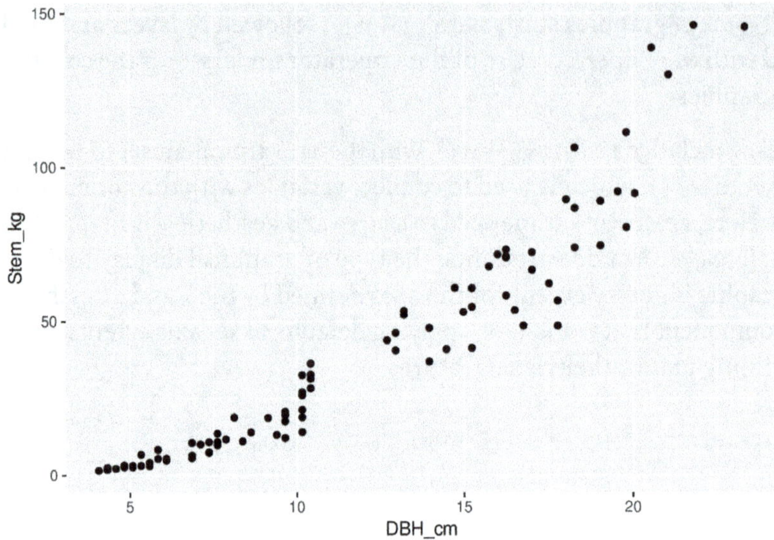

Looking at the `geom_point()` manual page, you'll see the first two arguments are `mapping` and `data`. By default, the values for these arguments are inherited from the initial call to `ggplot()`.[5] Alternatively, values for `mapping` (and even `data`) can be specified within `geom_point()`, and the same applies to other geometries. The code below produces the same output as above but specifies `mapping` within `geom_point()`.

```
ggplot(data = fef) +
  geom_point(mapping = aes(x = DBH_cm, y = Stem_kg))
```

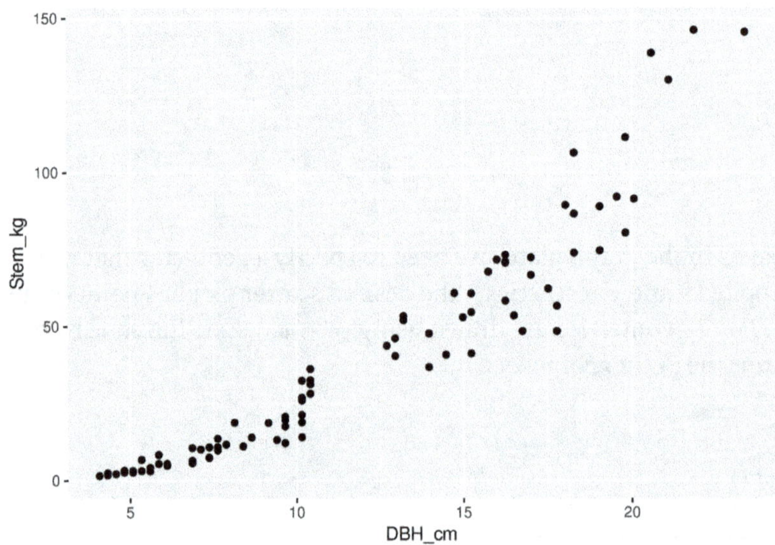

Return again to the `geom_point()` manual page and scroll down to the Aesthetics section. This section lists all the `aes()` arguments used to map variables to point geometry features.

[5]Access the manual page via `?ggplot2::geom_point()`.

Notice that x and y are required, but all other aesthetics are optional.

Let's map point color to `Species`. To do so, we include `color = Species` in the point geometry's `aes()` in the code below. Notice that in the resulting graphic, a color scale for species and an associated legend are automatically generated. `Species` is a categorical variable; hence, *scales* selects a default discrete color scale. These often very reasonable default colors and legend placement decisions facilitate efficient exploration of data. Later, we'll show how to override these and other defaults.

```
ggplot(data = fef) +
  geom_point(mapping = aes(x = DBH_cm, y = Stem_kg, color = Species))
```

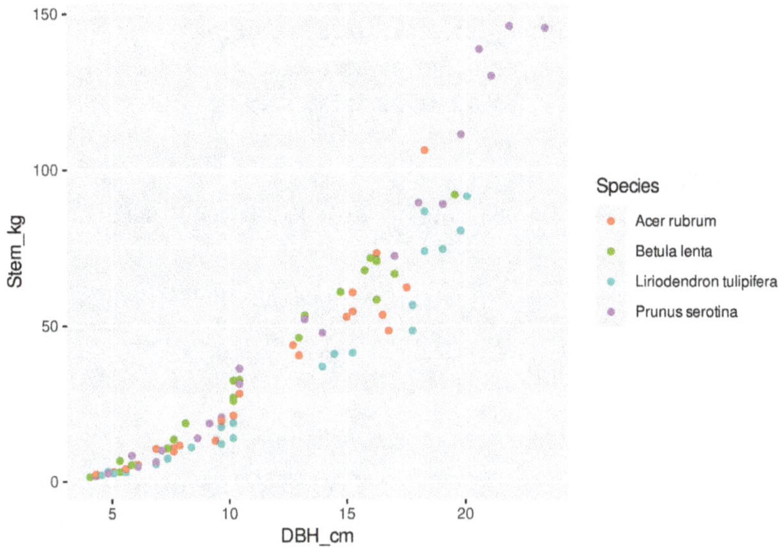

Recall from Section 4.3 that discrete variables (e.g., integers or characters) can be represented as factors. If you plan to represent integer or character variables as categories in your graphic, it's often useful to convert them to factors before sending them into `ggplot()`. One particularly motivating reason to convert discrete variables to factors is that it gives you finer control over how the categories are ordered and labeled in the resulting legends and other graphic elements.

As you'll notice in the graphic above, the default is to list species in alphabetical order within the legend. Suppose we would like them displayed in reverse alphabetical order. This can be done by converting `Species` to a factor and specifying the factor levels in the order we wish for them to appear in the legend.[6]

```
fef <- fef %>%
  mutate(Species = factor(Species,
```

[6]In Section 9.2.2, we show an alternative—and, in some situations, simpler—approach that uses scales to control discrete variable ordering in the graphic's elements.

```
                    levels = c("Prunus serotina",
                               "Liriodendron tulipifera",
                               "Betula lenta",
                               "Acer rubrum")))
fef %>%
  glimpse() # Check out the Species factor.
```

```
#> Rows: 88
#> Columns: 5
#> $ Species       <fct> Acer rubrum, Acer rubrum, Acer ~
#> $ DBH_cm        <dbl> 15.240, 17.526, 16.256, 16.510,~
#> $ Height_m      <dbl> 14.6304, 14.6304, 14.6304, 14.9~
#> $ Stem_kg       <dbl> 54.7, 62.3, 73.3, 53.6, 106.4, ~
#> $ Otherwoody_kg <dbl> 50.4, 60.8, 41.2, 52.3, 52.5, 2~
```

```
ggplot(data = fef) +
  geom_point(mapping = aes(x = DBH_cm, y = Stem_kg, color = Species))
```

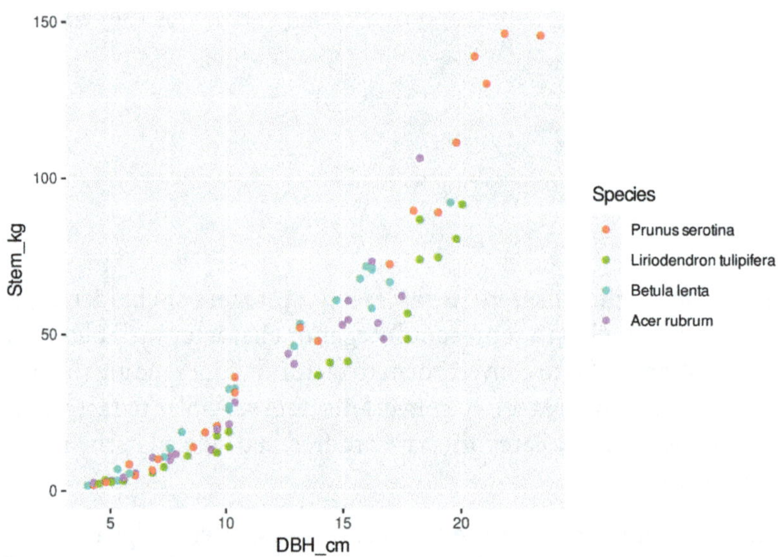

Next, let's map color to the continuous `Height_m` variable by setting `color = Height_m` in the point geometry's `aes()`.

```
ggplot(data = fef) +
  geom_point(mapping = aes(x = DBH_cm, y = Stem_kg, color = Height_m))
```

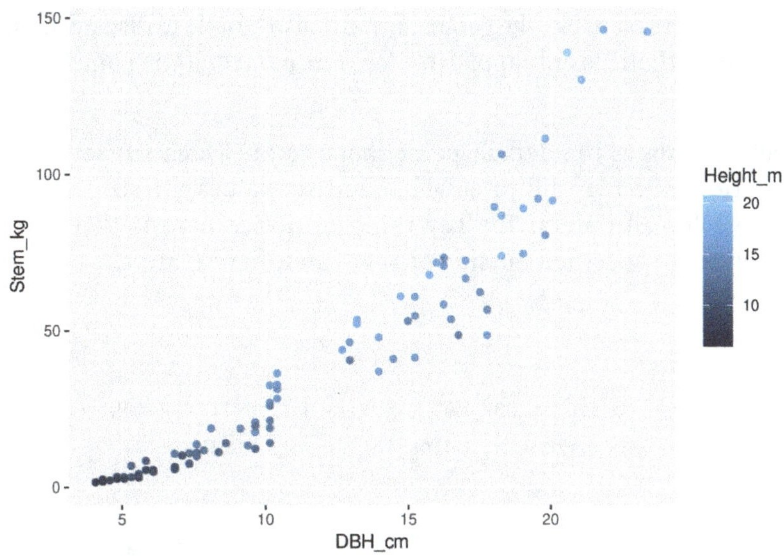

As you can see from the resulting legend colorbar, when the color aesthetic is mapped to a continuous variable (e.g., `Height_m`), *scales* assign each possible value in the variable's range to a color along a color gradient.[7] As expected, the graphic above shows that taller trees have larger DBH and stem weight.

As you've seen in the `geom_point()` manual page, there are many possible aesthetics to which we can map a variable. The aesthetic defaults are set up to handle *reasonable* mappings. Here, *reasonable* means that *scales* can make an interpretable mapping between the variable's type (i.e., categorical or continuous) and the aesthetic.

Above, we mapped species (categorical) to color and height (continuous) to color, and both resulted in reasonable mappings. Mapping species or height to the size aesthetic will also work by default[8]. The shape aesthetic is well suited for species, but it's not immediately apparent how to map shape to height, as suggested by the error thrown by the code below.

```
ggplot(data = fef) +
  geom_point(mapping = aes(x = DBH_cm, y = Stem_kg, shape = Height_m))
```

```
#> Error in `geom_point()`:
#> ! Problem while computing aesthetics.
#> i Error occurred in the 1st layer.
#> Caused by error in `scale_f()`:
#> ! A continuous variable cannot be mapped to the shape
#>   aesthetic.
#> i Choose a different aesthetic or use
#>   `scale_shape_binned()`.
```

[7]See what happens when you force `Height_m` to be treated as a discrete variable, i.e., set `color = factor(Height_m)`.

[8]Although representing species with point size is not very meaningful.

In addition to being reasonable, the geometry must also be able to display the aesthetic. For example, the fill aesthetic is only applicable for `geom_point()` if the point shape supports a fill aesthetic.[9]

The code below changes the default point shape to 21 (which supports color, fill, and stroke aesthetics). We map fill to `Height_m` and set values for size, color, and stroke to be the same for all points. The key thing to notice here is that size, color, and stroke aesthetics are specified outside of `aes()` and, hence, are not mapped to a data variable.

```
ggplot(data = fef) +
  geom_point(mapping = aes(x = DBH_cm, y = Stem_kg, fill = Height_m),
             shape = 21, size = 5, color = "black", stroke = 2)
```

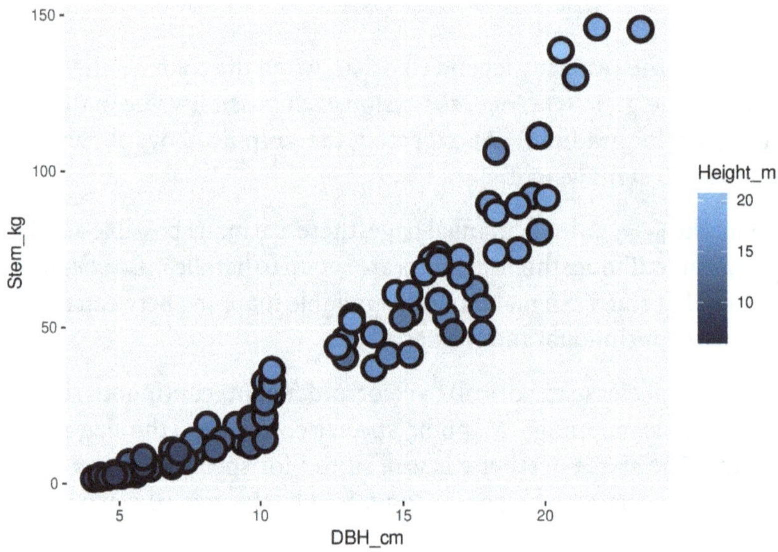

Next, let's map height to color and species to shape. Note that we also change the size of the points to a static value.

```
ggplot(data = fef) +
  geom_point(mapping = aes(x = DBH_cm, y = Stem_kg, color = Height_m,
                           shape = Species), size = 2)
```

[9]See `vignette("ggplot2-specs")` to learn about each geometry's aesthetics. For example, find the *Point* section in the vignette and read about the various values for `shape` and how `color`, `fill`, and `stroke` may—or may not—apply to each shape.

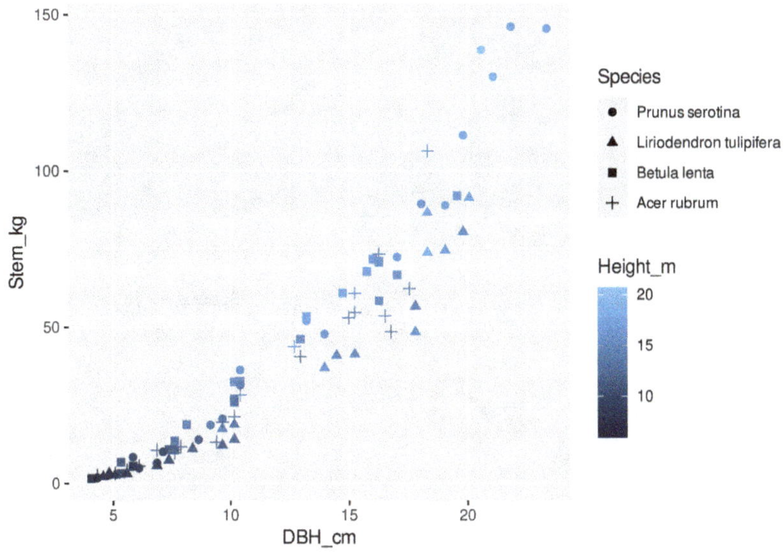

While the figure above compactly communicates several different relationships—via color and shape scales—consideration should always be given to the audience. It takes time and thought to create clear and concise graphics that efficiently communicate information. Always keep your audience in mind. What are the key points you want to communicate and are those points clearly portrayed in the graphic? How hard does the audience have to work to understand what you're trying to communicate? Would multiple graphics more clearly communicate key information? In some cases, portraying multiple scales in a single graphic is the perfect way to communicate information; in other cases, it might be preferable to offer scale-specific graphics.

The `ggplot2` package is a tool for a broad range of tasks. At one extreme, the package offers functionality to produce publication-quality, sophisticated, and beautiful graphics; at the other extreme, it's a tool for rapid data exploration used to identify errors, features, patterns, and possible avenues for analysis. In the latter extreme, there's an audience of one—yourself—and the focus is on efficient data exploration using as little code as possible.

You can create new variables directly within `aes()`. For example, say we wanted to compare *Prunus serotina* to other species in the scatter plot. Math and logical operators are evaluated on variables in `aes()` prior to assigning them to the aesthetic. This is illustrated in the code below, where the logical `Species == "Prunus serotina"` result is assigned to `color`, and a discrete scale is rendered.[10]

```
ggplot(data = fef) +
  geom_point(mapping = aes(x = DBH_cm, y = Stem_kg,
                           color = Species == "Prunus serotina"),
             size = 3)
```

[10]Try this with a continuous variable; e.g., identify all trees over 15 meters using `aes(x = DBH_cm, y = Stem_kg, color = Height_m > 15)`.

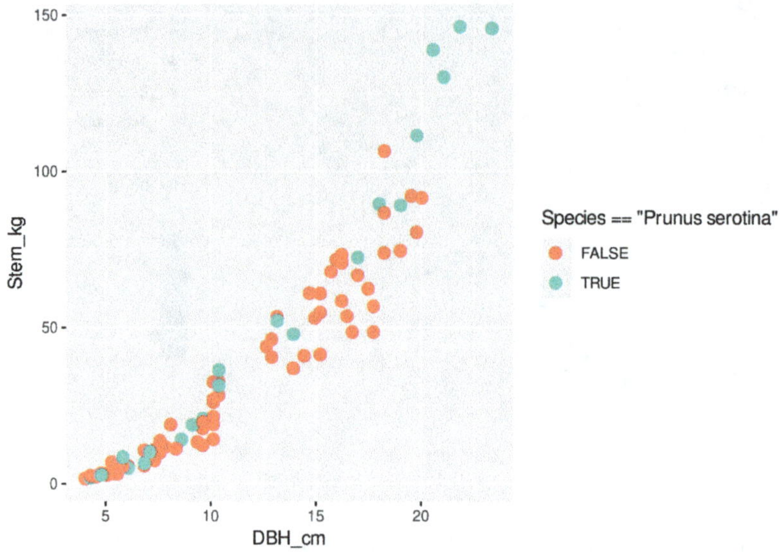

The scatter plot above makes it instantly clear that *Prunus serotina* was sampled fairly evenly across the DBH range, while other species were not sampled beyond about 20 cm DBH and 125 kg stem weight. Notice the efficiency of creating new variables within `aes()` compared with creating a new variable and adding it to `fef` via `mutate()` prior to calling `ggplot()`. The ability to create new variables within `aes()` on the fly, combined with good defaults for a graphic's numerous visual features, is key to efficient EDA.

Additional geometry layers are added with the `+` operator and inherit data and aesthetic mappings by default. For example, let's add a smooth line geometry layer over the points. The code below uses `geom_smooth()` to add the desired line. There are many steps happening behind the scenes here. For example, the `stats` package is fitting a statistical model to the `x` and `y` mappings to estimate the smooth trend line with an associated measure of uncertainty. Here, the trend line appears in the default color blue, with the confidence interval band in gray.

```
ggplot(data = fef, mapping = aes(x = DBH_cm, y = Stem_kg)) +
  geom_point() +
  geom_smooth()
```

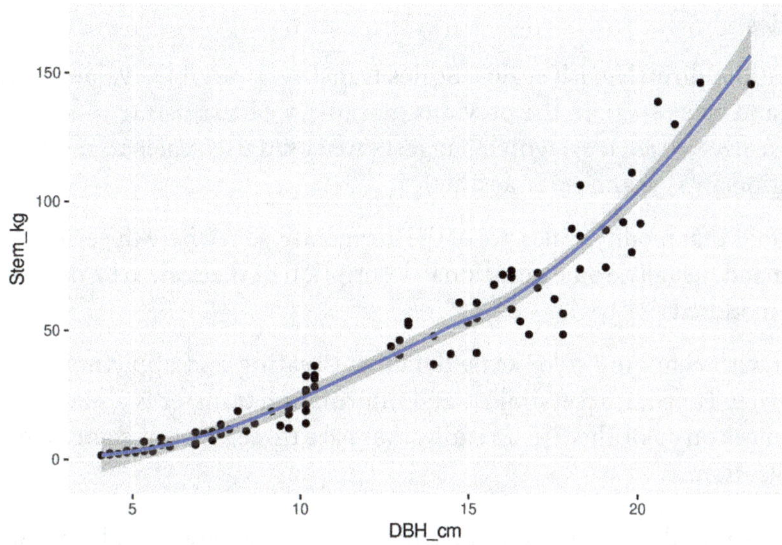

Notice the smooth line above is layered over the points. This layer order occurs because `geom_smooth()` is added after `geom_point()`—simply switch the layer order if you'd like the points to appear over the line.

The code below places the smooth line beneath the points, adds the species `color` aesthetic, and removes the confidence interval band by setting the `se` argument in `geom_smooth()` to `FALSE`. Notice the `color` aesthetic is now applied to both the point and line geometries, such that a separate smooth line is produced for each species.

```
ggplot(data = fef,
       mapping = aes(x = DBH_cm, y = Stem_kg, color = Species)) +
  geom_smooth(se = FALSE) +
  geom_point()
```

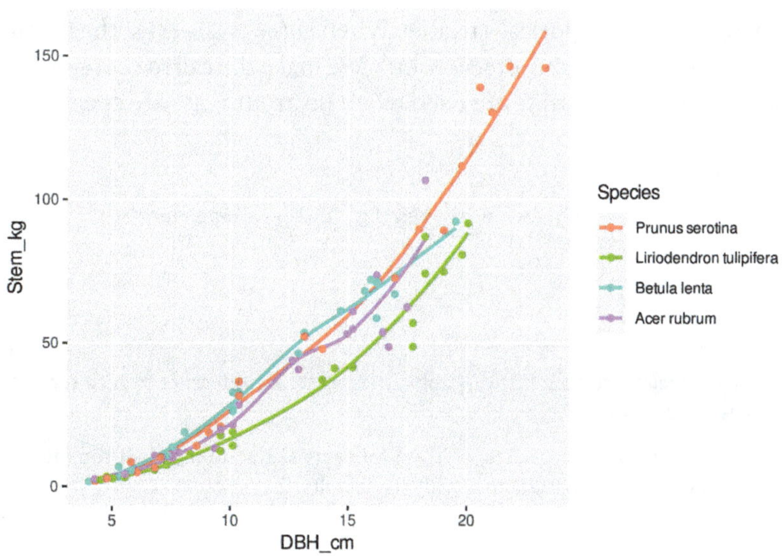

9.2.2 Scales

Let's work more directly with scales. Scales translate a variable's values to visual representation (and vice versa). In the previous example, we had `mapping = aes(x = DBH_cm, y = Stem_kg, color = Species)`, which suggests we could use scales to modify the default representation for `x`, `y`, and `color` aesthetics.

Most functions that modify scales are of the form `scale_*()`, where the asterisk is the name of the scale and, usually, some additional description of the scale type or how you intend for it to be modified.[11]

Let's begin with changing color scale defaults. Creating and applying color palettes to maximize visual appeal, accessibility, and information transfer is a science in itself (see, e.g., references on color theory). Broadly, there are three types of color palettes used for data visualization.

- **Qualitative** used for discrete unordered numeric or categorical variables. Discrete values are paired with a color, and all colors receive the same interpretive weight. Colors are typically chosen to be distinct from one another to emphasize the lack of natural ordering and improve visual differentiation among values.
- **Sequential** used for continuous numeric or discrete ordered numeric or categorical variables. Continuous numeric values are represented with a color gradient depicting a continuum from high to low or low to high. Similarly, discrete ordered values are paired with colors in a color gradient, with colors chosen to depict the incremental value change from one discrete value to the next.
- **Divergent** used for continuous numeric or discrete ordered numeric or categorical variables that increase and decrease from a neutral or central value. Values are represented using two diverging color gradients. The gradients start at a specified central value and extend to two extremes.

`Species` is an unordered categorical variable. When `color = Species`, the default color scale is implicitly set with `scale_color_discrete()`.[12] We make the call to `scale_color_discrete()` explicit below; again, the defaults are used when no arguments are specified.

```
ggplot(data = fef,
       mapping = aes(x = DBH_cm, y = Stem_kg, color = Species)) +
  geom_point() +
  scale_color_discrete()
```

---

[11]Take a look in the *Scales* section at https://ggplot2.tidyverse.org/reference to get a sense of the functions used to modify aesthetic scales.

[12]In general, there is an equivalent `scale_fill_*()` for every `scale_color_*()`. If you map a variable to the `fill` aesthetic, use `scale_fill_discrete()` to modify its defaults.

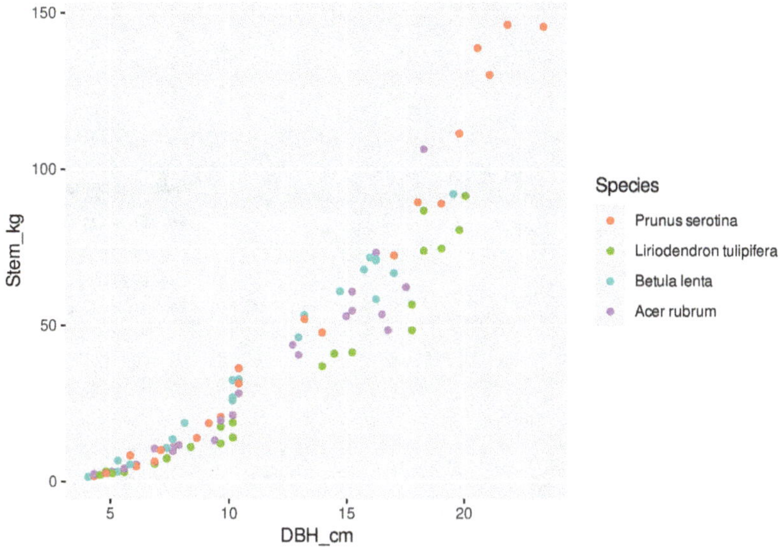

As illustrated in the code and output above, defaults are used if no arguments are passed into the scale function. Let's take a look at the `scale_color_discrete()` manual page, via `?ggplot2::scale_color_hue`, to get a sense of defaults that can be changed.[13]

Scroll down in the manual page and read about the `name` argument. All scale functions have a `name` argument, and it's used to provide a better title for the given aesthetic in the graphic. The default for `name` is the aesthetic's variable name. For example, the variable name mapped to `color` is used as the color legend title, and variables mapped to `x` and `y` are used for the respective axis titles (you can see this in the graphic above where `x` and `y` titles are *DBH_cm* and *Stem_kg*, respectively, and the color legend title is *Species*).

The code below changes the color scale name from its default to "Scientific name".

```
ggplot(data = fef,
       mapping = aes(x = DBH_cm, y = Stem_kg, color = Species)) +
  geom_point() +
  scale_color_discrete(name = "Scientific name")
```

[13]Note when you go to the `scale_color_discrete()` manual page via `?ggplot2::scale_color_discrete`, you learn that `scale_color_discrete()` is actually calling `scale_color_hue()`, hence the reason we point you to the `scale_color_hue()` manual page.

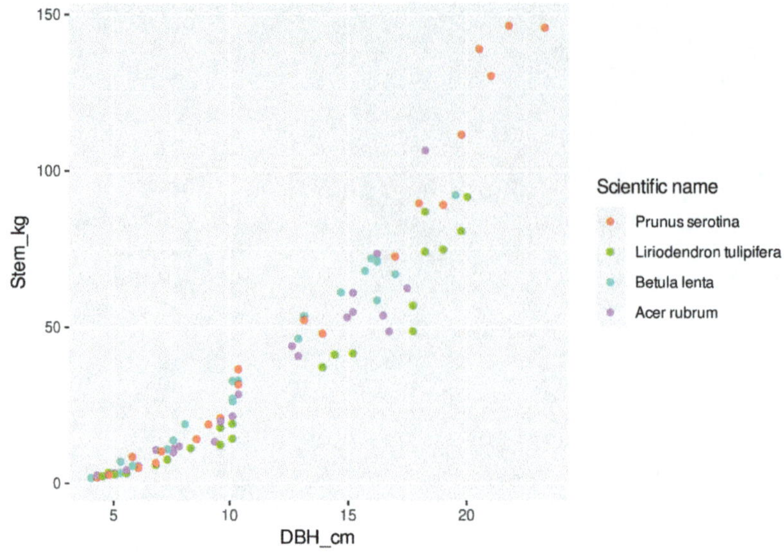

Next, let's modify the order in which species appear in the legend. We saw earlier that order can be changed through the factor data type specification, but it can also be changed using the scale function's `limits` argument. Returning to the manual page, notice that an optional value for `limits` is "a character vector that defines possible values of the scale and their order."

We specify our desired species order below in the vector assigned to `limits`. For fun, we also remove the legend title by setting `name` to `NULL` (a title is not really needed here, as species names and associated colors make the legend's role self-evident).

```
ggplot(data = fef,
       mapping = aes(x = DBH_cm, y = Stem_kg, color = Species)) +
  geom_point() +
  scale_color_discrete(name = NULL,
                       limits = c("Acer rubrum", "Betula lenta",
                                  "Liriodendron tulipifera",
                                  "Prunus serotina"))
```

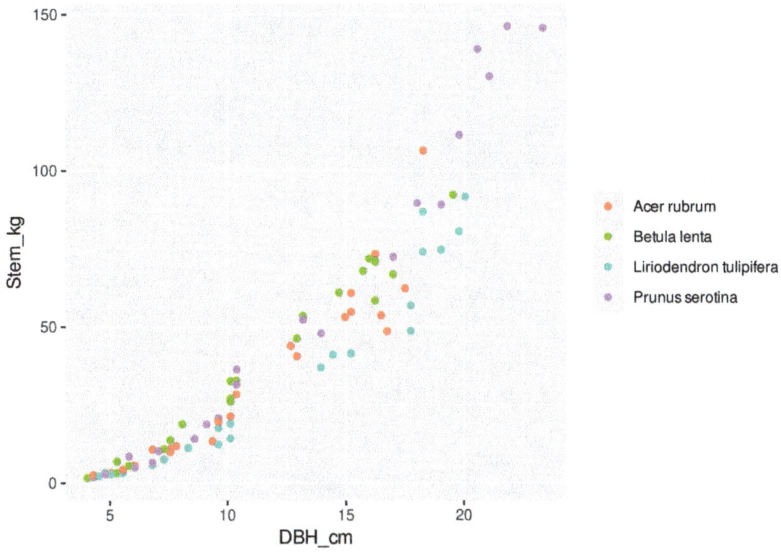

There are specialized scale functions available to simplify changing defaults and to provide expanded options for creating interesting and informative features. For example, in the code below, `scale_color_discrete()` is replaced with `scale_color_manual()`, which is used to set a qualitative color palette we define in `spp_col`.[14]

Our chosen color palette is passed to `scale_color_manual()` via the `values` argument. Additionally, we again use the scale's `name` argument to change the default title "Species" to "Scientific name."

```
spp_col <- c("Acer rubrum" = "firebrick", "Betula lenta" = "plum",
             "Liriodendron tulipifera" = "aquamarine",
             "Prunus serotina" = "steelblue")

ggplot(data = fef,
       mapping = aes(x = DBH_cm, y = Stem_kg, color = Species)) +
  geom_point() +
  scale_color_manual(name = "Scientific name", values = spp_col)
```

[14]Recall, you can see R's built-in colors by running `colors()` on the console.

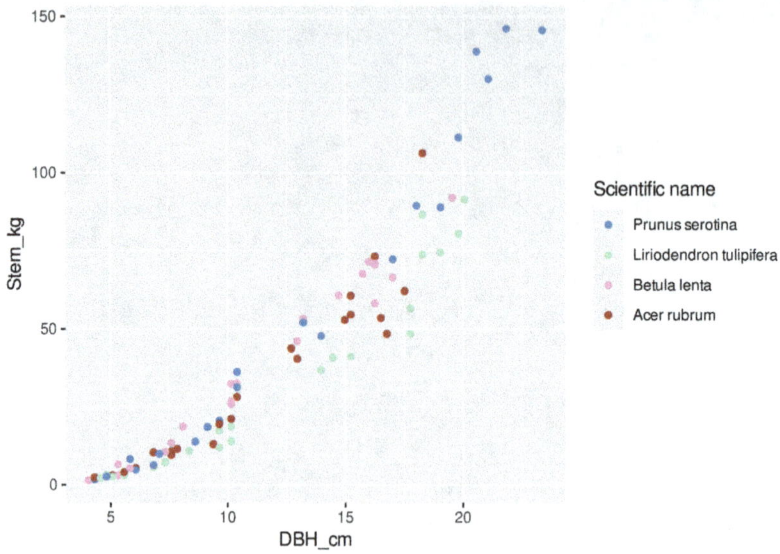

These colors are pretty nice; however, we might be better off leaving color palette selection to the experts. There are several functions within `ggplot2` that provide thoughtfully designed color palettes.

The code below uses `scale_color_brewer()` to provide palettes selected by Cynthia Brewer, a highly accomplished cartographer well known for her ColorBrewer palettes.[15]

The `palette` argument within `scale_color_brewer()` accepts a palette name. For example, we selected the *Set2* palette in the code below.[16]

```
ggplot(data = fef,
       mapping = aes(x = DBH_cm, y = Stem_kg, color = Species)) +
  geom_point(size = 3) +
  scale_color_brewer(name = "Latin name", palette = "Set2")
```

[15]See https://colorbrewer2.org for a visual tour of ColorBrewer palettes. The web application allows for visualization of different palettes and indicates which are colorblind-friendly and print well in black and white.

[16]See `?ggplot2::scale_color_brewer` for the list of palette names.

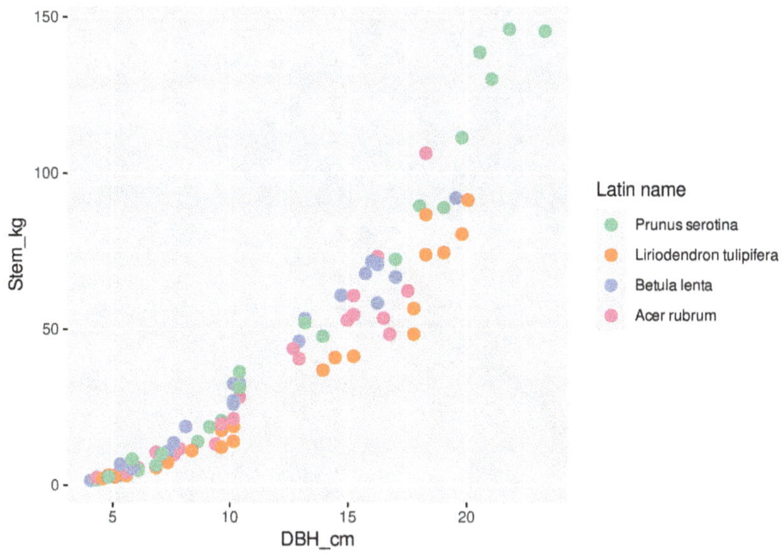

There are lots of exciting color palettes within `ggplot2` and available through R packages created by color enthusiasts. For example, we enjoy using palettes from the American filmmaker Wes Anderson's movies via the `wesanderson` package, illustrated below.[17]

```
library(wesanderson)

ggplot(data = fef,
       mapping = aes(x = DBH_cm, y = Stem_kg, color = Species)) +
  geom_point(size = 3) +
  geom_smooth(se = FALSE) +
  scale_color_manual(name = "Latin name",
                     values = wes_palette("AsteroidCity1"))
```

[17]You will need to install the `wesanderson` package by running `install.packages("wesanderson")` in your console before calling `library(wesanderson)`.

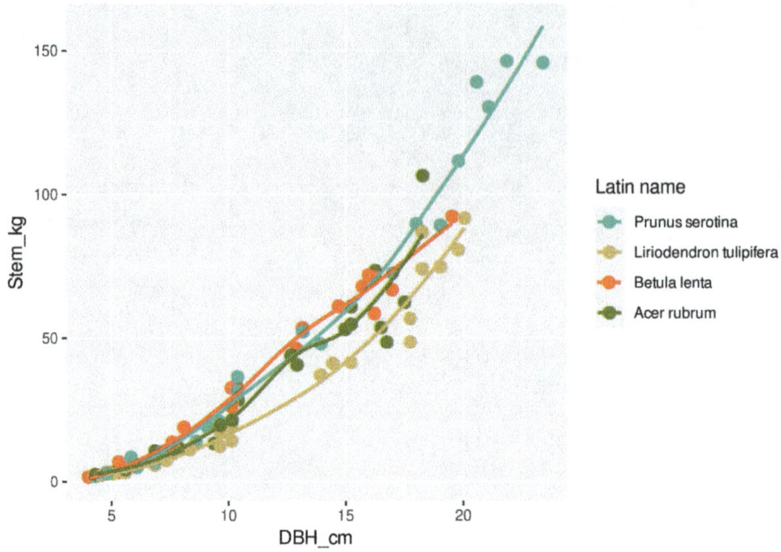

When a color aesthetic is mapped to a continuous variable, the variable's values are associated with colors in a sequential or divergent gradient. We saw such a mapping previously when `color = Height_m`, and the default colorbar gradient went from dark blue (small height values) to light blue (large height values).

We can select different built-in color gradients or create our own using a `scale_*()` function. For example, there are several excellent built-in continuous and divergent ColorBrewer palettes to choose from via `scale_color_brewer()` or `scale_fill_brewer()`. Alternatively, create your own using `scale_*_gradient()` for a two-color gradient (low-high), `scale_*_gradient2()` for a diverging color gradient (low-mid-high), or `scale_*_gradientn()` for an n-color gradient, where `*` is either `color` or `fill`.

The example code below uses `scale_fill_gradient()` to create a two-color gradient from hex color `#E28A00` to `#0201A5`.[18] Like many R functions, colors can be specified as quoted R color names or hex colors.

```
ggplot(data = fef,
       mapping = aes(x = DBH_cm, y = Stem_kg, fill = Height_m)) +
  geom_point(size = 5, shape = 21, color = "black") +
  scale_fill_gradient(low = "#E28A00", high = "#0201A5")
```

[18]A color hex code is a combination of alphanumeric characters beginning with the pound symbol that uniquely represents a color's mix of red, green, and blue. It's the most common way to digitally represent colors.

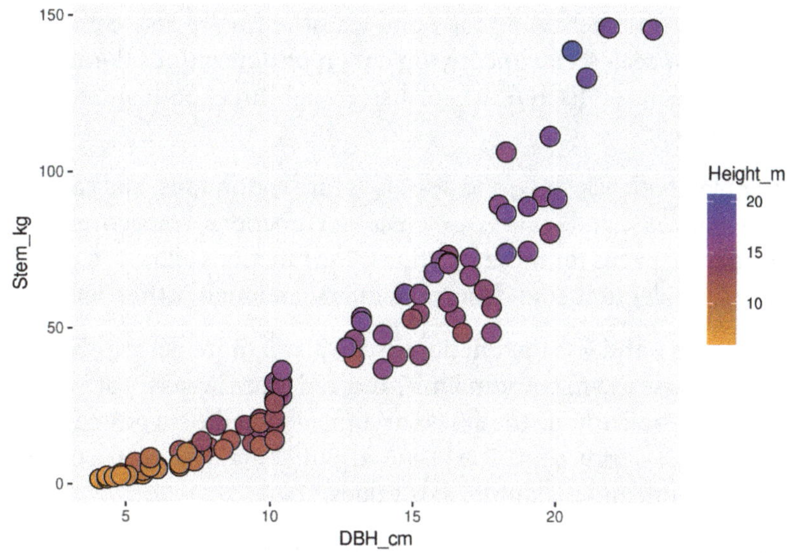

Just for fun, we'll try the `scale_fill_gradientn()` function, which builds a color gradient using any number of colors. The code below generates 10 colors from the `wesanderson` package *Zissou1* palette. In the code, we also change the colorbar name to "Height (m)" and add breaks at 7, 13.5, and 20 cm.

```
ggplot(data = fef,
       mapping = aes(x = DBH_cm, y = Stem_kg,fill = Height_m)) +
  geom_point(size = 5, shape = 21, color = "black") +
  scale_fill_gradientn(
    name = "Height (m)",
    breaks = c(7, 13.5, 20),
    colors = wes_palette(name = "Zissou1", n = 10, type = "continuous")
    )
```

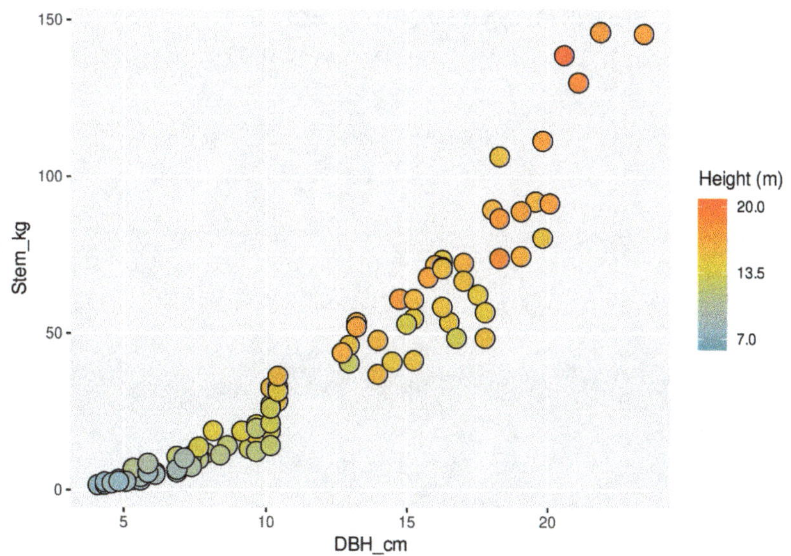

Changes to default scale behaviors for x and y aesthetics are also controlled with scale functions. Like color scales, the underlying data type determines the scale function. An axis is either continuous or discrete, where discrete subsumes both ordered and unordered categorical data types.

For our scatter plot, both DBH_cm (x) and Stem_kg (y) are continuous, and can be modified via the scale_x_continuous() and scale_y_continuous() functions, respectively. Take a moment to look at the manual pages for these functions. The functions allow you to change the axis name, tick mark break number and location, limits, and many other defaults.

The code below uses x and y scale functions to display a more descriptive name, force an origin at zero, increase the maximum limit, and add more breaks. For illustration, x-axis breaks are provided explicitly via the breaks argument, and the suggested number of y-axis breaks via the n.breaks argument. The limits argument takes a vector of length two that specifies the minimum and maximum axis values.

The default plot includes a bit of padding on the axis ends to ensure data are positioned away from the axis limits. You can change this default behavior using the expand argument, and can use the expansion() function to generate values to give to expand. We specify zero multiplicative and additive padding at the axis lower limit (i.e., the origin), and default padding at the axis upper limit (see ?ggplot2::expansion).

```
ggplot(data = fef,
       mapping = aes(x = DBH_cm, y = Stem_kg, color = Species)) +
  geom_point() +
  scale_x_continuous(name = "DBH (cm)",
                     breaks = seq(0, 26, 2), limits=c(0, 26),
                     expand = expansion(mult = c(0, 0.05),
                                        add = c(0, 0))) +
  scale_y_continuous(name = "Stem weight (kg)",
                     n.breaks = 6, limits = c(0, 160),
                     expand = expansion(mult = c(0, 0.05),
                                        add = c(0, 0)))
```

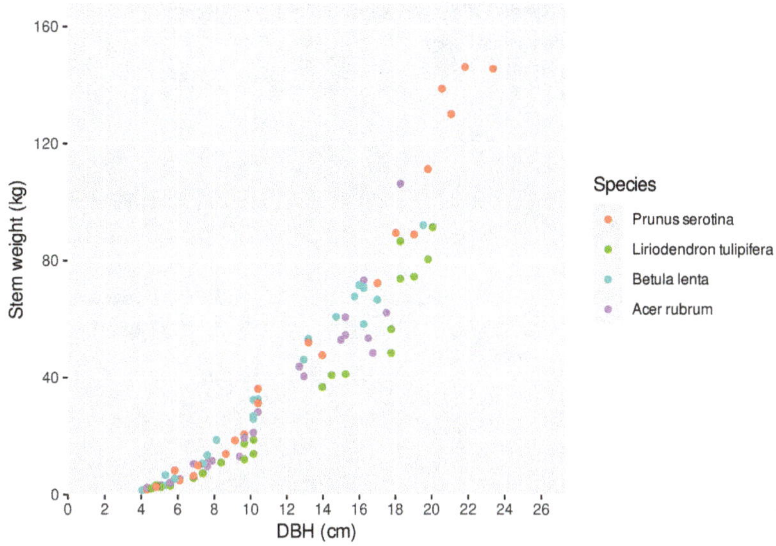

`xlim()` and `ylim()` provide shortcuts for specifying axis limits (e.g., we could use `xlim(0, 26)` in place of `scale_x_continuous(limits = c(0, 26))`). More generally, the `lims()` function is a shortcut for adjusting limits for any aesthetic.

Similarly, shortcuts are provided for axis scale names via `xlab()` and `ylab()`, and a general function `labs()` is used for any aesthetic. The code below uses `labs()` to change axis and color names. The `labs()` function can also be used to add descriptive text to other parts of the graphic, such as a title, subtitle, and caption, as illustrated below.[19]

```
ggplot(data = fef,
       mapping = aes(x = DBH_cm, y = Stem_kg, color = Species)) +
  geom_point() +
  labs(x = "DBH (cm)", y = "Stem weight (kg)", color = "Latin name",
       title = "Relationship between DBH and stem weight by species",
       subtitle = "USDA Fernow Experimental Forest",
       caption = "Data source: www.fs.usda.gov")
```

[19]The `ggtitle()` function can also be used to add a title, subtitle, and caption.

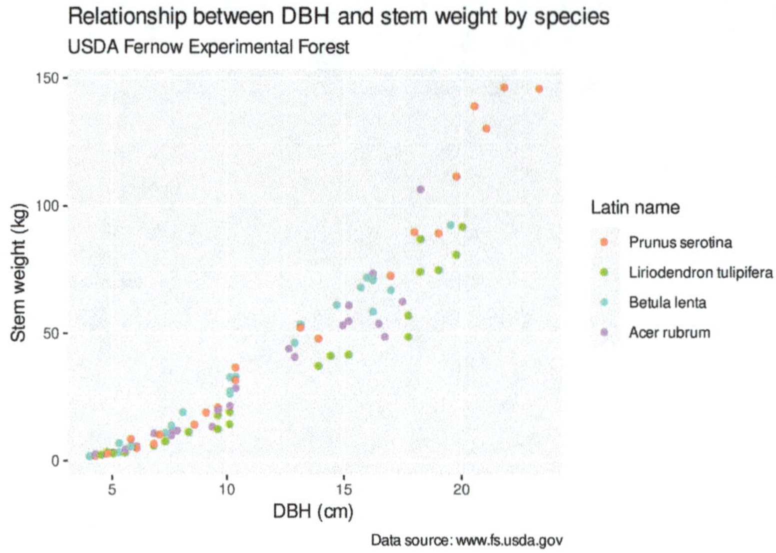

We often want to zoom into a region of a graphic to highlight a feature or get a better look at a relationship. There are two approaches to zooming.

1. Adjust axis scale limits to the data range of interest. This approach, however, can have unintended consequences and is not recommended. The unintended consequences occur because data outside an axis scale limit are removed from the data used to create the graphic, i.e., stats are computed using the visual data subset.

2. Set limits on the coordinate system. Setting limits on the coordinate system will zoom into the defined region (like you're looking at it with a magnifying glass) and will not change the underlying data. In our current example, we're using the Cartesian coordinate system, limits of which are set using `coord_cartesian()`.

The next two bits of code and output illustrate the zooming approaches and the unintended consequences of setting axis scale limits.

The smooth line in the graphic below shows the relationship between DBH and stem weight (along with an associated confidence interval) using the entire `fef` dataset.

```
ggplot(data = fef, mapping = aes(x = DBH_cm, y = Stem_kg)) +
  geom_point() +
  geom_smooth()
```

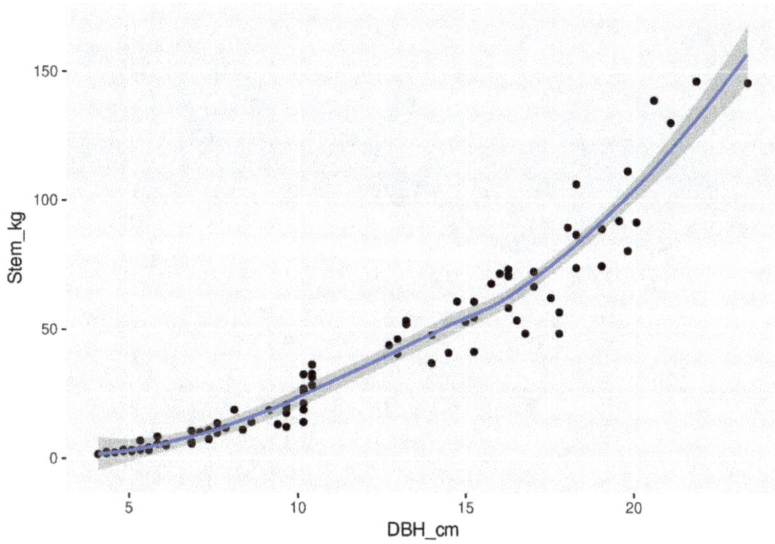

Say we want to get a better look at the relationship for trees with DBH between 18 and 24 cm. Using the first approach, we adjust axis scale limits using `xlim()` (we could also use the `limits` argument in `scale_x_continuous()`).

As the warning below the code suggests, this approach removes 74 `fef` rows. These removed rows are trees with `DBH_cm` outside the 18 to 24 cm range.[20] Hence, the smooth line is fit to only those trees within the limits, and the result looks very different from the relationship of interest in the graphic above.

```
ggplot(data = fef, mapping = aes(x = DBH_cm, y = Stem_kg)) +
  geom_point() +
  geom_smooth() +
  xlim(18, 24)
```

```
#> Warning: Removed 74 rows containing non-finite outside the scale
#> range (`stat_smooth()`).
```

```
#> Warning: Removed 74 rows containing missing values or values outside
#> the scale range (`geom_point()`).
```

[20]In fact, all 74 trees have DBH less than 18 cm, which you can prove to yourself via `sum(fef$DBH_cm < 18)` or `fef %>% filter(DBH_cm < 18) %>% nrow()`.

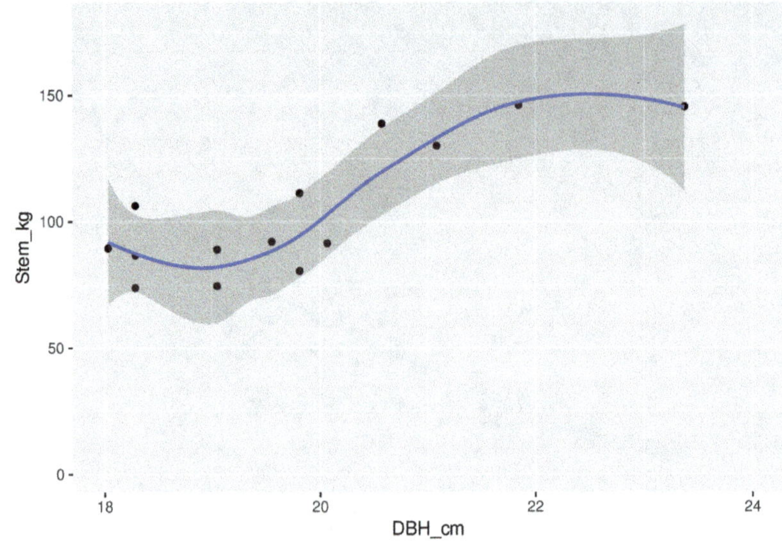

The code below uses the second and preferred approach. This is simply a zoom into the 18 to 24 cm region of the graphic and uses the full `fef` dataset to estimate the smooth line and its confidence interval.

```
ggplot(data = fef, mapping = aes(x = DBH_cm, y = Stem_kg)) +
  geom_point() +
  geom_smooth() +
  coord_cartesian(xlim = c(18, 24))
```

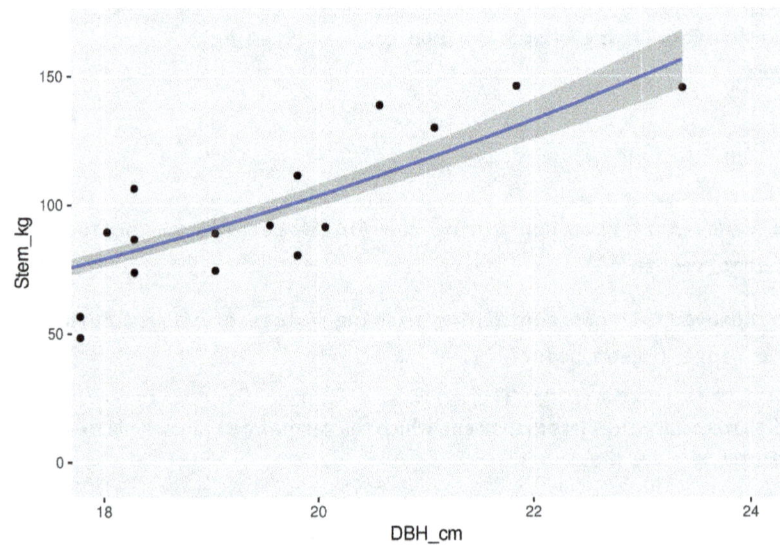

9.2.3 Guides and themes

There are many other features to consider when making visually appealing, informative, and accessible graphics. These include legend layout, size, placement, orientation, and font type, size, and placement, as well as the graphic's general look and feel. Fine-tuning scale defaults is accomplished via scale *guides*, whereas manipulating the graphic's long list of non-data elements is done via *themes*.

A scale's `guide` argument requires a guide object created using a guide function such as `guide_axis()`, `guide_legend()`, or `guide_colorbar()`. These and similar `guide_*()` functions can be specified in their respective scale functions or more generally for any scale within `guides()`. We most frequently use `guide_legend()`, `guide_colorbar()`, and similar legend-type guides.

For example, let's make the colorbar a bit taller and narrower via the `barheight` and `barwidth` arguments. See `?ggplot2::guide_colorbar()` for arguments to make these and many other changes.

```
ggplot(data = fef,
       mapping = aes(x = DBH_cm, y = Stem_kg, color = Height_m)) +
  geom_point() +
  scale_color_continuous(name = "Height (m)",
                         guide = guide_colorbar(barheight = 15,
                                                barwidth = 1))
```

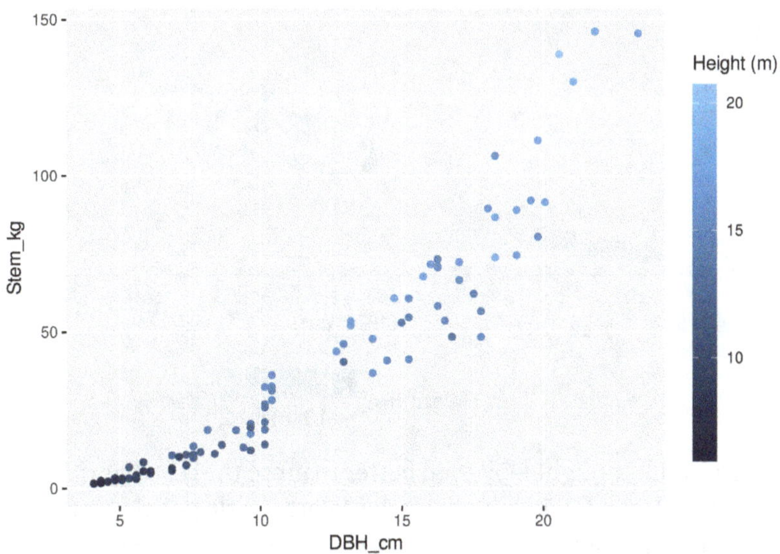

Arguments passed to the `theme()` function control a wide array of non-data graphic elements and are where we spend much of our time getting the graphic to look "just right." Take a look at `?ggplot2::theme` to get a sense of the defaults you can change.

There are also several ready-made complete themes designed to set a suite of theme defaults—see `?ggplot2::theme_gray`. We typically choose one of these complete themes, then further modify its defaults via `theme()`.

In the following example, we aim for a fairly polished-looking scatter plot by adding useful labels to the scales and choosing the `theme_bw()` complete theme, which creates a classic black-on-white look. We then change a long list of the theme's defaults.

As you'll see, many `theme()` arguments require an object returned by a function. For example, the `text` argument needs an object returned by the `element_text()` function (see `?ggplot2::element_text` for ways text can be modified).[21]

```
ggplot(data = fef,
       mapping = aes(x = DBH_cm, y = Stem_kg, color = Height_m)) +
  geom_point(size = 3) +
  labs(color = "Height (m)", x = "DBH (cm)", y = "Stem weight (kg)") +
  theme_bw() +
  theme(text = element_text(size = 14), # Increase all text elements.
        axis.text = element_text(size = 14), # Increase axis tick labels.
        legend.text = element_text(size = 14), # Increase legend labels.
        panel.grid.minor = element_blank(), # Remove minor grid lines.
        legend.direction = "horizontal", # Make the colorbar horizontal.
        legend.position = "bottom" # Move the colorbar to the bottom.
        )
```

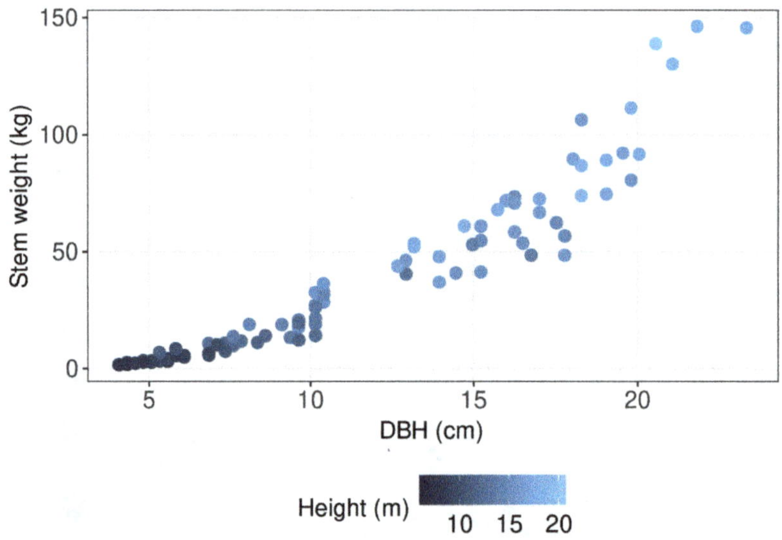

This looks pretty good. It might look even better to move the colorbar title from the left to the top center of the legend. You can adjust the colorbar title position and centering using the color scale guide, as shown below (same code as above with `scale_color_continuous()` added).

```
ggplot(data = fef,
       mapping = aes(x = DBH_cm, y = Stem_kg, color = Height_m)) +
  geom_point(size = 3) +
  labs(color = "Height (m)", x = "DBH (cm)", y = "Stem weight (kg)") +
```

[21]Also see the `base_size` argument in any `theme_*()` function to quickly manipulate the default text size of a graphic.

```
  theme_bw() +
  theme(text = element_text(size = 14),
        axis.text = element_text(size = 14),
        legend.text = element_text(size = 14),
        panel.grid.minor = element_blank(),
        legend.direction = "horizontal",
        legend.position = "bottom"
        ) +
  scale_color_continuous(guide = guide_colorbar(title.position = "top",
                                                title.hjust = 0.5))
```

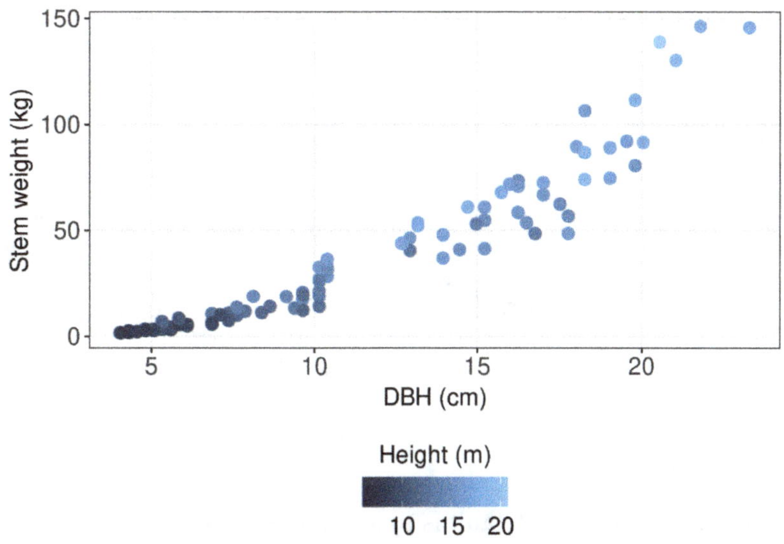

As you learn ggplot2, figuring out which graphic components to modify to achieve your desired change can be challenging (and perhaps a bit frustrating). However, Google, user forums, manual pages, and the ever-expanding set of ggplot2 references and vignettes will be key to developing proficiency and building a codebase for future graphics.

9.2.4 Facets

Figure 1.1 in Section 1.2.1 shows the relationship between DBH and component weight for four different tree components and four different species. Each of the 16 panels in this graphic shows a unique component and species combination. A facet is one side of something with many sides, hence the naming for `facet_*()` functions that create these multi-panel graphics. The `facet_grid()` function forms a matrix of panels defined by row and column faceting variables and is most useful when you have two discrete variables and all combinations of the variables exist in the data. The `facet_wrap()` function is useful when you have one variable with multiple levels.

Creating a facet graphic is as easy as adding `facet_grid()` or `facet_wrap()` to your `ggplot()` expression and identifying the faceting variable or variables using the `vars()` function, as illustrated below.

```
ggplot(data = fef, mapping = aes(x = DBH_cm, y = Stem_kg)) +
  geom_point() +
  geom_smooth() +
  facet_wrap(vars(Species))
```

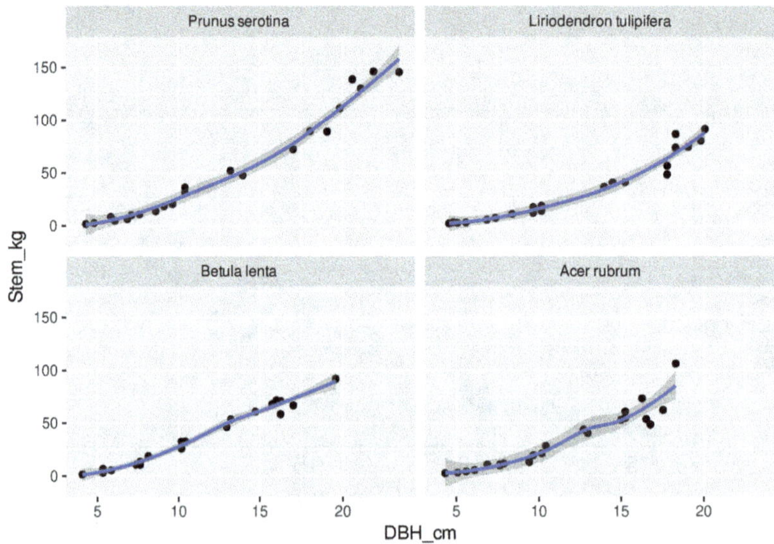

As you can see above, the facet default is to share axis scales across all panels. You can allow each panel to have its own scales, computed from its data subset, by adding `scales = "free"` (or `"free_x"` or `"free_y"` for axis-specific control), as illustrated below.

```
ggplot(data = fef, mapping = aes(x = DBH_cm, y = Stem_kg)) +
  geom_point() +
  geom_smooth() +
  facet_wrap(vars(Species), scales = "free")
```

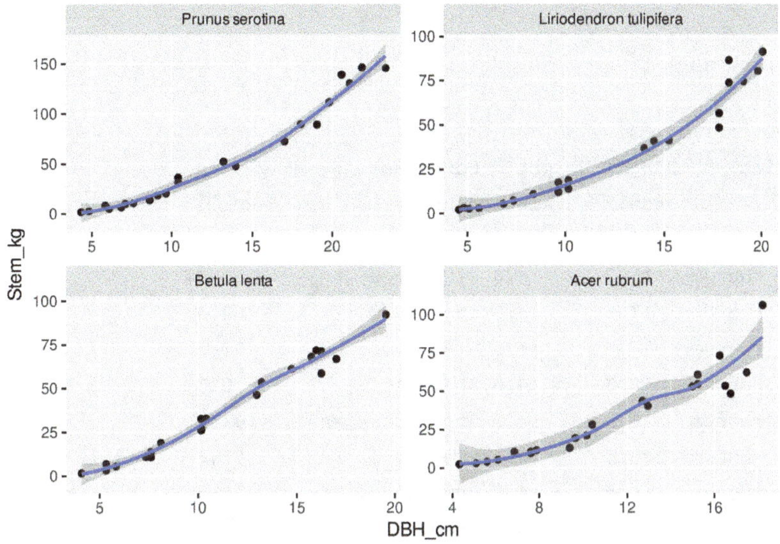

The `fef` data subset we're working with includes weight measurements for stem (`Stem_kg`) and other woody material (`Otherwood_kg`) components (see below).

```
fef
```

```
#> # A tibble: 88 x 5
#>    Species     DBH_cm Height_m Stem_kg Otherwoody_kg
#>    <fct>        <dbl>    <dbl>   <dbl>         <dbl>
#>  1 Acer rubrum  15.2     14.6     54.7          50.4
#>  2 Acer rubrum  17.5     14.6     62.3          60.8
#>  3 Acer rubrum  16.3     14.6     73.3          41.2
#>  4 Acer rubrum  16.5     14.9     53.6          52.3
#>  5 Acer rubrum  18.3     15.5    106.           52.5
#>  6 Acer rubrum   7.87    12.2     11.7           2.5
#>  7 Acer rubrum   5.08     9.30     3.2           2
#>  8 Acer rubrum  10.4     15.2     28.3          10.1
#>  9 Acer rubrum   6.10     8.53     5.5           2.5
#> 10 Acer rubrum   6.86    12.3     10.6           4.5
#> # i 78 more rows
```

Say we want to add facets for these two tree component weights. To do so, we'll need to make the messy `fef` data tidy (it's messy because the component weights aren't held in a single column—they're in wide format). Referring to Section 8.1, we'll tidy up `fef` by moving component weights from wide to long format as follows.

```
fef_long <- fef %>%
  pivot_longer(cols = c("Stem_kg", "Otherwoody_kg"),
               names_to = "Component",
               values_to = "Weight_kg")
```

```
fef_long # Take a look.
```

```
#> # A tibble: 176 x 5
#>    Species      DBH_cm Height_m Component     Weight_kg
#>    <fct>         <dbl>    <dbl> <chr>             <dbl>
#>  1 Acer rubrum    15.2     14.6 Stem_kg            54.7
#>  2 Acer rubrum    15.2     14.6 Otherwoody_kg      50.4
#>  3 Acer rubrum    17.5     14.6 Stem_kg            62.3
#>  4 Acer rubrum    17.5     14.6 Otherwoody_kg      60.8
#>  5 Acer rubrum    16.3     14.6 Stem_kg            73.3
#>  6 Acer rubrum    16.3     14.6 Otherwoody_kg      41.2
#>  7 Acer rubrum    16.5     14.9 Stem_kg            53.6
#>  8 Acer rubrum    16.5     14.9 Otherwoody_kg      52.3
#>  9 Acer rubrum    18.3     15.5 Stem_kg           106.
#> 10 Acer rubrum    18.3     15.5 Otherwoody_kg      52.5
#> # i 166 more rows
```

Now we're ready to add facets for component! Notice in the code below, the `y` in `aes()` is now set to `Weight_kg`, and we specify `Component` as the additional faceting variable in a call to `facet_grid()`.

```
ggplot(data = fef_long, mapping = aes(x = DBH_cm, y = Weight_kg)) +
  geom_point() +
  geom_smooth() +
  facet_grid(vars(Species), vars(Component))
```

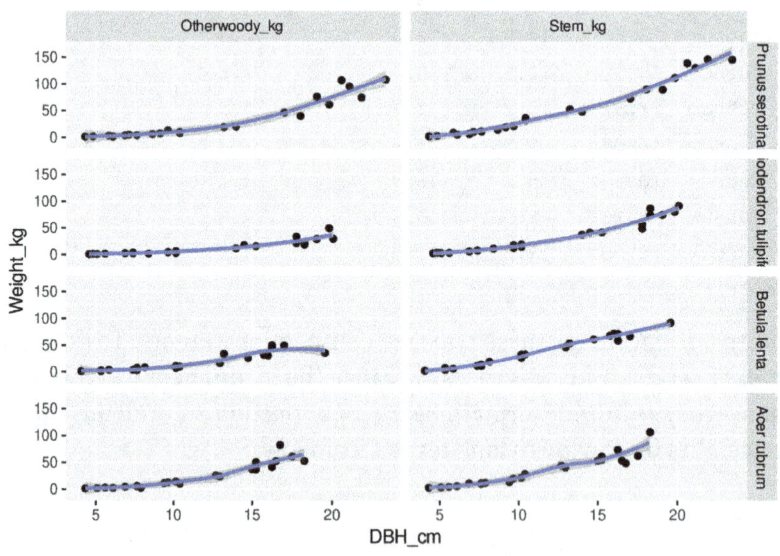

Looks pretty good, but some final tweaks are needed. Notice the row label for *Liriodendron tulipifera* is cut off—that needs fixing. Also, let's add better labels to the axes and columns, and reverse the column order (it feels right to have stem come before otherwoody).

Changing the graphic's column order and labels is most easily accomplished by converting the `Component` faceting variable to a factor, which is done below using a call to `factor()` within `mutate()`. The order in which columns appear in the graphic is determined by the order of the factor levels, and the column labels will be the factor values or, if specified, the factor labels. In the code below, both the `levels` and `labels` arguments are specified in the `factor()` function. Defining the factor with `labels` effectively replaces `Stem_kg` with `Stem` and `Otherwoody_kg` with `Otherwoody` in the newly formed factor.

```
fef_long <- fef_long %>%
  mutate(Component = factor(Component,
                            levels = c("Stem_kg", "Otherwoody_kg"),
                            labels = c("Stem", "Otherwoody")))
fef_long %>%
  glimpse() # Check out the Component factor.
```

```
#> Rows: 176
#> Columns: 5
#> $ Species   <fct> Acer rubrum, Acer rubrum, Acer rubr~
#> $ DBH_cm    <dbl> 15.240, 15.240, 17.526, 17.526, 16.~
#> $ Height_m  <dbl> 14.6304, 14.6304, 14.6304, 14.6304,~
#> $ Component <fct> Stem, Otherwoody, Stem, Otherwoody,~
#> $ Weight_kg <dbl> 54.7, 50.4, 62.3, 60.8, 73.3, 41.2,~
```

Getting the row names to wrap, and not be cut off, is best accomplished using the `facet_grid()` `labeller` argument, as illustrated below. Here, the `labeller` argument takes an object created using the `labeller()` function, which specifies how the species names should be wrapped onto multiple lines after a given width is exceeded. Column labels can also be specified using the `labeller` argument, but it's easier to specify the labels when forming the `Component` factor.

```
ggplot(data = fef_long, mapping = aes(x = DBH_cm, y = Weight_kg)) +
  geom_point() +
  geom_smooth() +
  facet_grid(vars(Species), vars(Component),
             labeller = labeller(Species =
                                   label_wrap_gen(width = 10,
                                                  multi_line = TRUE)))
```

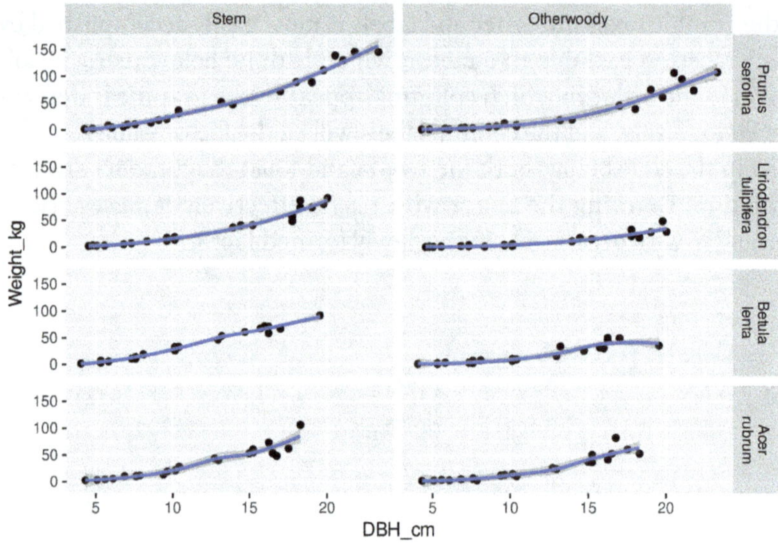

As with other topics covered in this chapter, we provide an incomplete tour of facets. These are extremely useful tools for exploring your data and communicating information. We'll explore additional facet functionality in subsequent sections.

9.2.5 Building graphics incrementally

As you've seen, constructing a graphic is an incremental process that will involve trial and error. It can be useful to build a base graphic to which you add layers and modifying functions. The base graphic and subsequent incremental additions can be saved as a `ggplot` object. Below, we define a base graphic and assign it to `base_plot`, then update `base_plot` incrementally once we're happy with the additions. Notice the graphic isn't plotted until you run `base_plot` on its own (which is equivalent to running `print(base_plot)`).

```
base_plot <- ggplot(data = fef, mapping = aes(x = DBH_cm, y = Stem_kg)) +
  geom_point(size = 5, shape = 21, mapping = aes(fill = Height_m))

base_plot # Implicit for print(base_plot).
```

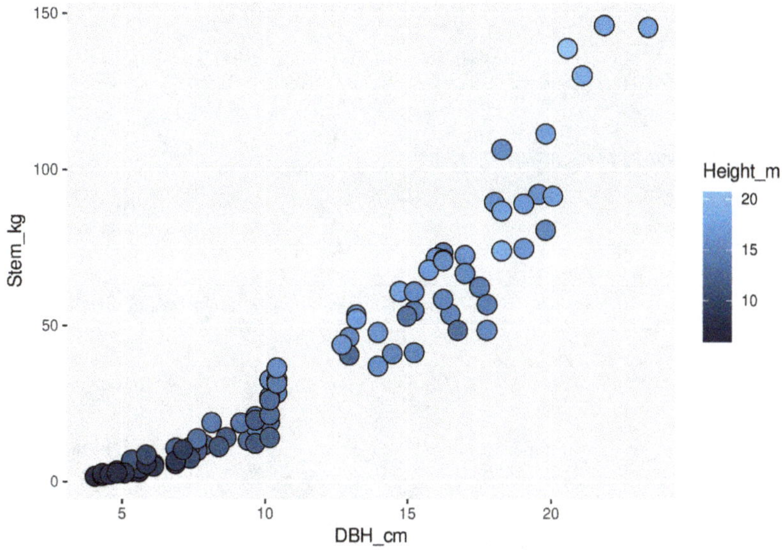

Now you can redefine or modify the point geometry layer, and add more layers and modifying functions to `base_plot`. Below, we try out a color gradient to see if we like it.

```
# Don't like the colors.
base_plot + scale_fill_gradient(low = "blue", high = "green")
```

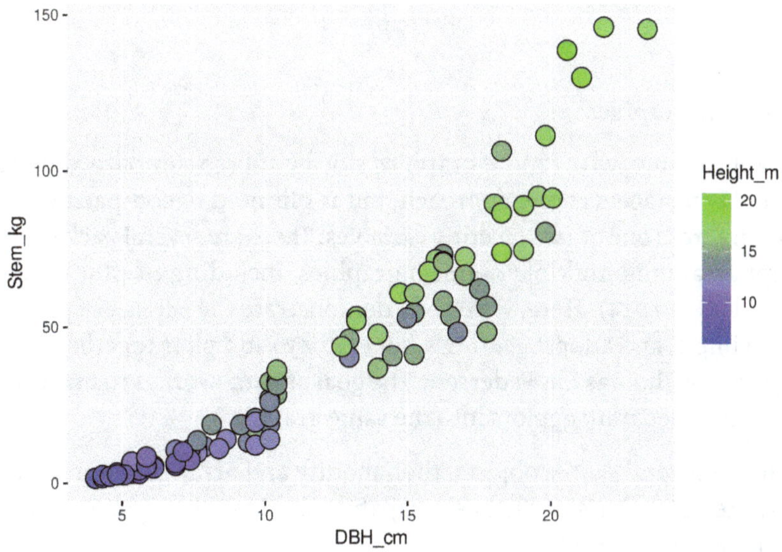

Try some different colors.

```
# Better colors, save it and move on.
base_plot <- base_plot +
  scale_fill_gradient(low = "white", high = "seagreen")
```

Now adjust scales and themes to your liking.

```
base_plot <- base_plot +
  labs(fill = "Height (m)", x = "DBH (cm)", y = "Stem (kg)") +
  theme_bw() +
  theme(text = element_text(size = 14))

base_plot # Looks good!
```

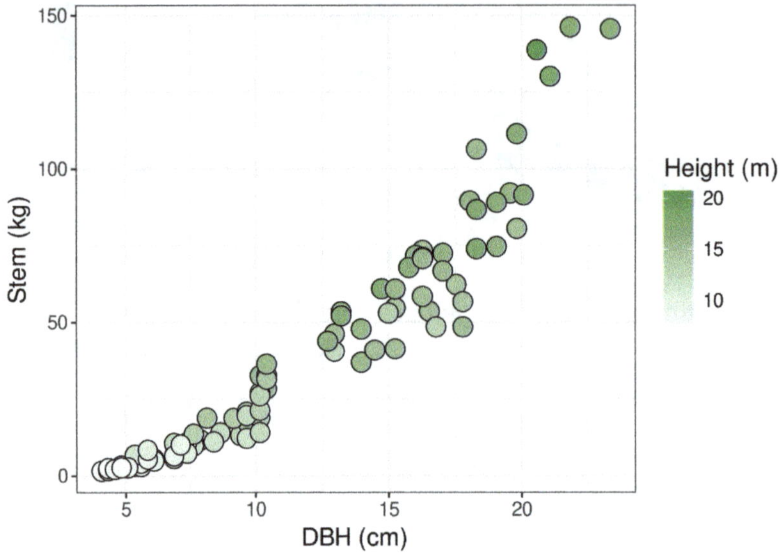

9.2.6 Arranging graphics

We often want to show two or more graphics side by side to compare and communicate information. Using facets is one approach, but it's limited to comparing the same data representation across one or two faceting variables. There are several packages that provide functions for arranging multiple `ggplot()` graphics, including `cowplot` (Wilke 2024) and `patchwork` (Pedersen 2024). Here, we briefly demonstrate the `patchwork` package, because it's versatile, simple, and adopts `ggplot2`'s `+` operator to add plots together. In the words of the package author Thomas Lin Pedersen, "the goal of patchwork is to make it ridiculously simple to combine separate ggplots into the same graphic."

Below, we make several `ggplot` objects, then modify and arrange them in different ways to demonstrate `patchwork`'s most basic capabilities. Install the `patchwork` package via `install.packages("patchwork")`.

```
p_1 <- ggplot(data = fef,
              mapping = aes(x = DBH_cm)) +
  geom_histogram() +
  ggtitle("Plot 1")

p_2 <- ggplot(data = fef,
```

```
              mapping = aes(x = DBH_cm, y = Stem_kg)) +
  geom_point() +
  ggtitle("Plot 2")

p_3 <- ggplot(data = fef_long,
              mapping = aes(x = DBH_cm, y = Weight_kg)) +
  geom_point() +
  facet_wrap(vars(Component)) +
  ggtitle("Plot 3")

p_4 <- ggplot(data = fef,
              mapping = aes(x = DBH_cm, y = Stem_kg, color = Species)) +
  geom_point() +
  geom_smooth() +
  ggtitle("Plot 4")
```

Add two graphics together using the + operator.

```
library(patchwork)

p_1 + p_2
```

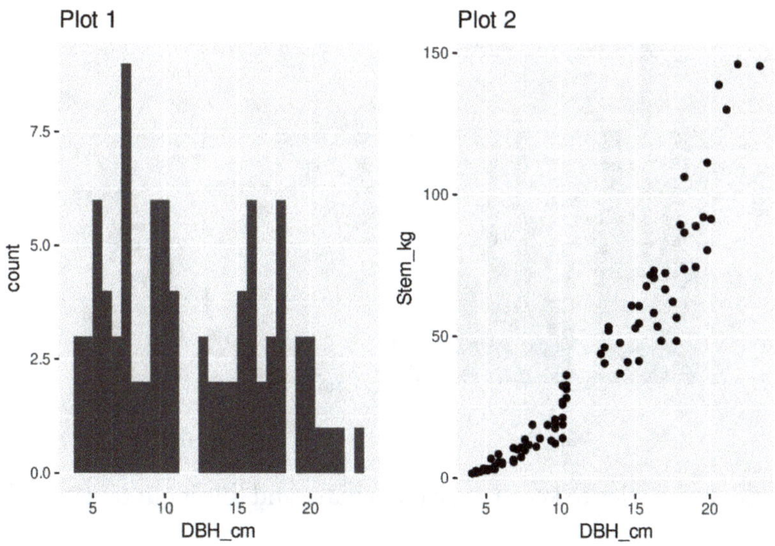

When adding graphics together, the last added `ggplot` is active and will receive any added layers or modifying functions.

```
p_1 + p_2 + labs(subtitle = 'This appears in the right plot')
```

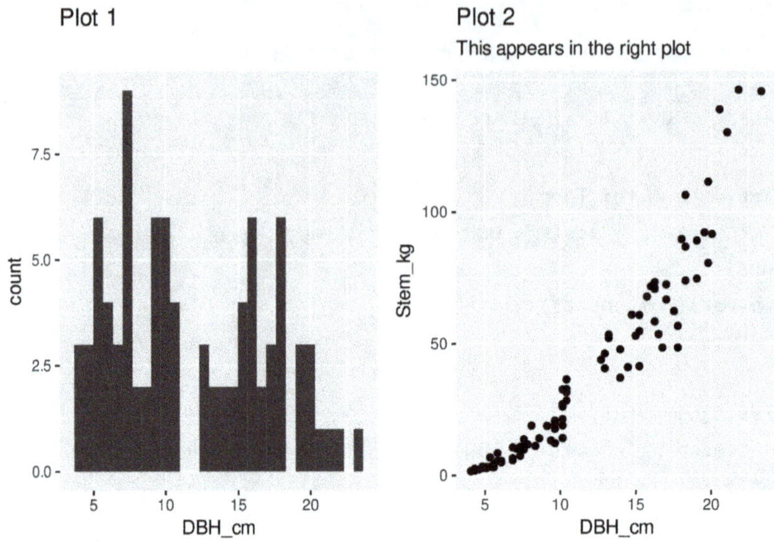

Parentheses can be used to control order of operations—for example, use them to add layers or modifying functions to a `ggplot` that's not the last `ggplot`.

```
(p_1 + labs(subtitle = 'This appears in the left plot')) + p_2
```

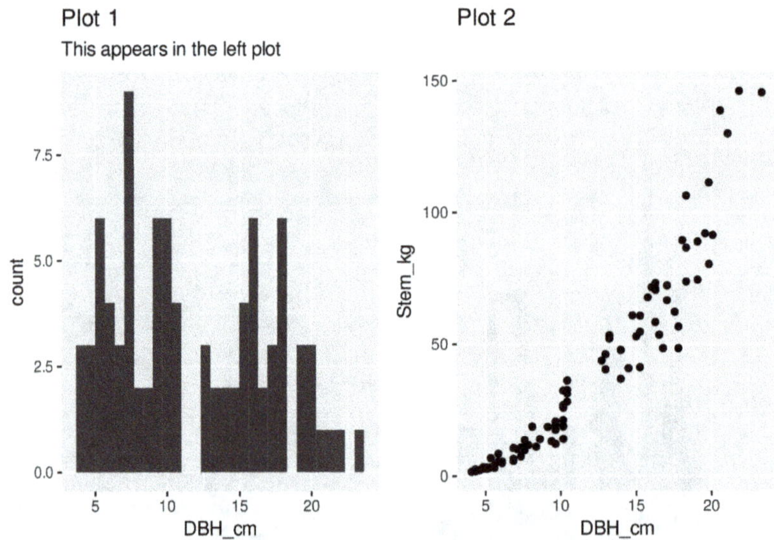

By default, `patchwork` tries to keep the layout square and fills it out by row order.

```
p_1 + p_2 + p_3 + p_4
```

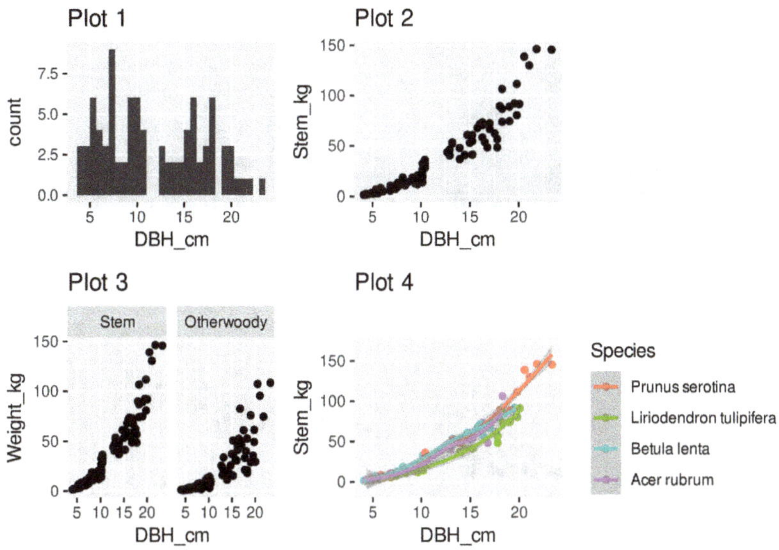

Control the layout by adding the `plot_layout()` function.

```
p_1 + p_2 + p_3 + p_4 + plot_layout(nrow = 2, byrow = FALSE)
```

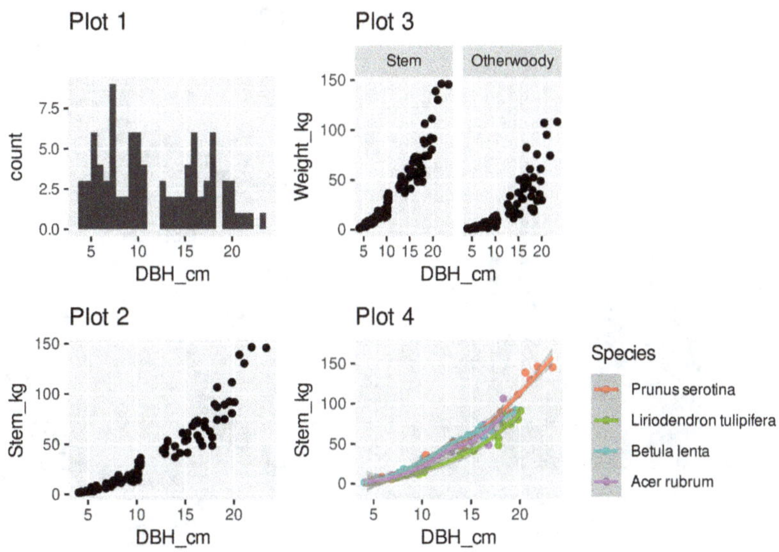

In addition to controlling placement via `plot_layout()`, you can use the `|` operator to place plots beside each other and the `/` operator to stack them.

```
p_1 / p_2
```

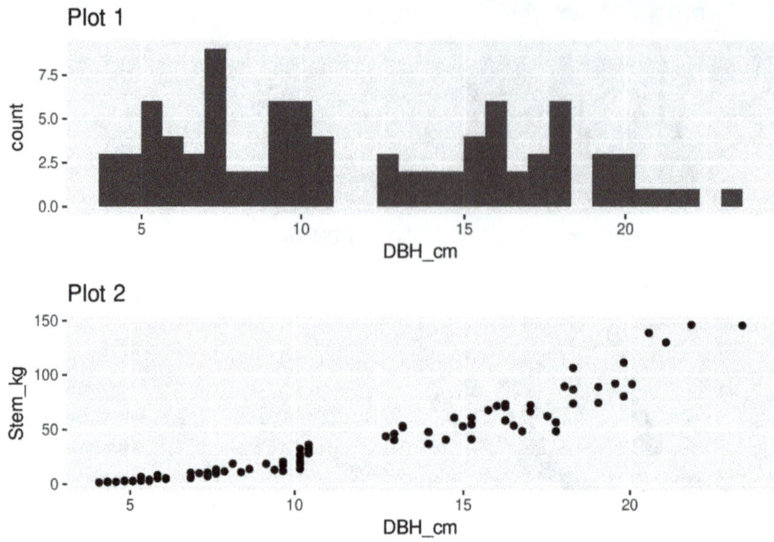

Using parentheses with the placement operators lets you create nested layouts.

```
p_1 | (p_2 / p_3)
```

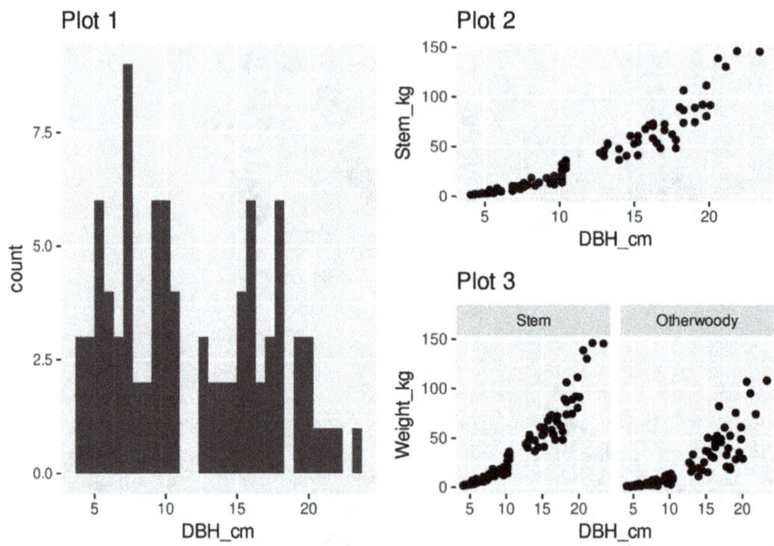

This is a very brief overview of some of `patchwork`'s capabilities. See its manual pages, vignette via `vignette("patchwork")`, and online reference at https://patchwork.data-imaginist.com for much more, including how to add annotation, collect legends into one place, remove duplicate legends, create overlays and insets, and further control layout and alignment.

9.2.7 Saving graphics

We often want to export our graphics for use in an external document or share them with colleagues. There are several ways to save graphics in a variety of file formats. One approach is to use ggsave(), see ?ggplot2::ggsave(). The plot argument default in ggsave() is to save the most recent graphic; however, you can also give plot a ggplot object, like the ones created in the previous section. The device argument lets you choose among a variety of vector (e.g., "eps", "ps", "pdf", "svg") or raster (e.g., "jpeg", "tiff", "png", "bmp", "wmf") formats.[22]

9.3 Exploring different graphics

We explore ggplot2 further using the FACE dataset introduced in Section 1.2.3. Recall, the Aspen FACE was designed to assess the effect of increasing tropospheric ozone (O_3) and carbon dioxide (CO_2) concentration on structure and functioning of northern forest ecosystems.

As noted previously, ggplot2 functions are designed to work with tidy data; therefore, we'll use the tidy FACE tibble face_long created at the end of Section 8.2. You can either recreate this tibble by following the code in Section 8.1 or read it into face_long from the "datasets/FACE_long.csv" file using the code below.

```
face_long <- read_csv("datasets/FACE_long.csv")
face_long # Take a look.
```

```
#> # A tibble: 9,955 x 6
#>      Rep Treat Clone    ID  Year Height_cm
#>    <dbl> <dbl> <dbl> <dbl> <dbl>     <dbl>
#>  1     1     1     8    45  2001        NA
#>  2     1     1     8    45  2002        NA
#>  3     1     1     8    45  2003        NA
#>  4     1     1     8    45  2004        NA
#>  5     1     1     8    45  2005        NA
#>  6     1     1   216    44  2001       547
#>  7     1     1   216    44  2002       622
#>  8     1     1   216    44  2003       715
#>  9     1     1   216    44  2004       716
#> 10     1     1   216    44  2005       817
#> # i 9,945 more rows
```

Here, the Rep, Treat, and Clone columns record experimental design information, ID records a unique tree identification number, Year records the year in which the tree's height was measured, and Height_cm holds the recorded height. As mentioned previously, the FACE data are longitudinal, meaning observational units—trees in this case—are measured over

[22]Vector files use point, line, curve, and polygon elements, whereas raster files depict these elements using shaded pixels. Vector images are often preferred for publication-quality graphics because they can be edited, scaled without loss of definition, and provide crisper detail.

time. Most longitudinal data are collected to understand how observational units respond to different treatments. The FACE was designed to look at how tree growth responds to elevated tropospheric O_3 and CO_2. We'll explore this question using a series of graphics.

We begin by adding a bit more information to `face_long`. As the name suggests, the `Treat` variable records each tree's treatment. Trees were assigned to one of four treatments for the study's duration. The treatments are coded 1 (control), 2 (elevated CO_2), 3 (elevated O_3), and 4 (elevated CO_2 and O_3). Let's add this more descriptive information to `Treat` by converting the column to a factor and associating treatment numeric code levels with treatment labels. We'll also convert `Year` to a factor.

```
face_long <- face_long %>%
  mutate(Treat = factor(Treat, levels = 1:4,
                              labels = c("Control", "Elevated CO2",
                                         "Elevated O3",
                                         "Elevated CO2 and O3")),
         Year = factor(Year))

face_long # Take a look at the new factors.
```

```
#> # A tibble: 9,955 x 6
#>      Rep Treat   Clone    ID Year  Height_cm
#>    <dbl> <fct>   <dbl> <dbl> <fct>     <dbl>
#>  1     1 Control     8    45 2001         NA
#>  2     1 Control     8    45 2002         NA
#>  3     1 Control     8    45 2003         NA
#>  4     1 Control     8    45 2004         NA
#>  5     1 Control     8    45 2005         NA
#>  6     1 Control   216    44 2001        547
#>  7     1 Control   216    44 2002        622
#>  8     1 Control   216    44 2003        715
#>  9     1 Control   216    44 2004        716
#> 10     1 Control   216    44 2005        817
#> # i 9,945 more rows
```

Let's get a little better sense of these data. We can see from the output above that `face_long` has 9955 rows and there are some `NA` `Height_cm` values. Let's figure out the number of trees, number of missing height measurements, and number of trees in each treatment (recall, trees are assigned to only one treatment).

```
face_long %>%
  summarize(n_tree = n_distinct(ID), # Trees in the study.
            n_na_heights = sum(is.na(Height_cm))) # Number of NA.
```

```
#> # A tibble: 1 x 2
```

```
#>   n_tree n_na_heights
#>    <int>        <int>
#> 1   1991         1344
```

```
face_long %>%
  group_by(Treat) %>%
  summarise(n_trees = n_distinct(ID)) # Trees in each treatment.
```

```
#> # A tibble: 4 x 2
#>   Treat                n_trees
#>   <fct>                  <int>
#> 1 Control                  498
#> 2 Elevated CO2             497
#> 3 Elevated O3              497
#> 4 Elevated CO2 and O3      499
```

Some working hypotheses that motivated the FACE study were that elevated atmospheric CO_2 could increase tree growth rate, while increased O_3 could damage leaf tissue and decrease growth rate. Let's see if the data show a treatment effect on height growth over time.

Begin by plotting each tree's growth over years as a line, and color the line by treatment. Given what we learned about aesthetics and geometries in the previous section, the code below might be a logical starting point.

```
ggplot(data = face_long,
       mapping = aes(x = Year, y = Height_cm, color = Treat)) +
  geom_line()
```

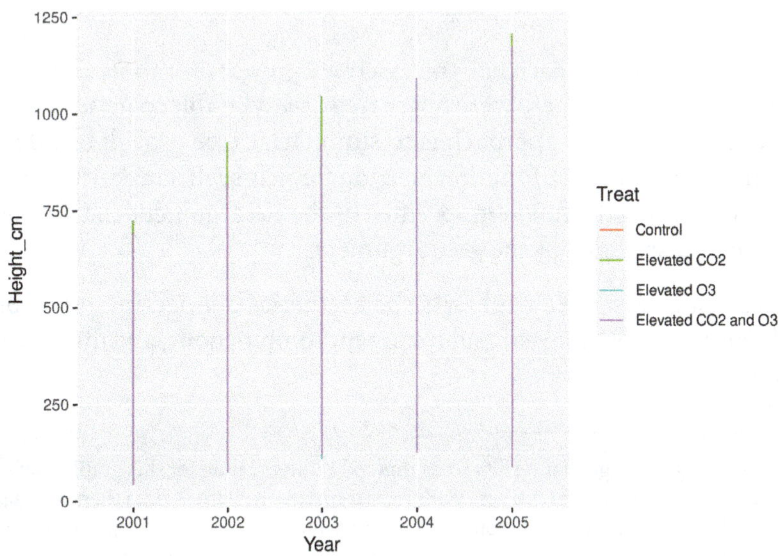

While the graphic isn't what we're aiming for, it's close. It's actually drawing a line connecting all height measurements within year and treatment. As it stands, there's nothing telling `geom_line()` that sets of height measurements are associated with individual trees over years. This information is provided using the `group` aesthetic (see `?ggplot2::aes_group_order`) via `group = ID` in `aes()` below.

```
ggplot(data = face_long,
       mapping = aes(x = Year, y = Height_cm, color = Treat,
                     group = ID)) +
  geom_line()
```

```
#> Warning: Removed 1336 rows containing missing values or values
#> outside the scale range (`geom_line()`).
```

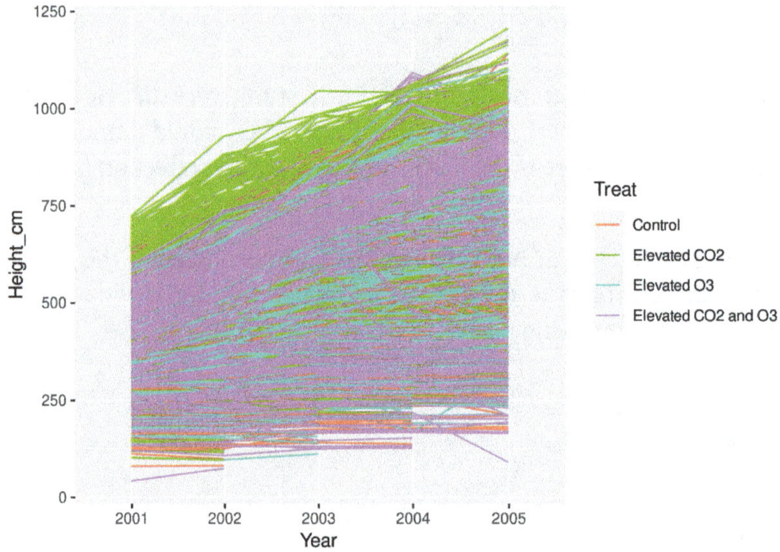

Okay, this is looking better.[23] Each line traces a tree's growth over time, and color shows the tree's treatment. Perhaps if there were fewer trees, we'd be able to notice some treatment effect. So perhaps the better approach is to summarize tree growth trends within each treatment. The `stat_summary()` function is made for just such a task—it summarizes `y` at discrete values of `x` using the function specified by the `fun` argument and displays the result using the geometry specified by the `geom` argument.

The call to `stat_summary()` below inherits the aesthetics defined in `aes()`, computes the mean height measurement for year and treatment combinations, and displays the result as a line geometry.

[23] Notice the warning message that results from this code. The warning for this graphic says there are 1336 rows removed due to missing, i.e., `NA`, values. We know from exploring these data there are 1344 `NA` `Height_cm` values and no missing years. So, we might wonder why the warning doesn't say 1344 rows removed. The answer is in the `geom_line()` manual page under the missing value handling section, which notes that `NA`s at the beginning or end of lines are removed without a warning. This accounts for the disparity of 8 `NA`s removed.

```
ggplot(data = face_long,
       mapping = aes(x = Year, y = Height_cm, color = Treat,
                     group = Treat)) +
  stat_summary(geom = "line", fun = mean)
```

```
#> Warning: Removed 1344 rows containing non-finite outside the scale
#> range (`stat_summary()`).
```

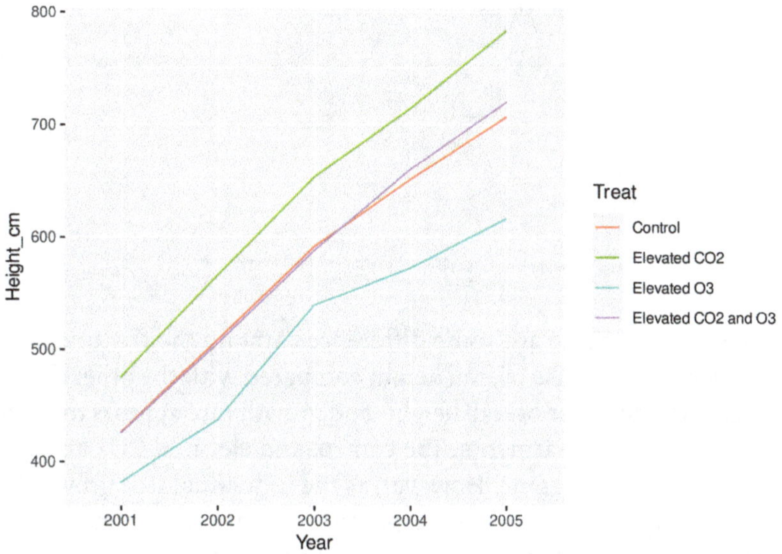

Each line in the graphic traces mean growth within each treatment over years.[24]

It's instructive to take this one step farther. Let's draw the individual trees below the mean growth lines. We'll make the individual tree lines semi-transparent so they don't dominate the scene. To do so, we keep the `x`, `y`, and `color` aesthetics in `ggplot()`, map `geom_line()` to `group = ID`, and map `stat_summary()` to `group = Treat`. Setting transparency is done via the `alpha` aesthetic.

```
ggplot(data = face_long,
       mapping = aes(x = Year, y = Height_cm, color = Treat)) +
  geom_line(mapping = aes(group = ID), alpha = 0.05) +
  stat_summary(geom = "line", fun = mean, mapping = aes(group = Treat),
               linewidth = 1)
```

```
#> Warning: Removed 1344 rows containing non-finite outside the scale
#> range (`stat_summary()`).

#> Warning: Removed 1336 rows containing missing values or values
```

[24]Notice the `ggplot()` warning says it removed 1344 rows containing non-finite values. These rows correspond to the 1344 `NA` `Height_cm` values.

```
#> outside the scale range (`geom_line()`).
```

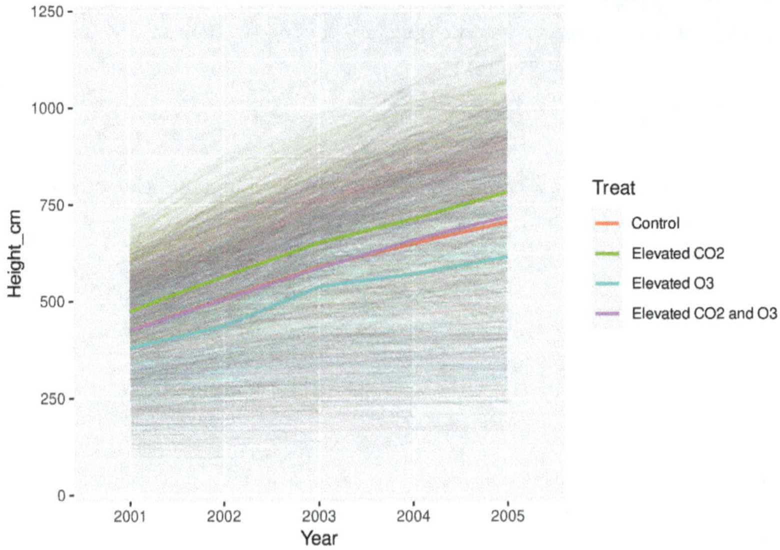

The mean lines suggest there are some differences among the treatments. Specifically, elevated CO_2 seems to increase overall height compared with the other three treatments. Elevated O_3 seems to decrease overall height, and growth rate appears to decrease beyond 2003 relative to the other treatments. The control and elevated CO_2 and O_3 height and growth trends are about the same. However, as the individual tree growth lines suggest, there's large variability in individual tree response to the various treatments. This variability needs to be taken into account to draw statistically valid conclusions about treatment effect. Formal analysis of the FACE data and results from similar studies have shown O_3's negative growth effects on tree seedlings. However, long-term positive growth effects of elevated CO_2 haven't been clearly demonstrated, especially in the presence of elevated O_3 and other environmental factors that negatively impact growth; see, e.g., McLaughlin et al. (2007), Karnosky et al. (2007), and Davis et al. (2022).

9.4 Visualizing distributions

There are a number of geometries used to display data distributions. The appropriate `ggplot2` geometry for the job depends on whether the distribution is defined by one or more variables, data types of the variables, and how variables are to be summarized.

In this section, we focus on a limited set of geometries most often used to summarize forestry data distributions. In particular, we use geometries that create bar charts to visualize forest inventory data by summarizing a continuous variable (e.g., volume, basal area) or a discrete variable (e.g., occurrence) by one or more categorical variables, e.g., DBH class, species. We begin by illustrating the geometries using a very simple toy dataset, then provide an example using the Elk County timber cruise dataset introduced in Section 1.2.2.

Below is a toy dataset comprising six trees and their associated species code (`Spp`), DBH (`DBH_cm`), and biomass (`Bio_kg`).[25] We also add a 2 cm DBH class variable with intervals $[18, 20)$, $[20, 22)$, $[22, 24)$, and $[24, 26]$ using `mutate()` and the handy `cut_width()` function, introduced in Section 8.3, and place it to the right of the `DBH_cm` column.

```
tbl <- tibble(Spp = c("ACRU", "ACRU", "FAGR", "FAGR", "BEPA", "BEPA"),
              DBH_cm = c(24, 22, 19, 24, 20, 18),
              Bio_kg = c(272, 221, 173, 306, 176, 138))

# Add DBH class.
tbl <- tbl %>%
  mutate(tbl,
         DBH_2cm_class = cut_width(DBH_cm, width = 2,
                                   center = 19, closed = "left"),
         .after = DBH_cm)
# Take a look.
tbl
```

```
#> # A tibble: 6 x 4
#>   Spp   DBH_cm DBH_2cm_class Bio_kg
#>   <chr>  <dbl> <fct>          <dbl>
#> 1 ACRU      24 [22,24]          272
#> 2 ACRU      22 [22,24]          221
#> 3 FAGR      19 [18,20)          173
#> 4 FAGR      24 [22,24]          306
#> 5 BEPA      20 [20,22)          176
#> 6 BEPA      18 [18,20)          138
```

There are two types of bar chart geometries that differ in how bar height is computed.

1. `geom_bar()` makes bar height proportional to the number of discrete value occurrences, i.e., counts, in the given variable (or, if the `weight` aesthetic is supplied, the sum of the weight values). These counts (or sums) can be partitioned by other discrete variables. The `geom_bar()` function uses the `stat_count()` stat function, which counts discrete value occurrences.

2. `geom_col()` makes bar height proportional to the value in the given variable. These sums can be partitioned by one or more discrete variables. The `geom_col()` function uses the `stat_identity()` stat function, which leaves values unchanged.

The code below uses `geom_bar()` to show tree count by `DBH_class` and tree count by species and `DBH_class`. We place the resulting bar charts side by side to facilitate comparison.

---

[25] Species codes are from the PLANTS database (USDA 2021) with FAGR (*Fagus grandifolia*), ACRU (*Acer saccharum*), and BEPA (*Betula papyrifera*), and biomass was estimated from DBH using equations provided in Jenkins et al. (2002).

```
ggplot(tbl, aes(x = DBH_2cm_class)) +
  geom_bar()

ggplot(tbl, aes(x = DBH_2cm_class, fill = Spp)) +
  geom_bar()
```

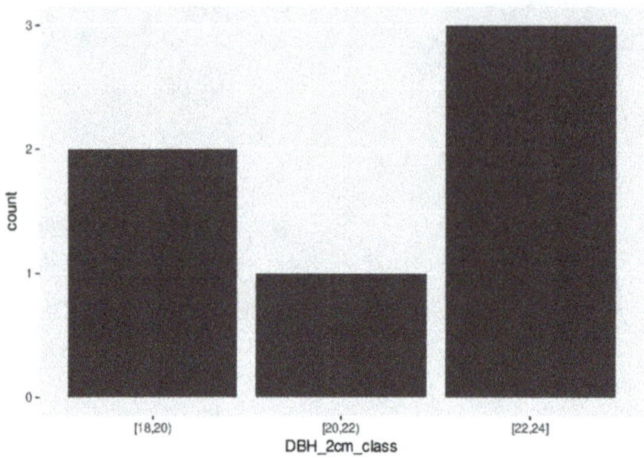

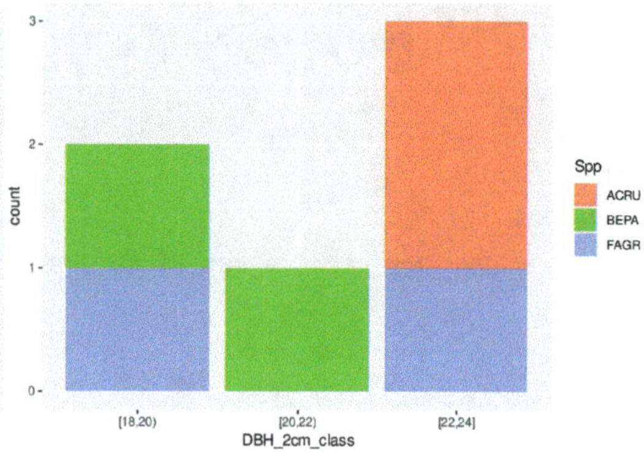

Notice in the right bar chart above, the default behavior is to stack species counts, i.e., for the $[18, 20)$ cm DBH class there's one FAGR and one BEPA, and in the $[22, 24]$ cm DBH class there are two ACRU and one FAGR. This default behavior is controlled using *position adjustment* of elements within a layer. Three adjustments apply primarily to bars:

- `position_stack()` stacks overlapping bars on top of each other (default),
- `position_fill()` stacks overlapping bars, scaling so the top is always at 1,
- `position_dodge()` places overlapping bars side by side.

Each position adjustment is illustrated below.

```
bars <- ggplot(tbl, aes(x = DBH_2cm_class, fill = Spp)) +
  xlab(NULL) +
  ylab(NULL) +
  theme(legend.position = "none")

bars + geom_bar() # Default, same as geom_bar(position = "stack")

bars + geom_bar(position = "fill")

bars + geom_bar(position = "dodge")
```

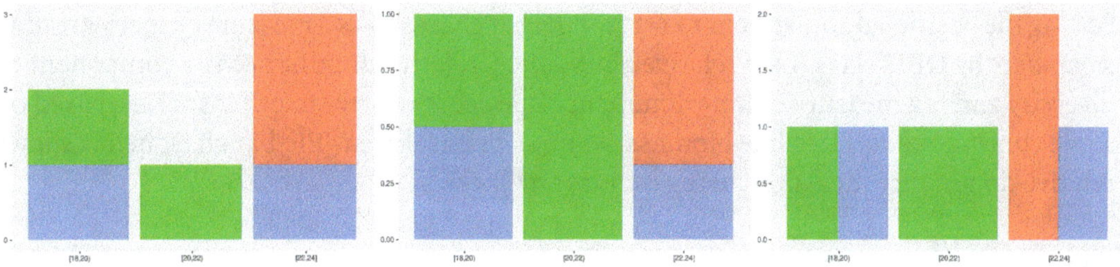

All three position adjustments above have use in inventory summaries. Select the one that best conveys the information you wish to communicate. For example, the middle bar chart (i.e., `position = "fill"`) shows species *relative abundance* by DBH class. Relative abundance is a component of diversity and is a measure of how common or rare a species is relative to other species. This chart shows FAGR and BEPA are equally abundant in the $[18, 20)$ cm class, BEPA dominates the $[20, 22)$ cm class, and ACRU is more abundant than FAGR in the $[22, 24]$ cm classes.

Next, let's use `geom_col()` to make two bar charts that show total biomass by `DBH_class` and biomass by species and `DBH_class`. Notice, `geom_col()` requires a `y` aesthetic, which is summed by `x` aesthetic class and any other discrete aesthetic provided to `color` or `fill`. We again take the default `position = "stack"` for each bar chart and place the charts side by side to facilitate comparison.

```
ggplot(tbl, aes(x = DBH_2cm_class, y = Bio_kg)) +
  geom_col()

ggplot(tbl, aes(x = DBH_2cm_class, y = Bio_kg, fill = Spp)) +
  geom_col()
```

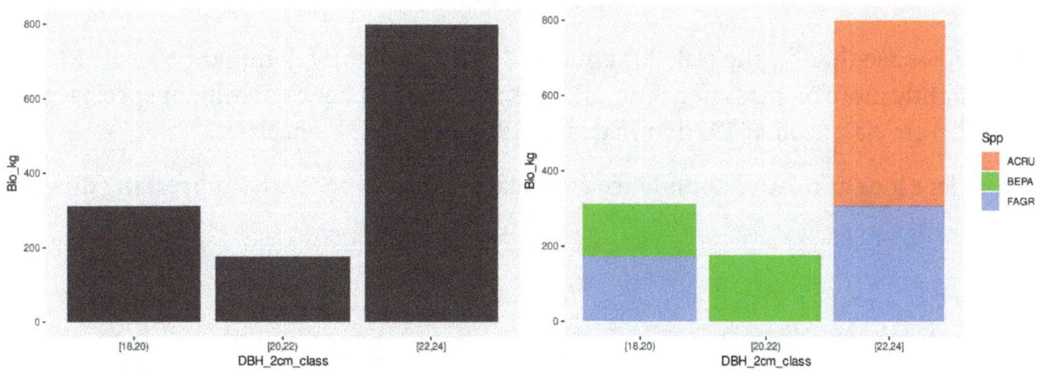

The figures above look similar to those from `geom_bar()`, but now the bar height is the sum of `Bio_kg`.[26]

[26]As an aside, you might notice some horizontal white lines in the bars. They're artifacts from how the data are internally summarized and from omitting the `color` aesthetic (which determines the bar segment's outline color). So the lines you're seeing are the default `color = NA` aesthetic. To remove those pesky lines—so as not to confuse your reader—match the `color` and `fill` argument values, e.g., in the first chart above add `color = "grey35"` to `geom_col()`, and for the second chart add `color = Spp` to `aes()`. Hopefully this default will change in

Setting the positional adjustment to `fill` for the bar chart shows a measure of species *relative dominance* by DBH class. Like relative abundance, relative dominance is a component of diversity and is a measure of how dominant a species is relative to other species, based on some continuous variable like biomass, volume, or basal area. The bar chart below shows relative dominance measured using biomass by species.

```
ggplot(tbl, aes(x = DBH_2cm_class, y = Bio_kg, fill = Spp)) +
  geom_col(position = "fill")
```

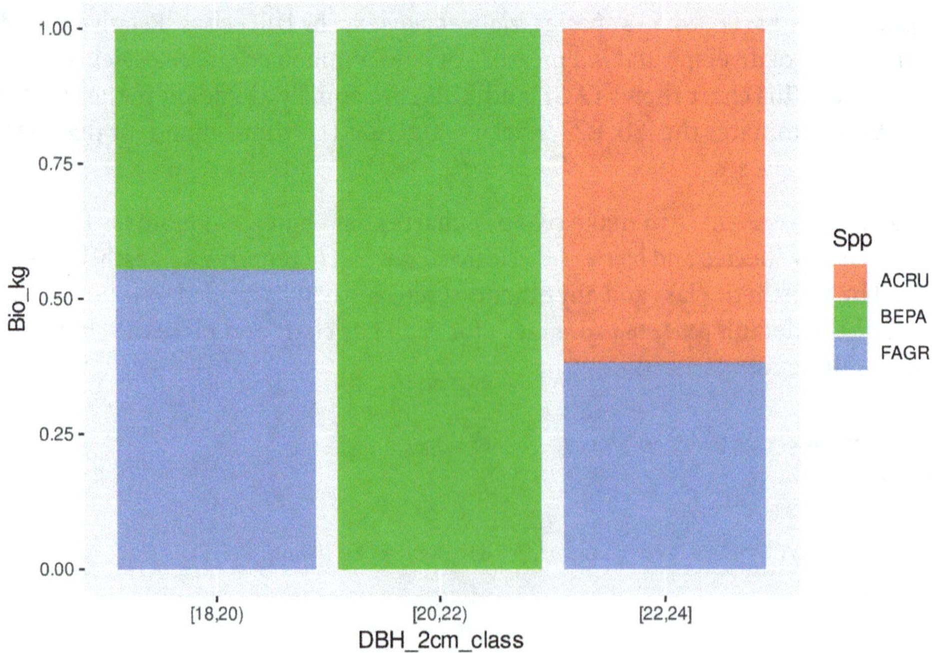

Above, we see that for the $[18, 20)$ cm class, FAGR is slightly dominant over BEPA (i.e., has slightly more biomass per acre), BEPA is the only and hence dominant species in the $[20, 22)$ cm class, and ACRU dominates FAGR in the $[22, 24]$ cm class.

Now, let's look at relative abundance and relative dominance without breaking down by DBH class.

```
ggplot(tbl, aes(x = Spp)) +
  geom_bar()

ggplot(tbl, aes(x = Spp, y = Bio_kg)) +
  geom_col()
```

---

future `ggplot2` versions.

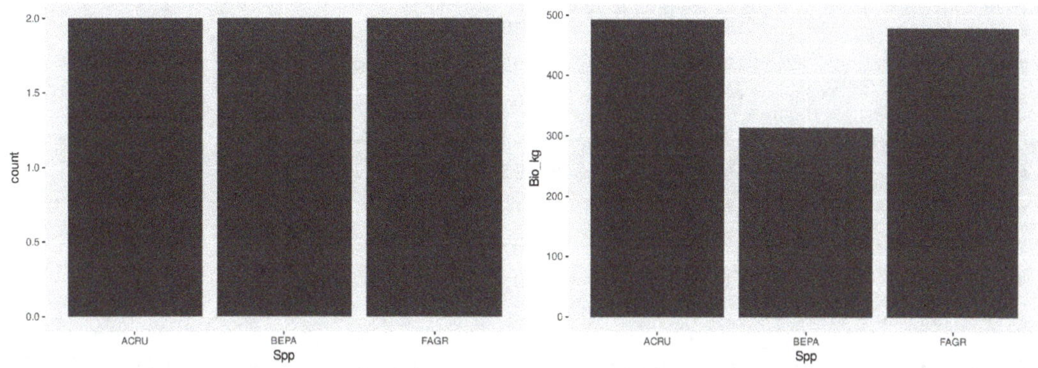

The left bar chart shows that species are equally abundant, because there are two of each. The right plot shows ACRU has slightly more biomass than FAGR and is hence the dominant species, based on this measure of dominance.

9.4.1 Elk County timber cruise bar charts

Let's return to the Elk County timber cruise dataset introduced in Section 1.2.2 and developed in more detail at the beginning of Section 8.5. Much of what is covered here builds on stand and stock tables covered in Section 8.3—in fact, several graphics we create are simply visual representations of tabular summaries created previously.

Our initial focus is on one-dimensional distributions of continuous variables summarized by one or more categorical variables. Begin by reading the per acre estimates for acceptable growing stock into `elk_ags` and adding the DBH class factor (`DBH_2in_range`), which more explicitly defines the two-inch DBH class used when the data were recorded in the field.[27] Refer to the beginning of Section 8.5 for a description of these data and column definitions. Here, per acre number of trees (`Trees_ac`), basal area (`BA_ac`), and number of 16-foot logs (`Logs_ac`) are estimated for each observed species and two-inch DBH class combination.[28]

```
elk_ags <- read_csv("datasets/elk_spp_dbh_estimates.csv")
elk_ags <- elk_ags %>%
  mutate(DBH_2in_range = cut_width(DBH_2in, width = 2,
                                   boundary = 1, closed = "left"),
         .after = DBH_2in)
elk_ags
```

```
#> # A tibble: 43 x 6
#>    Species DBH_2in DBH_2in_range Trees_ac BA_ac Logs_ac
#>    <chr>     <dbl> <fct>            <dbl> <dbl>   <dbl>
#>  1 Beech        12 [11,13)          1.33  1.04    1.33
#>  2 Beech        14 [13,15)          0.585 0.625   0.682
```

[27]You could use `DBH_2in` in the subsequent bar charts, but defining the DBH classes as factors that use interval notation eases interpretation.

[28]Methods to generate these estimates from tree measurement data will be covered in Chapter 12.

```
#>  3 Beech        16 [15,17)        0.448  0.625   0.746
#>  4 Beech        20 [19,21)        0.0955 0.208   0.143
#>  5 Black ~      12 [11,13)        1.06   0.833   1.06
#>  6 Black ~      14 [13,15)        0.974  1.04    1.07
#>  7 Black ~      16 [15,17)        0.895  1.25    1.42
#>  8 Black ~      18 [17,19)        0.118  0.208   0.236
#>  9 Black ~      20 [19,21)        0.0955 0.208   0.191
#> 10 Black ~      12 [11,13)        0.796  0.625   0.796
#> # i 33 more rows
```

Below, we make two sets of bar charts: one set for number of trees per acre and the other for basal area per acre. In each set, we show the variable's distribution across DBH classes, with and without species information. To get a per acre estimate for each variable, you must sum the estimates within the desired class, which is accomplished using the `geom_col()` function. We make adjustments to labels, fill color, and theme elements. We also place the legend inside the plot area to save space. Per the footnote a few pages back, we match the `color` to the `fill` aesthetic to get rid of the pesky horizontal line artifacts that appear in the bars.

```
ggplot(elk_ags, aes(x = DBH_2in_range, y = Trees_ac)) +
  geom_col(color = "grey35") +
  xlab("DBH class (in)") +
  ylab("Trees per acre") +
  theme_bw() +
  theme(text = element_text(size = 14), # Increase font size.
        axis.title.x = element_text(vjust = -1)) # Move label down.

ggplot(elk_ags, aes(x = DBH_2in_range, y = Trees_ac,
                    fill = Species, color = Species)) +
  geom_col() +
  xlab("DBH class (in)") +
  ylab("Trees per acre") +
  scale_fill_brewer(name = NULL, type = "qual", palette = "Paired") +
  scale_color_brewer(name = NULL, type = "qual", palette = "Paired") +
  theme_bw() +
  theme(text = element_text(size = 14), # Increase font size.
        axis.title.x = element_text(vjust = -1), # Move label down.
        legend.position = 'inside', # Move legend into plot area.
        legend.position.inside = c(0.85,0.65)) # Location in plot for legend.
```

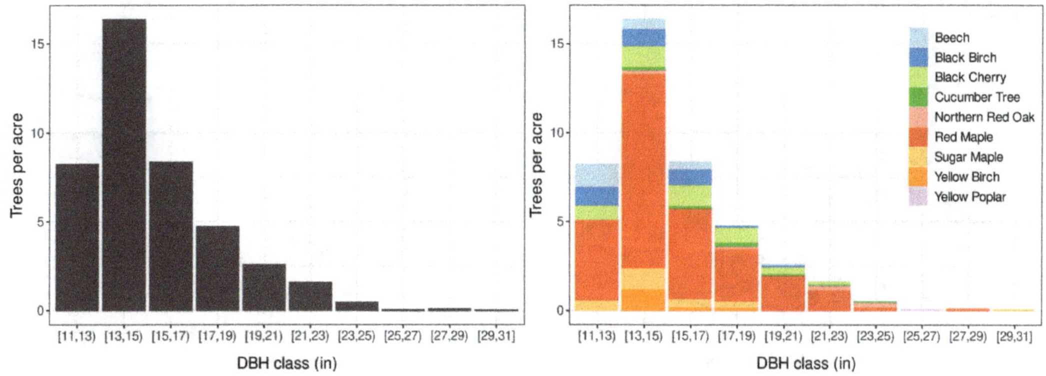

The right bar chart above is a visual representation of the stand table created in Exercise 8.6. Comparing this table with the bar charts above, you can see column totals equal total bar heights for each DBH class. Also, for a given DBH class, species values in the table equal the corresponding bar segment height. Both stem density (i.e., trees per acre) representations show red maple is the dominant species across DBH classes, and particularly so in the $[13, 15)$ in class. Rare species are northern red oak and yellow poplar.

```
ggplot(elk_ags, aes(x = DBH_2in_range, y = Logs_ac)) +
  geom_col(color = "grey35") +
  xlab("DBH class (in)") +
  ylab("Logs per ac") +
  theme_bw() +
  theme(text = element_text(size = 14), # Increase font size.
        axis.title.x = element_text(vjust = -1)) # Move label down.

ggplot(elk_ags, aes(x = DBH_2in_range, y = Logs_ac,
                    fill = Species, color = Species)) +
  geom_col() +
  xlab("DBH class (in)") +
  ylab("Logs per ac") +
  scale_fill_brewer(name = NULL, type = "qual", palette = "Paired") +
  scale_color_brewer(name = NULL, type = "qual", palette = "Paired") +
  theme_bw() +
  theme(text = element_text(size = 14), # Increase font size.
        axis.title.x = element_text(vjust = -1), # Move label down.
        legend.position = 'inside', # Move legend into plot area
        legend.position.inside = c(0.85,0.65)) # Location in plot for legend.
```

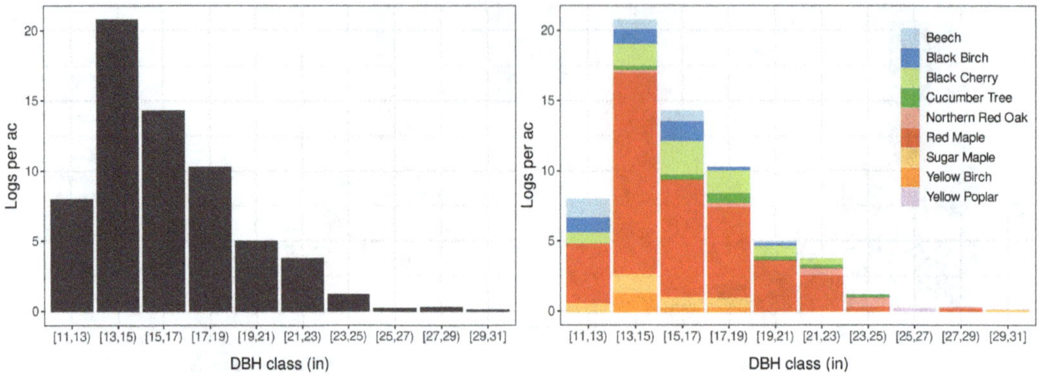

The right bar chart above is a visual representation of the stock table created in Exercise 8.8. Again, comparing this table with the bar charts above, you can see column totals equal total bar heights for each DBH class, and bar segment heights in the right chart correspond to values in the stock table. Given the abundance of red maple in the stand, it's not surprising the right chart above shows the species also carries the most log volume across DBH classes. Of course, similar charts can be created for other continuous variables, such as biomass, basal area, and monetary value if log prices by species are available.

In this section, we took a brief look at the `geom_bar()` and `geom_col()` functions used to create bar charts when the `x` aesthetic is discrete and known (e.g., DBH class or species). Said differently, here we know the bins (DBH class) to partition the variable(s) of interest (e.g., number of trees or biomass). Oftentimes, however, our interest is in getting a sense of how a variable's values are distributed across its value range, and binning serves only to allow for smoothing across the graphic's columns. That is, we want to understand value frequency across the variable's value range. Such understanding is most often sought for continuous variables, where there are no natural bins. In such cases, `geom_histogram()`, see `?ggplot2::geom_histogram()`, might be more appropriate, as illustrated below.

For example, let's look at the distribution of basal area across the Elk County property. This time, however, let's use the tree measurement data available in `"datasets/PA/Elk_trees.csv"`, which is read into `elk` below. We also do a bit of data prep work, including computing basal area from `DBH` (in) and adding it to `elk` as `BA` (ft^2), selecting a few columns we want to work with, removing any rows with an `NA`, and duplicating measurements based on the forester's tree count (i.e., to save time, trees of the same species, DBH, and number of logs are counted and recorded in the `Tree_count` column).

```
elk <- read_csv("datasets/PA/Elk_trees.csv")

# Compute BA and select a few variables we'll work with.
elk <- elk %>%
  mutate(BA = 0.005454 * DBH^2) %>%
  select(Species_name, Product_name, BA, Tree_count)
```

```
# Remove any row with an NA.
elk <- elk %>%
  drop_na()

# Use uncount() to duplicate rows according to tree count.
elk <- elk %>%
  uncount(Tree_count)
```

Okay, let's get a sense of the basal area distribution across the property. We also add an x-axis label with a math superscript for feet squared using the `expression()` function.

```
ggplot(elk, aes(x = BA)) +
  geom_histogram(fill = "steelblue", color = "black") + # A little color.
  xlab(expression(paste("Basal area ", (ft^{2})))) + # Nice math label.
  ylab("Number of trees")
```

```
#> `stat_bin()` using `bins = 30`. Pick better value with
#> `binwidth`.
```

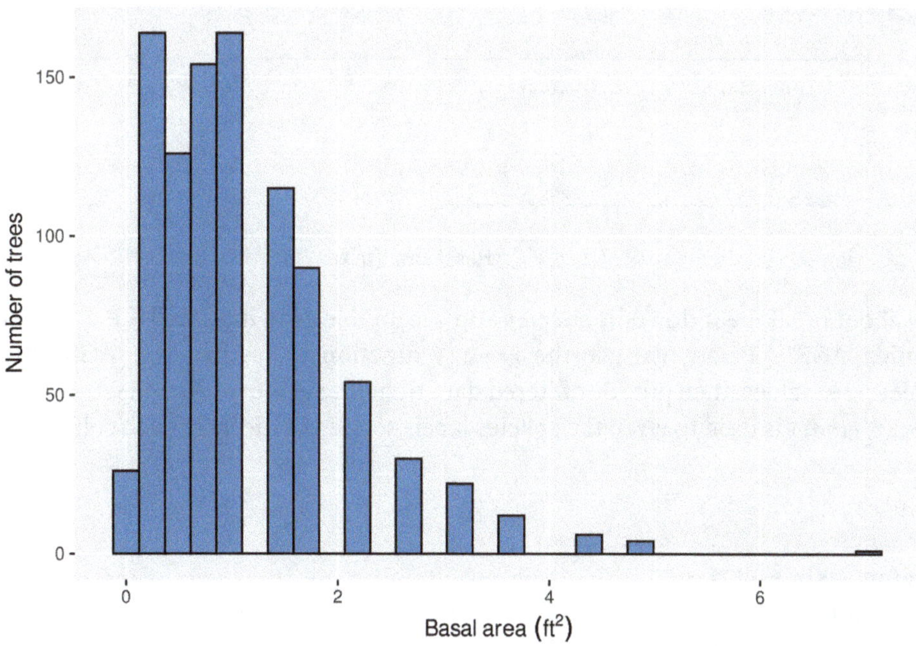

First, notice the message produced by the code above regarding bins. Binwidth is the x-axis interval width over which the variable's values are counted. `geom_histogram()` tries to guess a reasonable binwidth, which in this case resulted in 30 bins across the x-axis, but as the message suggests, you can set your own width. The histogram above shows most trees have a basal area less than 1.5 (ft^2) and at least one with a basal area greater than 6 (ft^2).

Now, let's break the basal area distribution down by species.

```
ggplot(elk, aes(x = BA)) +
  geom_histogram(fill = "steelblue", color = "black") +
  xlab(expression(paste("Basal area ", (ft^{2})))) +
  ylab("Number of trees") +
  facet_wrap(vars(Species_name))
```

```
#> `stat_bin()` using `bins = 30`. Pick better value with
#> `binwidth`.
```

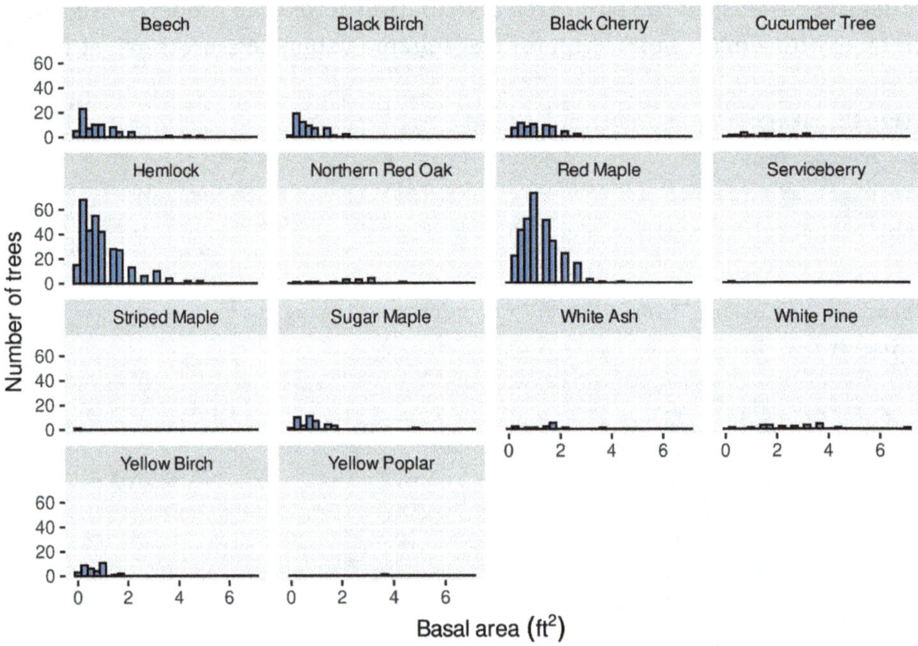

How about breaking it down by species and acceptable growing stock (i.e., `Product_name` includes "AGS")? Below, we use the `grepl()` function to test for the "AGS" string in `Product_name` values, then pipe our filtered data right into `ggplot()`. The `labeller` argument in `facet_grid()` is used to wrap the species labels so they fit nicely in each column.

```
elk %>%
  filter(grepl("AGS", Product_name)) %>%
  ggplot(aes(x = BA)) +
    geom_histogram(fill = "steelblue", color = "steelblue") +
    xlab(expression(paste("Basal area ", (ft^{2})))) +
    ylab("Number of trees") +
    theme(strip.text = element_text(size = 6)) + # Reduce facet label size.
    facet_grid(vars(Product_name), vars(Species_name),
               labeller = labeller(Species_name =
                                     label_wrap_gen(width = 8,
                                                    multi_line = TRUE)))
```

```
#> `stat_bin()` using `bins = 30`. Pick better value with
#> `binwidth`.
```

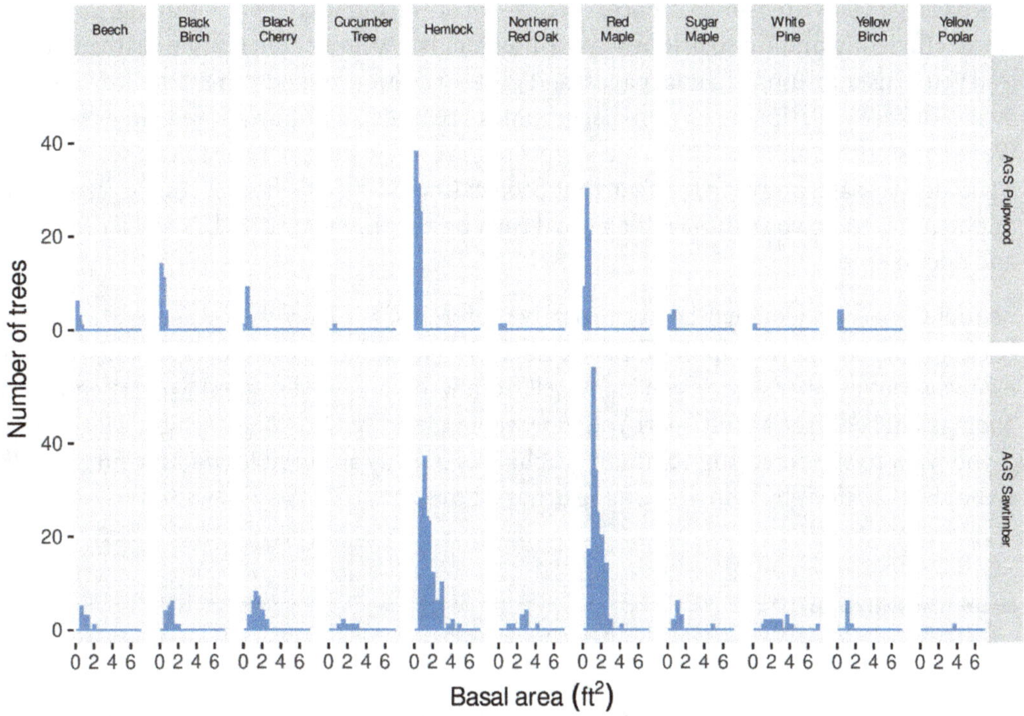

We've looked at three geometries for visualizing one-dimensional distributions, i.e., `geom_col()`, `geom_bar()`, and `geom_histogram()`. While we've only scratched the surface of how these functions can be used and modified, the applications covered here will get you going in the right direction for summarizing most forest inventory data you'll encounter. Additional learning resources are provided in the next section, and we'll continue to use these and related `ggplot2` functions throughout the book.

9.5 Summary

This chapter offered a brief introduction to the `ggplot2` package. We defined Hadley Wickham's grammar of graphics and how grammar components are strung together to create graphics. In Section 9.2, we applied some grammar components to create and modify scatter plots and illustrated geometry, mapping, stats, scale, guide, and theme features. Here too, we presented `ggplot2` as a dual-use tool for efficient EDA and for creating beautiful, informative graphics.

Section 9.4 focused on bar charts most often used to summarize and convey information from forest inventories. In particular, we looked at how stand and stock tables are visualized. We also defined relative abundance and relative dominance, and how these measures of stand diversity are easily computed and visualized.

Our aim for this chapter was to point you in the right direction as you set out to make `ggplot2` graphics. Like other topics covered in this book, becoming comfortable and proficient with this package takes time and patience. There are many resources to help you along. Often, the key to efficiently finding solutions to your problem is a well-written Google search that will lead you to user forums, manual pages, and `ggplot2` references and vignettes. Growing your own codebase of scripts used to make graphics becomes an important resource over time. We're constantly going back into old scripts to grab a few lines of code to create or modify a geometry, scale, or guide for a current project (including this book)! This is yet another opportunity to do your future self a good turn by keeping your codebase well organized and documented.

As mentioned previously, the authoritative reference for learning to use `ggplot2` is the book "*ggplot2*: Elegant Graphics for Data Analysis" (Wickham, Navarro, et al. 2023) and the extensive online reference. If you learn well from webinars, we'd suggest `ggplot2` co-author Thomas Lin Pedersen's multi-part YouTube webinar entitled "*Plotting Anything with ggplot2*." If you want to start creating common graphics as quickly as possible, we'd recommend *The R Graphics Cookbook* by Winston Chang (Chang 2018).[29]

9.6 Exercises

Exercises 9.1 through 9.3 build on the Elk County timber cruise bar charts created in Section 9.4.1. Below, we recreate the `elk_ags` tibble, which contains the per acre estimates for acceptable growing stock and the DBH two-inch class factor.

```
elk_ags <- read_csv("datasets/elk_spp_dbh_estimates.csv")
elk_ags <- elk_ags %>%
  mutate(DBH_2in_range = cut_width(DBH_2in, width = 2,
                                   boundary = 1, closed = "left"),
         .after = DBH_2in)
elk_ags
```

```
#> # A tibble: 43 x 6
#>    Species DBH_2in DBH_2in_range Trees_ac BA_ac Logs_ac
#>    <chr>     <dbl> <fct>            <dbl> <dbl>   <dbl>
#>  1 Beech        12 [11,13)         1.33   1.04    1.33
#>  2 Beech        14 [13,15)         0.585  0.625   0.682
#>  3 Beech        16 [15,17)         0.448  0.625   0.746
#>  4 Beech        20 [19,21)         0.0955 0.208   0.143
#>  5 Black ~      12 [11,13)         1.06   0.833   1.06
#>  6 Black ~      14 [13,15)         0.974  1.04    1.07
#>  7 Black ~      16 [15,17)         0.895  1.25    1.42
```

[29] A free online version of this book is available at https://r-graphics.org.

```
#>  8 Black ~       18 [17,19)         0.118  0.208   0.236
#>  9 Black ~       20 [19,21)         0.0955 0.208   0.191
#> 10 Black ~       12 [11,13)         0.796  0.625   0.796
#> # i 33 more rows
```

Exercise 9.1. Write code using `elk_ags` to reproduce the following bar chart that visualizes relative stand density in the Elk County timber cruise dataset.

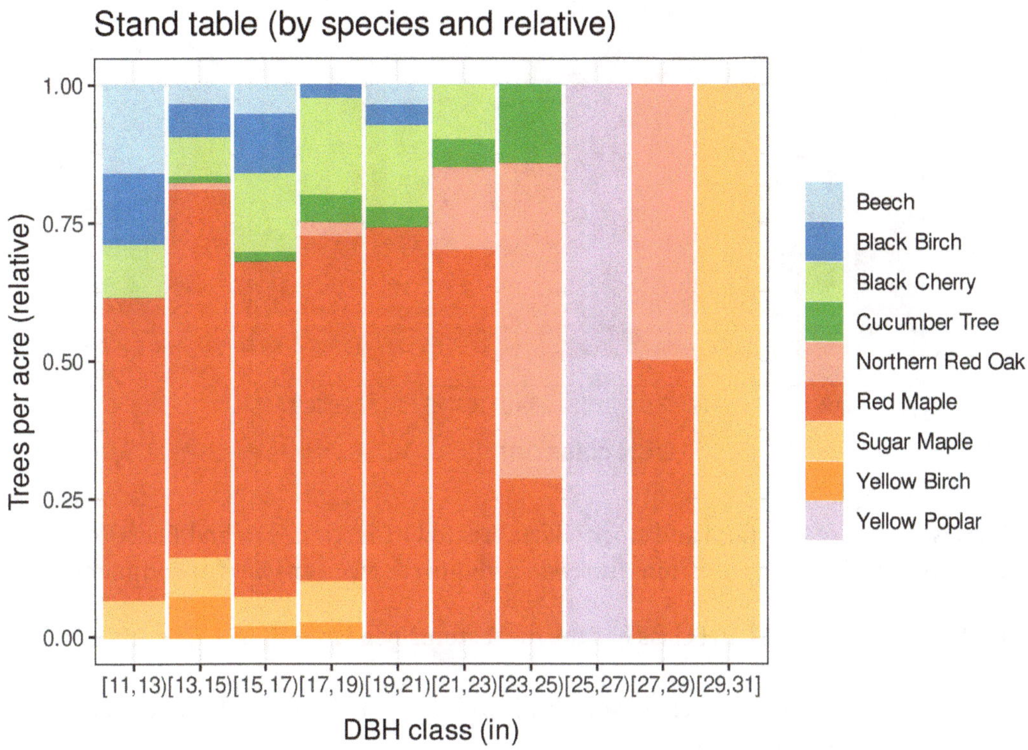

Exercise 9.2. Write code using `elk_ags` to reproduce the following bar chart that is a visual representation of the relative dominance stock table by species in the Elk County timber cruise dataset.

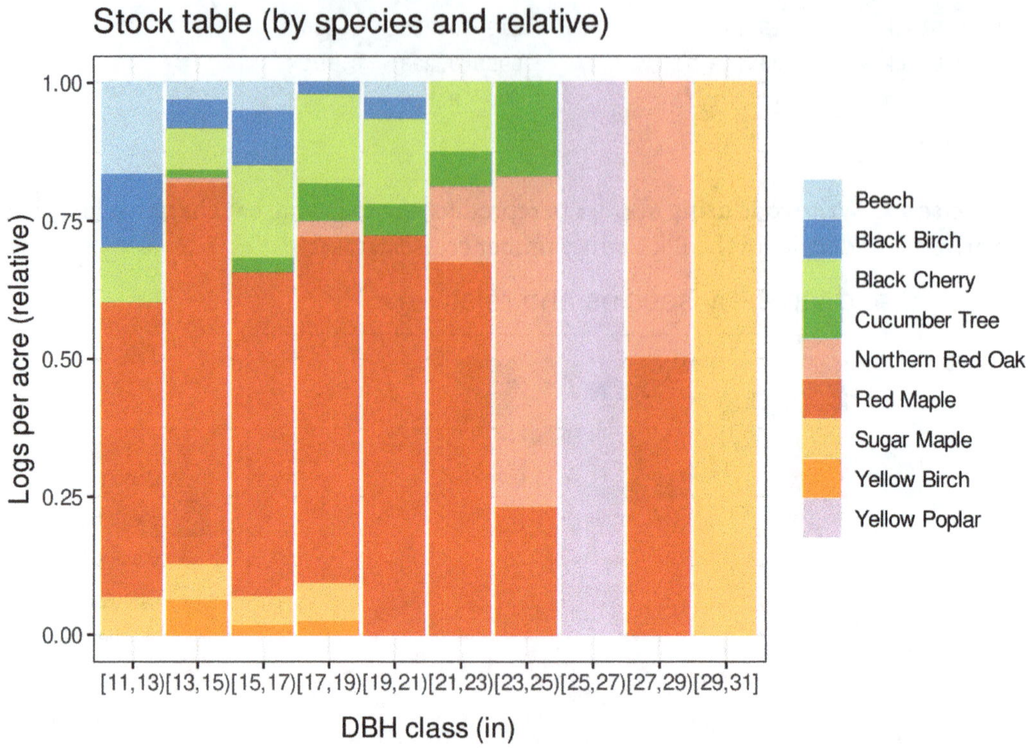

Exercise 9.3. Use the stand and stock charts created in Section 9.4.1 and Exercises 9.1-9.2 to answer the following questions (no code is required; write answers as comments).

a. Which species and DBH class has the most stems per acre?

b. In 25 years, which species do you think will have the largest relative density in the $[25, 27)$ inch DBH class?

c. Which species and DBH class has the most logs per acre?

d. Which species has the most logs per acre across DBH classes?

e. In 25 years, which species do you think will dominate (in logs per acre) across most DBH classes?

In Exercises 9.4-9.11, we'll use the aspen FACE dataset to visualize how increasing tropospheric ozone (O_3) and carbon dioxide (CO_2) influenced aspen diameter growth from 1997 to 2005. Note that while some exercises do not explicitly use `ggplot2`, they showcase a typical `tidyverse` pipeline where we manipulate and wrangle data using `dplyr`, tidy data using `tidyr`, then plot data using `ggplot2`. We load the complete FACE dataset below.

```
face <- read_csv("datasets/FACE/FACE_aspen_core_growth.csv")
face %>%
```

```
names()
```

```
#>  [1] "Rep"                "Treat"
#>  [3] "Clone"              "E Clone"
#>  [5] "Row"                "Col"
#>  [7] "ID #"               "1997Initial_Height"
#>  [9] "1997Initial_Diam"   "1997Final_Height"
#> [11] "1997Final_Diam"     "1998_Height"
#> [13] "1998_Diam"          "1999_Height"
#> [15] "1999_Diam"          "2000_Height"
#> [17] "2000_Diam"          "2001_Height"
#> [19] "2001_AvgDiam"       "2001_Diam@3cm"
#> [21] "2001_Diam@10cm"     "2002_Height"
#> [23] "2002_Diam@10cm"     "2003_Height"
#> [25] "2003_Diam@10cm"     "2003_DBH"
#> [27] "2004_Height"        "2004_Diam@10cm"
#> [29] "2004_DBH"           "2005_Height"
#> [31] "2005_Diam@10cm"     "2005_DBH"
#> [33] "2006_Height"        "2006_DBH"
#> [35] "2007_Height"        "2007_DBH"
#> [37] "2008_Height"        "2008_DBH"
#> [39] "Notes"              "Comment1"
#> [41] "Comment2"           "Comment3"
#> [43] "Comment4"           "Comment5"
#> [45] "Comment6"
```

Exercise 9.4. Using `dplyr` tools learned in Chapter 7 (and code from earlier in this chapter), convert the `Treat` variable to a factor with the following labels: "Control" (`Treat` = 1), "Elevated CO2" (`Treat` = 2), "Elevated O3" (`Treat` = 3), "Elevated CO2 and O3" (`Treat` = 4).

Exercise 9.5. Use `dplyr::select()` and the `contains()` helper function to select only the columns `Rep`, `Treat`, `Clone`, `ID #`, and those containing diameter measurements at 10 cm from 1997 to 2005. Note that the columns `2001_Diam@3cm` and `1997Initial_Diam` should not be included in the final dataset. Your new `face` tibble should match ours below.

```
#> Rows: 1,991
#> Columns: 14
#> $ Rep              <dbl> 1, 1, 1, 1, 1, 1, 1, 1, 1, 1~
#> $ Treat            <fct> Control, Control, Control, C~
#> $ Clone            <dbl> 8, 216, 8, 216, 216, 271, 27~
#> $ ID               <dbl> 45, 44, 43, 42, 54, 55, 56, ~
#> $ `1997Final_Diam` <dbl> 0.40, 0.90, 0.25, 0.78, 0.75~
#> $ `1998_Diam`      <dbl> 1.34, 3.11, 0.96, 3.03, 1.92~
#> $ `1999_Diam`      <dbl> 2.52, 5.19, 1.71, 5.03, 3.47~
#> $ `2000_Diam`      <dbl> NA, 5.960, 2.110, 5.775, 3.9~
```

```
#> $ `2001_AvgDiam`   <dbl> NA, 6.43, 2.21, 6.14, 4.05, ~
#> $ `2001_Diam@10cm` <dbl> NA, 6.29, 2.20, 5.92, 4.00, ~
#> $ `2002_Diam@10cm` <dbl> NA, 6.91, 2.51, 6.62, 4.04, ~
#> $ `2003_Diam@10cm` <dbl> NA, 7.83, 2.42, 7.58, 4.30, ~
#> $ `2004_Diam@10cm` <dbl> NA, 8.6, 2.6, 8.1, 4.4, 6.5,~
#> $ `2005_Diam@10cm` <dbl> NA, 9.1, 2.6, 8.7, 4.3, 6.7,~
```

Exercise 9.6. Use `pivot_longer()` from `tidyr` to convert the `face` dataset to long format. Save the new tibble as `face_long`. Use `glimpse()` to make sure your tibble matches ours below.

```
#> Rows: 19,910
#> Columns: 6
#> $ Rep       <dbl> 1, 1, 1, 1, 1, 1, 1, 1, 1, 1, 1, 1,~
#> $ Treat     <fct> Control, Control, Control, Control,~
#> $ Clone     <dbl> 8, 8, 8, 8, 8, 8, 8, 8, 8, 8, 216, ~
#> $ ID        <dbl> 45, 45, 45, 45, 45, 45, 45, 45, 45,~
#> $ Year_Type <chr> "1997Final_Diam", "1998_Diam", "199~
#> $ Diam_cm   <dbl> 0.40, 1.34, 2.52, NA, NA, NA, NA, N~
```

Exercise 9.7. Use `separate_wider_delim()` to extract only the `Year` from the `Year_Type` column. Hint, use `delim = '_'`. Again, make sure your updated `face_long` matches the results of our `glimpse()` below.

```
#> Rows: 19,910
#> Columns: 6
#> $ Rep     <dbl> 1, 1, 1, 1, 1, 1, 1, 1, 1, 1, 1, 1, 1~
#> $ Treat   <fct> Control, Control, Control, Control, C~
#> $ Clone   <dbl> 8, 8, 8, 8, 8, 8, 8, 8, 8, 8, 216, 21~
#> $ ID      <dbl> 45, 45, 45, 45, 45, 45, 45, 45, 45, 4~
#> $ Year    <chr> "1997Final", "1998", "1999", "2000", ~
#> $ Diam_cm <dbl> 0.40, 1.34, 2.52, NA, NA, NA, NA, NA,~
```

Exercise 9.8. Notice the `Year` column is a character variable because the records in 1997 were recorded as `1997Final`. Use `tidyr::parse_number()` to extract just the numeric values from `Year`. Also convert the resulting `Year` column to a factor.

Exercise 9.9. Use the `face_long` dataset to create the following plot, which shows the mean diameter at 10 cm in each year for each treatment. Hint, use the `stat_summary()` function to easily calculate means within each treatment level. We used the viridis discrete color theme (see `?scale_color_viridis_d`) and set the point size to 3.

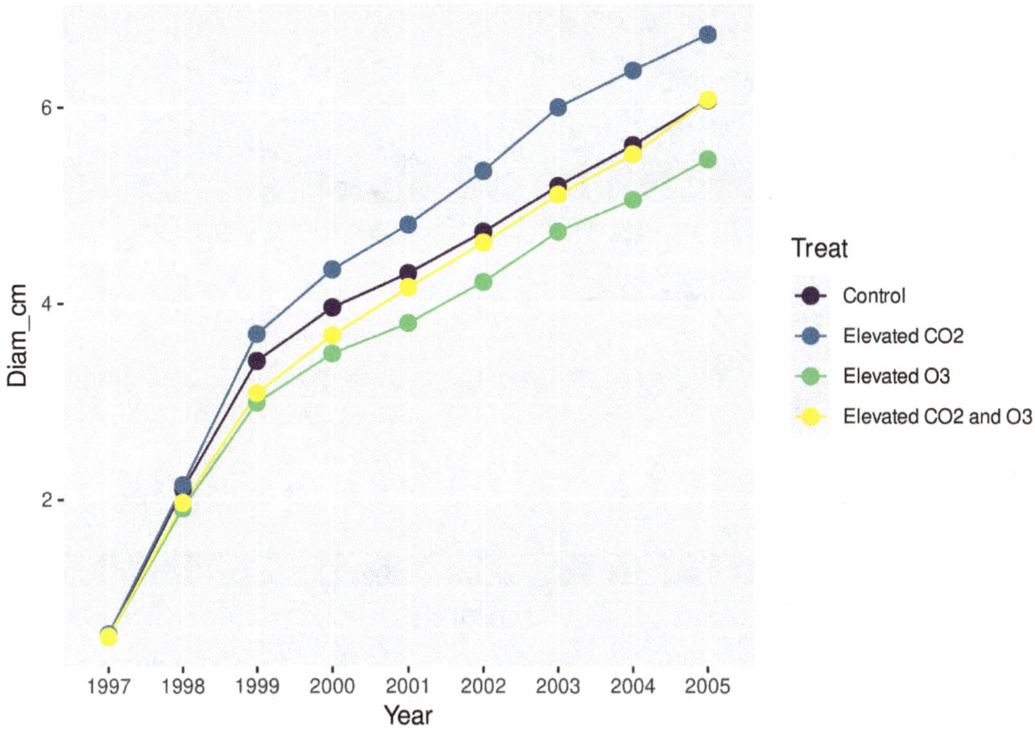

Exercise 9.10. Create the same plot as Exercise 9.9, but this time use a `dplyr` `group_by()` and `summarize()` to first calculate the means, and then do not use `stat_summary()`.

Exercise 9.11. Adapt your code from Exercise 9.9 to make the plot a bit more visually appealing by: (a) adjusting the axis labels; (b) removing the legend title and moving the legend below the plot; (c) changing the theme to `theme_bw()`; (d) setting the text size to 14; and (e) removing the panel grid from the plot theme. The resulting plot should look like the following.

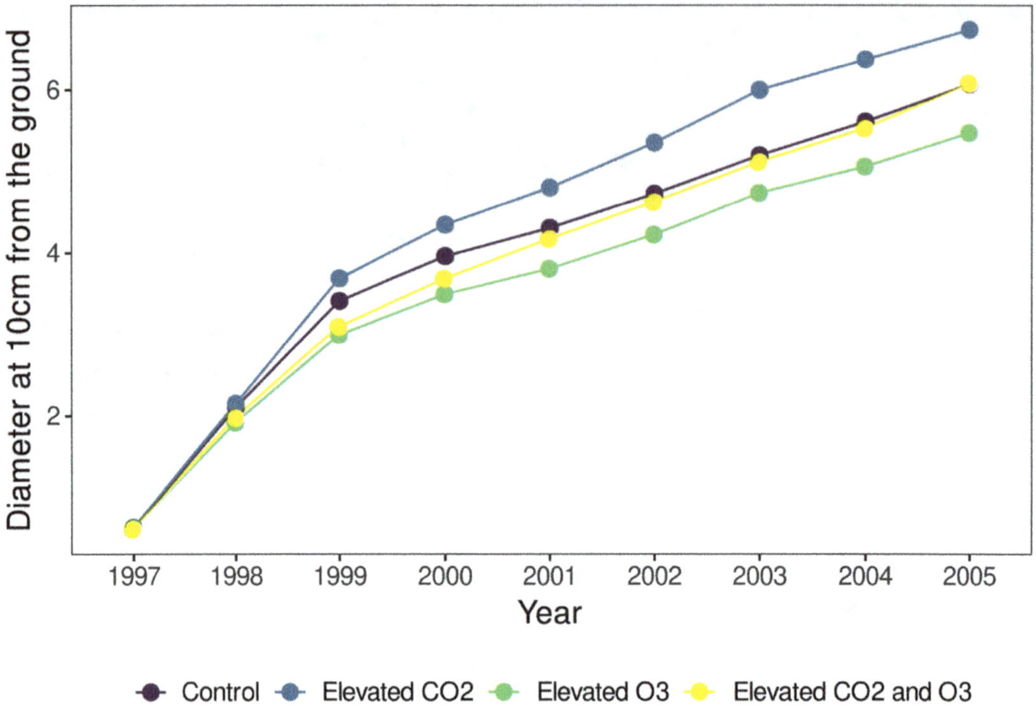

In Exercises 9.12-9.15, we'll work with the Minnesota tree growth dataset introduced in Section 6.1. In each exercise, we provide the final plot, and you need to write the code to generate the plot.

```
mn_trees <- read.csv("datasets/mn_trees_subset.csv")
mn_trees %>%
  glimpse()
```

```
#> Rows: 11,649
#> Columns: 8
#> $ stand_id <int> 1, 1, 1, 1, 1, 1, 1, 1, 1, 1, 1, 1, ~
#> $ plot_id  <int> 1, 1, 1, 1, 1, 1, 1, 1, 1, 1, 1, 1, ~
#> $ tree_id  <int> 1, 1, 1, 1, 1, 1, 1, 1, 1, 1, 1, 1, ~
#> $ year     <int> 1960, 1961, 1962, 1963, 1964, 1965, ~
#> $ species  <chr> "ABBA", "ABBA", "ABBA", "ABBA", "ABB~
#> $ age      <int> 1, 2, 3, 4, 5, 6, 7, 8, 9, 10, 11, 1~
#> $ rad_inc  <dbl> 0.930, 0.950, 0.985, 0.985, 0.715, 0~
#> $ DBH      <dbl> 2.1563, 2.3463, 2.5433, 2.7403, 2.88~
```

Exercise 9.12. Create a histogram that shows the diameter distribution of all trees in 2007. Use `dplyr::filter()` to only show trees in 2007, piping the result directly into `ggplot()`. We used a fill color of `"firebrick"` and `theme_classic()`.

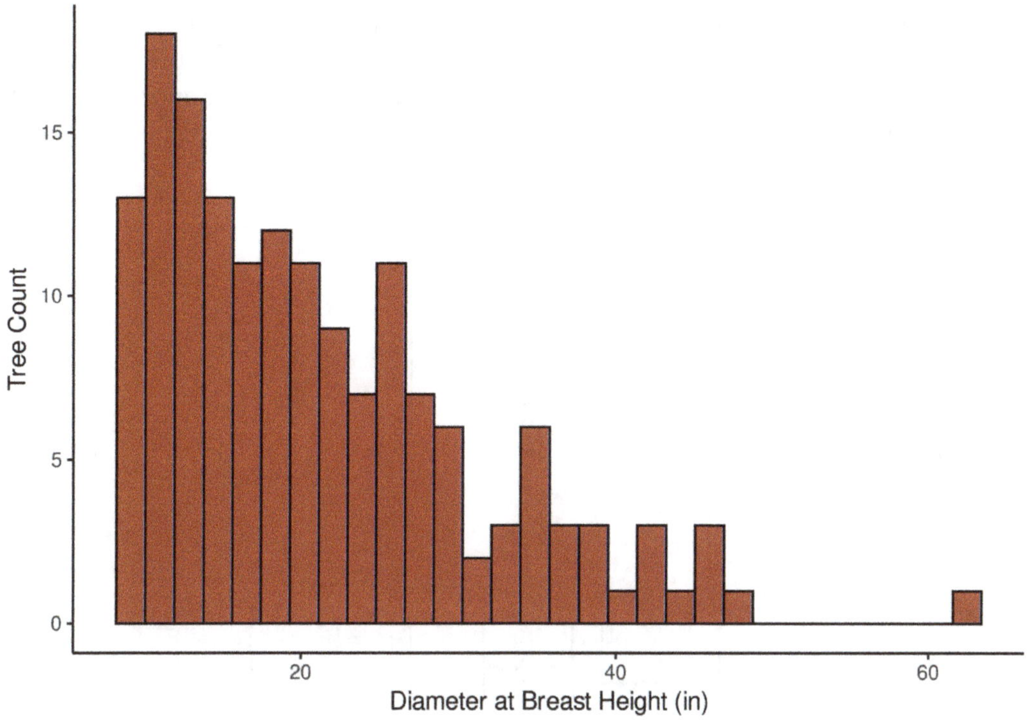

Exercise 9.13. Use `facet_wrap()` to create a histogram of the tree diameter distribution in 2007 separately for the five stands in the `mn_trees` dataset. Use `theme_bw()` and make the plot have two columns and three rows.

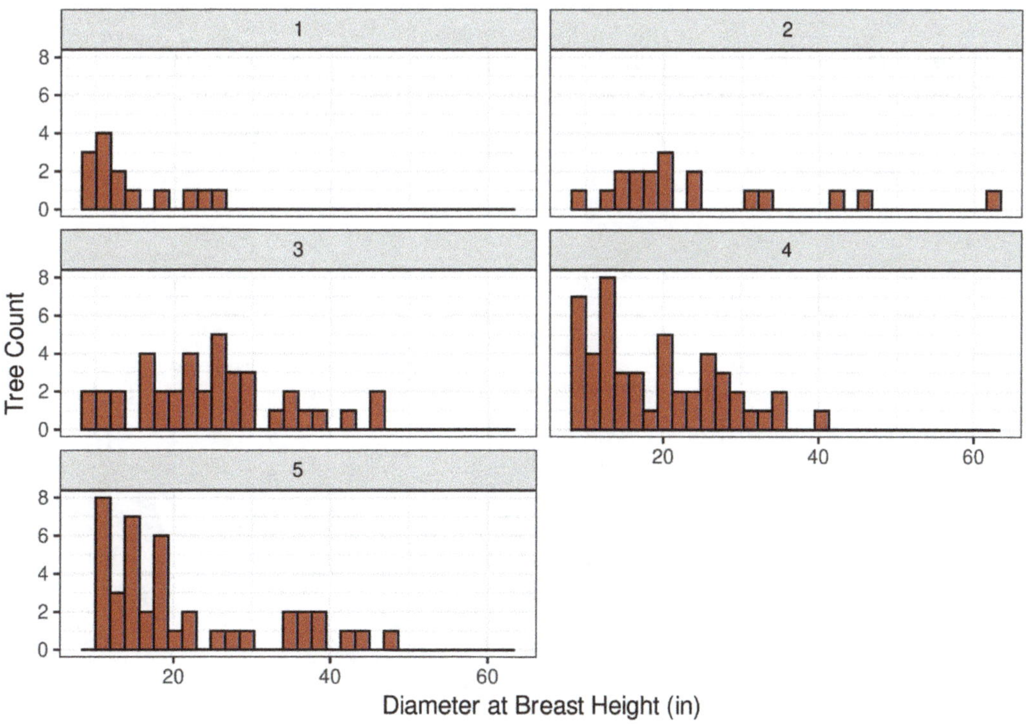

Exercise 9.14. `geom_boxplot()` produces box-and-whisker plots for a given variable across a set of discrete groups. Use `geom_boxplot()` to show the distribution of DBH values across the eight species in the `mn_trees` dataset. Use `"lightblue"` as the fill color and `"black"` as the border color.

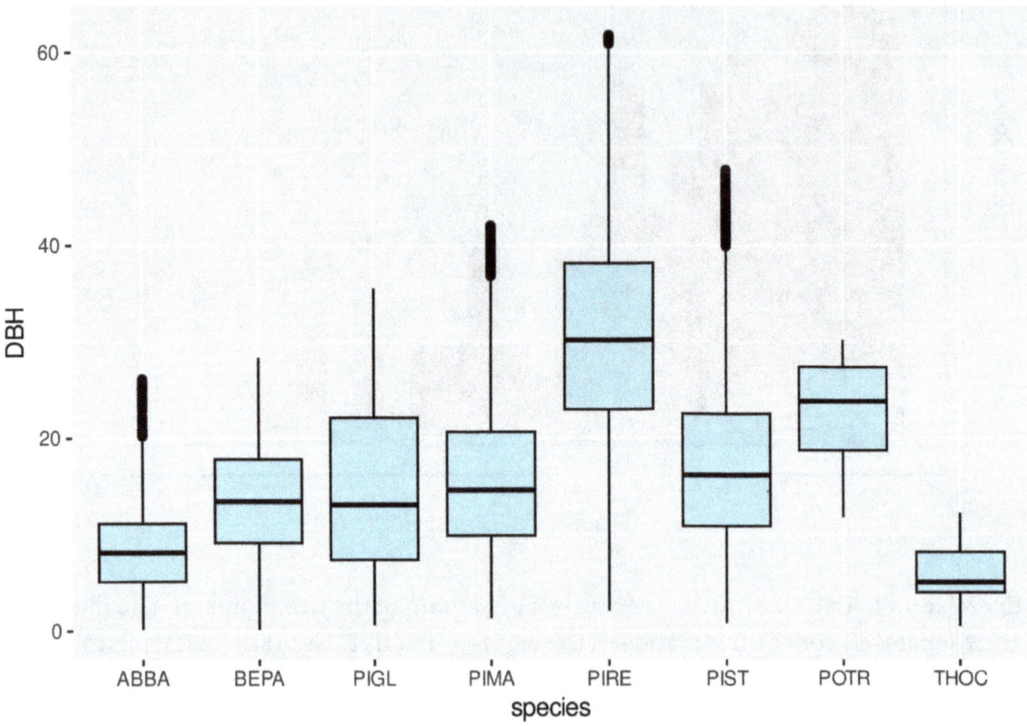

Exercise 9.15. Use the `coord_flip()` function to flip the coordinates of the boxplot created in Exercise 9.14. Also note the different theme, axis labels, title, and caption.

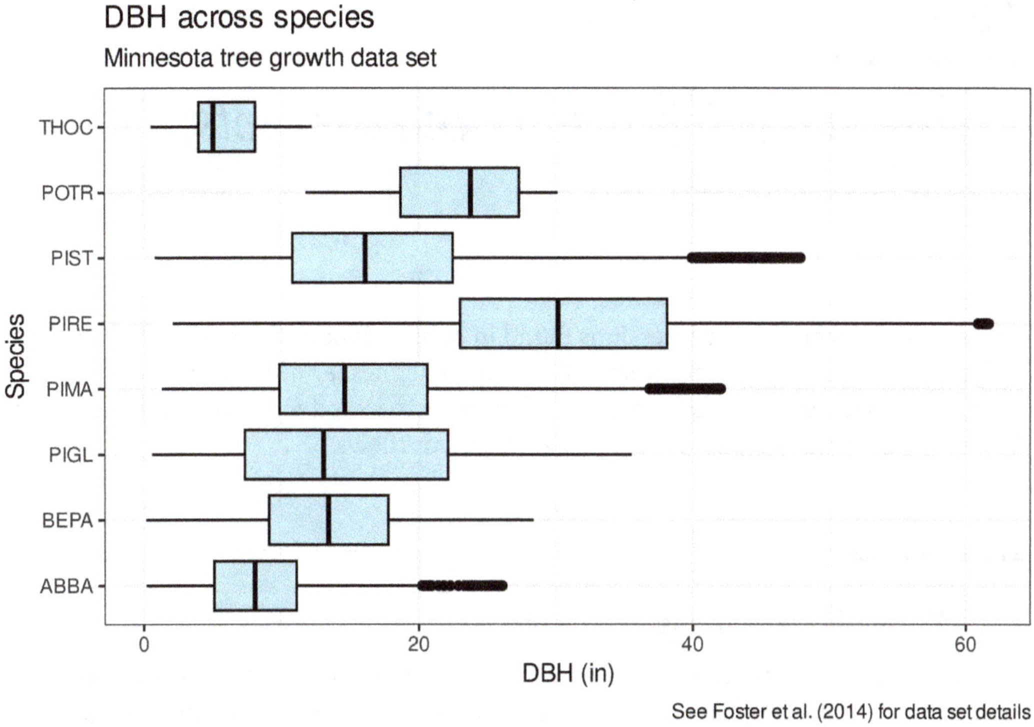
DBH across species
Minnesota tree growth data set
THOC
POTR
PIST
PIRE
PIMA
PIGL
BEPA
ABBA
Species
0
20
40
60
DBH (in)
See Foster et al. (2014) for data set details

10

Preliminary definitions and concepts

Translating mathematical expressions found in publications and reference material into code is a fundamental data analyst skill. This chapter defines terminology and mathematical notation used throughout the remainder of the book to introduce statistical concepts and how they can be translated into efficient analysis workflows.

10.1 Estimating population parameters

A central aim of data analysis, as covered in this book, is to infer (meaning to deduce or learn about) population characteristics of interest from a sample taken from the population. The *population* is an aggregate of units that belong to some defined group. The *units* are individuals or objects that collectively comprise the population. In some cases we're able to *observe* (meaning measure) all units in the population. Observing all population units is a *census* and means we can describe the population characteristic of interest in its entirety and without error (assuming no measurement error). Recall, the Harvard Forest dataset described in Section 1.2.5 is a census of the population, with units defined as all woody vegetation with a DBH of 1 cm and larger. In most practical cases, however, it's too time-consuming or expensive to observe all units in the population. In such cases, we settle for observing a *sample*, which means a subset of population units. If selected appropriately, this sample might ultimately provide a useful *estimate* of the population characteristic of interest.

We use the word *parameter* to mean a summary measure of a population characteristic. For example, a parameter could be an average, total, or proportion. Similarly, we use the word *statistic* to mean a summary measure of a sample characteristic, and like a parameter, a statistic could be an average, total, or proportion. Statistics are used to infer something about the population parameter of interest.

A characteristic that might vary from unit to unit is called a *variable*. Our observations, or measurements, of a variable in a sample of population units inform statistics, which serve as estimates of the population parameter of interest. The typical data analysis progression takes the following steps.

1. Define the population, population units, and population parameters of interest, then

DOI: 10.1201/9781003222699-10

2. Measure one or more variables on a sample of population units, and compute one or more statistics that serve as estimates of the population parameters. Typically, these statistics provide our best estimate of the parameter and quantify how certain we are in the estimate.

Here's an example to illustrate this new vocabulary. Say you want to know the total merchantable timber volume on a 50-acre property, where you define "merchantable timber volume" as volume from live trees 10.0 inches DBH or larger with at least one 16-foot sawlog. The *population* comprises all trees that meet your definition of merchantable timber volume on the 50-acre property. The population *parameter* of interest is total merchantable timber volume. The population *units* are the individual trees that meet your merchantable definition. The *variable* measured on each unit (i.e., tree) is its volume.[1]

Alternatively, and more typically in practice, we might define the population units as some fixed-area field plot. For example, we might divide the 50-acre property into 250 1/5-th acre non-overlapping plots. In this case, the *variable* is the cumulative volume of all trees that meet the population definition for merchantable timber on each 1/5-th acre plot. Regardless of whether our population unit is an individual tree or a group of trees on a plot, we likely won't be able to take a *census* of the property due to time or effort constraints, but rather we'll select a *sample* of population units for which we'll measure volume. Then, given the sample measurements, we compute a *statistic* that serves as our *estimate* for total merchantable timber volume on the property.

10.1.1 Types of variables

It's useful to connect the R data types introduced in Section 4.2.1 with those of variables we encounter in subsequent chapters. Following Figure 10.1, a variable is initially classified as either *quantitative* or *qualitative*.

A quantitative variable has values that give a notion of magnitude; that is, the values are a numerical measure. This numerical measure is either *continuous* or *discrete*. A quantitative continuous variable can take an infinite (not countable) number of possible values, i.e., an infinitely fine increment from one value to the next. Said differently, a continuous variable can have an arbitrary number of decimal places for a given value. Examples include height, weight, and volume. In comparison, a quantitative discrete variable can take a finite (countable) number of possible values when moving from one value to the next. The values are often (but not always) integers. Examples include age in whole years, number of trees, and number of fire events.

A qualitative variable (also referred to as a categorical variable) has values that represent different categories. Qualitative variables are either *nominal* or *ordinal*. A qualitative nominal variable takes values for which no ordering is possible or implied in the categories. For example, the variable species is nominal because there's no inherent or natural order in species names. Similarly, forest ownership type coded as federal, state, private, or tribal is

---

[1]Although in practice we typically measure DBH and height, then compute volume using an allometric equation like those briefly discussed in Section 1.2.1 and illustrated in Sections 5.3.2 and 7.14.3.

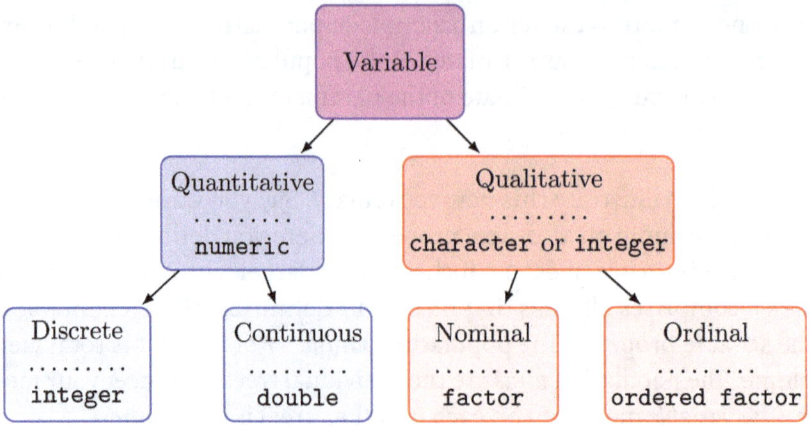

FIGURE 10.1: Variable types with corresponding R data types added below the dotted line.

nominal because there's no apparent ordering. In contrast, the categories of a qualitative ordinal variable have some natural ordering. For example, the variable tree canopy position can take the values suppressed, intermediate, co-dominant, or dominant, where these categories themselves imply an ordering. Further examples include disturbance severity code with categories low, medium, and high, or forest or ecosystem succession stages. Again, in these examples, there is a natural order to the categories.

10.2 Tools of the trade: Review of notation

Frank Freese was a prolific forest biometrician who worked for the USFS Southern Experiment Station in Asheville, North Carolina, and the Forest Products Lab in Madison, Wisconsin. He wrote several primers on applied statistics for practicing foresters. Despite being somewhat dated, Frank's works are clearly written, fun to read, and offer lots of worked examples that make them enduring resources for those interested in an efficient and accessible introduction to classical statistics and sampling methods. In one of his works entitled *Elementary Forest Sampling* (Freese 1962), he includes a section called "Tools of the Trade: Subscripts, Summations, and Brackets" that provides a refresher (or perhaps a first look) at more frequently used mathematical expressions in applied statistics and sampling. Following Frank's lead, we offer a similar section that covers common mathematical notation and its translation to R code.

Throughout the remainder of this book, Greek letters, e.g., μ, σ, α, β, η, represent population *parameters*.[2] Roman letters, e.g., x, y, z, represent *variables*. For example, we might say $x = 5$, which is a *scalar* (i.e., single value), or $y = (1, 4, 5)$, which is a *vector* of three values. A vector is a collection of one or more values organized such that the values can be referenced using an *index*—analogous to the R vector data object introduced in

[2]These are the letters mu, sigma, alpha, beta, and eta, respectively.

Section 4.2. A *sample statistic*, which is any quantity computed from values in a sample and used to estimate population parameters, will be represented using a Roman letter, sometimes decorated with a *bar*, *hat*, or some other symbol. For example, $\bar{y}$ is read as *y bar* and $\hat{R}$ as *R hat*.

10.2.1 Mathematical operators, subscripts, and nested data

You already know many mathematical operators: addition `+`, subtraction `-`, multiplication `*`, division `\`, exponentiation `^`, and assignment `=`. Like the comparison and logical operators introduced in Section 4.7, mathematical operators are applied to *operands*. For example, in the expression 1 + 2, there are two operands: the left operand is 1 and the right operand is 2. An operator is *binary* if it has two operands and *unary* if it has a single operand.

Although you might not have thought about it this way, the minus sign `-` can be unary (negation) or binary (subtraction). The unary form reverses the sign of the operand on its right; the binary form subtracts the right operand from the left.

Subscripts appear as one or more letters to the lower right side of a variable. For example, x_i is a variable x with subscript i. The subscript letter is a placeholder for an integer value used to index elements in a vector. Say $x = (4, 1, 5, 2, 8)$; that is, the variable x is a vector of length 5 with values 4, 1, 5, 2, and 8. If $i = 1$, then x_i equals 4—the first element in the vector. Similarly, when $i = 2$, x_i equals 1; when $i = 3$, x_i equals 5; and so on.

You'll see in later chapters that summing the elements of a vector is a recurring theme when computing sample statistics and other quantities of interest. Following from the previous example, the sum of the vector x could be written as

$$(x_1 + x_2 + x_3 + x_4 + x_5).$$

A more compact way to write this summation is

$$(x_1 + x_2 + ... + x_5),$$

where the ..., which is an ellipsis, means to continue the same pattern for x_3 and x_4. An even more compact way to express this summation is

$$\sum_{i=1}^{5} x_i,$$

where $\sum$ is a symbol that means add a `+` between the values of x_i each time i changes. The $i = 1$ below the summation symbol says subscript i starts at value 1 and increases by 1 until it equals the number above the summation symbol, which in this case is 5. In some books, you'll see the number on top of the summation symbol omitted, which should be taken to mean increment the subscript to the number of elements in the vector being summed.

Very often we use more than one subscript to reference values of a variable. For example, from Section 1.2.1, recall the FEF dataset comprises 88 trees sampled on 17 plots across two watersheds. More specifically, 44 trees were measured across 8 plots in watershed 1, and 44 trees were measured across 9 plots in watershed 2. We'll see later that it's helpful to use a separate subscript to index each level in these kinds of nested datasets.

TABLE 10.1: Data used for summation examples.

Plot (i)	Tree (j)	x	y
1	1	1.2	3
1	2	2.4	4
2	1	0.4	2
2	2	6.3	6
2	3	2.2	5

For now, let's just focus on the data in one of the two watersheds. Let x represent stem biomass and use subscript i for plot and j for tree. Using this notation, we can uniquely identify any tree biomass using the subscript $x_{i,j}$. Translating this notation into words, we might say $x_{i,j}$ is the biomass of the j-th tree in the i-th plot. Now, using these subscripts, say we want the total biomass of all trees across all plots, which can be written

$$\sum_{i=1}^{m}\sum_{j=1}^{n_i} x_{i,j},$$

where m is the number of plots and n_i is the number of trees in plot i. Notice the subscript on n is necessary because there might be a different number of trees in each plot, and the i says that n_i is plot-specific. The expanded equivalent of this summation is

$$\begin{aligned}
&x_{1,1} + x_{1,2} + \dots + x_{1,n_1}\\
&\quad + x_{2,1} + x_{2,2} + \dots + x_{2,n_2}\\
&\qquad\qquad \vdots\\
&\quad + x_{i,1} + x_{i,2} + \dots + x_{i,n_i}\\
&\qquad\qquad \vdots\\
&\quad + x_{m,1} + x_{m,2} + \dots + x_{m,n_m},
\end{aligned}$$

where the $\vdots$ indicates that each omitted row in the expression looks the same as the first two, except that they increment the i index until it reaches m.

Consider the example data given in Table 10.1. Using the same notation as the FEF biomass summation example above, these example data comprise $m = 2$ plots (indexed using i), with multiple trees measured in each plot (indexed using j). For plot $i = 1$, there are $n_i = 2$ trees measured, and for plot $i = 2$, there are $n_i = 3$ trees measured. The columns x and y represent two variables measured on each tree.

Using the indexing described above, the sum of the x values in Table 10.1 can be written as

$$\begin{aligned}
\sum_{i=1}^{m}\sum_{j=1}^{n_i} x_{i,j} &= x_{1,1} + x_{1,2} + x_{2,1} + x_{2,2} + x_{2,3}\\
&= 1.2 + 2.4 + 0.4 + 6.3 + 2.2\\
&= 12.5.
\end{aligned}$$

This subscript notation extends in a straightforward way to accommodate any number of different data levels combined with any mathematical operator (we focus on the summation operator here because it's used most frequently in the methods considered in this book). As an example of extending this notation, consider the watershed level of the FEF data and index watershed using subscript h, with H representing the number of watersheds. Tree biomass measurements x can now be indexed as the j-th tree, within the i-th plot, within the h-th watershed. The total biomass is now expressed as

$$\sum_{h=1}^{H}\sum_{i=1}^{m_h}\sum_{j=1}^{n_i} x_{h,i,j}. \tag{10.1}$$

Note that in the summation above we added the h subscript to m to acknowledge that there might be a different number of plots within each watershed.

A common characteristic of datasets we encounter is that tree measurements (i.e., *observations*) are *nested* in one or more levels. The FEF data are a perfect example: observations are nested in plots, plots are nested in watersheds, and the watersheds are nested in the entire FEF. The Elk County inventory data in Section 1.2.2 have tree observations nested in plots, which are nested in the forested property. The PEF data in Section 1.2.4 have tree observations nested in year, nested in inventory plot, nested in management unit, and the management units are nested in the entire PEF. The FACE data in Section 1.2.3 have a complex nesting structure, with aspen tree diameter observations nested in clone, nested in year, nested in experimental replicate, and nested in treatment. As illustrated in (10.1), subscripts provide useful notation to describe how observations are nested at different levels and the order in which different mathematical operators apply to the observations. This nesting becomes especially important in the programming and analysis chapters, where we apply different estimators and seek data summaries at various data levels.

As demonstrated in subsequent chapters, the `dplyr` `group_by()` and `summarize()` workflows, along with other tidyverse functions, are perfectly suited to efficiently analyzing the kind of nested data we encounter in practice.

10.2.2 Order of operations

Perhaps at one time you learned the acronym PEMDAS for Parenthesis, Exponents, Multiplication, Division, Addition, and Subtraction, along with its handy mnemonic Please Excuse My Dear Aunt Sally. Given two or more operations in a single expression, the order of the letters in PEMDAS tells us what to calculate first, second, third, and so on, until the calculation is complete. Confusing the order of operations is a common coding error when implementing the estimators considered in subsequent chapters—it's time well invested reviewing and practicing these foundational rules.

Consider the summation of n values of a variable x squared, expressed as

$$\sum_{i=1}^{n} x_i^2. \tag{10.2}$$

The two operations in (10.2) are raising each x_i to the exponent of 2 (i.e., squaring), and summing the n resulting values of x_i^2. Here, our handy acronym PEMDAS reminds us to apply exponents before summation, so we first square each x_i and then sum the resulting n values.

Now let's add parentheses around the summation in (10.2) and see how the order of operations changes

$$\left(\sum_{i=1}^{n} x_i\right)^2 . \tag{10.3}$$

Looking at (10.3) and applying PEMDAS, the order of operations is to first sum the n x values because they are inside parentheses, then square the *scalar* result. A scalar is a mathematical term meaning a single real number, which is what you're left with after the summation in (10.3).

Following the data structure in Table 10.1, consider the expression

$$\sum_{i=1}^{m}\left(\sum_{j=1}^{n_i} x_{i,j}\right)^2 , \tag{10.4}$$

where, recall, i indexes the m plots and j indexes the n_i trees on each plot. There are a few steps to evaluate (10.4). First, for each value of i, sum over the n_i values of $x_{i,j}$, then square the resulting scalar. Second, sum the m resulting scalars. Here's the expanded form:

$$\begin{aligned}\sum_{i=1}^{m}\left(\sum_{j=1}^{n_i} x_{i,j}\right)^2 &= \left(x_{1,1} + x_{1,2} + \ldots + x_{1,n_1}\right)^2 \\ &+ \left(x_{2,1} + x_{2,2} + \ldots + x_{2,n_2}\right)^2 \\ &\vdots \\ &+ \left(x_{m,1} + x_{m,2} + \ldots + x_{m,n_m}\right)^2 .\end{aligned} \tag{10.5}$$

Using data in Table 10.1, (10.5) equals 92.17.

From here forward, we'll drop the comma between subscript elements (e.g., $x_{i,j}$) to reduce visual clutter. The comma is handy when first introducing subscripts because it clearly separates multiple indices and helps explain subsetting. However, as equations grow longer the comma adds nothing to meaning and only clutters the notation, so we'll omit it to keep expressions cleaner and easier to read.

You might also encounter two or more variables with subscripts in the same expression. For example,

$$\sum_{i=1}^{n} x_i y_i = x_1 y_1 + x_2 y_2 + \ldots + x_n y_n, \tag{10.6}$$

where, following PEMDAS, multiplication of x_i and y_i comes before summation.

Here's another expression with two variables

$$\left(\sum_{i=1}^{n} x_i\right)\left(\sum_{i=1}^{n} y_i\right) = (x_1 + x_2 + \ldots + x_n)(y_1 + y_2 + \ldots + y_n), \tag{10.7}$$

where, because of the parentheses, we first sum the n values of x and y, then multiply the two resulting scalars.

10.2.3 Tools of the trade using R

As noted at the beginning of Section 10.2, a vector is a collection of one or more values organized so they can be referenced using an index. Since we're talking about mathematical operations here, it's implied the vector elements are numeric. We saw in Section 2.1.1 that a vector is created using the `c()` function. We use this function again to make a vector `x` using the x values in Table 10.1.

```
x <- c(1.2, 2.4, 0.4, 6.3, 2.2)
```

Recall from Section 4.2, we access vector elements using their position or index in square brackets, starting with 1 for the first element and ending with `length(x)` (equivalent to 5) for the last element. For example, `x[1]` is 1.2, `x[2]` is 2.4, and `x[5]` is 2.2. Notice these indexes follow the mathematical subscript notation introduced above and are equivalent to x_1, x_2, and x_5.

R applies unary operators to all elements in a vector. In the code below, we see how the unary negation operator changes the sign of all elements.

```
-x
```

```
#> [1] -1.2 -2.4 -0.4 -6.3 -2.2
```

Note the code above only changes the sign of the printed values, not the values in `x`. If you want to change the values in `x`, you need to assign the negated vector back to `x`.

```
x <- -x
x
```

```
#> [1] -1.2 -2.4 -0.4 -6.3 -2.2
```

```
# Then change it back to the original values.
x <- -x
x
```

```
#> [1] 1.2 2.4 0.4 6.3 2.2
```

A binary operator between two vectors of the same length takes the left and right operands from each element along the vectors (an elementwise operation). This is illustrated in the elementwise multiplication of the x and y variables from Table 10.1.

```
y <- c(3, 4, 2, 6, 5)
x*y
```

```
#> [1]  3.6  9.6  0.8 37.8 11.0
```

Here's a very important concept we initially touched upon in Sections 4.2.3 and 4.7. For elementwise operations involving two or more vectors, the vectors must be the same length. If R encounters two vectors of different lengths in a binary operation, it replicates (recycles) the smaller vector until it's the same length as the longest vector, then performs the operation.

For example, say we scale the vector `x` by 5, i.e., `5 * x`. Recall, R stores single values as a vector of length 1, so the 5 is replicated `length(x)` times before the elementwise multiplication.

Consider another example where the vector `z` (defined in the code below) is shorter than `x`. Since `x` is the longer vector (length 5), `z` is recycled until it matches the length of `x`, so `z` will be `1, 2, 1, 2, 1`. Because the length of `x` is not a multiple of the original length of `z`, the recycled `z` is truncated and R kindly prints a warning, but still returns the result, as illustrated below.

```
z <- c(1,2)
x+z
```

```
#> Warning in x + z: longer object length is not a
#> multiple of shorter object length
```

```
#> [1] 2.2 4.4 1.4 8.3 3.2
```

R follows the PEMDAS order of operations. Take a moment to study R's precedence of operators manual page, accessed via `?Syntax`. The manual page lists operators in precedence groups, from highest (applied first) to lowest (applied last). You'll see that the PEMDAS ordering is reflected in the manual page's precedence ordering, and in how these mathematical operators are positioned relative to other R operators and syntax.

The next lines of code implement (10.2) and (10.3), respectively. Notice the `sum()` function is equivalent to the $\sum$ operator, and exponentiation is done relative to the parentheses.

```
sum(x^2)
```

```
#> [1] 51.89
```

```
sum(x)^2
```

```
#> [1] 156.25
```

We implement (10.6) and (10.7) below.

```
sum(x * y)
```

```
#> [1] 62.8
```

```
sum(x) * sum(y)
```

```
#> [1] 250
```

Above, and in subsequent chapters, we often illustrate ideas using single vectors that represent measurements on a variable. However, in practice, these vectors are typically columns in a data frame. So let's extend the order of operations examples above to variables in a data frame. We begin by moving the data in Table 10.1 to a data frame using the `tibble()` function, introduced in Section 6.2.

```
plots <- tibble(plot_index = c(1, 1, 2, 2, 2),
                tree_index = c(1, 2, 1, 2, 3),
                x = c(1.2, 2.4, 0.4, 6.3, 2.2),
                y = c(3, 4, 2, 6, 5))
```

Now let's repeat the last few order of operations examples above using `dplyr` functions covered in Chapter 7.[3] The code below computes (10.2), (10.3), (10.6), and (10.7), all in one call to `summarize()`.

```
plots %>%
  summarize(`sum(x^2)` = sum(x^2),
            `sum(x)^2` = sum(x)^2,
            `sum(x*y)` = sum(x * y),
            `sum(x)*sum(y)` = sum(x) * sum(y))
```

```
#> # A tibble: 1 x 4
#>   `sum(x^2)` `sum(x)^2` `sum(x*y)` `sum(x)*sum(y)`
#>        <dbl>      <dbl>      <dbl>           <dbl>
#> 1       51.9       156.       62.8             250
```

Going back to subscripts and nested data. We'll often want to apply the same equation to subsets of data indexed by one or more qualitative variables that indicate group membership. For example, let's use the `plot_index` column in the `plots` data frame to apply equations to trees within each plot. This is easily done by using `plot_index` as a grouping variable, as illustrated below.

---

[3]In this example we're using non-syntactic names in `summarize()`, protected with backticks (see Section 3.2.2 for naming rules). Using non-syntactic names should generally be avoided, but it's useful here to match the mathematical expression with its resulting value.

```
plots %>%
  group_by(plot_index) %>%
  summarize(`sum(x^2)` = sum(x^2),
            `sum(x)^2` = sum(x)^2,
            `sum(x*y)` = sum(x * y),
            `sum(x)*sum(y)` = sum(x) * sum(y))
```

```
#> # A tibble: 2 x 5
#>   plot_index `sum(x^2)` `sum(x)^2` `sum(x*y)`
#>        <dbl>      <dbl>      <dbl>      <dbl>
#> 1          1        7.2       13.0       13.2
#> 2          2       44.7       79.2       49.6
#> # i 1 more variable: `sum(x)*sum(y)` <dbl>
```

Let's go back and look at (10.4) and its expanded form in (10.5). Recall i indexes plot and j indexes tree within plot, m is the number of plots, and n_i is the number of trees within each plot. For the data in Table 10.1, $m = 2$; for $i = 1, n_i = 2$; and for $i = 2, n_i = 3$.

The double summation can be completed using the following, where `within_plots` is the squared sum of `x` for each plot, and `across_plots` is the sum of the `within_plots` values.

```
plots %>%
  group_by(plot_index) %>%
  summarize(within_plots = sum(x)^2) %>%
  summarize(across_plots = sum(within_plots))
```

```
#> # A tibble: 1 x 1
#>   across_plots
#>          <dbl>
#> 1         92.2
```

As always, it can be instructive to run the code above incrementally and compare the output of each call to `summarize()` with the right-hand side of (10.5).

10.3 Summary

In this chapter, we covered essential data analysis vocabulary and statistical notation used throughout the remainder of the book. We began by defining concepts related to populations, samples, and variables. We described the typical data analysis progression: define the population, population units, and the population parameter of interest, then measure one or more variables on a sample of population units and use these measurements to compute a statistic that serves as an estimate of that parameter. We'll see more concrete examples of this workflow beginning in Chapter 11.

Next, we extended our discussion to quantitative and qualitative variables. Quantitative variables are numeric and defined as discrete (finite increment between values, e.g., an integer) or continuous (infinite increment between values, e.g., a number with decimals). Qualitative variables represent different categories and can be nominal (no order) or ordinal (ordered).

We then shifted to the notation used throughout the remainder of the book. We use Greek letters (e.g., μ, σ) to represent population parameters and Roman letters (e.g., x, y) to represent variables, and Roman letters—sometimes decorated with bars, hats, or other symbols—to represent statistics.

We reviewed mathematical operators and discussed subscripts and their role in summations and nested data. We stressed the importance of understanding how subscripts reference variable values in a nested dataset, as nested data are extremely common in both forestry and environmental sciences.

We finished by discussing the ever-important order of operations. While it might seem conceptually straightforward, we've spent precious hours (even days) tracking down bugs that turned out to be simple errors caused by overlooking PEMDAS.

With *tools of the trade* in hand, you're ready for Chapter 11, which develops the statistical foundations you'll use to make inferences about populations.

10.4 Exercises

With the exception of Exercise 10.10, all exercises should be completed *by hand*, i.e., without using R. Use the information in Table 10.1 to complete Exercises 10.1 through 10.9.

Exercise 10.1. How many tree measurements of variable x are there for Plot 1 and Plot 2?

Exercise 10.2. Evaluate the expression $\sum_{j=1}^{2} x_{ij}$, where $i = 1$, i.e. sum the x variable values for Plot 1.

Exercise 10.3. Evaluate the expression $\sum_{i=1}^{2} \sum_{j=1}^{n_i} x_{ij}$, where n_i is the total number of trees in each plot.

Exercise 10.4. Evaluate the expression $\sum_{i=1}^{2} \sum_{j=1}^{n_i} x_{ij}^2$, where n_i is the total number of trees in each plot.

Exercise 10.5. Evaluate the expression $\sum_{i=1}^{2} (\sum_{j=1}^{n_i} x_{ij})^2$, where n_i is the total number of trees in each plot.

Exercise 10.6. Evaluate the expression $\sum_{i=1}^{2} \sum_{j=1}^{n_i} x_{ij}y_{ij}$, where n_i is the total number of trees in each plot.

Exercise 10.7. Evaluate the expression $\sum_{i=1}^{2}\sum_{j=1}^{n_i} x_{ij}^2 y_{ij}$, where n_i is the total number of trees in each plot.

Exercise 10.8. Is $\sum_{i=1}^{n} x_i^2$ equal to $\left(\sum_{i=1}^{n} x_i\right)^2$?

Prove your answer using the x values, with i redefined to sequentially index the x values from $i = 1, \ldots, n$ where $n = 5$.

Exercise 10.9. Is $\sum_{i=1}^{n} x_i y_i$ equal to $\left(\sum_{i=1}^{n} x_i\right)\left(\sum_{i=1}^{n} y_i\right)$?

Prove your answer using the x and y values, with i redefined to sequentially index the x and y values from $i = 1, \ldots, n$ where $n = 5$.

Exercise 10.10. Evaluate the following expression first by hand (with the aid of a calculator), then using R:

$$\frac{(\exp(14) + 10) \times \sqrt{5}}{\ln(4) - 5 \times 10^2},$$

where exp is the exponential function and ln is the natural logarithm (i.e., $\log_e$), which correspond to `exp()` and `log()` in R.

11

Basic statistical concepts

This chapter lays out the basic statistical concepts used for making inferences about a population using sample information. We generalize and expand on these concepts to accommodate forestry applications in Chapters 12-18. As noted in Section 1.1, our focus is on an applied introduction to inferential methods using R and, as a result, we don't develop and present the methods' theoretical underpinnings. Here, and in subsequent chapters, we refer the reader to subject matter books on survey sampling to build a deeper understanding of the topics we touch upon. To this end, we provide a brief bibliography in Section 11.2.

11.1 Characterizing a population

As described in Section 10.1, we're interested in characterizing a population using one or more parameters. Parameters are generally divided into two categories. Measures of *central tendency* summarize the central point or typical value, and include the mean, median, and mode parameters. Measures of *dispersion* summarize the degree to which values are distributed around a central tendency and include variance, standard deviation, and range. These parameters are defined below. Following the notation laid out in Section 10.2, Greek letters represent parameters.

In this chapter, we define parameters under a *finite population* view, where the population consists of N discrete, identifiable units. This provides a straightforward and practical framework for introducing sampling ideas and forms the basis for most classical examples. The same foundational ideas extend to the *infinite population* view introduced in Chapter 12, which better reflects forest inventory applications.

11.1.1 Measures of central tendency

The population arithmetic *mean* is

$$\mu = \frac{1}{N}\sum_{i=1}^{N} y_i, \tag{11.1}$$

where N is the number of units in the population and y_i is the variable value on the i-th unit.

DOI: 10.1201/9781003222699-11

The arithmetic mean is the most common measure of central tendency in subsequent analyses. There are other types of means, such as quadratic and harmonic, used in some specific settings touched upon later in the book. Unless otherwise noted, when we refer to the "mean" or "average" we're talking about the arithmetic mean.

The mean is generally only useful to summarize quantitative variables. However, in a few special cases, such as proportions, we do compute a mean for qualitative variables (see Section 11.1.3 for details).

The population *median* is computed by first arranging the N values in ascending or descending order of magnitude. The median is the middle value in the ordered values, such that 50% of the values are less than the median and 50% of the values are greater than the median. If the y variable values are arranged by magnitude, then the median is

$$\eta = \begin{cases} y_{(N+1)/2} & \text{if } N \text{ is odd} \\ \frac{y_{N/2}+y_{N/2+1}}{2} & \text{if } N \text{ is even,} \end{cases} \tag{11.2}$$

where, if N is odd, the median is the value with index $(N+1)/2$ (top line in the equation)—otherwise, it's the mean of the two middle values (bottom line of the equation).

The population *mode* is the value(s) that occurs most frequently. Some populations have more than one mode.

11.1.2 Measures of dispersion

The population *variance* is a measure of dispersion with two equivalent expressions defined as

$$\begin{aligned} \sigma^2 &= \frac{1}{N-1}\sum_{i=1}^{N}(y_i-\mu)^2 \\ &= \frac{1}{N-1}\left(\sum_{i=1}^{N} y_i^2 - \frac{\left(\sum_{i=1}^{N} y_i\right)^2}{N}\right). \end{aligned} \tag{11.3}$$

Some prefer the second expression because it's easier to use in spreadsheet software. Both expressions are equally convenient to compute in R. When comparing (11.3) with the population variance defined in other references, you might notice we use $N-1$ in the denominator while other authors use N, i.e., our definition differs by a factor of $\frac{N-1}{N}$. Like Cochran (1977), Lohr (2019), and others, we chose this definition because it simplifies presentation of some sampling concepts. The alternative definitions lead to equivalent inference.

Because variance is in squared units, its interpretation isn't intuitive. For example, if y is measured in tons of forest biomass, describing tons of biomass squared isn't very useful. For this reason, we typically work with or report the population *standard deviation*, which is the square root of the variance and is in the same measurement units as the variable being summarized. The standard deviation is defined as

$$\sigma = \sqrt{\sigma^2} = \sqrt{\frac{1}{N-1}\sum_{i=1}^{N}(y_i-\mu)^2}. \tag{11.4}$$

Often we're interested in comparing the variation between two populations. If the populations have the same mean, then their standard deviations can be compared directly. However, if the populations have different means, then the standard deviations must first be mean standardized before comparison. The *coefficient of variation* (CV) is a mean-standardized measure of relative variability, commonly expressed in percent units, and is defined as

$$\text{CV} = 100\frac{\sigma}{\mu}, \tag{11.5}$$

for $\mu \neq 0$.

The *range* is perhaps the simplest summary of dispersion to compute. It's defined as the difference between the lowest and highest values in the data. If the values of y were sorted in ascending order of magnitude, the range would be $y_N - y_1$.

11.1.3 Totals and proportions

In many forestry applications, in addition to the population mean, we're interested in the population total, e.g., total volume, biomass, number of species, or forest area. The population total is

$$\tau = \sum_{i=1}^{N} y_i = N\mu, \tag{11.6}$$

where μ is defined in (11.1).

There are many instances where the population parameter of interest is a proportion. For example, we might want to know the proportion of forest on a mixed landcover property or the proportion of diseased individuals in a population. The population proportion parameter, denoted by p, is computed using (11.1) with a binary qualitative variable y taking values 0 or 1 to indicate the state of an observation. To compute a property's forested proportion, we set $y_i = 1$ if the i-th unit is forested and $y_i = 0$ otherwise. Similarly, $y_i = 1$ if the i-th individual is diseased and $y_i = 0$ otherwise. Using this binary coding of the variable, the population proportion is

$$p = \frac{1}{N}\sum_{i=1}^{N} y_i. \tag{11.7}$$

Although a population proportion can be viewed as the mean of a binary variable, it's conventional to denote this parameter directly by p rather than introducing additional notation.

The population variance of a proportion, obtained by applying (11.3) to a binary variable, is

$$\begin{aligned}
\sigma^2_p &= \frac{1}{N-1}\sum_{i=1}^{N}(y_i - p)^2 \\
&= \frac{\sum_{i=1}^{N} y_i^2 - 2p\sum_{i=1}^{N} y_i + Np^2}{N-1} \\
&= \frac{N}{N-1}p(1-p).
\end{aligned} \tag{11.8}$$

Notice we use the subscript p on σ_p^2 simply as a reminder that this variance corresponds to a population proportion.

11.1.4 Tabular and graphical representations of data distributions

Data analysis objectives commonly focus on estimating a limited number of parameters. In addition to the desired parameter estimates, a conscientious analyst includes a narrative documenting the analysis process. The narrative, written around exploratory data analysis (EDA) tables and figures, supports estimation methods, provides context for the estimates, and provides insight not conveyed by estimate values alone.

Frequency tables and associated graphs efficiently convey information about the data distribution. A frequency table organizes values into an easy to read and interpret format. A value's *frequency* is the number of times it occurs. The frequency table arranges values in ascending or descending order with their corresponding frequencies. Stand tables, as introduced in Section 8.3, are frequency tables and we saw in Section 9.4 how to translate a frequency table into a graphic by placing its values along the x-axis and corresponding frequency on the y-axis. For example, a *histogram* uses vertical bars or lines to indicate each value's frequency. Other graphics connect the frequencies using lines or capture the histogram's general shape using a smooth curve. See Section 11.1.5 for examples.

Skewness is a term used to describe a distribution's shape, specifically its asymmetry. A *positive-skewed* distribution, as illustrated in Figure 11.1, has a long tail in the positive direction. Vertical lines in Figure 11.1 identify the distribution's mean, median, and mode. The mean of a positive-skewed distribution is on the positive side of the distribution's peak. This is because the mean is "pulled" toward the relatively few, but large, values out on the distribution's tail. The median of a positive-skewed distribution is to the left of the mean and, unlike the mean, isn't sensitive to large extreme values on the distribution's tails. A *negative-skewed* distribution has a long tail in the negative direction, again illustrated in Figure 11.1. Like the positive-skewed distribution, the negative-skewed distribution mean is pulled off the peak. However, the mean in a negative-skewed distribution is pulled in the negative direction by the long tail. The median of a negative-skewed distribution is to the right of the mean.

The first row in Figure 11.1 illustrates unimodal distributions with positive and negative skew. We describe these distributions as unimodal because there's only one prominent mass of observations at and around the mode. Depending on the type of analysis we pursue, the description of a distribution's modality (i.e., the number of peaks it contains) can deviate from the mode's formal definition and be a qualitative description about the number of clearly defined peaks. The second row in Figure 11.1 illustrates a bimodal (two modes) and unimodal symmetric distribution. These distributions are symmetric about the mean. Notice for symmetric distributions, the mean and median are equal. Chapter 9 covers coding tools used to produce graphics like those in Figures 11.1.

The lower right distribution in Figure 11.1 illustrates the Normal distribution, also known as the Gaussian distribution. Due to its role in theoretical results used to quantify parameter estimate uncertainty, the *Normal distribution* is the most conspicuous probability

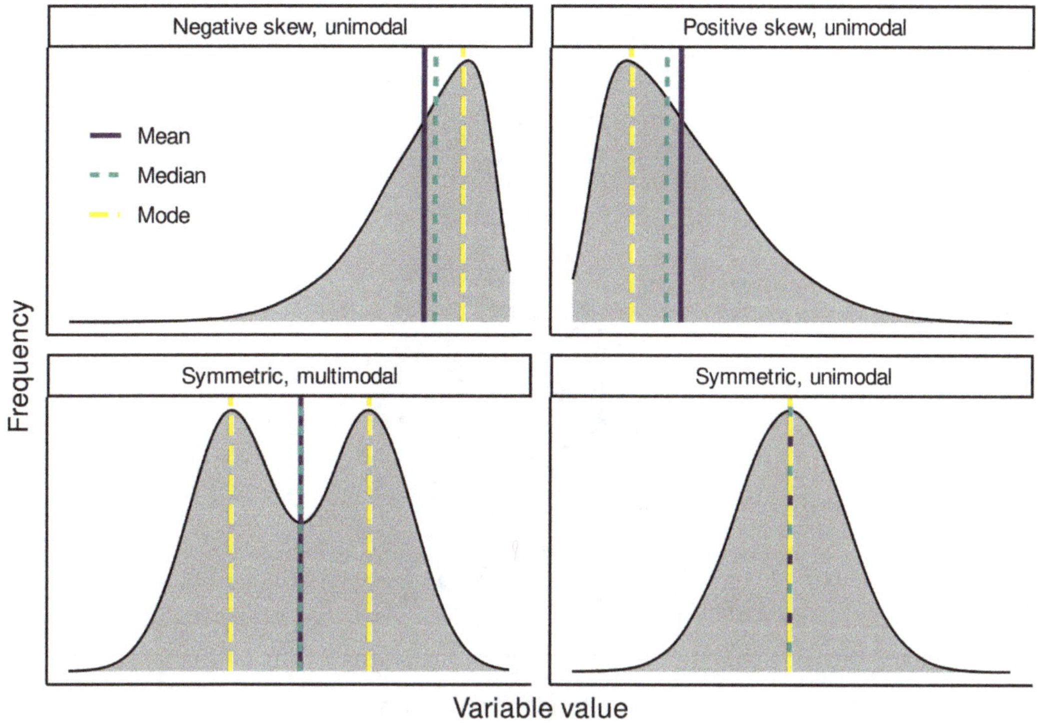

FIGURE 11.1: Distribution shapes and associated measures of central tendency.

distribution in statistics. The Normal distribution depicted in Figure 11.2 provides a bit more detail. As we've already seen, this distribution is unimodal and symmetric about the mean. Most data in a Normal distribution, about 68.3%, are within 1 standard deviation (i.e., -1σ to 1σ) of the mean, and about 95.5% are within 2 standard deviations (i.e., -2σ to 2σ) of the mean. The distribution's center and dispersion are defined by its mean and standard deviation parameters, respectively.

11.1.5 Illustrative analysis

Let's connect these data representations (i.e., tables and graphics) and parameters to an illustrative analysis of the Harvard Forest dataset introduced in Section 1.2.5. Recall this is a rare dataset in that it's a census of all trees in the 35 ha study area.[1] Figure 11.3 shows the 35 ha Harvard Forest ForestGEO plot partitioned into 140 non-overlapping square 0.25 ha units. The shape and size of these 140 units is just one way to delineate population units—selected here because it serves well for this illustration. Section 12.1 covers some considerations for defining population unit shape and size.

Our population parameter of interest, which we'll call μ, is the average biomass in metric tons per ha (Mg/ha). The variable recorded for each population unit, say y, is the biomass

[1]Perfect for building understanding about approaches for summarizing population parameters, then later for exploring parameter estimation methods.

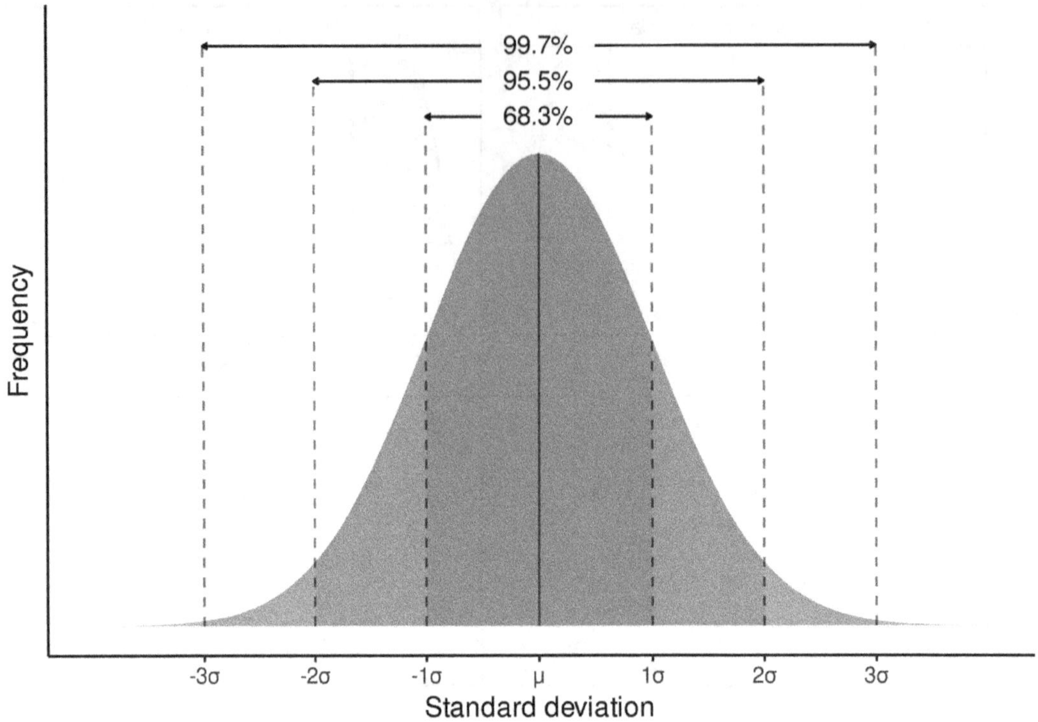

FIGURE 11.2: Normal distribution with shaded regions representing different percents of area under the curve.

Mg per 0.25 ha (this is the sum of tree biomass on each unit). It will, however, simplify life later if we record y as the biomass Mg per ha instead of biomass Mg per 0.25 ha. To do this, multiply biomass Mg per 0.25 ha unit by 4 (i.e., 1/0.25).[2] Figure 11.4 shows each unit's biomass (Mg/ha).

Table 11.1 is the frequency table for the data shown in Figure 11.4. The frequency values in this table are the number of population units with biomass (Mg/ha) within the given interval. The sum of frequency values equals the number of units in the population $N =$ 140.

The frequency distribution mode in Table 11.1 is the $(350, 450]$ interval, i.e., most population units have somewhere between 350 and 450 Mg/ha, and the majority of the units fall within the two middle intervals.

Figure 11.5 is a graphical representation of the frequency distribution summarized in Table 11.1. In fact, Figure 11.5 shows two different representations superimposed upon one another. The first is a histogram with rectangular bar width set to 25 Mg/ha. Except for the interval width, the histogram is a direct translation of the frequency table. Analogous

[2]This is called an expansion factor. We'll get a lot more practice using them in Chapter 12.

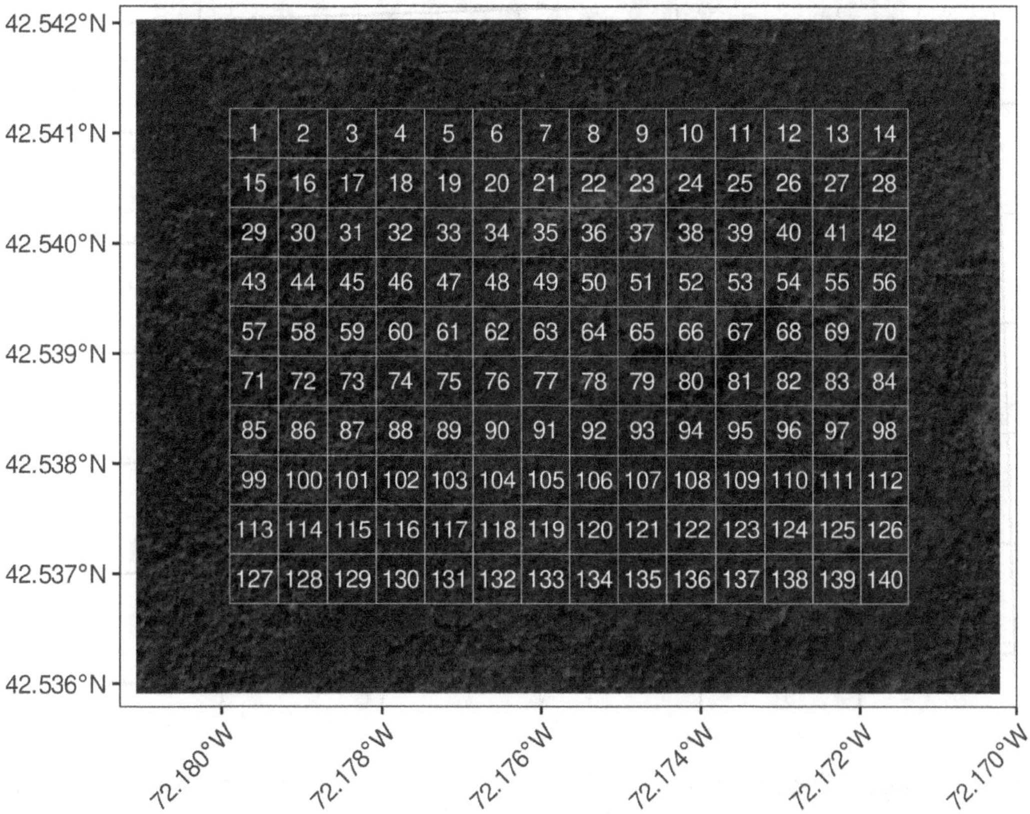

FIGURE 11.3: Harvard Forest comprising 140 0.25 (ha) square population units.

TABLE 11.1: Harvard Forest biomass (Mg/ha) frequency table for values shown in Figure 11.4.

	Biomass (Mg/ha)					
	[50,150]	(150,250]	(250,350]	(350,450]	(450,550]	(550,650]
Frequency	7	8	38	73	13	1

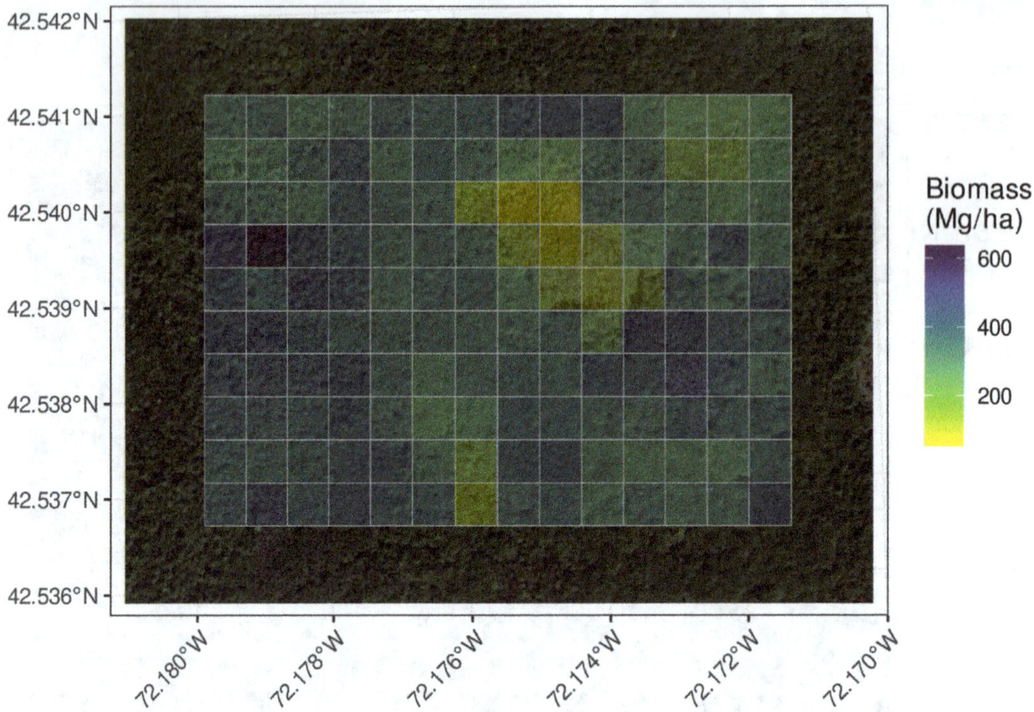

FIGURE 11.4: Harvard Forest biomass (Mg/ha) for each population unit.

to summing the frequency values in the table, the combined height of the histogram bars equals the number of population units $N = 140$. The second is a smoothed representation of the histogram shown as the blue curve with slight gray shading beneath. Unlike the frequency table and histogram, the area under this curve doesn't necessarily equal the number of units in the population. Rather, the curve is used to give a general impression of the distribution—smoothing over anomalies that might not be meaningful in the grand scheme of things.

Vertical lines in Figure 11.5 identify the distribution's mean, median, and mode. Following from Section 11.1.4, this distribution shows a slight negative skew, with the mean to the negative side of the distribution's peak and the median to the right of the mean. The mode, being the most frequently occurring value (or values), is clearly identified in Figure 11.5 as the tallest bar, at about 387.5 Mg/ha. This is a unimodal distribution because there's only one prominent mass of observations at and around the mode.

Let's compute the population parameter values shown in Figure 11.5. The code below uses `tidyverse` functions to read a file containing the population unit data shown in Figure 11.4 and assigns it to the `hf` data frame. We then select the 140 biomass measurements and assign them to a vector called `bio_mg_per_ha`. Given `bio_mg_per_ha`, we first compute the mean following (11.1) then use R's built-in `mean()` function.

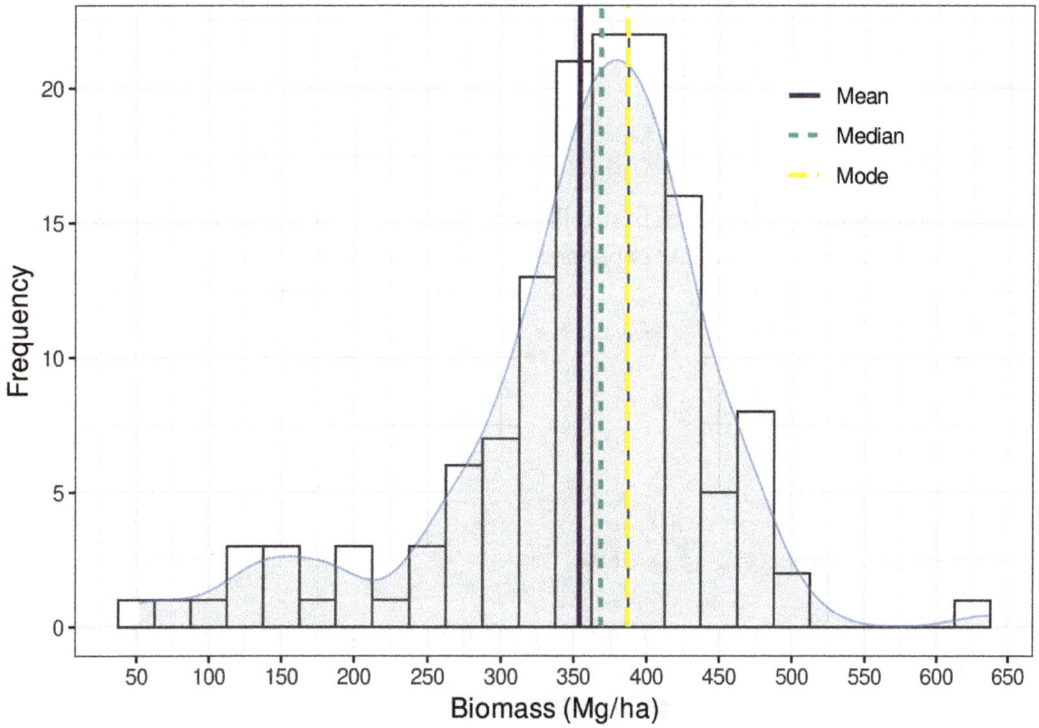

FIGURE 11.5: Harvard Forest biomass (Mg/ha) frequency histogram for values shown in Figure 11.4.

```
hf <- read_csv("datasets/HF/HF_pop_units.csv")
bio_mg_per_ha <- hf %>% pull(bio_mg_per_ha)

# Assign the vector length to N and check that it's what we expect.
N <- length(bio_mg_per_ha)
N
```

```
#> [1] 140
```

```
# Take a look at the first several measurements.
head(bio_mg_per_ha)
```

```
#> [1] 352.69 380.87 314.18 340.12 382.87 384.20
```

```
# Calculate the mean "by hand."
sum(bio_mg_per_ha) / N
```

```
#> [1] 354.15
```

```
# Calculate the mean using R's function.
mean(bio_mg_per_ha)
```

```
#> [1] 354.15
```

Now let's compute the median, first following (11.2) then using `median()`. The operators used to compute the median are explained in the text below the code.

```
# First test if the number of measurements is even or odd.
N %% 2 == 0 # No remainder, so it's even.
```

```
#> [1] TRUE
```

```
# Arrange the measurements in ascending order.
bio_mg_per_ha <- sort(bio_mg_per_ha)

# Given the even number of measurements, the median
# is the mean of the two middle elements.
(bio_mg_per_ha[N / 2] + bio_mg_per_ha[N / 2 + 1]) / 2
```

```
#> [1] 369.05
```

```
# Now the easy way.
median(bio_mg_per_ha)
```

```
#> [1] 369.05
```

There are a few things to notice in the code used to compute the median. The first line uses the modulus operator `%%` and equality operator `==` to test whether `N` is even or odd. The modulus operator returns the remainder from `N / 2`, i.e., the remainder of the division. We then test if the remainder is equal to zero using the `==` operator, which returns `TRUE` if the values on either side are equal and `FALSE` otherwise. The order of operations for these two operators is from left to right, such that `N %% 2` is evaluated before the logical test.[3] Next, as the name implies, `sort()` arranges the vector's values, with a default ordering from least to greatest. Finally, following (11.2), the median is the mean of vector values at index `N / 2` and `N / 2 + 1`.

Next, let's compute the population's variance using the two mathematically equivalent expressions in (11.3) then the standard deviation (11.4).

[3] If in doubt about the order of operations, you could always use parentheses to make it explicit for you and others who might read your code, e.g., `(N %% 2) == 0`. That said, we strongly recommend learning the order of operations for the various operators you'll encounter (see Sections 10.2.2 and 10.2.3).

```
# Define mu just for convenience.
mu <- mean(bio_mg_per_ha)
sum((bio_mg_per_ha - mu)^2) / (N - 1)
```

```
#> [1] 8340.1
```

```
# or
sigma_sq <- (sum(bio_mg_per_ha^2) -
               sum(bio_mg_per_ha)^2 / N) / (N - 1)
sigma_sq
```

```
#> [1] 8340.1
```

```
# Now the standard deviation.
sqrt(sigma_sq)
```

```
#> [1] 91.324
```

Returning to a point made in Section 10.2.3, notice in the second line of code above that `(bio_mg_per_ha - mu)^2` is a vector operation with `mu` being subtracted from each element in `bio_mg_per_ha`, then each resulting difference is squared before being summed via `sum()`.[4] Like the `mean()` and `median()` functions we saw previously, you can also use the built-in variance function `var()` and standard deviation `sd()` to compute (11.3) and (11.4), respectively.

11.2 Introduction to sampling methods

The following sections provide the basics for implementing and interpreting results from routine probability sampling inference. For a more complete treatment, we defer to other books that delve more deeply into application and theory. While somewhat dated, the extensive book by Johnson (2000) provides a balanced perspective on both theory and application. As noted previously, Freese (1962) is clearly written and covers basic statistical concepts and analysis for common sampling designs used in forestry. Shiver and Borders (1995) and Schreuder et al. (2004) provide a modernized and expanded version of Freese (1962), with additional material on forestry methods. The entertaining and informative book by Iles (2003) is a fun read, full of wit, wisdom, and practical suggestions. Gregoire and Valentine (2008) serves as a keystone reference bridging applied and rigorous theoretical treatments of forest survey sampling topics; their book is best approached by those with a solid understanding of probability theory who want to better understand the development

[4]It's often instructive to run lines of code like this from the inside out, i.e., first run `bio_mg_per_ha - mu`, then run `(bio_mg_per_ha - mu)^2`, and so on. If you're using RStudio you can highlight `bio_mg_per_ha - mu` then press the Run button or keyboard shortcut.

and properties of various estimators. Similarly, Mandallaz (2007) provides many theoretical insights with real-world implications, along with coverage of model-assisted, geostatistical, and optimal sampling schemes with applications in national forest inventory (NFI). From a purely sampling theory standpoint, Cochran (1977) and Claes-Magnus et al. (1977) are popular references among many practitioners. We particularly like Lohr (2019), who offers a well-balanced contemporary take on sampling design and analysis. For readers interested in a full understanding of how and why probabilistic sampling methods work, along with extensions to model-assisted inference, the classical books Claes-Magnus et al. (1977) and Särndal et al. (1992) are essential resources.

There are many lively discussions among experts on how to select a sample and use its information to draw conclusions about the population. Much of the literature noted above, and the majority of the material in this book, follows the classical design-based inference framework. *Design-based* inference relies on the known probability of selection in sampling. A potentially richer and more flexible paradigm, known as *model-based* inference, is capable of delivering equally valid and potentially more intuitive interpretations of uncertainty quantification—we encourage the reader to explore it (see, e.g., Ghosh and Meeden (1997), Ericson (1969), Little (2004), and Green et al. (2020)).

11.2.1 Characteristics of samples

In most practical cases, it's too time consuming and/or expensive to observe all units in the population. In such cases, we settle for observing a sample, meaning a subset of the population units. If selected appropriately, this sample can provide a useful estimate for the population parameter of interest.

A fundamental assumption is information gleaned from the sample holds for the population from which it was drawn. The ideal sample is an exact scaled-down version of the population; however, in practice this is never achieved.[5] Therefore, our aim is to select a sample that's representative of the population. This starts by recognizing two sources of bias to avoid. *Bias*, in this context, means our sample differs in a systematic way from the population. *Selection bias* occurs when some part of the population isn't included in the sample. For example, in an effort to avoid exercise, you establish your timber cruise plots for a woodlot along the side closest to the road—effectively excluding the interior from the sample. Another example is avoiding thickets when conducting tree regeneration surveys, preferring to install sample plots only where it's easy to walk—again potentially missing important areas of regeneration protected from deer browsing. These are examples of *convenience sampling* and are very common sources of selection bias.[6]

---

[5]Actually, it might be achieved, but we'd never know it because we don't observe the entire population and hence can't compare it with the sample.

[6]Many years ago, as an undergraduate, the first author was on a field crew conducting seedling regeneration counts on permanent long-term regeneration study plots. One stand we were working in was wall-to-wall hay-scented fern. On a few of the plots, the crew leader decided we should remove all the ferns that overtopped the seedlings to simplify the counting process. Clearly this was a convenience for us, but it likely biased subsequent remeasurements by changing seedling growth rates through removal of competition. To this day, I envision well-defined circles of advanced regeneration on those plots in a sea of ferns.

Measurement bias and personal bias are other common sources of bias we wish to avoid. *Measurement bias* occurs when an instrument provides a measurement that's systematically different from the true value, perhaps due to damage or miscalibration. *Personal bias* refers to a human error that results in a measurement that's systematically different from the true value. The classic forestry example of personal bias is when someone consistently takes tree DBH measurements at a height other than 4.5 ft (1.37 m) from the ground on the uphill side of the tree (Omule 1980; Luoma et al. 2017).

11.2.2 Probability samples

Here, we introduce classical techniques for drawing inference using probability sampling from a finite population. As defined in Section 11.1, a finite population contains a fixed, countable set of units. We begin with this setting because it's the one most commonly encountered in introductory statistics or survey sampling courses, and it provides a straightforward framework for building foundational sampling concepts.

In Chapter 12, we adapt these ideas to forest inventory applications. Instead of sampling from a fixed list of discrete units, we select sample locations within a forest and base inference on measurements made around those locations. This leads to the *infinite population* perspective used in forest inventory. While the way samples are laid out in the field differs from list-based sampling, the same core principles of probability sampling still apply.

Under *probability sampling*, each possible sample has a known probability of selection and each population unit must have a non-zero probability of being included in a sample. As this definition suggests, two different probabilities comprise probability sampling. The *sample probability* is the probability of selecting a particular sample (i.e., the specific set of population units observed), and *inclusion probability* is the probability of including a given population unit in the sample.

Probability sampling is the preferred approach to avoid the temptations and shortcomings of convenience sampling. The probability sample guarantees that each unit in the population could appear in the sample and provides a mechanism to assess the expected quality of a statistic calculated from the sample. A *statistic* is a quantity computed using values in a sample and used to estimate a population parameter. Although probability sampling can't guarantee that any given sample is representative of the population with respect to a parameter of interest, it does provide a way to quantify sampling error. *Sampling error* is the difference between a sample-based estimate for a population parameter and the true, but unknown, value of the parameter.

Selection of a probability sample from a finite population requires the existence of a sampling frame. For our current purposes, the *sampling frame* is a list of all units in the population from which we select sample units. At this point, it's helpful to clarify what we mean by a *unit* by distinguishing between observation units and sampling units. The *observation unit* is the physical object we actually measure. The *sampling unit* is the unit selected for measurement from the sampling frame.

To make this distinction concrete, recall the example at the end of Section 10.1, where we were interested in estimating total merchantable timber volume on a 50-acre property.

Here, the 1/5-th acre plot is the sampling unit, and the trees measured within each plot are the observation units. Unless stated otherwise, when we say *unit*, we mean *sampling unit*.

11.2.3 Sampling distribution and properties of estimators

Let's review. A *statistic* is a quantity computed from sample data and used to estimate a population parameter. We could consider several different statistics (i.e., computed from different algebraic expressions, or "rules") that might serve to estimate a particular population parameter. So, which statistics give "good" estimates? We might even ask which statistic gives the "best" estimate. Of course, these questions also force us to define "good" and "best."

In practice, we typically have one sample and hence one estimate. Because of sampling error, comparing a single estimate to the true value of the population parameter (if it were known) would tell us little about the quality of the statistic. Rather, to answer these and related questions, we introduce the idea of an *estimator*, which is a rule for calculating an estimate from sample data.

Given a sampling frame, we draw a probability sample of size n from a population. Using these sample data, the estimator yields one estimate for the population parameter of interest. Next, say we repeat the process using a different, but equally valid, probability sample of size n from the population. Because we're using a different sample, the second estimate is quite likely different from the first. By extension, using the estimator to compute an estimate for each of a large number of different samples of size n from the population, you'll end up with an equally large number of estimates. The distribution of these estimates is called the sampling distribution. The *sampling distribution* of a statistic is the distribution of that statistic computed from a large number of repeated samples, or all possible samples, of the same size drawn from a population. The sampling distribution depends on characteristics of the population, statistic computed, sampling procedure, and sample size.

We can learn about a statistic by comparing properties of its sampling distribution with the population parameter of interest. Specifically, we can revisit the questions posed earlier—"which statistic gives a good estimate" and "which statistic gives the best estimate"—by restating them as "which estimator gives a good estimate" and "which estimator gives the best estimate."

For a given population parameter, there are three key qualities an estimator is generally expected to meet to be considered "good." The estimator should be:

1. *unbiased*—its expected value equals the true parameter;
2. *consistent*—as sample size increases, estimates converge to the parameter;
3. *relatively efficient*—among competing estimators, it has the smallest sampling variance.

In later chapters, we'll introduce settings where the "best" estimator might not meet all three qualities but is still the most appropriate choice given the objectives, sampling design, and logistical constraints. In some cases, an unbiased estimator may not be available under the design. In others, unbiased estimators exist but are not the best choice. For example, an approximately unbiased estimator might be preferred if, e.g., it provides lower variance, improved stability in small samples, or greater efficiency under the sampling design.

Taken together, these estimator qualities provide a practical basis for defining "best" and a principled way of choosing among alternative estimators, depending on which qualities are most important under the given sampling design.

Let's illustrate some probability sampling concepts using a small population where all units have known values. We'll then propose an estimator for the population's mean, generate its sampling distribution, see if it's unbiased, compute its variance, and explore some of its other qualities.

Following notation in Section 10.2, population parameters are represented using Greek letters, and statistics are represented using Roman letters often decorated with a bar or hat above the letter, i.e., the symbols $^-$ and $\hat{}$, respectively. For example, as you'll see shortly, we use μ and $\bar{y}$ to represent the population mean parameter and its corresponding sample-based estimate, respectively.

The population has $N = 9$ units with the following values

$$\begin{array}{lll} y_1 = 5 & y_4 = 8 & y_7 = 5 \\ y_2 = 1 & y_5 = 9 & y_8 = 1 \\ y_3 = 1 & y_6 = 1 & y_9 = 5, \end{array}$$

where the subscript on variable y indexes the population unit. The population mean is

$$\mu = \frac{\sum_{i=1}^{N} y_i}{N} = \frac{5+1+1+8+9+1+5+1+5}{9} = 4. \tag{11.9}$$

The sampling frame comprises the population unit indexes 1 through 9. If we consider a sample size of $n = 2$ and select sample units without replacement (meaning a unit cannot appear more than once in a given sample, see Section 11.3) then 36 possible samples can be formed for this population.[7] As we'll see in Section 11.3, the procedure used to select these samples implies any given sample has a 1/36 probability of selection.

An estimator we could use for μ is the sample mean $\bar{y} = \frac{\sum_{i=1}^{n} y_i}{n}$. The sampling distribution of this estimator is generated by calculating $\bar{y}$ for each of the 36 samples. For example, given the sample comprising units {1, 4}, our estimate is

$$\bar{y} = \frac{y_1 + y_4}{n} = \frac{5+8}{2} = 6.5. \tag{11.10}$$

This calculation is then repeated for the remaining 35 samples, resulting in 36 estimates. Different samples can yield the same estimate value. For example, in addition to the sample comprising units {1, 4}, used in (11.10), samples with units {4, 7} and {4, 9} also result in $\bar{y} = 6.5$.[8]

---

[7]You can generate the sampling unit indexes for all 36 samples using `combn(9, 2)`.

[8]Take one of these samples and prove this fact to yourself.

TABLE 11.2: Sampling distribution of the sample mean statistic.

	$\bar{y}$						
	1	3	4.5	5	6.5	7	8.5
Frequency	6	12	4	7	3	3	1
Probability	$\frac{6}{36}$	$\frac{12}{36}$	$\frac{4}{36}$	$\frac{7}{36}$	$\frac{3}{36}$	$\frac{3}{36}$	$\frac{1}{36}$

The estimate value frequency among the 36 estimates is given in the first row of Table 11.2 (note, the first row sums to 36). As we just saw, three samples resulted in an estimate of 6.5, which is reflected in the frequency table.

Next, using the frequency and knowing the probability of sample selection, i.e., 1/36, we can calculate the probability of estimate value occurrence, which is given in the second row of Table 11.2 (note, this row sums to 1). This second row is the sampling distribution for the estimate values in the first row.[9] The sampling distribution histogram is plotted in Figure 11.6.[10] Looking at either Table 11.2 or Figure 11.6, the sampling distribution tells us there's a one-in-six chance (i.e., 6/36) of drawing a sample that results in $\bar{y} = 1$, a one-in-three chance (i.e., 12/36) of drawing a sample that results in $\bar{y} = 3$, and so on.

Given the sampling distribution, we can now assess if our estimator $\bar{y}$ is unbiased. This starts by computing the expected value of $\bar{y}$, denoted by $\mu_{\bar{y}}$. The *expected value* of a statistic is the weighted average of its m sampling distribution values, with the probability of each value as the weight. Using the $m = 7$ estimate values in Table 11.2, the estimator's expected value is

$$\begin{aligned} \mu_{\bar{y}} &= \sum_{j=1}^{m} P_j \bar{y}_j \\ &= \frac{6}{36}(1) + \frac{12}{36}(3) + ... + \frac{1}{36}(8.5) \\ &= 4, \end{aligned} \tag{11.11}$$

where j indexes the table columns, and P_j is the probability of estimating $\bar{y}_j$. The *estimation bias* of $\bar{y}$ is

$$\begin{aligned} \text{Bias}[\bar{y}] &= \mu_{\bar{y}} - \mu \\ &= 4 - 4 \\ &= 0. \end{aligned} \tag{11.12}$$

[9] More specifically, the sampling distribution in this example is a *discrete probability distribution*, meaning the estimator can take on only a finite set of distinct values, each with an associated probability, and these probabilities sum to 1.

[10] We use the rectangular bar histogram, e.g., Figure 11.5, to represent value intervals. The vertical line style in Figure 11.6 is appropriate to represent discrete values.

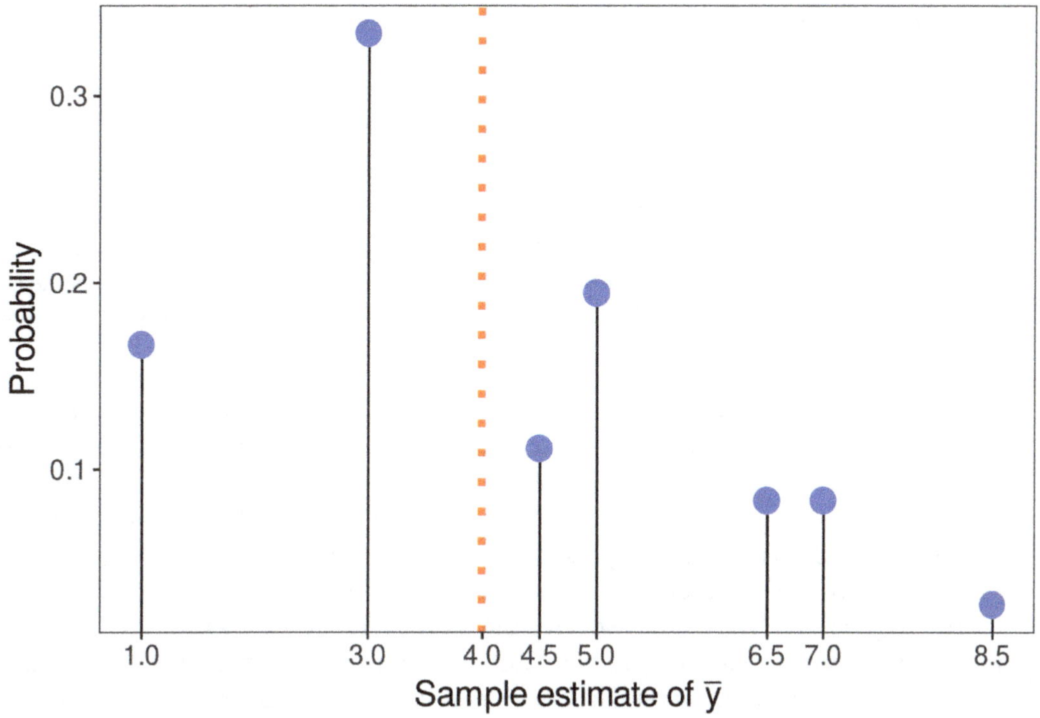

FIGURE 11.6: Sampling distribution of $\bar{y}$ with the population mean μ added as a vertical dotted line.

When the estimation bias is zero, as is the case here, we say the estimator is *unbiased*. Notice, unbiased doesn't mean the estimates are equal to the parameter μ, but rather the weighted average of the estimates equals μ. This is seen in Figure 11.6, where none of the estimates actually equal μ (i.e., due to sampling error), but rather their weighted average (11.11) equals μ.

The *variance* of the sampling distribution of $\bar{y}$ is

$$\begin{aligned}\sigma^2_{\bar{y}} &= \sum_{j=1}^{m} P_j(\bar{y}_j - \mu_{\bar{y}})^2 \\ &= \frac{6}{36}(1-4)^2 + \frac{12}{36}(3-4)^2 + ... + \frac{1}{36}(8.5-4)^2 \\ &= 3.89. \end{aligned} \tag{11.13}$$

A small variance means the estimates are more similar in value, and large variance means the estimates have greater differences in value.

We compute an estimator's variance for a few reasons. First, as noted previously, it's needed when comparing among alternative estimators using *relative efficiency*. Second, related to relative efficiency, we often seek unbiased estimators with low variance or high *precision*, where precision is the inverse of the variance. Third, variance is a component of *mean squared error* (MSE), which is a common measure of an estimator's *accuracy*. Smaller values of MSE indicate greater estimator accuracy. The MSE of the sampling distribution of $\bar{y}$ is

$$\begin{aligned} \text{MSE}[\bar{y}] &= \sigma_{\bar{y}}^2 + (\text{Bias}[\bar{y}])^2 \\ &= 3.89 + (0)^2 \\ &= 3.89. \end{aligned} \tag{11.14}$$

Notice, in (11.14), when the estimator is unbiased (i.e., estimation bias is zero), the MSE reduces to the estimator's variance. For some population parameters of interest and for some sampling procedures, we find ourselves working with biased estimators. In such settings, MSE is a useful criterion for choosing among alternative estimators.

Figure 11.7 provides a bit more intuition about the different ways to describe an estimator's characteristics. The figure illustrates various combinations of estimation bias and precision, which together result in different levels of accuracy (i.e., MSE). Within the context of evaluating these qualities of an estimator, you can think of the dots on the targets as being estimate values in a sampling distribution. The center of the scatter becomes less biased moving from the left targets to the right targets. Similarly, the scatter of points increases in precision (i.e., variance decreases) moving from the top targets to the lower targets. Since accuracy, as measured by MSE, is a combination of estimation bias and precision, each of the four targets differs in its accuracy. The upper left target has the lowest accuracy (i.e., biased and imprecise) and the lower right target has the highest accuracy (i.e., unbiased and precise). If these four targets represented sampling distributions of four different estimators, then we'd select the one that produced the lower right target.

As a very brief final example, let's see how a different estimator for the population mean μ performs relative to $\bar{y}$. Purely for illustration, we'll naively consider the *quadratic mean* as the alternative estimator. The quadratic mean is defined as

$$\bar{y}_Q = \sqrt{\frac{\sum_{i=1}^{n} y_i^2}{n}}. \tag{11.15}$$

Using the sampling distribution for $\bar{y}_Q$, we find the estimator is biased $\text{Bias}[\bar{y}_Q] = 0.53$, with variance $\sigma_{\bar{y}_Q}^2 = 4.38$, and mean squared error $\text{MSE}[\bar{y}_Q] = 4.66$. Clearly, compared with $\bar{y}_Q$, $\bar{y}$ is the preferred estimator for μ, given it's unbiased, has greater relative efficiency ($\sigma_{\bar{y}}^2 < \sigma_{\bar{y}_Q}^2$) and is more accurate ($\text{MSE}[\bar{y}] < \text{MSE}[\bar{y}_Q]$).

11.3 Simple Random Sampling

Simple random sampling is the most basic form of probability sampling and provides the basis for the more complex sampling designs explored in subsequent chapters. Simple random sampling can be conducted *with replacement* or *without replacement*, where in both cases the n units are selected independently and at random from the sampling frame. With replacement means the same sample unit can be included more than once in the sample.

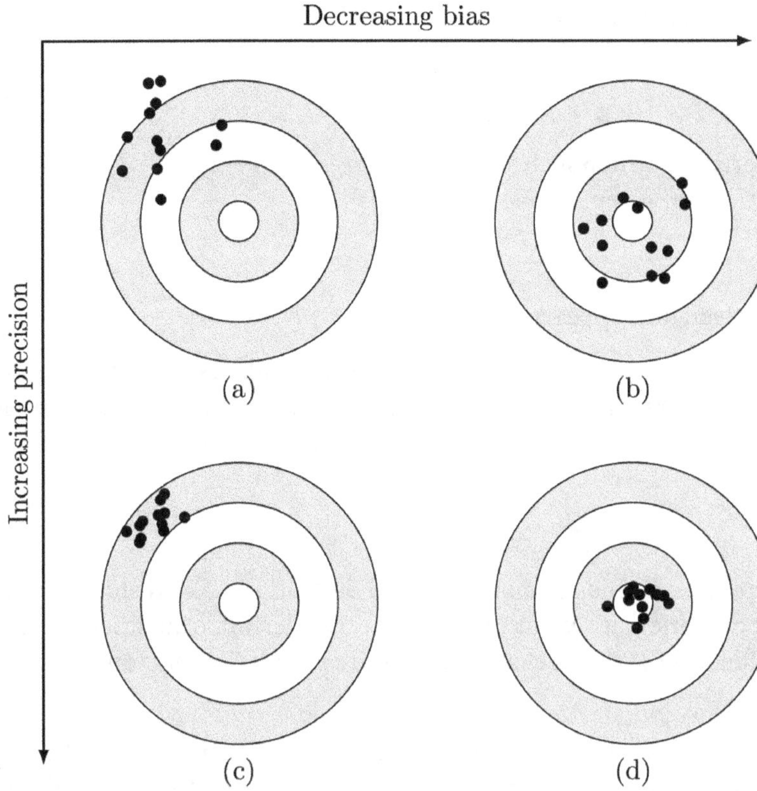

FIGURE 11.7: Illustration of bias, precision, and accuracy continuum. (a) Biased, imprecise, and inaccurate. (b) Unbiased, imprecise, and inaccurate. (c) Biased, precise, and inaccurate. (d) Unbiased, precise, and accurate.

Without replacement means the sample unit can be included only once in the sample, i.e., all units in the sample are distinct. The word *replacement* refers to the idea that we're replacing units selected into the sample back into the population—such that they could be selected again.[11]

11.3.1 Simple Random Sampling without replacement

In the finite population setting, we usually prefer using a simple random sample without replacement (SRS). A SRS of size n is selected so every possible combination of n distinct units in the population has the same probability of being selected as the sample.[12] There

[11]Recall, the sampling frame lists all units in the population, so, for this discussion, the sampling frame and population are synonymous.

[12]The term *simple* in simple random sampling refers to this equal-probability structure, which greatly simplifies the mathematics of estimation, even though implementing SRS in the field is often not simple.

are $\binom{N}{n}$[13] equally likely samples of size n,

$$\binom{N}{n} = \frac{N!}{n!(N-n)!}, \tag{11.16}$$

where ! denotes the factorial expansion such that $n! = n\cdot(n-1)\cdot(n-2)\cdot(n-3)\cdot\ldots\cdot 3\cdot 2\cdot 1$. The probability of selecting any given sample of n units is

$$\frac{n!(N-n)!}{N!}, \tag{11.17}$$

and the probability any given unit appears in the sample is n/N.[14]

As we'll illustrate in Section 11.3.5, given a sampling frame that indexes the N population units, a set of n random integers selected from 1 to N are typically used to identify sample units. To estimate the population mean μ (11.1) using SRS, we use the sample mean defined as

$$\bar{y} = \frac{\sum_{i=1}^{n} y_i}{n}. \tag{11.18}$$

We explored some properties of this estimator for μ in Section 11.2.3 and showed that, for one small example population, it's unbiased and has other desirable qualities. A more general result by Cornfield (1944) shows that for any finite population, $\bar{y}$ is an unbiased estimator for μ.

Recall from Section 11.2.3, the variance of $\bar{y}$ is computed using (11.13) and measures the variability among estimates for μ computed from different samples. Under SRS, this variance is reexpressed in terms of the population variance σ^2, given in (11.3), as

$$\sigma^2_{\bar{y}} = \left(1 - \frac{n}{N}\right)\frac{\sigma^2}{n}, \tag{11.19}$$

which will prove useful very shortly.

The $1 - \frac{n}{N}$ term in (11.19) is called the *finite population correction* factor. Its role is to reduce the variance of $\bar{y}$ as the *sampling fraction* n/N increases.

In estimation problems, we're almost always looking for ways to reduce $\sigma^2_{\bar{y}}$ (or, equivalently in practice, ways to reduce its estimate given in (11.21)). Intuitively, under SRS, as our sample size n increases, we learn more about the population and, hence, there should be less variability among our estimates. The extreme is when n equals N, at which point there's no sampling variability because the only eligible sample is the entire population—and the FPC ensures $\sigma^2_{\bar{y}}$ appropriately equals zero.

In many applications, the population is large relative to the sample. In such cases, $\sigma^2_{\bar{y}}$ is reduced more by the sample size n in the denominator of the second term in (11.19) than by the sampling fraction. As a general rule, if the sampling fraction—expressed as the

---

[13] This is called a *combination*, which is a way to select n items from a set of N items where the order of selection does not matter. The notation $\binom{N}{n}$ is read as "N choose n."

[14] For example, when $N = 9$ and $n = 2$, $\binom{9}{2} = 36$, which matches the number of samples used to construct the sampling distribution in Section 11.2.3. In practice, this calculation isn't something you need to do to select a sample; it's included here to explain where equal-probability results come from.

percent of the population observed—is less than 2%, the FPC has little impact and can be dropped from (11.19) and subsequent equations to simplify life (Cochran 1977). As we'll cover in Chapter 12, many forest inventory settings assume an infinite population, in which case the FPC is not appropriate.

In practice, we don't know the population variance σ^2 used in $\sigma^2_{\bar{y}}$ (11.19); rather, we estimate it using the sample variance defined as

$$s^2 = \frac{1}{n-1}\sum_{i=1}^{n}(y_i - \bar{y})^2. \tag{11.20}$$

Then, given s^2, the unbiased estimator for $\sigma^2_{\bar{y}}$ is

$$s^2_{\bar{y}} = \left(1 - \frac{n}{N}\right)\frac{s^2}{n}, \tag{11.21}$$

where, notice, this is the same expression as (11.19) except s^2 is substituted for σ^2. As was the case for $\bar{y}$ and μ, it can be shown that s^2 and $s^2_{\bar{y}}$ are unbiased estimators for σ^2 and $\sigma^2_{\bar{y}}$, respectively (Cornfield 1944).

The population standard deviation σ is estimated using the sample standard deviation defined as

$$s = \sqrt{s^2}. \tag{11.22}$$

Following from (11.21), we estimate the standard deviation of $\bar{y}$ as

$$s_{\bar{y}} = \sqrt{s^2_{\bar{y}}} = \sqrt{\left(1 - \frac{n}{N}\right)\frac{s^2}{n}}. \tag{11.23}$$

This statistic is so special and used so often that it has its own name; it's called the *standard error of the mean*. Given one sample from the population, the standard error of the mean is our best estimate for the standard deviation of $\bar{y}$'s sampling distribution and—as we'll see in Section 11.3.4—is key to quantifying uncertainty for the estimated population mean and other parameters of interest.

If the FPC isn't needed, then the estimate for the standard error of the mean (11.23) simplifies to

$$s_{\bar{y}} = \sqrt{\frac{s^2}{n}} = \frac{s}{\sqrt{n}}. \tag{11.24}$$

11.3.2 Simple Random Sampling with replacement

A simple random sample with replacement (SRSWR) of size n from a population of N units is seen as selecting n *independent* samples of size 1. Here, independent means the selection of any given unit doesn't change the probability any unit is selected. To generate a SRSWR of size n, one unit is randomly selected from the population with probability $1/N$, representing the first sample unit. Then the sample unit is returned to the population (i.e., replaced), and the second unit is selected with probability $1/N$, then the unit is returned to the population. This process repeats until n units are selected into the sample. Note this sample may include the same sample unit multiple times.

When sampling units are selected using SRSWR, the population mean is estimated using $\bar{y}$ (11.18), and population variance is estimated using s^2 (11.20). Due to the replacement, even if our sample size equals the population size, there's no guarantee (in fact it's highly unlikely) that all population units will be observed; hence, there's no notion of a FPC when estimating the standard error of the mean. The standard error of the mean $s_{\bar{y}}$ under SRSWR is computed using (11.24).

When the sampling fraction is large, the SRS standard error of the mean estimator (11.23) is smaller than the SRSWR estimator (11.24). The gain in efficiency comes from the FPC, which reduces the sampling variance under SRS. When the sampling fraction is small, the FPC has little effect, and the two approaches yield nearly identical standard errors.

11.3.3 Estimating totals and proportions

Given the sample mean (11.18), an unbiased estimator of the population total τ (11.6) is

$$\hat{t} = N\bar{y}. \tag{11.25}$$

Following from (11.19), the sampling distribution variance of $\hat{t}$ is

$$\sigma^2_{\hat{t}} = N^2\sigma^2_{\bar{y}}. \tag{11.26}$$

The unbiased estimator for $\sigma^2_{\hat{t}}$ is

$$s^2_{\hat{t}} = N^2 s^2_{\bar{y}} \tag{11.27}$$

and its standard error is

$$s_{\hat{t}} = \sqrt{N^2 s^2_{\bar{y}}} = N s_{\bar{y}}, \tag{11.28}$$

where, depending on the size of the sampling fraction or whether sampling is conducted using SRS or SRSWR, $s_{\bar{y}}$ is given by either (11.23) or (11.24).

The development of estimators for means, totals, and their associated dispersion in the preceding sections focused on quantitative variables. In many cases, however, we want to make inferences about qualitative population parameters such as proportions. Given a sample of size n, the unbiased estimator for the population proportion parameter p (11.7) is

$$\hat{p} = \frac{\sum_{i=1}^{n} y_i}{n}, \tag{11.29}$$

where, recall, y is a binary variable taking values 0 or 1.

Uncertainty quantification for $\hat{p}$ proceeds from its sampling distribution. Substituting the population variance σ^2 in (11.19) with the population proportion variance σ^2_p given in (11.8) and simplifying yields the variance of $\hat{p}$

$$\sigma^2_{\hat{p}} = \left(\frac{N-n}{N-1}\right)\frac{p(1-p)}{n}. \tag{11.30}$$

In practice, $\sigma^2_{\hat{p}}$ is unknown and must be estimated. This begins with estimating the sample variance as

$$s^2 = \frac{1}{n-1}\sum_{i=1}^{n}(y_i - \hat{p})^2 = \frac{n}{n-1}\hat{p}(1-\hat{p}). \tag{11.31}$$

Substituting s^2 into (11.19) and simplifying yields the estimator for the variance of $\hat{p}$

$$s^2_{\hat{p}} = \left(1 - \frac{n}{N}\right)\frac{\hat{p}(1-\hat{p})}{n-1}, \tag{11.32}$$

and the corresponding estimate of the standard error

$$s_{\hat{p}} = \sqrt{\left(1 - \frac{n}{N}\right)\frac{\hat{p}(1-\hat{p})}{n-1}}. \tag{11.33}$$

As before, the FPC can be dropped from (11.33) when it's not warranted or appropriate.

11.3.4 Confidence intervals

Consider the quote from Frank Freese in his introduction to statistical methods in forestry.

> We have it on good authority that "you can fool all of the people some of the time." The oldest and simplest device for misleading folks is the barefaced lie. A method that is nearly as effective and far more subtle is to report a sample estimate without any indication of its reliability.
>
> — Freese (1967)

In other words, we should accompany each estimate with a reliability statement that answers the question, "How confident are you in the estimate?" Ideally, the statement offers a range of values around the estimate and the probability that the range includes the parameter being estimated. This range is called a confidence interval. A *confidence interval* is determined by an upper and lower value, referred to as the *confidence limits*.

We incorporate probability by defining intervals with specific confidence levels. A *confidence level* is typically defined as $1 - \alpha$, where α is referred to as a *significance level*.[15] A confidence interval with confidence level $1 - \alpha$ is referred to as a $100(1 - \alpha)\%$ confidence interval. For example, a 95% confidence interval has a confidence level of 0.95 ($\alpha = 0.05$, or 5%, which are the same thing).

Three ingredients: 1) an estimate for the population parameter; 2) the standard error of the estimate; and 3) a t-value are combined to form a confidence interval, generically written as

$$\text{Estimate} \pm \text{t}_{\text{df},1-\frac{\alpha}{2}} \cdot (\text{Standard Error}), \tag{11.34}$$

where $\pm$ is the plus-minus symbol. The t-value is a *critical value* taken from a t distribution—not to be confused with the estimator $\hat{t}$ used elsewhere in this book for population totals. The product of the t-value and the standard error, to the right of the $\pm$ in (11.34), is called the *margin of error* (or confidence interval half-width).

---

[15] You may recall from previous experience the use of α in hypothesis testing—this is the same α we describe here.

Say we want to estimate the population mean μ using a sample collected under SRS. We've already seen that the sample mean $\bar{y}$ (11.18) is an unbiased estimator for μ. Further, we learned how to compute the standard error of the mean $s_{\bar{y}}$ (11.23). The third ingredient, $\mathrm{t}_{\mathrm{df},1-\frac{\alpha}{2}}$, comes from a t distribution and allows us to associate a given confidence level with our confidence interval.

The t-value depends on the confidence level $(1-\alpha)$ through the term $\left(1-\frac{\alpha}{2}\right)$, as well as the sample size n through the term $\mathrm{df}=n-1$. The `qt()` function returns the t-value for a given α and sample size n. Using these three ingredients, a confidence interval for the population mean is

$$\bar{y} \pm \mathrm{t}_{n-1,1-\frac{\alpha}{2}} \cdot s_{\bar{y}}. \tag{11.35}$$

For a given standard error $s_{\bar{y}}$, decreasing α increases the width of the confidence interval. For example, a 99% confidence interval ($\alpha = 0.01$) is wider than an 80% confidence interval ($\alpha = 0.2$). Changing α doesn't change the precision of the estimate, only our level of confidence that the interval covers the parameter being estimated. See Figure 11.9 in Section 11.3.5 for illustration.

While we focus primarily on estimating the mean, we can generate confidence intervals for any population parameter of interest. Confidence intervals always take the form described in (11.34). For example, if we're interested in the population total τ, the confidence interval is

$$\hat{t} \pm \mathrm{t}_{n-1,1-\frac{\alpha}{2}} \cdot s_{\hat{t}}, \tag{11.36}$$

where $\hat{t}$ and $s_{\hat{t}}$ are defined in (11.25) and (11.28), respectively.

11.3.4.1 Confidence interval interpretation

Say we're interested in estimating the average depth of the duff layer in a forest stand with its 95% confidence interval. The duff is the organic material layer between the uppermost soil mineral horizon and the litter layer. A SRS of n = 20 soil cores yielded a sample mean $\bar{y}$ = 8 (in) and standard error of the mean $s_{\bar{y}}$ = 0.5 (in). To construct a 95% confidence interval, the corresponding t-value is found using the code below.

```
n <- 20
alpha <- 0.05
t <- qt(p = 1 - alpha / 2, df = n - 1)
t
```

```
#> [1] 2.093
```

Then, following (11.35), the confidence interval is

$$8 \pm 2.093(0.5)$$

or (8-1.05 , 8+1.05) = (6.95 , 9.05) (in).

The key to interpreting a confidence interval is to recognize the confidence level is the percent of intervals—where each interval is computed using a different sample from the population—that include the population parameter being estimated. Said differently, for a 90% confidence level, if we collected a large number of different samples of size n from the population, we'd expect 90% of the intervals computed from the samples to contain (cover or include) the parameter value being estimated.

Continuing with the duff layer example, it's correct to state "we're 95% confident the interval (6.95 , 9.05) (in) contains the average duff depth." It's **not** correct to state "there's a 95% probability, or chance, the average duff depth is between (6.95 , 9.05) (in)."

The distinction between these statements is subtle but important. The correct statement references a level of confidence that the interval contains the true average duff depth. The incorrect statement suggests the interval has a probability of including the average duff depth. A single confidence interval (which is typically all we're able to compute because we only have one sample) either includes or doesn't include the population parameter—there's no probability of inclusion. There is, however, a probability that a portion of confidence intervals, computed from repeated sampling, contain the population parameter. This idea is demonstrated in Section 11.3.5.

11.3.4.2 Conditions for confidence intervals

For a confidence interval to accurately capture the given confidence level, three conditions must be met.

1. *Randomness*: the population units are selected into the sample using a probability sampling method, e.g., SRS or SRSWR.
2. *Normality*: the estimator's sampling distribution is approximately normal; see; e.g., Section 11.2. Because we typically have one sample from the population, we can't construct the sampling distribution to test this condition. Rather, we rely on theoretical results and sample data EDA. Three cases result in an approximately normal sampling distribution.
    a. The condition is met if the population from which we draw the sample is normally distributed. In practice, however, we usually don't know the population's distribution.
    b. The condition is met if the sample size is sufficiently large. A theoretical result from the *Central Limit Theorem* states that regardless of the population's distribution, the sampling distribution will be approximately normal if the sample size is sufficiently large. What is "sufficiently large" depends on the degree to which the population's distribution deviates from a normal distribution—the greater the deviation from normal (e.g., multiple modes and not symmetric), the larger the sample size needed to ensure a normally distributed sampling distribution. For an approximately normal population distribution, some results show that a sample size n of 25 or 30 is sufficient; see, e.g., Hogg et al. (2015).
    c. The condition might be met even if our sample is small, but we can ensure the population's distribution is unimodal, symmetric, continuous, and doesn't have outliers. An *outlier* is an observation that lies an abnormal distance from other values in the distribution. While we're unlikely to observe the population, we might

try and infer its shape and the presence of outliers from the sample. If the sample data are approximately symmetric and don't have outliers, then we might assume the population's distribution is approximately normal. However, this assumption and associated EDA results should be clearly stated when reporting the confidence interval.

3. *Independence*: the sampling units are selected independent of one another. Under SRS, sampling units aren't independent because selecting a unit into the sample changes the selection probability of non-selected units (i.e., the first unit selected into the sample has selection probability 1/N, the second unit has selection probability 1/(N-1), and so on). However, if the sampling fraction is less than ~10%, we can assume the independence condition is met.

If these three conditions are met, the resulting confidence interval should be accurate and conclusions reliable.[16]

11.3.4.3 Estimate precision

The question arises, "how do we increase the precision of our estimate?" Or, equivalently, for a fixed confidence level, "how do we narrow the confidence interval?" To answer these questions, consider again the standard error of the mean formula (11.24), ignoring the FPC for now.

The standard error depends on two components: the population variance σ^2 (estimated by the sample variance s^2) and the sample size n. In most applications, once the population and sampling units are defined, σ^2 is essentially fixed—it reflects the natural variability among units. In contrast, n is fully under our control. So, to increase precision and narrow the confidence interval, we increase the sample size.

Of course, there's no free lunch: larger samples typically require more time, money, and resources. While the population variance can be influenced indirectly during study design (e.g., by how sampling units are defined), we assume that structure is already in place. Our focus here is on improving precision through sample size. We'll explore how to determine the sample size needed for a desired level of precision in Section 11.3.6.

11.3.5 Illustrative analysis

Continuing with the Harvard Forest analysis, let's use a SRS of size $n = 30$ to estimate a few of the population parameters computed in Section 11.1.5. In particular, we're interested in estimating average biomass (Mg/ha), μ, and total biomass, τ, along with their associated confidence intervals.

Given the $N = 140$ population units delineated in Figure 11.4, (11.16) tells us the number of distinct samples of size $n = 30$ that could be drawn is

$$\frac{140!}{30!(140-30)!} \tag{11.37}$$

[16] See Exercises 11.2-11.5 for an exploration of the consequences of failing to meet these three conditions.

or

```
# Recall, bio_mg_per_ha was previously extracted from the
# Harvard Forest dataset and is a vector of biomass (Mg/ha)
# measured on all population units.
N <- length(bio_mg_per_ha)
n <- 30
factorial(N) / (factorial(n) * factorial(N - n)) # Or using choose(N, n).
```

```
#> [1] 3195451784096803839625217966080
```

which is an astonishingly large number! This calculation is shown only to emphasize how many equally valid samples exist under SRS; in practice, we draw and analyze just one realized sample. Importantly, following the discussion in Sections 11.2.3 and 11.3.4, each distinct sample could yield a different, but equally valid, estimate and associated confidence interval for the parameter of interest.

Let's draw one of the possible samples. Selecting the sample begins with the sampling frame, which is just a list of all population units from which we select sample units. Here, the sampling frame consists of a vector of integers from 1 to 140 that index the population units and is illustrated in Figure 11.3. Our task is to select n indices at random and without replacement from this sampling frame.

This is accomplished using `sample.int()`, which returns n integers from 1 to N. When consulting the `sample.int()` manual page, notice the `TRUE` or `FALSE` value passed to the function argument `replace` determines if selection is SRSWR (with replacement) or SRS (without replacement), respectively. Note that the argument names in `sample.int()` differ from our notation: the argument `n` corresponds to the population size N, and `size` corresponds to the sample size n. This requires a small amount of cross-walking between notation and code.

The vector of integers returned from `sample.int()` is assigned to `sample_index`, which we'll use to select the corresponding population units into the sample.

```
set.seed(seed = 1) # Sets the seed for the RNG in sample.int.
sample_index <- sample.int(n = N, size = n, replace = FALSE)
sample_index
```

```
#>  [1]  68 129  43  14  51  85  21 106  74   7  73  79
#> [13] 135  37 105  89 110 101 127  34 133 125  44 139
#> [25]  33  84  35  70 132  42
```

There are a few things to notice in the code above. `sample.int()` uses a random number generator (RNG), which, as the name suggests, generates random integers within the range of values we prescribe. Just for fun, run `sample.int(n = 140, size = 30, replace = FALSE)` a few times in your console. You'll see that each time the function is called it returns a different series of random integers.

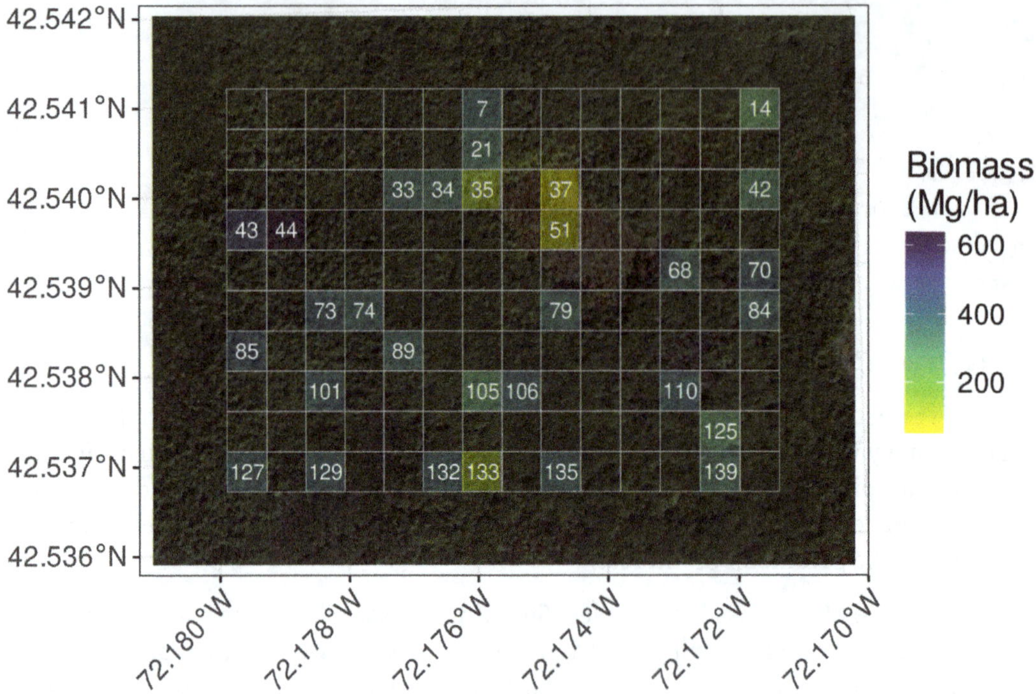

FIGURE 11.8: Harvard Forest biomass (Mg/ha) sampling units selected using Simple Random Sampling without replacement.

`set.seed()` sets the RNG's state, called the *seed*, used to generate the sequence of random numbers. Setting the RNG's seed via `set.seed(seed = 1)` before calling `sample.int()` ensures you get the same result each time you run the code. This is important when we wish to reproduce analyses or research results that rely on RNGs. The specific value of the seed is arbitrary; what matters is using the same seed to reproduce results.

Using the sampling unit indices stored in `sample_index`, we create a new vector called `bio_mg_per_ha_n30` containing the biomass measured on the sampled units. This vector represents our observed data, denoted generically as y, and is the only information from the population used to compute the subsequent estimates.

```
bio_mg_per_ha_n30 <- bio_mg_per_ha[sample_index]
```

Figure 11.8 shows the 30 sampling units selected via `sample.int()`. Biomass values (Mg/ha) are encoded by color, while the numbers shown within selected units identify the sample indices.

Let's start by estimating the population mean μ using the sample mean (11.18), assigned to `y_bar`.

```
y_bar <- sum(bio_mg_per_ha_n30) / n  # or mean(bio_mg_per_ha_n30)
y_bar
```

```
#> [1] 351.74
```

This estimate looks very close to the true population value of $\mu = 354.15$ (Mg/ha). However, as noted in Section 11.3.4, we should also provide a statement of uncertainty for the estimate. In this case, uncertainty is quantified using a confidence interval with a stated confidence level.

Let's construct a 95% confidence interval for the mean. From Section 11.3.4, recall the three key ingredients are: 1) an estimate for the population parameter; 2) the standard error of the estimate; 3) a t-value. We have the first ingredient, `y_bar`. The second ingredient starts with computing the sample variance s^2 defined in (11.20).

```
s_sq <- 1 / (n - 1) * sum((bio_mg_per_ha_n30 - y_bar)^2)
# or var(bio_mg_per_ha_n30)
```

The standard error of the mean is then computed using (11.23) and stored as `s_y_bar`.

```
s_y_bar <- sqrt((1 - n / N) * s_sq / n)
```

The final ingredient is the t-value, which comes from `qt()` given a probability of $1 - \frac{\alpha}{2}$ and degrees of freedom $n - 1$.

```
alpha <- 0.05
t <- qt(p = 1 - alpha / 2, df = n - 1)
```

Using these three ingredients, the code below follows (11.35) to compute the margin of error and the resulting lower and upper confidence limits.

```
y_bar_mar_err <- t * s_y_bar # Margin of error.
y_bar_lower <- y_bar - y_bar_mar_err # Lower limit.
y_bar_upper <- y_bar + y_bar_mar_err # Upper limit.

round(y_bar_mar_err, 2)
```

```
#> [1] 39.19
```

```
round(y_bar_lower, 2)
```

```
#> [1] 312.55
```

```
round(y_bar_upper, 2)
```

```
#> [1] 390.93
```

There are several ways to report an estimate with associated uncertainty. Here are a few example statements:

- We estimate 351.74 Mg/ha of biomass with a 95% margin of error $\pm$ 39.19.
- We estimate 351.74 Mg/ha biomass with a 95% confidence interval from 312.55 to 390.93.
- Biomass (Mg/ha) was estimated to be 351.74 (312.55, 390.93), with a 95% confidence interval in parentheses.

Because we know the population mean for these data, we can see our sample is one of the approximately 95% of possible samples that yield a confidence interval containing $\mu =$ 354.15.

Given that we've already computed $\bar{y}$ and its standard error $s_{\bar{y}}$, we can now estimate total biomass (11.25), its standard error (11.28), and a 95% confidence interval. These quantities are computed in the code below. Just for fun, the last line of code uses `paste0()` to make a character string that includes the estimate and 95% confidence interval. As the name suggests, `paste0()` pastes together comma-separated numeric or character objects and returns the resulting character string.[17]

```
A_ha <- N / 4 # Forest area in ha (i.e., 35 ha).
t_hat <- A_ha * y_bar # Estimate for the total.
s_t_hat <- A_ha * s_y_bar # Standard error of the total estimate.
t_hat_mar_err <- t * s_t_hat # Margin of error, using previous t value.
t_hat_lower <- t_hat - t_hat_mar_err # Lower limit.
t_hat_upper <- t_hat + t_hat_mar_err # Upper limit.

paste0(round(t_hat, 2), " (", round(t_hat_lower, 2), ", ",
       round(t_hat_upper, 2), ") (Mg/ha)")
```

```
#> [1] "12310.93 (10939.15, 13682.7) (Mg/ha)"
```

```
# Or equivalently.
paste0(round(A_ha * y_bar, 2), " (", round(A_ha * y_bar_lower, 2),
       ", ", round(A_ha * y_bar_upper, 2), ") (Mg/ha)")
```

```
#> [1] "12310.93 (10939.15, 13682.7) (Mg/ha)"
```

---

[17]There is a more general function called `paste()` with an argument `sep`, the value of which separates the terms, e.g., `paste(":", ")", sep="-")` results in :-).

In the code above, the total biomass estimate `t_hat` and associated standard error `s_t_hat` are computed by scaling `y_bar` (Mg/ha) and `s_y_bar` (Mg/ha) by the forest area `A_ha`.[18]

Let's step through the use of `A_ha` in this code. Recall the N population units are each 0.25 ha. However, back in Section 11.1.5, we scaled biomass observations on all units by 4 so that $\bar{y}$ and $s_{\bar{y}}$ were expressed on the desired per ha basis, i.e., Mg/ha. This change in area basis means we can't simply scale $\bar{y}$ (Mg/ha) and $s_{\bar{y}}$ (Mg/ha) by N to get the total $\hat{t}$ and standard error $s_{\hat{t}}$, as directed by (11.25) and (11.28). Rather, so that units of $\bar{y}$ and $s_{\bar{y}}$ cancel out correctly when computing the total (Mg), we scale $\bar{y}$ and $s_{\bar{y}}$ by `A_ha`, the total area in ha. It's important to keep track of units. There's an implicit assumption in (11.25) and (11.28) that units of N match the units of $\bar{y}$ and $s_{\bar{y}}$.

Next, let's reinforce a few key points about confidence intervals using an illustration. Figure 11.9 shows biomass confidence intervals estimated using 100 different samples of size $n = 30$ drawn using SRS from the Harvard Forest population. For each sample, we computed an 80%, 95%, and 99% confidence interval. The 100 resulting intervals, for each confidence level, are drawn as horizontal lines in Figure 11.9 and colored blue if they cover the population biomass value $\mu = 354.15$ (represented as the vertical line) and red otherwise. The empirical (meaning observed) interval coverage is 78%, 95%, and 99% for the 80%, 95%, and 99% confidence intervals, respectively. Recall, the desired confidence level is only theoretically achieved when an exceptionally large number of samples is considered, and the population and samples meet the conditions laid out in Section 11.3.4. Therefore, the empirical results based on 100 samples might not match their respective confidence levels exactly.

In Figure 11.9, notice for any given sample, the trend in interval width moving from the 80%, 95%, to 99% confidence level—as the confidence level increases, so too does the interval width. This trend is not because an estimate's precision is increasing (its precision stays the same regardless of confidence level); rather, wider intervals are needed to ensure the desired level of coverage. Each sample has the same precision as estimated by the standard error of the mean $s_{\bar{y}}$. As noted in Section 11.3.4.3, once the sampling design is set, the only practical way to increase an estimate's precision is to increase the sample size n, which in turn decreases the standard error of the mean.

11.3.6 Sample size

We start our sample size discussion with another insightful quote from Frank Freese.

> Samples cost money. So do errors. The aim in planning a survey should be to take enough observations to obtain the desired precision—no more, no less.
>
> — Freese (1967)

---

[18] Notice, we could save some lines of code by just scaling the confidence limits for `y_bar` by `A_ha` to get the limits for the total biomass.

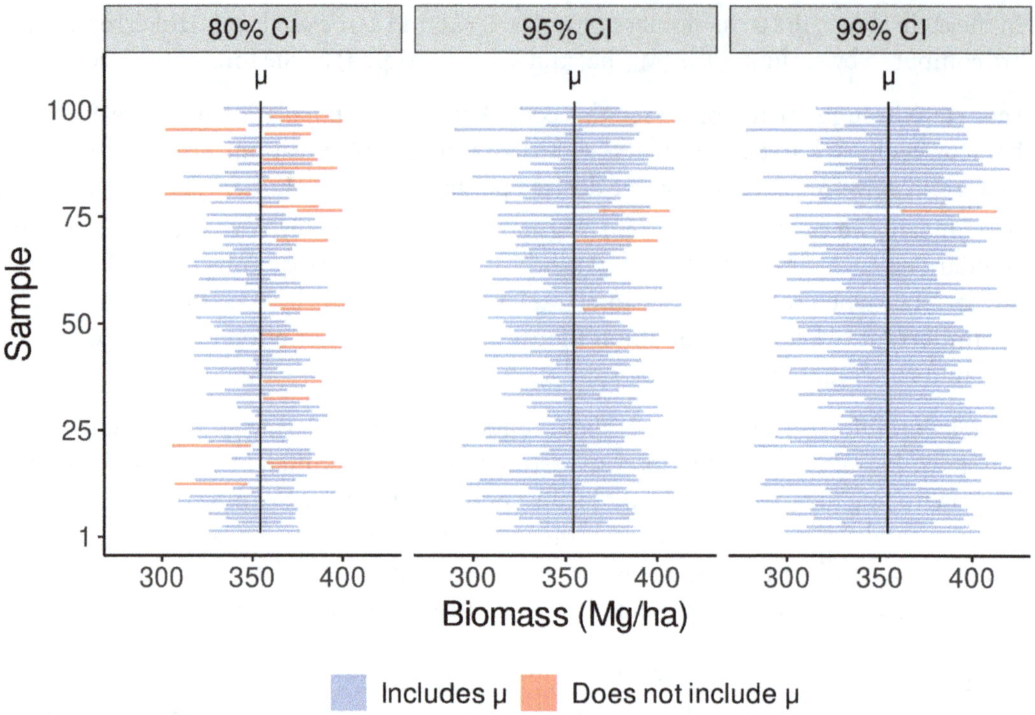

FIGURE 11.9: Harvard Forest biomass (Mg/ha) confidence intervals from 100 samples of size 30 selected using Simple Random Sampling without replacement.

In other words, when designing a sampling effort, our desired precision should determine the sample size to collect. This requires forethought about how precise an estimate needs to be to meet survey objectives. For example, do we need an estimate to be within $\pm$ 5 tons of biomass per acre of the true value, or is an estimate of $\pm$ 10 tons sufficient? As we'll see, the difference between the sample size needed to achieve a precision of $\pm$ 5 versus $\pm$ 10 tons per acre could be large, so we should be sure the additional effort and expense is warranted.

An approach to determine sample size comes directly from the notions developed in Section 11.3.4 for confidence intervals. Recall, the product of the t-value and standard error, to the right of $\pm$ in (11.34), is the margin of error. Notice how (11.34) reads like the biomass example in the preceding paragraph, where we have an estimate $\pm$ a possible error of 5 tons per acre.

To compute a sample size, we choose the value to the right of the $\pm$ and call it the *allowable error*. As developed below, in addition to specifying the allowable error, we must set the probability of meeting the allowable error. Similar to constructing confidence intervals, this probability is expressed as a confidence level. Therefore, if our survey objectives require biomass per acre be within $\pm$ 5 tons of the true value with high confidence, we state the sample size problem as: *what is the minimum sample size needed to achieve an allowable error of* $\pm$ *5 tons with a 99% confidence level?*

The confidence level interpretation follows from the confidence interval. For example, say we find the sample size n needed to meet some specified allowable error at a confidence level 90%. If we were able to collect a large number of different samples each of size n, then we'd expect 90% of the estimates computed from the samples to be within the desired allowable error of the parameter value.

To reduce clutter, we'll suspend using the subscript notation associated with the t-value, but remember that it has $n-1$ degrees of freedom and confidence level $1-\alpha$. Consider again the margin of error introduced in Section 11.3.4. Following (11.34) and using $s_{\bar{y}}$ that includes the FPC (11.23), the margin of error for the population mean is

$$\begin{aligned} E &= \mathrm{t}\, s_{\bar{y}} \\ &= \mathrm{t}\sqrt{\left(1-\frac{n}{N}\right)\frac{s^2}{n}}. \end{aligned} \tag{11.38}$$

For the sample size calculation, E is the *allowable error* that we set based on survey objectives. Solving (11.38) for the sample size n gives us

$$n = \frac{\mathrm{t}^2\, s^2 N}{(E^2 N + \mathrm{t}^2\, s^2)}. \tag{11.39}$$

Below we define each term in (11.39) within the context of estimating sample size:

- n is the minimum sample size required to estimate the population mean μ to within $\pm E$ for a specified confidence level.
- t is the same t-value encountered before, which requires a confidence level $(1-\alpha)$ and the degrees of freedom $(n-1)$. Notice there's a circularity issue here—n appears on both sides of the equation. We'll introduce an algorithm to calculate n despite this issue.
- s^2 is the sample variance. Because we've not taken the sample, this must come from somewhere else, e.g., prior experience working with a similar population or small pilot study.
- E is the allowable error, which is the amount of error we can tolerate in the estimate computed using the resulting sample. Note that E is in the measurement units of the estimate.

Using this notation, we can say with probability $(1-\alpha)$ the absolute value of the difference between an estimate $\bar{y}$ based on n observations and the true parameter value μ is less than or equal to E with probability $(1-\alpha)$.[19]

If we're in a setting where it's appropriate to ignore the FPC, then the sample size calculation simplifies to

$$n = \frac{s^2\, \mathrm{t}^2}{E^2}. \tag{11.40}$$

Thus, given a variance estimate s^2, allowable error E, and confidence level (i.e., $1-\alpha$), we can estimate the minimum sample size n. The only hitch is we need a way to address

[19] Or stated more compactly $|\bar{y}-\mu| \leq E$, where $||$ takes the absolute value of the quantity between them.

the circularity issue caused by n appearing on both sides of the equation in sample size formulas (11.39) or (11.40).[20] We do this using an iterative algorithm outlined below.

1. Compute the $t_{n_0-1,1-\frac{\alpha}{2}}$ value using an educated guess for the sample size n_0 and your chosen α. Then apply the sample size formula and call the result n_1. Round n_1 up to the nearest integer.
2. Apply the sample size formula using $t_{n_1-1,1-\frac{\alpha}{2}}$. Call the result n_2. Round n_2 up.
3. Compare n_2 and n_1.
    a. If $n_2 = n_1$, n_2 is the minimum desired sample size and there are no further steps.
    b. If they're not equal, return to Step 2 and calculate the t-value using n_2. Call the result n_3 and round it up to the nearest integer.
4. Repeat Steps 2 and 3, incrementing the subscript on n for each iteration. You stop at iteration i when $n_i = n_{i-1}$ or when these values differ by an acceptably small amount across multiple iterations.

Even if your initial guess n_0 is way off, it should take only a few iterations to satisfy the condition described in Step 4.

In many cases, survey objectives call for the allowable error expressed as a percent of the parameter of interest. For example, what is the minimum sample size required to be within 10% of the true parameter at some confidence level? We do this by expressing the sample size formula in terms of the CV (11.5) and allowable error as a percent, denoted by $E\%$. In this case we restate (11.39) as

$$n = \frac{t^2(CV)^2N}{((E\%)^2N + t^2(CV)^2)} \tag{11.41}$$

and (11.40) as

$$n = \frac{(CV)^2\,t^2}{(E\%)^2}, \tag{11.42}$$

where the CV is expressed in percent units and is estimated as $100\,(s/\bar{y})$, either from a small pilot study sample or from prior experience working in similar forest types.

11.3.6.1 Illustrative analysis

Let's try a sample size calculation using the Harvard Forest data. In Section 11.3.5, our estimates were based on an arbitrary sample size of $n = 30$, which resulted in a 95% confidence interval margin of error of 39.19 biomass (Mg/ha). Now, say we want to be within $E = 50$ (Mg/ha) of the true value with a 95% confidence level.[21] The sample size is computed using (11.39) because the allowable error E is expressed in the same units as the estimate (Mg/ha), and a FPC is included since the sample was drawn from a finite population and the sampling fraction is expected to exceed 2%.

[20] Recall n enters the right-hand side of (11.39) or (11.40) through the t-value's degrees of freedom.

[21] Test your intuition. Given the margin of error we saw previously using $n = 30$, should the new sample size be more or less than 30?

Looking at (11.39), let's consider the terms we know. We know the population size $N = 140$, and we specified the allowable error $E = 50$ (Mg/ha). We also specified a 95% confidence level (i.e., $\alpha = 0.05$), which partially informs the t-value. The other piece of information needed for a t-value—and following the iterative algorithm above—is n_0, which we'll naively guess to be 20. The only unknown in (11.39) is the sample variance s^2, which we can get using a small pilot sample, previous inventory estimate, or data from a similar forest. Let's use 8000 for s^2.[22] Now we have everything needed to implement the iterative sample size calculation, which is presented in the code below.

```
N <- 140
E <- 50 # Biomass (Mg/ha).
alpha <- 0.05 # For 95% confidence level.
s_sq <- 8000 # An educated guess.
n_0 <- 20 # An educated guess.

# Step 1 (initialize)
t_0  <- qt(p = 1 - alpha / 2, df = n_0 - 1)
n_1 <- t_0^2 * s_sq * N / (E^2 * N + t_0^2 * s_sq)
n_1 <- ceiling(n_1) # Rounds up to the whole integer.
n_1
```

```
#> [1] 13
```

```
# Step 2 (first iteration)
t_1  <- qt(p = 1 - alpha / 2, df = n_1 - 1)
n_2 <- t_1^2 * s_sq * N / (E^2 * N + t_1^2* s_sq)
n_2 <- ceiling(n_2)
n_2 # Step 3, n_2 does not equal n_1, do Step 2 again.
```

```
#> [1] 14
```

```
# Step 2 (second iteration)
t_2  <- qt(p = 1 - alpha / 2, df = n_2 - 1)
n_3 <- t_2^2 * s_sq * N / (E^2 * N + t_2^2* s_sq)
n_3 <- ceiling(n_3)
n_3 # Step 3, n_3 equals n_2, STOP.
```

```
#> [1] 14
```

Notice how quickly the iterative algorithm converged to $n = 14$. Now, just for fun, we took a SRS of size 14 and computed a $\bar{y} = 347.41$. Given the true value $\mu = 354.15$, the absolute value of the difference $|\bar{y} - \mu| = 6.74$ (Mg/ha), which means our one sample is in 95% of the possible samples that result in an estimate falling within the desired allowable error.

---

[22]This is pretty close to the population variance of $\sigma^2 = 8340$, which, again, we hardly ever know in practice.

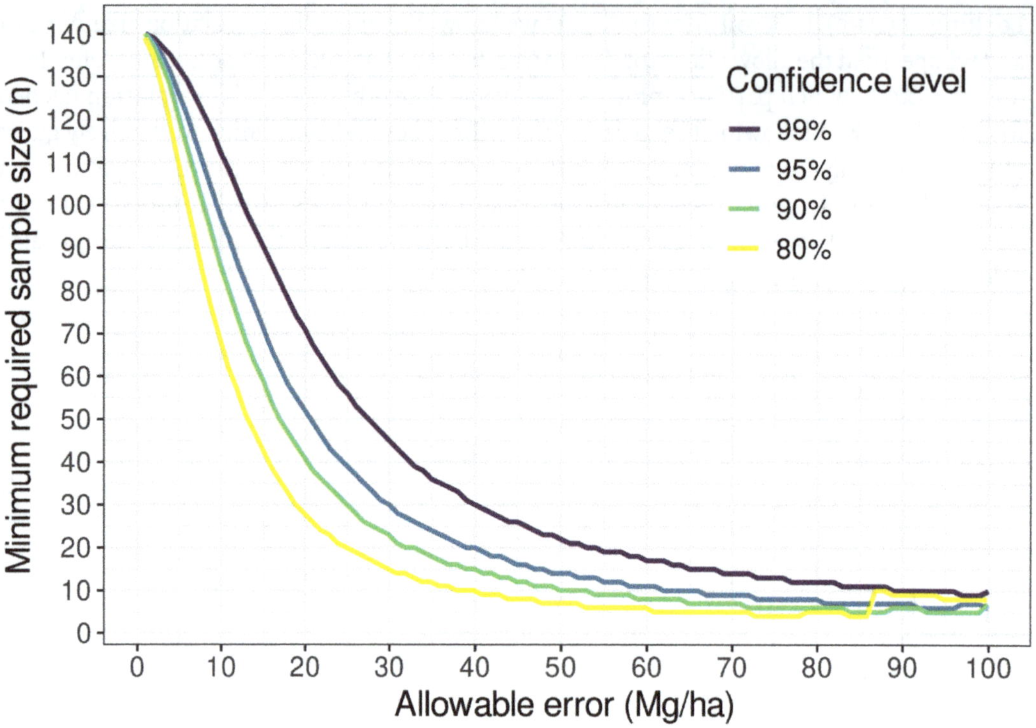

FIGURE 11.10: Effect of allowable error (E) and confidence level on minimum sample size for Harvard Forest biomass (Mg/ha) estimates.

Ideally the allowable error and confidence level are guided by survey objectives. However, in reality, limited financial resources and time often require that we increase the allowable error E and/or decrease the confidence level, both of which reduce the sample size. Figure 11.10 shows the relationship between sample size, allowable error, and confidence level for the Harvard Forest biomass analysis. In this figure, each curve corresponds to a different confidence level. Notice for a fixed allowable error, a decrease in confidence level results in a smaller necessary sample size. For example, if we want 99% of the samples to yield a mean biomass estimate within $E = 50$ (Mg/ha) of the true value, then we need a sample of at least $n = 22$; whereas, if we're content with only 80% of the possible estimates being within $E = 50$ (Mg/ha) of the true value, then only $n = 7$ observations are required.

11.3.7 Relative variation and estimator precision

When comparing variability or uncertainty across samples or estimators, it's helpful to standardize dispersion relative to the scale of the estimate. A common approach is to use a *relative measure of variation*, defined as

$$100 \cdot \frac{\text{measure of dispersion}}{\text{estimate}}. \tag{11.43}$$

Multiplying by 100 in (11.43) expresses the result in percent units, which is how relative variation is conventionally reported in practice.

This expression is used in two common contexts and interpretations.

1. When the numerator is the *sample standard deviation* and the denominator is the *sample mean*, the result is called the *sample coefficient of variation* (CV). This statistic is useful when comparing variability across groups or populations, especially when variables are measured on different scales or in different units. The *sample CV* reflects *natural variability* in the observed data (e.g., how variable tree height is among individual trees).

2. When the numerator is the *standard error of an estimator* (e.g., for a mean or total) and the denominator is the *estimate itself*, the result is referred to as the *relative standard error* (RSE), or equivalently, the *coefficient of variation of the estimator*. In forestry applications, this quantity is often called the *standard error in percent*. The *RSE* reflects *sampling variability*—how much an estimate would vary among repeated samples drawn under the same design.

All else equal, lower values are better in both contexts, indicating less variability (CV) or greater precision (RSE). These metrics are especially useful when comparing competing estimators, evaluating sampling strategies, or interpreting results for estimates on different scales.

Another related metric is the *percent improvement in precision* (or *percent reduction in variance*) between two estimators of the same quantity. Suppose $\hat{\theta}_A$ and $\hat{\theta}_B$ are two unbiased estimators with standard errors s_A and s_B, respectively, then,

$$\text{Percent improvement} = 100\left(1 - \frac{s_B^2}{s_A^2}\right). \tag{11.44}$$

This value is positive when estimator B is more precise (has lower variance) than A.

When estimators target different quantities or are expressed on different scales, comparing their relative variation can be helpful. One option is the *relative CV ratio*

$$\text{Relative CV ratio} = \frac{\text{CV}_{\hat{\theta}_B}}{\text{CV}_{\hat{\theta}_A}}. \tag{11.45}$$

Values below 1 suggest estimator B has lower relative uncertainty than A.

We'll return to these ideas in subsequent chapters, where we compare multiple estimators and sampling designs using these metrics.

11.3.8 Advantages and disadvantages of SRS

We've seen that estimators applied to samples selected using SRS and SRSWR yield unbiased estimates for population parameters such as the mean, total, variance, standard deviation, and measures of dispersion key to inference, such as the standard error of the mean. Ensuring these unbiased estimates is the central advantage of SRS and SRSWR. There are, however, several disadvantages listed below that make these selection methods impractical for many forest inventories.

1. Implementing a method to select sampling locations from the sampling frame can be cumbersome.
2. Traveling between and finding sampling locations' spatial coordinates in the field can be very time consuming.
3. Distributing sampling locations at random doesn't ensure a representative sample from a population with strong spatial patterns, e.g., by chance, clustering of sample locations might miss important spatial variability in population characteristics.
4. Using prior or ancillary information about population characteristics can improve efficiency, and SRS and SRSWR alone can't incorporate such information.

In Chapters 14 through 18, we'll introduce sampling designs that can provide inference comparable to, or in some cases more efficient than, SRS without the disadvantages noted above.

11.4 Summary

We introduced the idea of a population and how it's characterized using different summary measures called parameters. Parameters that describe the center point or typical value are called measures of central tendency, and parameters that describe the spread of values about the central tendency are called measures of dispersion. Without the luxury of observing the entire population, we settle for a sample of observations drawn from the population. A statistic is a characteristic of a sample used to estimate a parameter. A frequency table and its graphical histogram representations are EDA tools used to explore and communicate characteristics of either a population or sample.

Two probability sampling approaches, SRS and SRSWR, were introduced to guide selection of population units into a sample. A probability sample (1) avoids selection bias and (2) provides statistically defensible uncertainty quantification for mean and variance statistics. We also discussed how personal and measurement bias can cause a well-devised probability sample to poorly represent the population, and hence must be vigilantly guarded against.

We introduced the notation of an estimator and how it's used to generate a statistic's sampling distribution, which in turn is used to assess expected qualities of the statistic (e.g., bias, accuracy, precision). We illustrated how the sample mean statistic's sampling distribution (i.e., the standard error of the mean) is estimated and used to compute confidence intervals for mean, total, and proportion estimates. Given an allowable error and confidence level, the confidence interval mechanics were then extended to sample size calculations.

For quick reference, Table 11.3 summarizes statistics to estimate the population mean, total, and proportion using SRS, along with their respective standard errors. Recall, when the sampling fraction n/N is less than 2%, the FPC, i.e., $(1 - \frac{n}{N})$, can generally be dropped

TABLE 11.3: Statistics and associated standard errors for inference about population parameters.

Population	Estimate	Standard error of estimate
Variance σ^2	$s^2 = \frac{\sum_{i=1}^{n}(y_i-\bar{y})^2}{n-1}$	
Mean μ	$\bar{y} = \frac{\sum_{i=1}^{n} y_i}{n}$	$s_{\bar{y}} = \sqrt{\left(1-\frac{n}{N}\right)\frac{s^2}{n}}$
Total τ	$\hat{t} = N\bar{y}$	$s_{\hat{t}} = N s_{\bar{y}}$
Proportion p	$\hat{p} = \frac{\sum_{i=1}^{n} y_i}{n}$	$s_{\hat{p}} = \sqrt{\left(1-\frac{n}{N}\right)\frac{\hat{p}(1-\hat{p})}{n-1}}$

from the calculations. Statistics for SRSWR are the same as those given in Table 11.3, but with the FPC removed.

We demonstrated the inverse relationship between confidence level and margin of error, and between sample size and allowable error. A distinction was drawn between changing the confidence level and changing an estimate's precision. Once the sampling design is set, the only substantive way to increase an estimate's precision is to increase the sample size. Changing the confidence level affects the confidence interval's width to ensure that the prescribed percentage of intervals, computed from repeated sampling, contain the true parameter value. We also emphasized how the idea of repeated sampling underpins the interpretation of both confidence intervals and sample size calculations.

In the next chapter, we connect general statistical survey sampling topics covered here with forestry applications. There are several additional considerations that must be taken into account when applying the statistical tools developed here to forest settings. For example, the list sampling frame used to select sampling units will comprise an infinite set of sampling locations within the forest population's boundary. Additionally, the distinction between the sampling unit and the measurement unit becomes more important. In most forestry applications, trees are the measurement unit and summaries of their various measurements are the values associated with the sampling unit. So, rules will need to be established for how to select which trees are measured at each sampling unit and how their values should be summarized. The next chapter also forms the basis for the efficient sampling designs presented in later chapters.

11.5 Exercises

We'll use the Harvard Forest dataset for several exercises. Recall, as illustrated in Figure 11.3, this population comprises 140 0.25 ha units. Measurements on population units are held in the `"HF_pop_units.csv"` file. The code below reads in this file and assigns it to the `hf` data frame.

```
hf <- read_csv("datasets/HF/HF_pop_units.csv")

hf %>% glimpse()
```

```
#> Rows: 140
#> Columns: 3
#> $ trees_per_ha     <dbl> 3296, 2188, 2268, 2112, 1796, 1760~
#> $ bio_mg_per_ha    <dbl> 352.69, 380.87, 314.18, 340.12, 38~
#> $ prop_gr_30cm_ha <dbl> 0.038835, 0.104205, 0.058201, 0.07~
```

`dplyr::glimpse()` provides a compact description of `hf`. The output tells us `hf` has 140 observations (rows) and 3 variables (columns). The first column named `trees_per_ha` holds the number of trees on each 0.25 ha unit, expressed on a per ha basis (i.e., we've already applied the expansion factor). You're familiar with the second column `bio_mg_per_ha`, which holds the aboveground biomass (Mg/ha) on each unit and is the same data used for illustration throughout this chapter. The third column named `prop_gr_30cm_ha` holds the proportion of trees greater than 30 cm in DBH on each unit.

The next two lines of code extract `trees_per_ha` and `prop_gr_30cm_ha` from the data frame and set them as vectors for use in subsequent exercises.

```
# Extract the trees per ha and proportion columns.
trees_per_ha <- hf %>% pull(trees_per_ha)
prop_gr_30cm_ha <- hf %>% pull(prop_gr_30cm_ha)
```

Exercise 11.1. Using the `trees_per_ha` variable, compute each of the following parameters. First compute the parameters *by hand* (i.e., using R but following the mathematical expressions given in this chapter), then use R's built-in functions to check your answers.

a. mean
b. median
c. variance
d. range
e. standard deviation
f. total

Exercise 11.2. Use the base R `hist()` function or the `ggplot2` package to create a histogram similar to Figure 11.5 for `trees_per_ha`. Add a main title as well as x-axis and y-axis labels to the histogram. Make sure your x-axis label has the correct units. Ensure the number of histogram x-axis bins is sufficient to get a good sense of the data distribution. Using your histogram, with support from your work in Exercise 11.1, answer the following questions.

a. Is this a unimodal or multimodal distribution?

b. Is the distribution skewed? If so, is it negatively skewed or positively skewed?

c. How do the mean and median values computed in Exercise 11.1 support your conclusion about the distribution's skew?

d. What is the approximate value of the distribution's mode?

Exercise 11.3. In Section 11.3.5 we estimated the Harvard Forest's mean and total biomass along with the estimates' 95% confidence intervals. Following the code in Section 11.3.5, draw a SRS sample of size $n = 25$ from `trees_per_ha`. Set your RNG seed to 1, i.e., `set.seed(seed = 1)`, prior to creating your `sample_index` vector using `sample.int()`.

Use this sample to estimate the following.

a. Mean number of trees per ha and associated 90% confidence interval.

b. Total number of trees and associated 90% confidence interval.

Using your estimates and the population measurements in `trees_per_ha`, answer the following questions.

c. Does the population mean number of trees per ha and total number of trees fall within the confidence intervals you computed?

d. Say you took a large number of samples each of size n=25 drawn using SRS and a unique RNG seed. Then, with each sample, you computed an estimate for the mean number of trees per ha and associated 90% confidence interval. Approximately what percent of the resulting confidence intervals would include the true mean number of trees per ha?

Exercise 11.4. Consider the three conditions for valid confidence intervals listed in Section 11.3.4. Does the SRS of n=25 in Exercise 11.3 meet these conditions? List the conditions that are not met and describe why the sample does not meet the given condition. Hint, there is at least one condition that is not met. Also, you'll note the illustrative biomass analysis we present in this chapter does not meet one of the conditions either.

Exercise 11.5. We generated 100,000 samples of size n=25 from the Harvard Forest `trees_per_ha` vector of population measurements. For each sample, we estimated the mean number of trees per ha, computed the 95% confidence interval, and determined whether or not the 95% confidence interval contained the population mean. By definition, we expect about 95% of these confidence intervals to contain the population mean. However, we found only 89% containing the population mean.

Consider your answers to Exercises 11.2–11.4 and provide a likely reason why we found this result. What does this result suggest about situations where conditions required to compute confidence intervals are not met?

Exercise 11.6. Which is the most accurate interpretation of a 95% confidence interval for the mean resulting from probability sampling?

a. There is a 0.95 probability the confidence interval contains the population mean.

b. If the confidence interval was computed for each of a large number of equal sized samples, about 95% of intervals would cover the population mean.

c. The confidence interval contains about 95% of the population's distribution.

Explain why the answers you didn't choose are not correct interpretations.

Exercise 11.7. Say, a researcher is interested in estimating the mean (tons/ac) and total (tons) root mass for a particular invasive species within a 10 acre forest. They defined population units by tessellating the forest into 1/10000 acre non-overlapping squares. From the resulting sampling frame they use SRS to select n=5 sampling units that were subsequently measured and found to have 0.002, 0.001, 0.0, 0.0015, and 0.001 tons of the species's root mass. Given this information, answer the following questions.

a. How many population units comprise the 10 acre forest (i.e., what is N)?

b. What expansion factor will convert tons per sampling unit to tons per acre (i.e., what is the expansion factor)?

c. What is the mean root mass estimate expressed in tons/ac (i.e., $\bar{y}$)?

d. What are the lower and upper 80% confidence interval bounds for $\bar{y}$ found in (c)?

e. What is the total root mass estimate expressed in tons (i.e., what is $\hat{t}$)?

f. What are the lower and upper 80% confidence interval bounds for $\hat{t}$ found in (e)?

Exercise 11.8. Consider some population that has a mean of 14 cm. Say you took a SRS from this population and computed a 95% confidence interval with lower bound of 11.5 cm and upper bound of 12.3 cm. Can this confidence interval be correct? Explain your answer.

Exercise 11.9. Say you were hired by a land conservancy to inventory one of their forest properties. The conservancy is interested in knowing the amount of forest biomass on the property so they can trade it on a carbon market (approximately 50% of tree biomass is carbon). Upon completing your work, you provide the mean biomass (tons/ac) along with a 90% confidence interval in your well written and nicely formatted inventory report. After reading your report, the conservancy's representative says the biomass confidence intervals are much too wide and they need a more precise estimate for the property's biomass. To remedy the issue, they ask you to recompute the confidence interval using a 99% confidence level.

a. Given you'd like to do more work for the conservancy in the future, how would you gently and clearly explain the error in their request to increase estimate precision by increasing the confidence level used for the confidence interval? Part of your explanation should cover how to actually increase an estimate's precision.

b. When discussing alternatives to increase an estimate's precision, the conservancy's representative asks you if there is a better estimator you can use to estimate the population mean. Given your knowledge of sampling and the SRS design, you feel confident the sample mean is the best estimator to use. What reasons would you give to the representative to defend the use of the sample mean over alternative estimators?

c. Being so impressed with your explanation, the representative asked you to inventory a newly acquired property. With a more informed understanding of sample-based inference, they ask for a biomass estimate within 10% of the population mean with 95% confidence. It turns out, the new property is adjacent to the property you just inventoried and the two forests have similar species composition and structure. Given the biomass estimates from your previous inventory $\bar{y} = 60$ (tons/ac) and $s = 20$ (tons/ac), what sample size n will be needed for the new property to meet the desired level of precision? Use the sample size formula that does not include the FPC for these calculations.

Exercise 11.10. A forester working on the Manistee National Forest in Michigan undertook a small study to determine which size of fixed-area plot should be used to inventory wood volume on a portion of the forest. They considered 1/10, 1/5, and 1/4 acre plots. A small preliminary SRS of size $n_0 = 12$ plots was used to compute the CV for the three plot sizes. The CV and other values needed for this analysis are given in Table 11.4. The column labeled *Cost ($/plot)* is the labor cost to measure each plot size.

Which plot size is most cost effective if the forester wishes to produce estimates within 10% of the population mean with a 90% confidence level? To answer this question:

1. Fill in the column labeled *n plots* with the necessary sample size for each plot size, using the correct sample size formula. Assume there is no FPC adjustment. To calculate the sample size, first notice that the CV values in Table 11.4 are reported as percentages. Then decide which sample size formula is appropriate given this—either (11.40) or (11.42).

2. Fill in the *Total cost* column with the inventory cost, which equals the necessary sample size (*n plots*) times the per plot cost (*Cost ($/plot)*).

The plot size with the smallest total cost is the most cost-effective option. Based on your calculations, which plot size would you choose?

Exercise 11.11. Say you just landed your dream job as head arborist for a city's newly formed Shade Tree Commission. In your first week on the job, the Commissioner's Office asked you to provide a work budget for the city's largest neighborhood by the following Monday morning. Given the very short turnaround and the fact that there were no recent street tree assessments to work from, you decided to conduct a quick survey over the weekend to

TABLE 11.4: An example dataset.

n_0 plots	Plot size	CV	Cost ($/plot)	n plots	Total cost
12	0.10	75	16	?	?
12	0.20	60	24	?	?
12	0.25	53	29	?	?

estimate the proportion of neighborhood blocks with trees that require some attention. Using a list of the neighborhood's 204 blocks, you selected a sample of 30 blocks at random and without replacement. For each of the 30 blocks in your windshield survey, you recorded a 1 if work was required and 0 if no work was required.

The comma delimited survey result file found in `"datasets/neighborhood_survey.csv"` has $n = 30$ rows and two columns. The first column `block_index` holds the blocks' sampling unit index, and the second column `needs_work` is a binary variable where a value of 1 means the block needs work and 0 otherwise. Using these data, complete the following analyses to support your budget request.

1. Estimate the proportion of blocks that need work along with a 95% confidence interval.
2. From your previous work experience, you know that, on average, it costs $5,500 for each block that needs work. Using this average cost and your results from part 1, provide a cost estimate along with a 95% confidence interval for the entire neighborhood.

12

Estimating forest parameters

Building on statistical concepts developed in Chapter 11, this chapter covers the basic steps and considerations needed to estimate forest parameters from a sample.[1] In contrast to the finite population framework introduced earlier, this chapter adopts an infinite population perspective that better reflects how forest inventories are conducted in practice. Rather than sampling from a fixed list of discrete units, we place sample locations within the forest and summarize measurements made around each location. This shift shapes how sampling frames are defined, how units are selected, and how measurements are summarized.

The chapter closes with a section that embeds forest parameter estimation within a highly efficient `tidyverse` workflow, leveraging `dplyr` and `tidyr` functions covered in Chapters 7 and 8. We generalize and expand on these concepts and programming tools to accommodate more advanced forestry applications presented in subsequent chapters.

12.1 Sampling units for forestry applications

Collecting a sample used to estimate forest parameters is typically accomplished in three steps. First, sample locations, selected using some random mechanism, are placed within the forest. Second, a rule is used to select measurement trees around each location. Third, for each location, tree measurements are expanded to the desired per-unit-area basis and summarized to represent one observation within the sample. The most common rules for selecting measurement trees are *plot sampling* and *point sampling*.

Plot sampling is applied in many forestry applications, particularly in long-term monitoring efforts. A fixed-area plot is positioned at each sampling location, and trees within the plot boundary are measured. A sampling location can define the center for a circular plot, a corner for a rectangular plot, or the position for more complex plot shapes and configurations such as plot clusters or nested plots of different sizes. Under plot sampling, a tree's selection probability is proportional to the area of the plot used for sampling.

Point sampling is also a common approach to tree selection and a popular method for cruising timber. Under point sampling, the probability of selecting a tree for measurement is conditional on some function of its size. For this reason, the approach is often called probability proportional to size (PPS) sampling. Depending on the survey objectives, compared

[1]We use forest generically to represent a bounded land area holding the population of interest.

DOI: 10.1201/9781003222699-12

with plot sampling, point sampling can offer greater efficiency from both a field-effort and statistical standpoint. In general, there are gains in efficiency when trees are preferentially selected for measurement based on a size attribute related to the parameter of interest. For example, if we're interested in estimating volume, it's more efficient to select measurement trees proportional to their basal area (because basal area and volume are strongly related). However, if we're interested in estimating tree density (i.e., number of trees per unit area) or growth rates, plot sampling would be the preferred approach.

The next section connects forestry applications with concepts from Chapter 11. Specifically, we introduce the areal sampling frame, which takes the place of the list sampling frame introduced in Section 11.2.2. For the applications covered here, the areal sampling frame is used for both plot and point sampling. Then, because it's conceptually easier, we introduce plot sampling, expansion factors, boundary effects, and other related topics. This is followed by point sampling and how expansion factors and other sampling considerations change under this tree selection rule.

12.2 Areal sampling frame

Recall from Section 11.2.2 that SRS and related probability sampling methods require a sampling frame from which the sample is selected. In the examples presented in the previous chapter, the sampling frame is a list of all population units from which sample units are randomly selected. For instance, in the Harvard Forest biomass analysis described in Section 11.3.5, the frame was constructed by overlaying a grid of equal-sized squares across the forest extent (Figure 11.3). This grid delineated the N population units, with indices 1 through N, and sampling units were selected by drawing n random integers in that range. This is a valid and intuitive approach to sample selection.

Geographic information systems (GIS) and R's spatial data packages (see, e.g., Pebesma and Bivand (2023) and Brunsdon and Comber (2019)) can be used to tessellate—or grid—the geographic extent of a population using equal-area, non-overlapping shapes such as rectangles, as in the Harvard Forest example. Creating such grids is straightforward at a computer. However, logistical challenges, such as locating sampling unit boundaries and measuring trees within selected units in the field, often make these approaches impractical in forestry applications.

A more common approach in forestry is to use an *areal sampling frame*. This frame allows selection of any point location defined by geographic coordinates (e.g., longitude and latitude) within the population extent, usually outlined by one or more boundary polygons. Sampling locations are points placed using a random mechanism within this area. Each observation is a per-unit-area summary of the variable at that location, computed from all trees selected by the sampling rule, and treated as the value at that location.

Selecting discrete sampling units from a finite *list sampling frame* differs from selecting points from the infinitely many locations within an area. Fortunately, much of the statistical theory needed for valid inference is shared between list and areal frames. However, using an areal frame brings additional survey design, data collection, and analysis considerations, which are covered in this and subsequent chapters.

Using an areal frame for SRS introduces both statistical and logistical challenges. When sampling locations are randomly distributed, plots may overlap spatially—particularly with fixed-area plots if locations fall too close together. The same trees can then be measured more than once, increasing redundancy and reducing statistical efficiency. This doesn't violate SRS assumptions or the validity of design-based estimators, but it can yield less informative samples. Logistically, it can also be inefficient to navigate to a random set of widely dispersed locations in the field, especially in forested or rugged terrain. To address these issues, forest inventories typically use systematic designs in which sampling locations are spaced evenly across the population extent. To preserve key probability sampling properties, the systematic grid is positioned using a random mechanism. See Figure 1.2 for illustration and Chapter 14 on systematic sampling for further details.

Despite these practical considerations, we continue to use SRS in this chapter for both plot and point sampling because the concepts introduced—defining sampling units, expanding tree-level measurements, and summarizing observations—are foundational and largely design-agnostic. Moreover, several of the more complex designs introduced in subsequent chapters adopt or extend estimators originally developed under SRS. The ideas we establish here provide the grounding needed to understand and implement the more advanced forest sampling strategies that follow.

12.2.1 Finite population correction

While we emphasized the role of the *finite population correction* in Section 11.3, it's rarely applied in forest inventory. The FPC is relevant when sampling is conducted from a list sampling frame, where the population is finite, its size (i.e., N) is known, and each unit is discrete and identifiable. In this case, the correction reduces the standard error because, once selected, a unit is no longer eligible for reselection.

In contrast, with an areal sampling frame the number of possible sampling locations is effectively infinite, and there is no predefined finite set of discrete units. A sampling location can occur anywhere within the geographic extent. Because of this, the conditions required for applying the FPC don't hold, and the correction is not used when estimating standard errors for samples selected from an areal frame.

12.3 Plot sampling

As noted at the beginning of this chapter, *plot sampling* is a common approach for collecting forest inventory data used to estimate population parameters. Fixed-area plots, often circular or rectangular, are established at sampling locations selected from an areal sampling frame. All trees within the plot boundary that belong to the population of interest are then measured. The criteria for determining whether a tree is "in" the plot are typically defined in the field protocol—for example, a standing tree is measured if the center of its stem at DBH falls within the plot boundary.

The survey objectives define the population(s) of interest. For example, in a timber cruise, we might measure only marketable trees—live stems of a certain DBH, species, and log quality. In an Emerald Ash Borer (EAB) forest health assessment, we might measure live and dead ash species. In a regeneration survey, we might measure seedlings and saplings of desirable species.

In practice, we're almost always interested in estimating parameters for multiple populations. These populations often differ in ways that guide how they're sampled. For example, tree density (number of trees per unit area) can vary greatly among populations. Suppose we're assessing both regeneration potential and timber volume in a naturally regenerated, uneven-aged stand. These two stand components—regeneration and overstory—can be viewed as distinct populations. Such stands typically contain far more seedlings and saplings than large overstory trees on a per-area basis. Thus, an ideal large plot size for overstory trees would likely result in excessive time spent measuring small regeneration stems, while a small plot suited for regeneration would include too few overstory trees to reliably estimate timber volume with a feasible number of plots.

The solution is to co-locate plots sized appropriately for each population of interest, often by nesting smaller plots within larger ones. This might look like one or more 1/1000-th acre regeneration plots nested within a 1/20-th acre overstory plot. Smaller nested plots are often referred to as subplots.

When using plot sampling, each tree's selection probability is proportional to the area of the plot in which it's measured. More specifically, this probability equals the plot area divided by the geographic extent of the population. These ideas are further developed in Section 12.3.1 and illustrated in Section 12.3.6. They're key to expanding tree measurements to a per-unit-area basis (e.g., per acre or per hectare) and computing plot-level summaries that represent one observation within the sample.

12.3.1 Expansion factors and plot-level summaries

In most forestry applications, estimates are expressed on a standard per-unit-area basis (i.e., acre or hectare). This means that at some point in the analysis we need to scale tree measurements to the desired per-unit-area basis. For example, recall from Section 11.1.5 that we were interested in estimating the average biomass (Mg/ha) in the Harvard Forest. However, because our sampling units (i.e., plots) were 0.25 ha, we first multiplied the total tree biomass measured on each 0.25-ha unit by 4 (i.e., 1/0.25) to express it on a per-hectare basis. Here, the value 4 is referred to as an *expansion factor*. An *expansion factor* is a value used to scale measurements from one area basis to another. In the Harvard Forest analysis, the individual tree measurements were summarized to the 0.25-ha plot in advance, and the expansion factor was then applied to the plot-level summaries to arrive at the desired per-hectare expression.

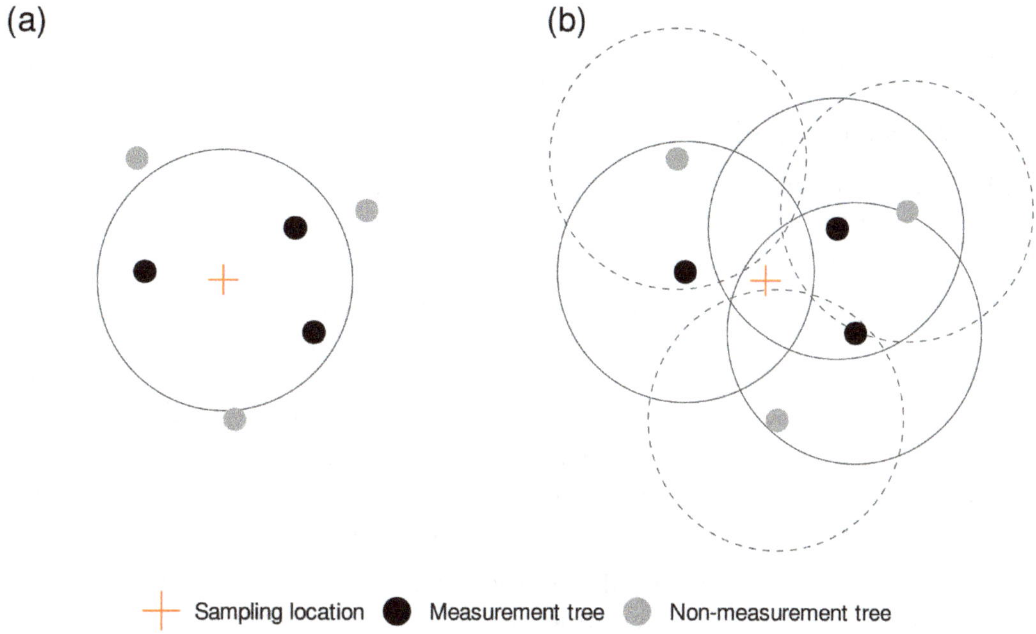

FIGURE 12.1: Two perspectives of measurement tree selection using a fixed-area circular plot. (a) Plot-centered, where the circle identifies the plot boundary. (b) Tree-centered, where circles identify each tree's inclusion zone with solid and dashed lines corresponding to measurement and non-measurement trees, respectively.

To better understand expansion factors, it's helpful to consider two perspectives for measurement tree selection. In the preceding section, our description of measurement tree selection takes a *plot-centered* perspective—that is, as illustrated in Figure 12.1(a), a plot boundary is established at a sampling location and trees inside the plot are measured. Alternatively, we can take a *tree-centered* perspective, in which an imaginary plot identical in shape and size to the one intended for the sampling location is established using each tree's location. The tree's imaginary plot is called its *inclusion zone* and is illustrated in Figure 12.1(b).[2] All trees with inclusion zones containing the sampling location are measured. Notice in Figure 12.1 that both perspectives result in the same set of measurement trees.

The j-th tree's measurements are scaled to a per-unit-area basis using the tree expansion factor, or tree factor (TF) for short, defined as

$$\text{TF}_j = \frac{\text{unit area}}{\text{inclusion zone area}_j}. \tag{12.1}$$

[2]The plot defining each tree's inclusion zone is created by replicating the plot intended for the sampling location, rotating it 180°, and positioning it at the tree's location; see Chapter 7 in Gregoire and Valentine (2008) for further explanation and examples. The inclusion zone rotation is not apparent in Figure 12.1(b) because the circular plot is symmetric about the sampling location.

In most applications, the unit-area numerator is 43,560 (ft^2/ac) or 10,000 (m^2/ha), and the denominator is the j-th tree's inclusion zone area in ft^2 or m^2. Regardless of the measurement system or chosen base units, TF is the number of trees per unit area that the given measurement tree represents.

As illustrated in Figure 12.1, under plot sampling a tree's inclusion zone equals the plot area used to sample the population to which the tree belongs. Hence, following (12.1), all trees measured on a given plot size will have the same TF.

We're careful to note "population" because it's often the case that different populations are measured using different plot sizes within the same survey effort. For example, following the description of overstory and regeneration plots in Section 12.3, a tree measured on the 1/20-th-acre (0.05 ac or 2,178 ft^2) fixed-area overstory plot will have a 0.05-acre inclusion zone and represent TF = 43,560/2,178 = 20 trees per acre. Similarly, a seedling or sapling measured on the 1/1,000-th-acre (0.001 ac or 43.56 ft^2)[3] subplot will have a 0.001-acre inclusion zone and represent TF = 43,560/43.56 = 1,000 trees per acre.

A per-unit-area expansion for the j-th tree's continuous or binary measurement (e.g., basal area, volume, biomass, logs, live/dead) is

$$y_j\,(\text{units/unit area}) = y_j\,(\text{units/tree}) \cdot \text{TF}_j\,(\text{trees/unit area})\,, \tag{12.2}$$

where y_j represents the tree-level measurement in the given units.

For example, following (12.2), suppose the j-th tree measured on a 0.05-acre plot had a DBH of 10 inches. Given $\text{TF}_j = 20$ trees per acre, its basal area per acre is

$$\begin{aligned}\text{BA}_j\left(\text{ft}^2/\text{acre}\right) &= \left(c \cdot \text{DBH}_j^2\right)\left(\text{ft}^2/\text{tree}\right) \cdot \text{TF}_j\,(\text{trees/acre})\\ &= 0.005454 \cdot 10^2 \cdot 20\\ &= 10.908,\end{aligned}$$

where $c = 0.005454 = \pi/(144 \cdot 4)$.[4, 5]

Next, suppose that same tree had a volume v_j of 10.4 ft^3; then, again following (12.2), its expanded volume is

$$\begin{aligned}v_j\left(\text{ft}^3/\text{acre}\right) &= v_j\left(\text{ft}^3/\text{tree}\right) \cdot \text{TF}_j\,(\text{trees/acre})\\ &= 10.4 \cdot 20\\ &= 208.\end{aligned}$$

The individual tree measurements, expanded to a per-unit-area basis, are then combined to form a *plot-level summary*. For the i-th sampling location, the general expression for a

[3] Often referred to as a milacre plot.

[4] Calculation of basal area from DBH and derivation of the constant c under English and metric systems are in Section 5.3.1.

[5] Here and throughout the remainder of the book, we intentionally retain more digits than would normally be reported. This makes intermediate calculations easier to follow and helps verify results using independent calculations or software. In practice, final estimates would be rounded to a sensible number of significant digits.

per-unit-area plot-level summary is

$$y_i = \sum_{j=1}^{m_i} y_{ij} \cdot \mathrm{TF}_{ij}, \tag{12.3}$$

where y_{ij} represents the j-th tree's measurement in the chosen units, TF_{ij} is its tree factor, and m_i is the number of measurement trees on plot i.[6]

Applying this general formula to specific variables, the plot-level summary for number of trees per unit area is

$$\mathrm{Trees}_i = \sum_{j=1}^{m_i} \mathrm{TF}_{ij}, \tag{12.4}$$

where the per-tree quantity y_{ij} in (12.3) is implicitly 1 for a counted tree. For plot sampling, this summation reduces to $m_i \cdot \mathrm{TF}$, where TF is the constant tree expansion factor for all trees measured on the plot.[7]

Similarly, the i-th sampling location's per-unit-area summaries for other variables are obtained by summing the per-unit-area contributions of its measured trees. For example, following (12.3), the plot-level summaries for basal area and volume per unit area are

$$\mathrm{BA}_i = \sum_{j=1}^{m_i} c \cdot \mathrm{DBH}_{ij}^2 \cdot \mathrm{TF}_{ij}, \tag{12.5}$$

and

$$v_i = \sum_{j=1}^{m_i} v_{ij} \cdot \mathrm{TF}_{ij}. \tag{12.6}$$

In this notation, Trees_i, BA_i, and v_i are simply instances of y_i from Section 11.3, with i indexing sample observations. These per-plot summaries serve as the observations for subsequent sample-based estimation.

12.3.2 Sample-based estimates

For n sampling locations, the corresponding n per-unit-area plot-level summaries of a given variable (see Section 12.3.1) serve as the sample observations used to estimate population parameters via the estimators described in Section 11.3. Following the notation introduced in Section 11.3, the n plot-level summaries are denoted y_i, for $i = 1, 2, \dots, n$, and are used to compute the sample mean, standard error of the mean, and confidence intervals using (11.18), (11.24), and (11.35), respectively.

---

[6] By standard order-of-operations rules (reviewed in Section 10.2), each tree measurement y_{ij} is first multiplied by its tree factor TF_{ij}, and the resulting products are then summed over trees on plot i. When in doubt, it's always worth pausing to check the implied order.

[7] You might ask, "If all trees are measured on the same plot area, why complicate life with this j subscript for tree-specific TF?" There are a several good reasons. First, as noted earlier, it's common to have more than one plot size in a given inventory, e.g., a plot for overstory trees and a nested subplot for regeneration, so the dataset includes trees with different TFs. Second, as developed later in Section 12.4, under point sampling a tree's inclusion zone depends on its size (e.g., its basal area), so TF differs by tree. Third, using a tree-specific TF supports an efficient workflow developed in Section 12.6 that works for both plot and point sampling.

These plot-level summaries—one per sampling location—form the essential link between tree-level measurements and population-level inference.

Continuing the example from Section 12.3.1, per-acre sample-mean estimates for tree density, basal area, and volume are

$$\overline{\text{Trees}} = \frac{\sum_{i=1}^{n} \text{Trees}_i}{n}, \tag{12.7}$$

$$\overline{\text{BA}} = \frac{\sum_{i=1}^{n} \text{BA}_i}{n}, \tag{12.8}$$

and

$$\bar{v} = \frac{\sum_{i=1}^{n} v_i}{n}. \tag{12.9}$$

These are simply variable-specific instances of $\bar{y}$, and their standard errors and confidence intervals follow directly from (11.24) and (11.35).

Estimates for population totals (e.g., total number of trees, basal area, or volume) are obtained by scaling each per-unit-area mean by the total forest area. This matches the approach in Section 11.3.3, where the population size N is replaced by the forest area A, expressed in the same units as the per-unit-area estimates. Computation of means and totals for several parameters are illustrated in Section 12.3.6.

The tree-centered perspective developed above continues to offer practical benefits. First, as shown in Section 12.3.1, it aids in introducing tree expansion factors. Second, from a practical computing standpoint, it supports a single analytical workflow (Section 12.6) that yields population parameter estimates for both plot and point sampling data. Third, it clarifies how forest sampling based on an areal sampling frame connects with broader survey sampling theory (see, e.g., Grosenbaugh (1958), Oderwald (1981), Roesch et al. (1993), and Gregoire and Valentine (2008)). Fourth, as explored in subsequent sections, it helps us recognize and address statistical issues that can arise during implementation.

12.3.3 Boundary effects

The validity of plot and point sampling requires that the probability of selecting a tree for measurement is proportional to the tree's inclusion zone. If any portion of a tree's inclusion zone falls outside the forest boundary, then that tree has a lower than expected probability of selection. This reduced probability can result in biased estimates, referred to as *boundary slopover* bias. This bias can, and should, be corrected in the field using methods found in most forest measurement books (see, e.g., Burkhart et al. (2018), Kershaw et al. (2016), and Gregoire and Valentine (2008) for an in-depth tour of methods proposed to correct boundary slopover bias).

Below, we review the *mirage* and *walkthrough* methods. Both are easy to implement and common in practice. They effectively correct for those trees with inclusion zones that extend beyond the boundary. We introduce the methods using circular plots, but they can also be applied for non-circular plots.

The *mirage method* was proposed by Schmid (1969).[8] The method can be applied when the boundary near the plot in question is straight (or made up of straight segments with well-defined corners) and you are able to work outside the boundary. The method is implemented using the following steps.

1. Establish the plot center. Measure all trees that fall inside the plot and are also inside the boundary. If any portion of the plot is over the boundary, then implement Step 2.
2. Determine the bearing with the shortest distance d from the plot center to the boundary. Establish a "mirage" (i.e., temporary) plot center along the bearing at distance d outside the boundary. From that mirage plot center, measure all trees that fall inside the mirage plot and also inside the boundary. Add these mirage plot measurement trees to the measurement trees recorded in Step 1.

Figure 12.2 shows three fixed-area circular plots with increasing amounts of boundary overlap. Following Step 2 above, the mirage sampling locations are established at a distance d outside the boundary. The shaded regions within each sampling location plot show how many times each measurement tree should be recorded. Notice that trees falling within the overlap between the sampling location plot and the mirage plot are recorded twice.

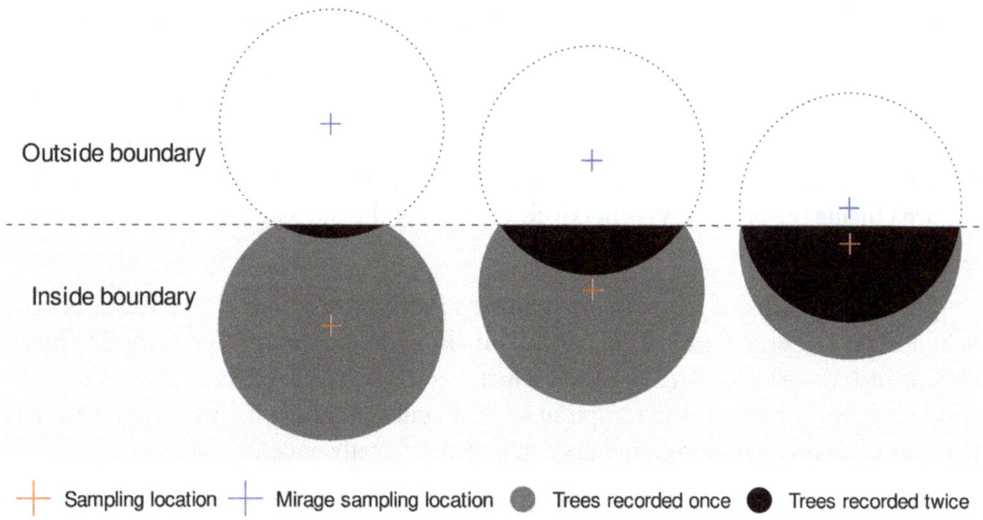

FIGURE 12.2: Three fixed-area circular plots with increasing boundary overlap. Boundary indicated by the dashed line. Mirage plots outlined with dotted line.

Consider Figure 12.3 to build some intuition for how the mirage method works. Like Figure 12.1, it shows measurement tree selection from both a plot- and tree-centered perspective. Figure 12.3(a) shows a plot with boundary overlap and two measurement trees. Following the mirage method, Tree 1 is recorded twice because it falls within the region where the

[8] Paul Schmid published his original mirage method development in German. Later, an English publication by Beers (1977) highlighted Schmid's method and proposed an extension for settings where the mirage method could not be implemented.

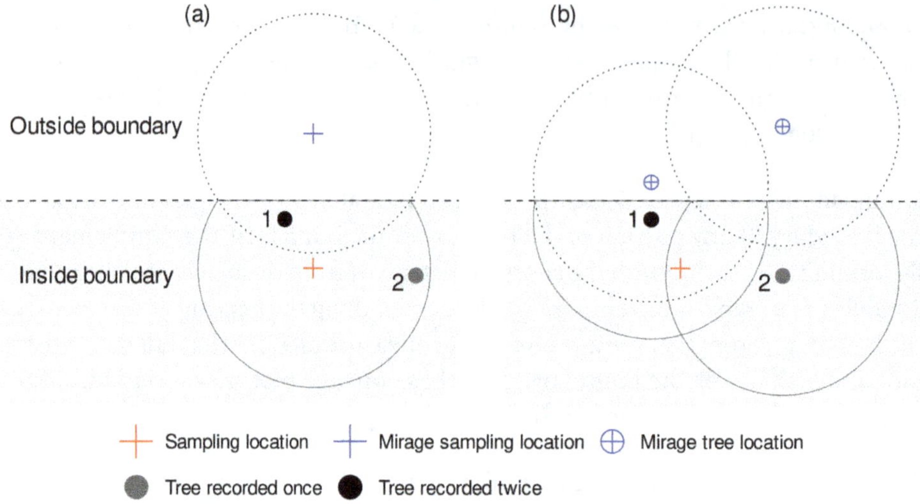

FIGURE 12.3: Plot-centered and tree-centered view of the mirage method using two example trees.

mirage plot overlaps the sampling location plot. Tree 2 is recorded only once because it doesn't fall within the overlap. Double counting Tree 1 effectively corrects for the fact that its inclusion zone is reduced by the portion extending beyond the boundary.

Given this notion, consider the corresponding tree-centered view in Figure 12.3(b), where a mirage tree inclusion zone has been established for each measurement tree. For a given tree, notice the inclusion zone area outside the boundary is equal to the mirage inclusion zone area inside the boundary—effectively correcting the inclusion zone area reduction, or making it whole.

It might help to think of the portion of a tree's inclusion zone area that extends beyond the boundary as being folded back onto itself along the boundary. The sampling location falls within Tree 1's inclusion zone and inclusion zone correction region; hence, that tree is recorded twice. For Tree 2, the sampling location falls within its inclusion zone but not its inclusion zone correction region; hence, it's recorded only once.

The *walkthrough method* was developed by Ducey et al. (2004). This method follows the same basic principles as the mirage method but is more general and works well even if the boundary near the plot in question is not straight, or if you are not able to work outside the boundary. The method is implemented using the following steps.

1. Establish the plot center. Measure all trees that fall inside the plot and are also inside the boundary.
2. Identify each tree in the plot that is closer to the boundary than to the plot center. For each of these trees, take the bearing from the plot center to the tree and also measure the distance d from the plot center to the tree. Then, following this bearing, walk from the tree for distance d. The point at distance d from the tree (i.e., where you stop walking) is called the "walkthrough point." If the walkthrough point is outside the boundary, then the tree is recorded twice; otherwise, the tree is recorded only once. It doesn't matter

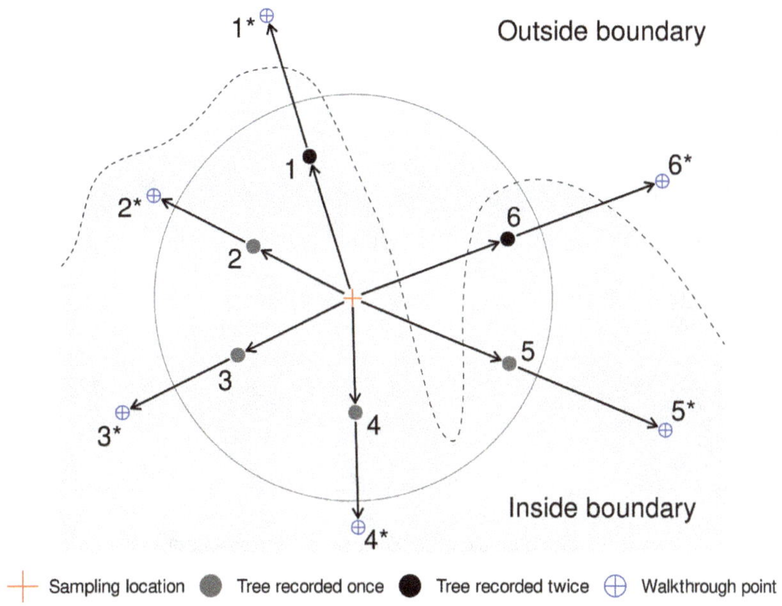

FIGURE 12.4: Fixed-area circular plot indicated by the solid line that overlaps a boundary indicated by the dashed line. Six trees fall in the plot. Arrows indicate direction and distance to walkthrough points denoted with asterisked tree number.

how many times you cross the boundary on the way to the walkthrough point—it's whether the walkthrough point is inside or outside the boundary that determines if the tree is recorded once or twice. It's only necessary to walk to the walkthrough point if it's not obvious whether the point is inside the boundary.

Consider the example in Figure 12.4. Here, a fixed-area circular plot center falls near a boundary. Six trees fall within the plot and boundary. Each tree's walkthrough point is also identified in the figure. In practice, only Trees 1, 2, 5, and 6 would likely be considered using the walkthrough method, since Trees 3 and 4 are clearly closer to the plot center than to the boundary. Trees 1 and 6 have walkthrough points outside the boundary, so they're recorded twice. The other trees have walkthrough points within the boundary and are recorded only once.

12.3.4 Slope corrections

Forest parameters are reported on a horizontal land area basis (e.g., tons per hectare or volume per acre). Horizontal areas are delineated on the horizontal plane. The land surface, with all its interesting terrain features, can be projected onto the horizontal plane; however, this projection results in some distortion that we must consider. The key issue here is that a fixed-area plot established on a slope (i.e., oblique plane) has a smaller area when projected onto the horizontal plane. Hence, tree factors computed for horizontal plot areas should not be applied to trees measured on sloped plots. Plots that fall on a slope require *slope correction* prior to use in estimation.

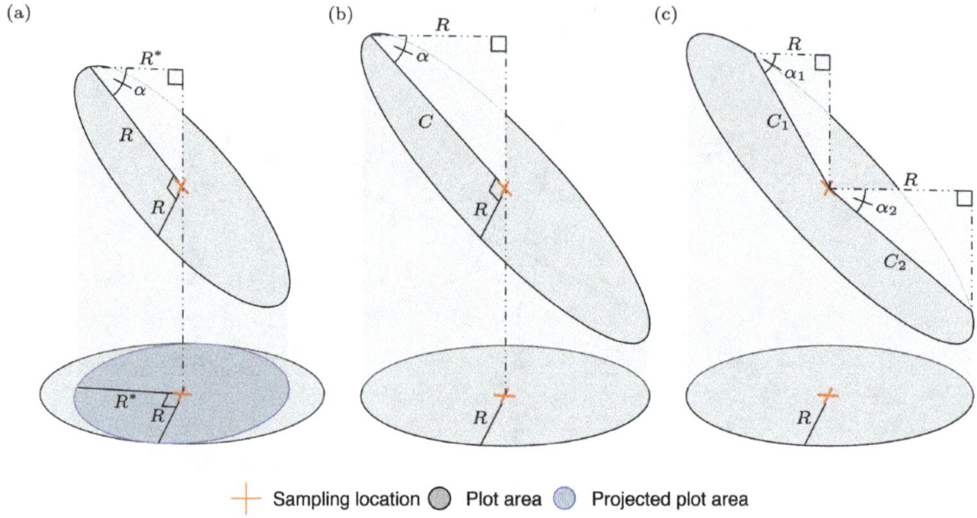

FIGURE 12.5: Illustration of how fixed-area plot dimensions change when projected between the horizontal (lower) and oblique (upper) planes. (a) Plot area when projecting a circular plot on an oblique plane onto the horizontal plane. (b,c) Examples used to find the oblique plane plot's radius and critical distance given R and slope angle α.

Let's build some intuition about the need for slope correction. Say we're using a fixed-area circular plot with radius R to conduct our cruise. If this circular plot is established on an oblique plane with slope α, the plot shape when projected onto the horizontal plane becomes an ellipse with major radius R, minor radius $R^* = \cos(\alpha)R$, and area $\pi R R^*$ (which is smaller than πR^2). This relationship is illustrated in Figure 12.5(a), where the upper sloped plot area is a circle with radius R that becomes an ellipse when projected down onto the horizontal plane. Importantly, the area of this projected ellipse is smaller than πR^2—hence, applying the tree factor computed assuming an area of πR^2 will cause downward bias in resulting estimates.

There are a few approaches for slope correction. The most general is to ensure that all measurements are made using horizontal distance, not the distance along the oblique plane. For example, if you had a circular plot that falls on an oblique plane with slope α, the plot shape becomes an ellipse with minor axis radius R perpendicular to the slope, major axis radius $C = R/\cos(\alpha)$ parallel with the slope, and area πRC. Once projected to the horizontal plane, this elliptical plot has the desired plot area of πR^2 (i.e., $C = R$ when $\alpha = 0$). This projection and associated dimensions are illustrated in Figure 12.5(b).

In practice, we don't actually lay out an elliptical slope plot—that would be painful. Rather, we check if each tree around the sloped plot center is within the corresponding horizontal plot radius. If a tree is within the horizontal plot radius, then it's a measurement tree.

Checking if a tree is "in" or "out" of the elliptical plot is straightforward once you recognize that it amounts to computing a right triangle's hypotenuse length with known angle α and adjacent side R, then comparing the hypotenuse length to the distance from the plot center to the tree in question. We'll call this hypotenuse C, the tree's *critical distance*. If the slope distance from the plot center to the tree is smaller than the tree's critical distance, then it's a measurement tree.

A tree's critical distance is determined by the slope between the plot center and the tree's location. The j-th tree has critical distance $C_j = R/\cos(\alpha_j)$, where α_j is the slope from the plot center to the center of the tree's stem (slope measured parallel to the ground at the tree's stem center). This slope can be found using a clinometer or laser rangefinder that is able to calculate slope.

For example, consider Figure 12.5(c), and say we'd like to compute a critical distance for a tree positioned along the direction indicated by line C_1. Using a clinometer, you find the slope α_1 is 30 degrees. The cruise protocol calls for a circular plot with $R = 26.33$ (ft) (the horizontal distance radius of a 1/20-th acre plot). The tree's critical distance is then $C_1 = R/\cos(\alpha_1) = 26.33/\cos(30) = 30.4$ (ft).[9] So, the tree must be within 30.4 (ft) of the plot center (measured along the slope) to be a measurement tree.

Now say the next tree is positioned along the C_2 line. Using a clinometer, you find the slope α_2 is 40 degrees. The tree's critical distance is $C_2 = R/\cos(\alpha_2) = 26.33/\cos(40) = 34.37$ (ft).

The slope correction approach described above is general and works even when the slope is not constant across the plot area. Many modern laser rangefinders report horizontal distance and hence automatically correct for slope. Using such a device, simply measure the horizontal distance from the plot center to each possible measurement tree. Those trees with horizontal distance (to the tree center) less than R are measurement trees. These results will match the time- and math-intensive approach of finding each tree's critical distance. This approach is also the most commonly used in national forest inventories and similar inventories where accuracy and repeatability are key.

There are two other approaches to slope correction often seen in practice.

1) *Radius adjustment* uses a larger circular plot on the slope that, when projected onto the horizontal plane, has area πR^2 (Bryan 1956).

2) *Tree factor adjustment* uses a circular plot with radius R on the slope but computes a plot-specific tree factor that accounts for the resulting smaller plot area when projected onto the horizontal plane (Beers 1969).

Both approaches require a measurement that captures the overall plot slope.

---

[9] R's trigonometric functions assume angles are in radians, not degrees. To reproduce this example in R using `cos()`, first convert the angle from degrees to radians with the `d2r()` function: `d2r <- function(d) { d * pi / 180 }`. So for the C_1 example: `26.33/cos(d2r(30))`. Other useful angle conversion functions are listed in Exercise 12.12.

To apply the radius adjustment approach, establish a circular plot on the slope with radius $R^* = R/\sqrt{cos(\alpha)}$. This circular plot defined by radius R^* on the slope will be an ellipse with area πR^2 when projected onto the horizontal plane. This works because the inclusion zone area of the projected ellipse equals that of the intended plot on the horizontal plane. This approach might be a good option for a timber cruise as it allows you to keep a constant radius, which greatly simplifies identifying measurement trees using a tape from the plot center. Foresters using this approach often rely on a table that provides a scaling factor for R given different slope ranges.

For example, if your cruise protocol calls for a circular plot with $R = 26.33$ (ft) (horizontal distance) and the plot falls on a $\alpha = 30$ (degree) slope, the adjusted radius is $R^* = R/\sqrt{\cos(\alpha)} = 28.3$ (ft) or using the equivalent scaling factor $R^* = r^* R$ where $r^* = 1/\sqrt{cos(\alpha)} = 1.07$. To create your own scaling factor table to take into the field, compute r^* for a range of slopes (see Exercise 12.12).

To apply the tree factor adjustment approach, establish a circular plot on the slope with radius R as if it were on the horizontal plane. Because the area of this slope plot is smaller than πR^2 when projected onto the horizontal plane, the tree factor for measurement trees needs to increase accordingly. Specifically, the tree factor for measurement trees on the slope plot is $\text{TF}^* = \text{TF}/\cos(\alpha)$, where $\text{TF} = \text{unit area}/(\pi R^2)$ (following (12.1)) or equivalently $\text{TF}^* = \text{unit area}/(\pi \cos(\alpha) R^2)$, where the denominator is the area of the horizontally projected ellipse, i.e., the blue area in Figure 12.5(a).

This approach is also easy to implement, but it requires that you record all plot slopes so you can compute corresponding tree factors after the cruise. As an example, say your horizontal plot radius is $R = 26.33$ (ft), which results in a $\text{TF} = 43560/(\pi 26.33^2) = 20$. The tree factor for trees measured on the oblique plot with a slope of 30 (degrees) and $R = 26.33$ (ft) is $\text{TF}^* = \text{TF}/\cos(\alpha) = 23.09401$.

12.3.5 Effect of plot size on variance

Plot size, i.e., the area covered by the sampling unit, affects variance and, in turn, inferences that involve variance such as confidence intervals and sample size calculations. For many variables measured in forestry applications, the variance of measurements collected on small plots is greater than the variance of measurements collected on larger plots. For example, the variance in biomass measurements (Mg/ha) taken on 0.25 ha plots will be larger than the variance in biomass measurements (Mg/ha) on 0.5 ha plots.

Large plots tend to have smaller variance because they average over small-scale horizontal forest structure caused by factors such as competition, harvesting, windthrow, disease, or fire. The effect of plot size on variance diminishes as stand structure becomes more uniform. For example, we expect a plantation to have a weaker plot size to variance relationship compared with structurally complex mixed-species uneven-aged stands.

Freese (1962) offers the following equation to approximate the relationship between plot size and variance.

$$s_2^2 = s_1^2\sqrt{\frac{a_1}{a_2}}, \tag{12.10}$$

where plot size a_1 has variance s_1^2 and plot size a_2 has variance s_2^2. One can replace the variances in (12.10) with their corresponding coefficients of variation (i.e., $CV_1 = s_1^2/\bar{y}_1$ and $CV_2 = s_2^2/\bar{y}_2$).

For example, following (12.10), if the variance of pulpwood volume per acre on 1/4-acre plots is $s_1^2 = 50$, the variance in pulpwood volume per acre on a 1/10-acre plot is approximately

$$
\begin{aligned}
s_2^2 &= 50\sqrt{\frac{0.25}{0.1}} \\
&= 79.06.
\end{aligned}
$$

Although plot size is often chosen based on experience or an existing field sampling protocol, the relationship between plot size and expected variance can provide an approach to select plot size and sample size based on allowable error and cost constraints. Cost comes into play here because "time is money"—to reach the same standard error, there are trade-offs between sampling more small plots versus fewer large plots.

For example, if travel time between many small plots is large relative to the time needed to take measurements on fewer large plots, one would favor a sampling design where fewer large plots are selected.

12.3.6 Illustration

Figure 12.6 shows a toy dataset that we'll use to illustrate estimation via plot sampling. The figure shows a forest area (1.32 ac) that delineates tree populations of interest. The forest is divided into two stands (Stand 1 is 0.64 ac and Stand 2 is 0.68 ac), and within each a distinction is made between overstory and regeneration trees. This partitioning allows us to define at most four populations: Stand 1 overstory trees, Stand 1 regeneration trees, Stand 2 overstory trees, and Stand 2 regeneration trees.

These are simulated data, meaning we created them, and as a result we have measurements on all trees, i.e., a census. These measurements allow us to compute the population parameter values of interest that are given in Table 12.1. For instructional purposes, we can compare these parameter values to their corresponding estimates—something we can't do when sampling a real population.[10]

The sample comprises trees measured on six overstory plots with accompanying regeneration subplots. Overstory and regeneration plots are circular fixed-areas with 24 ft and 6.8 ft radii, respectively. Plot sampling locations were selected at random (i.e., SRS) from each stand's areal sampling frame and serve as the overstory plot centers. Regeneration subplots were located 12 ft east of overstory plot centers.[11]

[10] Like the Harvard Forest census dataset, it's useful to learn about different estimators and associated computations using a census because it allows you to check your resulting estimates against the truth. Clearly, this is never a luxury we have when sampling a real population.

[11] Because much of the field activity occurs at the overstory plot center (e.g., holding the tape while checking in/out trees and setting permanent plot markers), regeneration subplots are typically offset from this high-traffic area to avoid trampling seedlings before they can be measured.

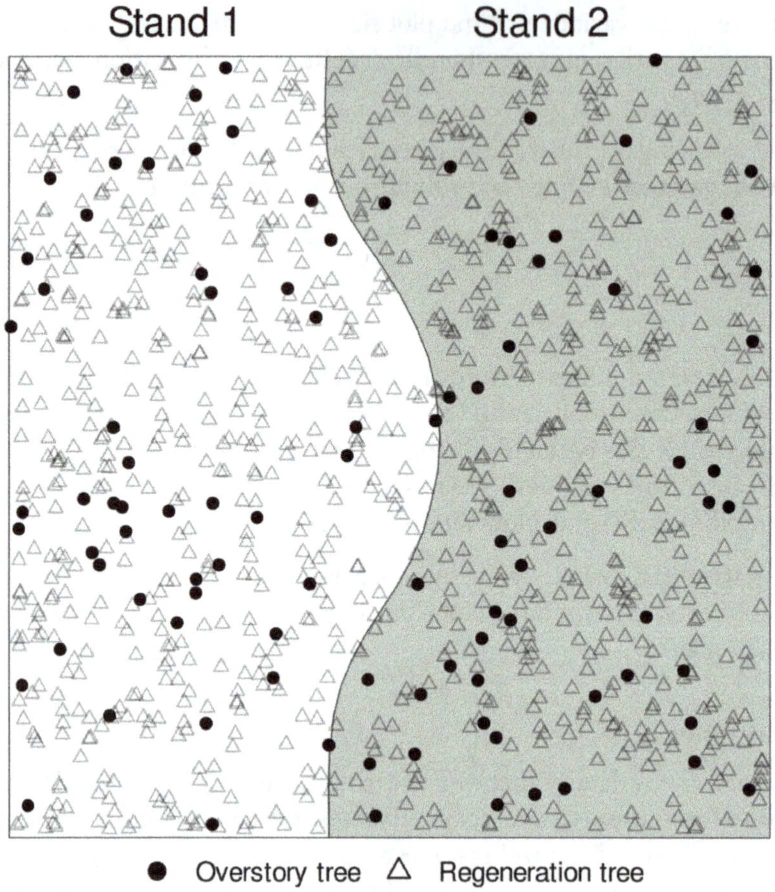

FIGURE 12.6: Toy dataset used to illustrate sample-based estimation.

TABLE 12.1: Parameters for the population shown in Figure 12.6.

	Overstory			Regeneration
Stand	Trees/ac	Basal area (ft^2/ac)	Volume (ft^3/ac)	Trees/ac
1	78.210	64.506	1809.93	782.10
2	73.206	27.907	643.04	732.06

On overstory plots, species, DBH (in), and height (ft) for all trees with DBH ≥ 2 in were measured. Additionally, volume (ft^3) was estimated for measurement trees using allometric equations provided in Honer (1967). On regeneration subplots, the number of trees by species with height greater than 2 ft and DBH < 2 in were recorded.

This illustration steps through the calculations used to estimate number of trees, basal area, and volume per acre, as well as their stand totals. To keep the focus on these steps, we'll initially limit our analysis to the overstory plot data in Stand 1. Later in Section 12.6, we

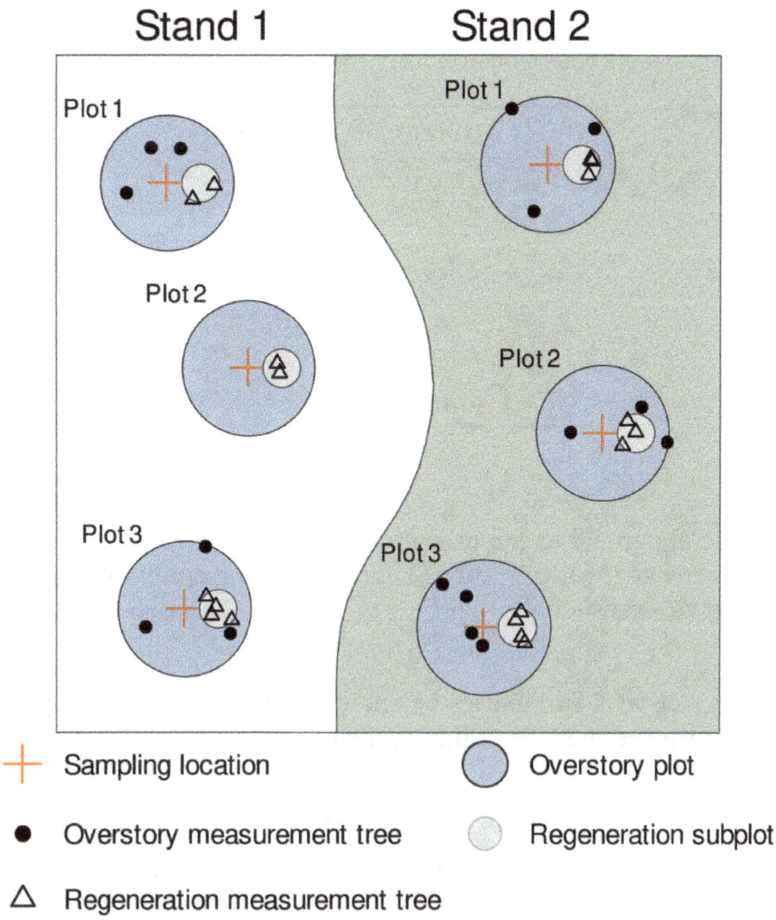

FIGURE 12.7: A sample from the populations shown in Figure 12.6. The sample comprises three fixed-area circular overstory plots randomly located within each stand. Regeneration subplots are nested within overstory plots.

develop an efficient workflow for both stands and apply it to estimate parameters for both the overstory and regeneration layers.

Table 12.2 provides overstory tree measurements for the three Stand 1 plots. Notice in Figure 12.7, there are no overstory trees on Plot 2 in Stand 1. It's critical that the absence of trees at a sampling location be included in subsequent estimates—meaning a plot with no trees is part of the sample, reflects a characteristic of the population, and needs to be included as a zero when computing population parameter estimates.

We include a line for Plot 2 in Table 12.2 with zero DBH and volume values to remind us to include these values in subsequent computations.

We divide the estimation process into two steps, which were covered in Sections 12.3.1 and 12.3.2, respectively.[12] First, compute plot-level summaries for each variable following

[12]Get used to these steps—they're a recurring theme in forest inventory data analysis.

TABLE 12.2: Stand 1 overstory tree measurements for inventory data shown in Figure 12.7. Zero values indicate plots where no overstory trees were measured.

Plot	DBH (in)	Volume (ft^3)
1	11.3	17.8
1	9.8	14.5
1	10.7	17.9
2	0.0	0.0
3	14.8	33.6
3	15.4	36.6
3	13.1	28.9

(12.3). The plot-level summaries comprise each plot's expanded and summarized tree measurements expressed on a per-unit-area basis (these are the sample observations). Second, use the plot-level summaries to compute the desired population parameter estimates (see Section 12.3.2).

Computing the plot-level summaries begins with calculating the TF used by all overstory trees measured on the 24 ft radius circular fixed-area plot. Following (12.1), the TF is

$$\begin{aligned}\text{TF (trees/ac)} &= \frac{\text{unit area}\left(\text{ft}^2/\text{ac}\right)}{\text{inclusion zone area}\left(\text{ft}^2/\text{tree}\right)}\\ &= \frac{43560}{\pi \cdot R^2} = \frac{43560}{\pi \cdot 24^2}\\ &= 24.07219,\end{aligned}$$

where R is the plot radius. This tree factor tells us that each tree measured on an overstory plot represents 24.07 trees per acre.

Next, we compute trees per acre for each of the $n = 3$ plots. Referring to Table 12.2 to get the number of trees on each plot and following (12.4), the plot-level trees per acre are

$$\begin{aligned}\text{Plot 1: } &\sum_{j=1}^{3} \text{TF}_j = 3 \cdot \text{TF} = 72.21656 \text{ (trees/ac)},\\ \text{Plot 2: } &\text{TF} \cdot 0 = 0 \text{ (trees/ac)},\\ \text{Plot 3: } &\sum_{j=1}^{3} \text{TF}_j = 3 \cdot \text{TF} = 72.21656 \text{ (trees/ac)},\end{aligned}$$

where j is the tree index. Notice in the calculations above, because the TF is a constant we pull it out of the summation and drop its subscript.

Using DBH measurements in Table 12.2 and following (12.5), the plot-level basal area per acre is

$$\begin{aligned}
\text{Plot 1: } \sum_{j=1}^{3} c \cdot \text{DBH}_j^2 \cdot \text{TF}_j &= c \cdot (11.3^2 + 9.8^2 + 10.7^2) \cdot \text{TF} \\
&= 44.4048 \left(\text{ft}^2/\text{ac}\right), \\
\text{Plot 2: TF} \cdot 0 = 0 \left(\text{ft}^2/\text{ac}\right), \\
\text{Plot 3: } \sum_{j=1}^{3} c \cdot \text{DBH}_j^2 \cdot \text{TF}_j &= c \cdot (14.8^2 + 15.4^2 + 13.1^2) \cdot \text{TF} \\
&= 82.42499 \left(\text{ft}^2/\text{ac}\right),
\end{aligned}$$

where $c = 0.005454$.[13] Using volume measurements in Table 12.2 and following (12.6), the plot-level volume per acre is

$$\begin{aligned}
\text{Plot 1: } \sum_{j=1}^{3} v_j \cdot \text{TF}_j &= (17.8 + 14.5 + 17.9) \cdot \text{TF} \\
&= 1208.42369 \left(\text{ft}^3/\text{ac}\right), \\
\text{Plot 2: TF} \cdot 0 = 0 \left(\text{ft}^3/\text{ac}\right), \\
\text{Plot 3: } \sum_{j=1}^{3} v_j \cdot \text{TF}_j &= (33.6 + 36.6 + 28.9) \cdot \text{TF} \\
&= 2385.55355 \left(\text{ft}^3/\text{ac}\right),
\end{aligned}$$

where v_j is the j-th tree's volume.

For reference, we collected the plot-level summaries of number of trees, basal area, and volume per acre computed above into Table 12.3. Given the n=3 sample observations in Table 12.3, you're back in familiar territory for computing SRS means, standard deviations, standard errors, and confidence intervals—all covered in Sections 11.3 and 12.3.2.

TABLE 12.3: Stand 1 overstory plot-level summary for inventory data shown in Figure 12.7.

Plot	Trees/ac	Basal area (ft^2/ac)	Volume (ft^3/ac)
1	72.2166	44.4048	1208.4237
2	0	0	0
3	72.2166	82.425	2385.5535

The code below computes each variable's mean (11.18), standard error of the mean (11.24), and 90% confidence interval. As discussed in Section 12.2.1, when sampling locations are selected using an areal sampling frame, we don't apply the FPC when computing the standard error of the mean.

[13] Calculation of basal area from DBH and derivation of the constant c under English and metric systems are in Section 5.3.1.

As you've already seen in Chapter 11, we try to align variable names in code with the statistical notation. Moving forward, we append `_bar` to indicate a sample mean (e.g., `y_bar` corresponds to $\bar{y}$) and prefix `s_` to denote a standard deviation or standard error (e.g., `s_y_bar` corresponds to $s_{\bar{y}}$). We won't always be perfectly consistent, but we include this naming where it helps connect the code to the notation. For example, in the code below we append an `_i` index to indicate plot-level summaries (e.g., `trees_i`, `ba_i`, and `vol_i` correspond to (12.4), (12.5), and (12.6), respectively).

```
n <- 3 # Sample size.

t <- qt(p = 1 - 0.1 / 2, df = n - 1) # t-value for 90% CI.

# Trees per acre.
trees_i <- c(72.2166, 0, 72.2166)
trees_bar <- mean(trees_i)
s_trees_bar <- sd(trees_i) / sqrt(n)
ci_trees <- c(trees_bar - t * s_trees_bar,
              trees_bar + t * s_trees_bar)

trees_bar # Mean number of trees per acre.

#> [1] 48.144

ci_trees # Confidence interval for mean number of trees per acre.

#> [1] -22.146 118.435

# Basal area per acre.
ba_i <- c(44.4048, 0, 82.425)
ba_bar <- mean(ba_i)
s_bar_ba <- sd(ba_i) / sqrt(n)
ci_ba <- c(ba_bar - t * s_bar_ba,
           ba_bar + t * s_bar_ba)

ba_bar # Mean basal area per acre.

#> [1] 42.277

ci_ba # Confidence interval for basal area per acre.

#> [1] -27.271 111.824
```

```
# Volume per acre.
vol_i <- c(1208.4237, 0, 2385.5535)
vol_bar <- mean(vol_i)
s_bar_vol <- sd(vol_i) / sqrt(n)
ci_vol <- c(vol_bar - t * s_bar_vol,
            vol_bar + t * s_bar_vol)

vol_bar # Mean volume per acre.
```

```
#> [1] 1198
```

```
ci_vol # Confidence interval for volume per acre.
```

```
#> [1] -812.91 3208.90
```

The means and associated confidence intervals computed in the code above are collected in Table 12.4.

TABLE 12.4: Stand 1 overstory mean estimates along with associated confidence intervals (CI) for inventory data shown in Figure 12.7 and recorded in Table 12.2.

	$\bar{y}$	90% CI
Trees/ac	48.144	(-22.146, 118.435)
Basal area (ft^2/ac)	42.277	(-27.271, 111.824)
Volume (ft^3/ac)	1197.992	(-812.913, 3208.898)

Stand 1 total and associated confidence interval for each variable can be computed using (11.25) and (11.28). These estimators for the total are modified by replacing population size N with stand area in acres $A = 0.64$ (as demonstrated in the Harvard Forest biomass analysis toward the end of Section 11.3.5).

Total estimates are given in Table 12.5. Importantly, notice that obtaining totals and associated confidence intervals is as easy as scaling mean per-acre estimates given in Table 12.4 by A—a simple and direct computation.

TABLE 12.5: Stand 1 overstory total estimates along with associated confidence intervals (CI) for inventory data shown in Figure 12.7 and recorded in Table 12.2.

	$\hat{t}$	90% CI
Trees	30.812	(-14.173, 75.798)
Basal area (ft^2)	27.057	(-17.454, 71.568)
Volume (ft^3)	766.715	(-520.264, 2053.695)

Notice the lower confidence interval bounds for the mean and total parameter estimates are negative. This should seem odd—after all, quantities like tree density, basal area, and volume are strictly positive. How can we have negative trees? Is such a thing possible? No. Is such a thing statistically possible? Yes.

This apparent contradiction arises from the limitations of normal approximation methods in small samples or high-variance settings. For further discussion on how to interpret such results—and why they should prompt careful reflection and defensive programming practices—see Section 12.5.

12.4 Point sampling

Like plot sampling, introduced in Section 12.3, point sampling is a rule for selecting measurement trees around a sampling location. The Austrian forester Walter Bitterlich introduced point sampling in the forestry literature under the German name "Winkelzählprobe," which translates into English as "angle count sampling" (Bitterlich 1952). Bitterlich named his method angle count sampling because basal area per unit area can be estimated by simply counting trees selected using an angle from a sampling location (Bitterlich 1952, 1984).

Bitterlich's ingenious method is most commonly used in timber cruises because of its ease of application, time efficiency, and flexibility to meet a range of inferential objectives. The time efficiency and flexibility come from the fact that you can match information collection effort with the desired level of inference. For example, if you're only interested in estimating mean basal area per unit area, then a simple count (i.e., tally) of measurement trees across sampling locations is all that's needed. If you'd like to estimate a confidence interval for mean basal area per unit area, simply keep track of the number of trees tallied at each sampling location. Estimating other parameters such as volume, biomass, or trees per unit area typically requires additional measurements such as DBH.[14]

Shortly after Bitterlich's 1952 publication, the American forester Lewis R. Grosenbaugh helped popularize the method, which he coined *point sampling*, among American foresters (Grosenbaugh 1952; Grosenbaugh and Stover 1957).[15] Grosenbaugh (1958) generalized point sampling to estimate parameters beyond basal area, e.g., volume, biomass, and density, and connected the method to existing PPS sample survey theory. A few years later, Palley and Horwitz (1961) further detailed point sampling's theoretical underpinnings by providing statistical proofs that mean and variance estimators are unbiased for several common sampling designs.

Under point sampling, a tree's inclusion zone area—and hence its selection probability—is a function of a chosen characteristic. Here, the characteristic that determines a tree's

[14] Later in the book, we introduce variants such as Big BAF sampling (Section 16.5), where DBH is measured on only a subset of the counted trees to support estimation of additional parameters, while retaining the basic point-sampling framework.

[15] Bitterlich's angle count method is also known as Bitterlich sampling, prism sampling, plot-less sampling, variable radius plot sampling, variable plot sampling, and plotless cruising.

selection probability is typically related to its size, e.g., basal area or height. This is why point sampling is a PPS method.

Timber cruises are typically conducted to estimate stand value and, for most stands, timber value is concentrated in large diameter trees. From a time efficiency standpoint, it's logical that we spend our valuable field time measuring those trees that contribute most to parameters of interest, e.g., timber volume with associated standard error. This is the primary motivation and result of point sampling: when selection probability scales with tree size, large trees are measured more frequently than small trees.

This section focuses on *horizontal point sampling*, which uses a tree's basal area to determine its selection probability. It's called horizontal point sampling because the forester projects the discerning angle horizontally from the sampling location toward each tree. Related sampling methods such as vertical point sampling, horizontal line sampling, and vertical line sampling are summarized in Kershaw et al. (2016). We focus on horizontal point sampling because it's the most common in forest inventory and because its general concepts are transferable to related sampling methods.

Figure 12.8(a) illustrates a horizontal point sampling measurement tree selection rule. From a sampling location, the forester projects a known and fixed angle θ toward each tree's

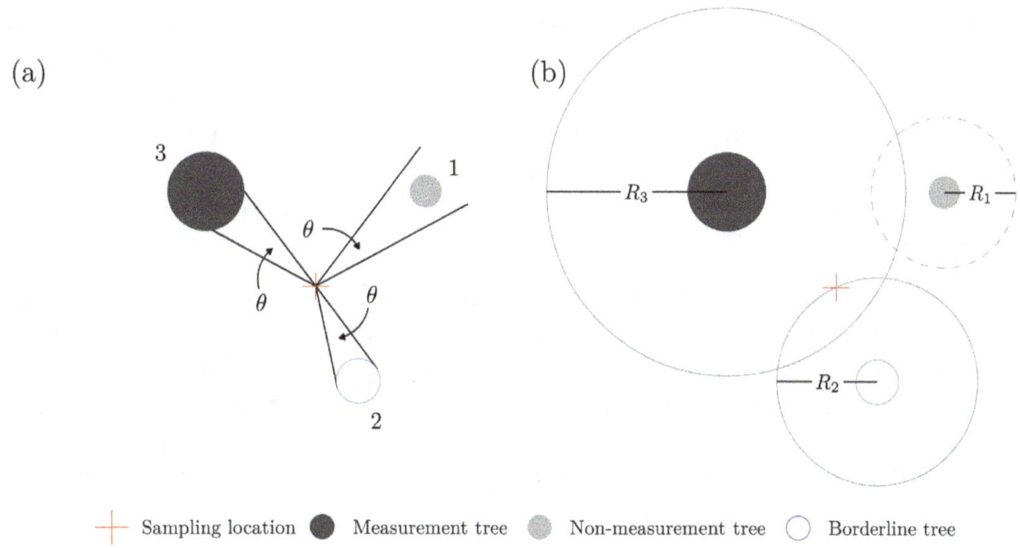

FIGURE 12.8: Point sampling example with three trees around a sampling location. Trees are identified using filled circles with diameter equal to the tree's DBH. The orange cross indicates the sampling location. (a) Fixed and known angle θ is projected from the sampling location and used to identify measurement (3), borderline (2), and non-measurement (1) trees. (b) Unfilled circles identify each tree's inclusion zone with R_1, R_2 and R_3 being the inclusion zone radius for Trees 1, 2, and 3, respectively. Inclusion zones delineated with solid lines identify measurement and borderline trees. Inclusion zones delineated with dashed lines identify non-measurement trees.

bole at breast height. Given a tree's DBH and distance from the sampling location (i.e., the angle's vertex), the angle may appear:

1) wider than the tree bole, in which case the tree is not measured;
2) the same width as the tree bole (i.e., the angle's rays are tangent to the bole), in which case the tree is considered a *borderline* tree and a measurement determination will take some more effort, as we'll describe below;
3) narrower than the tree bole, in which case the tree is measured.

If we apply this rule to the three trees in Figure 12.8(a), Tree 1 is not measured, Tree 2 is a borderline tree, and Tree 3 is measured. Here again, depending on the inferential objectives, "measured" might imply that you simply tally (i.e., add the tree to a tree count) or take detailed measurements of interest (e.g., DBH, height).

The angle gauge, relascope, and wedge prism are the most common tools used in practice to project the angle θ from the sampling location to discern a tree's measurement status (i.e., is the tree "in" or "out" of the tally). The angle gauge consists of a small piece of metal of known width, held a known distance from the forester's eye, which is positioned over the sampling location; see, e.g., Fehrmann (2024). The relascope, developed by Walter Bitterlich specifically for point sampling, is a sophisticated and multipurpose tool. The forester sights through the relascope's eyepiece with their eye positioned over the sampling location. The relascope automatically corrects for slope (see Section 12.4.4) and allows the user to choose among various angles to maximize cruise efficiency (see Section 12.4.3). The wedge prism, introduced by Bruce (1955), is a ground and calibrated prism that refracts light at the desired angle. In contrast to the angle gauge and relascope, where the forester's eye is the vertex of the angle at the sampling location, the wedge prism itself must be positioned directly over the sampling location. See Burkhart et al. (2018) or similar practical guides to learn how these tools are used.

Let's begin building some intuition about point sampling by connecting measurement tree selection using an angle with a tree's inclusion zone as developed in Section 12.3. Recall that under plot sampling, a tree is selected for measurement if its inclusion zone contains a sampling location (see, e.g., Figure 12.1(b)). Also, under plot sampling, all trees have the same inclusion zone area that is equal to the plot area used in the survey.[16] Because all trees then have the same inclusion zone, they have equal probability of selection for measurement and, following (12.1), the same tree factor.

Under horizontal point sampling, a tree's inclusion zone area—and thus its probability of selection—is a function of DBH. As shown in Figure 12.8(b), the circular inclusion zone defined by the tree's inclusion zone radius (R) has area proportional to DBH^2, and therefore proportional to basal area. In other words, larger trees have larger inclusion zones, making them more likely to be selected.

[16]This statement holds when a tree's inclusion zone lies fully within the boundary. When any portion of the zone falls outside the boundary, the tree has a reduced probability of selection, leading to boundary slopover bias and the need for corrective methods (Section 12.3.3).

A tree's inclusion zone radius (R) is called its *limiting distance*. This is because if the distance between the sampling location and the tree center is less than or equal to the tree's limiting distance, then the tree is measured. If the distance between the sampling location and the tree center is greater than the tree's limiting distance, then the tree is not measured.

This relationship can be seen in Figure 12.8(b). Here, in particular, notice the borderline tree's limiting distance, R_2, equals the distance between the sampling location and the tree's center. Figure 12.9 provides a closer look at a borderline tree. At the borderline, the tree is the maximum distance from the sampling location and still measured. Notice in Figure 12.9, if the tree's DBH is smaller or the tree is farther from the sampling location, the angle would be wider than the tree—hence, it would not qualify for measurement.

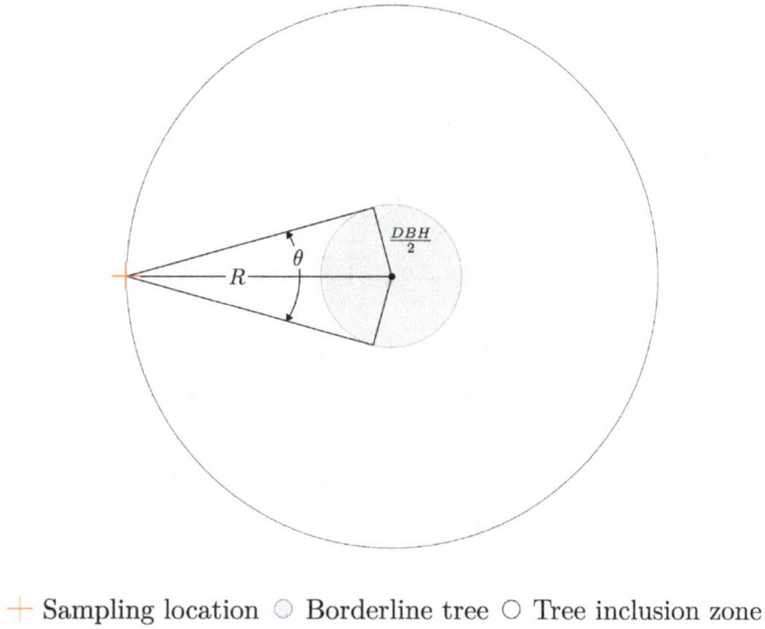

FIGURE 12.9: Horizontal point sampling example of a borderline tree.

When projecting the angle from the sampling location, if there's uncertainty about a tree's status (i.e., is it a measurement tree or not), then additional measurements are needed to make a correct determination. This is done by measuring the tree's DBH, computing its limiting distance, and checking that distance against the distance from the sampling location to the tree's center.

Even without dense undergrowth or other factors that impair line of sight, it's often difficult—even for experienced foresters—to identify borderline trees using an angle gauge, prism, or relascope. See, for example, work by Iles and Fall (1988), who assessed professional cruisers' skill in discerning borderline trees using a wedge prism.[17]

[17]Interestingly, they found that even though errors were made by professional cruisers in calling borderline trees as "in" or "out," the errors tended to cancel over the entire cruise. Those errors that didn't cancel out (i.e.,

Incorrect status determination can introduce bias, so it's worth the extra time to conduct the limiting distance calculation and measurement needed for a correct determination.

For a fixed angle θ, the relationship between a tree's DBH and its inclusion zone radius R is determined by geometry. Specifically, once DBH and R are expressed in the same length units, their ratio depends only on the chosen angle.

When DBH is measured in inches and R is measured in feet, converting DBH to feet gives

$$k = \frac{\text{DBH (in)}}{12\, R\,\text{(ft)}}. \tag{12.11}$$

Similarly, when DBH is measured in centimeters and R is measured in meters, converting DBH to meters gives

$$k = \frac{\text{DBH (cm)}}{100\, R\,\text{(m)}}. \tag{12.12}$$

In (12.11) and (12.12), the conversion factors (12 or 100) simply rescale DBH into the same length units as R, so the ratio compares two quantities expressed in the same units. Regardless of the measurement system used, this ratio depends only on the chosen angle and can be written as

$$k = 2\sin\left(\frac{\theta}{2}\right). \tag{12.13}$$

Equivalently, the inclusion zone radius can be viewed as a fixed multiple of DBH, with the multiplier determined entirely by the angle used in point sampling.

If k is known and a tree's DBH is measured, the corresponding limiting distance can be computed directly. When DBH is measured in inches, the inclusion zone radius in feet is

$$R\,\text{(ft)} = \frac{\text{DBH (in)}}{12\, k}. \tag{12.14}$$

Likewise, when DBH is measured in centimeters, the inclusion zone radius in meters is

$$R\,\text{(m)} = \frac{\text{DBH (cm)}}{100\, k}. \tag{12.15}$$

The expressions above clarify the geometry of point sampling, but they're not especially convenient for routine field use. In the next section, we build on these results to develop field-useful expansion factors and point-level summaries for per-unit-area estimation. We then return to the field mechanics in Section 12.4.3 and provide practical formulas for computing the limiting distance R.

12.4.1 Expansion factors and point-level summaries

Recall from Section 12.3.1, the tree factor is the number of trees per unit area a given measurement tree represents. For the j-th measurement tree, (12.1) defines the tree factor

bias) were typically due to inaccurate slope corrections and line of sight issues. Inexperienced cruisers tended to make more non-canceling errors. The authors strongly recommend measuring all potential borderline trees.

as

$$\text{TF}_j = \frac{\text{unit area}}{\text{inclusion zone area}_j}, \tag{12.16}$$

where, depending on your measurement system, the unit area numerator is 43560 (ft^2/ac) or 10000 (m^2/ha), and the denominator is the j-th tree's inclusion zone area in ft^2 or m^2.[18]

For horizontal point sampling, the inclusion zone is circular with area equal to πR_j^2. Because R_j is proportional to DBH_j, the tree factor is inversely proportional to DBH_j^2.

Using radius R_j defined in (12.14), the TF for the English system when DBH is measured in inches is

$$\text{TF}_j = \frac{43560}{\pi R_j^2} = \frac{43560}{\pi\left(\text{DBH}_j/(12k)\right)^2} = \frac{43560k^2}{(\pi/144)\,\text{DBH}_j^2},$$

then scaling the numerator and denominator by 1/4 gives us the more revealing and convenient

$$\text{TF}_j = \frac{(1/4)43560k^2}{(1/4)\,(\pi/144)\,\text{DBH}_j^2} = \frac{10890k^2}{0.005454\cdot\text{DBH}_j^2} = \frac{10890k^2}{\text{BA}_j}. \tag{12.17}$$

where BA_j is the j-th tree's basal area.

Similarly, using the definition of R_j from (12.15), the TF for the metric system when DBH is measured in centimeters is

$$\text{TF}_j = \frac{10000}{\pi R_j^2} = \frac{10000}{\pi\left(\text{DBH}_j/(100k)\right)^2} = \frac{10000k^2}{(\pi/10000)\,\text{DBH}_j^2},$$

then scaling the numerator and denominator by 1/4 gives us

$$\text{TF}_j = \frac{(1/4)10000k^2}{(1/4)\,(\pi/10000)\,\text{DBH}_j^2} = \frac{2500k^2}{0.00007854\cdot\text{DBH}_j^2} = \frac{2500k^2}{\text{BA}_j}. \tag{12.18}$$

As developed in Section 12.3.1, (12.2) defines the per-unit-area expansion factor for the j-th tree's continuous or binary variable measurement (e.g., basal area, volume, biomass, logs, live/dead) as

$$y_j\,(\text{units/unit area}) = y_j\,(\text{units/tree})\cdot\text{TF}_j\,(\text{trees/unit area}), \tag{12.19}$$

where y_j represents the tree-level variable measurement in the given units.[19]

It's instructive to first consider a tree's basal area expansion, which is called the *basal area factor* (BAF). Following (12.19) and (12.17), the English system BAF (with units included in parentheses) is

$$\begin{aligned}\text{BAF}_j\left(\text{ft}^2/\text{acre}\right) &= \text{BA}_j\left(\text{ft}^2/\text{tree}\right)\cdot\text{TF}_j\,(\text{trees/acre})\\ &= \cancel{\text{BA}_j}\cdot\left(\frac{10890k^2}{\cancel{\text{BA}_j}}\right)\\ &= 10890k^2.\end{aligned} \tag{12.20}$$

---

[18]While any area unit can be used, we focus on acres and hectares because they're most commonly used in forestry applications.

[19]Equation (12.19) is identical to (12.2); we repeat it here for convenience.

and following (12.19) and (12.18), the metric system BAF is

$$\begin{aligned}\text{BAF}_j\ (\text{m}^2/\text{ha}) &= \text{BA}_j\ (\text{m}^2/\text{tree}) \cdot \text{TF}_j\ (\text{trees/ha})\\ &= \cancel{\text{BA}_j} \cdot \left(\frac{2500k^2}{\cancel{\text{BA}_j}}\right)\\ &= 2500k^2.\end{aligned} \tag{12.21}$$

Notice in (12.20) and (12.21) that BAF depends **only** on the constant k. As discussed earlier, k is determined by the angle θ used to conduct the cruise.[20] Therefore, given θ, the basal area per unit area represented by each measurement tree is constant. This is a remarkable result! It means that each counted tree represents the same basal area per unit area, regardless of its diameter or its distance from the sample point.

By extension, this means you don't need to measure a tree's DBH to determine its contribution to estimating basal area per unit area—all you need to know is whether the tree is a measurement tree or not (i.e., is the tree "in" or "out")!

For example, say your chosen angle θ results in a BAF = 10 (ft^2/acre). From the result above, we know that each measurement tree (regardless of its DBH) represents 10 ft^2 of basal area per acre. If your chosen angle θ results in a BAF = 4 (m^2/ha), then each measurement tree represents 4 m^2 of basal area per hectare (again, regardless of its DBH).

To expand measurements for variables other than basal area, we follow (12.19), which relies on the tree factor and therefore requires a DBH measurement. Using the results in (12.20) and (12.21) and substituting into (12.17) and (12.18), respectively, the tree factor can be written compactly in terms of the BAF as

$$\text{TF}_j = \frac{\text{BAF}}{\text{BA}_j}. \tag{12.22}$$

For example, say you're using a BAF = 10 (ft^2/acre) and the j-th measurement tree has a DBH of 16 (in) and volume of 45.8 (ft^3).[21] Calculating the j-th tree's volume per acre is a two-step process.

First, compute the tree's tree factor using (12.22) as follows.

$$\begin{aligned}\text{TF}_j\ (\text{trees/acre}) &= \frac{\text{BAF}\left(\text{ft}^2/\text{acre}\right)}{\text{BA}_j\left(\text{ft}^2/\text{tree}\right)}\\ &= \frac{10}{0.005454 \cdot 16^2}\\ &= 7.16217.\end{aligned}$$

[20] Remember, you choose the angle θ, or perhaps it's chosen for you and specified in the survey protocol.

[21] Notice we dropped the j subscript on BAF. This is because a single BAF is typically used for a given population—more on this topic in Section 12.4.3.

Then, following (12.19), scale the tree's volume by its tree factor

$$\begin{aligned} v_j\left(\text{ft}^3/\text{acre}\right) &= v_j\left(\text{ft}^3/\text{tree}\right) \cdot \text{TF}_j\,(\text{trees/acre}) \\ &= 45.8 \cdot 7.16217 \\ &= 328. \end{aligned}$$

So, that one measurement tree represents a volume of 328 ft^3/acre.

We refer to the process of expanding and summing tree measurements at each sampling location as the *point-level summary*—the point-sampling analog to the plot-level summary introduced in Section 12.3.1.

For the i-th sampling location, the point-level summary for basal area per unit area is

$$\text{BA}_i = m_i \cdot \text{BAF}, \tag{12.23}$$

where m_i is the number of trees counted using the given BAF.

For other variables, the general expression for a point-level summary follows (12.3). For example, trees per acre are given by (12.4) and volume by (12.6). The only difference from plot sampling is that, under point sampling, the tree factor is computed using the basal-area-based form in (12.22).

12.4.2 Sample-based estimates

For n sampling locations, the corresponding n per-unit-area point-level summaries of a given variable (see Section 12.4.1) are the sample observations—that is, the y_i for $i = 1, 2, \dots, n$—used to estimate population parameters via the estimators provided in Section 11.3.

Point sampling follows the same sample-based estimation framework described in Section 11.3, with one exception. As noted at the beginning of Section 12.4, point sampling is unique in that it allows you to match data collection effort with the desired level of inference. The minimum data collection effort is called a *continuous tally*, meaning a count of measurement trees is kept across the n sampling locations (no additional information is recorded—not even how many measurement trees were observed at each sampling location).

At the end of a continuous tally cruise, you have the total number of measurement trees m, which is used to compute the mean basal area per-unit-area estimate as

$$\overline{\text{BA}} = \frac{m \cdot \text{BAF}}{n}. \tag{12.24}$$

If you'd like to estimate a confidence interval for mean basal area per unit area, record the number of trees tallied at each sampling location and compute the point-level basal area per unit area (12.23). Recognize these values as your y_is and follow the steps in Section 11.3.

Following Section 12.4.1, all variables except basal area require tree factors to compute their point-level summaries.[22] Again following the notation in Section 11.3, these n point-level summaries are your y_is, used to compute the sample mean, standard error of the mean, and confidence intervals via (11.18), (11.24), and (11.35), respectively. The same formulas used for plot sampling apply here when point-sampling tree factors are used in place of plot-based ones.

Estimates for totals (e.g., total basal area or volume) in the cruised population follow directly from the per-unit-area estimates as described in Section 11.3.3, with the simple change of replacing the population size N with the forest area A, expressed in the same units as the per-unit-area means.

Computation of basal area and other parameter estimates under point sampling is illustrated in Section 12.3.6.

12.4.3 Selecting a basal area factor and computing limiting distances

In practice, a horizontal point sampling cruise is described in terms of the BAF, not the angle θ that determines the BAF. For example, you might say "the cruise was conducted using a 40 BAF angle gauge (English units)," or perhaps "a 3 BAF wedge prism (metric units)."

While there are no specific rules for selecting a BAF, in practice you should select one that yields an average of 4 to 10 measurement trees per point. BAFs that yield average tree counts outside this range are not wrong—they're just less efficient.

Given a desired average number of trees per point and a rough estimate for the stand's basal area per unit area (e.g., obtained via prior experience with similar stands or from a quick preliminary cruise of a few points), select the BAF instrument (e.g., angle gauge or wedge prism) you have that's closest to the value computed using

$$\frac{\text{stand basal area per unit area}}{\text{desired tree count per point}}. \tag{12.25}$$

For example, say you'd like an average of 7 measurement trees per point. Looking at the stand, you come up with a rough basal area estimate of 150 ft^2/ac. Then, using (12.25), a reasonable BAF to cruise the stand would be $150/7 = 21.43$, so select a BAF of 20 (which is a standard instrument size).

Based on the guidance above or an established survey protocol, a single BAF is typically selected for each population in a stand. To keep roughly the same number of measurement trees per point, you might feel tempted to change the BAF from point to point—don't do this. Changing the BAF from point to point leads to errors and bias; see, e.g., Wensel et al. (1980) and Iles and Wilson (1988).

You can, however, use a different BAF for different populations within a stand, especially if the populations differ substantially in their average DBH and/or basal area. For example,

---

[22] There are a few exceptions where point-level summaries can be obtained without tree factors—see Section 12.4.6.

you might choose one BAF for large diameter sawtimber and a separate BAF for small diameter pulpwood. Of course, in such settings, it's critical to record the BAF used for each population.

Given a BAF, you can find the constant k, the tree's inclusion zone radius (R) given its DBH, and the R/DBH ratio called the *horizontal distance multiplier* (HDM), which is useful for quickly figuring out a tree's limiting distance. For the English system, with DBH in inches and BAF in ft^2/ac, these are

$$k = \sqrt{\frac{\text{BAF}}{10890}}, \tag{12.26}$$

$$R\,(\text{ft}) = \frac{\text{DBH}}{12k}, \tag{12.27}$$

$$\text{HDM}\left(\frac{\text{ft}}{\text{in}}\right) = \frac{R}{\text{DBH}} = \frac{1}{12k}, \tag{12.28}$$

and for the metric system, with DBH in centimeters and BAF in m^2/ha, these are

$$k = \sqrt{\frac{\text{BAF}}{2500}}, \tag{12.29}$$

$$R\,(\text{m}) = \frac{\text{DBH}}{100k}, \tag{12.30}$$

$$\text{HDM}\left(\frac{\text{m}}{\text{cm}}\right) = \frac{R}{\text{DBH}} = \frac{1}{100k}. \tag{12.31}$$

The HDM is a convenient value to have with you in the field to determine a tree's limiting distance. As noted previously, due to competing vegetation or other complicating factors, it might be difficult to discern if a given tree is a measurement tree. In such cases, you'll need to compare the tree's limiting distance to the distance between the sampling location and the tree's center.

Using the HDM, a tree's limiting distance is computed as

$$R = \text{HDM} \cdot \text{DBH}, \tag{12.32}$$

where the HDM has units of distance per unit diameter.

For example, say you're conducting a cruise using a metric 4 BAF (i.e., $k = 0.04$) and the tree in question has a DBH of 30 cm. Then, using (12.32) with the HDM from (12.31), its limiting distance is

$$\begin{aligned} R &= \text{HDM} \cdot \text{DBH} \\ &= \left(\frac{1}{100 \cdot 0.04}\right) \cdot 30 \\ &= 7.5\,\text{m}. \end{aligned}$$

Or say you have a borderline tree with a DBH of 18 inches and you're using an English 10 BAF (i.e., $k = 0.0303$). Then, using (12.32) with the HDM from (12.28), its limiting distance is

$$\begin{aligned} R &= \text{HDM} \cdot \text{DBH} \\ &= \left(\frac{1}{12 \cdot 0.0303}\right) \cdot 18 \\ &= 49.5 \text{ ft.} \end{aligned}$$

If you're lucky enough to be working in metric units with DBH in cm and BAF in m^2/ha, the limiting distance in (12.30) collapses to an easy-to-remember identity,

$$R\,(\text{m}) = \frac{\text{DBH (cm)}}{2\sqrt{\text{BAF}}}, \tag{12.33}$$

which is algebraically identical to (12.30) after substituting k from (12.29). Unfortunately, the English-unit equivalent doesn't work out as cleanly and isn't nearly as easy to remember or use in the field.[23]

While any angle θ could be used for a cruise, we generally select ones that provide convenient BAFs to work with (i.e., whole numbers). A few such BAFs and their corresponding angle, constant k, and HDM are provided in Tables 12.6 and 12.7 for the English and metric systems, respectively.[24] Such tables are useful to have with you in the field so you can quickly compute a tree's limiting distance given its DBH and HDM.

TABLE 12.6: Commonly used English system basal area factors (BAFs) and associated angle, constant, and horizontal distance multiplier (HDM).

English units			
BAF (ft^2/acre)	Angle $\theta°$	Constant k	HDM (ft/in)
5	1.2277	0.02143	3.89
10	1.7363	0.03030	2.75
15	2.1266	0.03711	2.25
20	2.4556	0.04285	1.94
25	2.7455	0.04791	1.74
30	3.0076	0.05249	1.59
35	3.2486	0.05669	1.47
40	3.4730	0.06061	1.38
50	3.8831	0.06776	1.23
60	4.2539	0.07423	1.12

[23] In English units, the analogous simplification is $R = \frac{\sqrt{10890}}{12} \cdot \frac{\text{DBH}}{\sqrt{\text{BAF}}} \approx \frac{8.696 \cdot \text{DBH}}{\sqrt{\text{BAF}}}$, which is far less tidy than the metric identity. This is why English-unit practice typically relies on tabulated HDMs, or computes k via (12.26) and then plugs it into (12.27) to get R.

[24] If you want to reproduce the constant k in Tables 12.7 and 12.6 in R using (12.13), convert the angle θ from degrees to radians with the `d2r()` function: `d2r <- function(d) { d * pi / 180 }`. Recall from Section 12.3.4 that R's trigonometric functions assume angles are in radians, not degrees.

TABLE 12.7: Commonly used metric system basal area factors (BAFs) and associated angle, constant, and horizontal distance multiplier (HDM).

Metric units			
BAF (m^2/ha)	Angle $\theta°$	Constant k	HDM (m/cm)
1	1.1459	0.02000	0.500
2	1.6206	0.02828	0.354
3	1.9849	0.03464	0.289
4	2.2920	0.04000	0.250
5	2.5626	0.04472	0.224
6	2.8072	0.04899	0.204
7	3.0322	0.05292	0.189
8	3.2416	0.05657	0.177
9	3.4383	0.06000	0.167
10	3.6243	0.06325	0.158

12.4.4 Boundary overlap and slope corrections

Like plot sampling, point sampling requires corrections for boundary and slope effects. The need for these corrections is the same as those for plot sampling, namely, selection probability is lower for trees that have their inclusion zone partially outside the forest boundary, and horizontal distance, not slope distance, should be used to determine if a point center falls within a tree's inclusion zone. The main difference between boundary overlap and slope correction approaches for plot sampling versus horizontal point sampling is that under horizontal point sampling a tree's inclusion zone is a function of its DBH (whereas under plot sampling, a tree's inclusion zone and hence selection probability is a function of plot area).

Under horizontal point sampling, if a measurement tree has any portion of its inclusion zone outside the forest boundary, then some approach to correcting the boundary effect should be used. In such cases, either the mirage or walkthrough method introduced in Section 12.3.3 can be applied. An in-depth look at these and other viable correction methods is presented in Kershaw et al. (2016) and Gregoire and Valentine (2008).

As in plot sampling, measurements taken to identify measurement trees are assumed to be on the horizontal plane. The relascope automatically corrects for slope while sighting possible measurement trees, so no additional effort is required to apply the correction. Slope correction using a wedge prism can be done by rotating the prism about the line of sight to match the slope angle along which you are sighting; see USFS (1996) or a similar field methods guide for details.

When using an angle gauge or other devices without a straightforward slope correction, you can convert horizontal limiting distance to slope limiting distance. This calculation is the same as that performed in Section 12.3.3 to compute C_1 and C_2 in Figure 12.5(c), but with plot radius R replaced by the j-th tree's limiting distance $R_j = \text{HDM} \cdot \text{DBH}_j$, following (12.32).

12.4.5 Illustration

This illustration uses the toy forest population described in Section 12.3.6 and shown in Figure 12.6. Here, point sampling locations are placed at the fixed-area plot centers used previously to illustrate plot sampling. The analysis presented in this section is the point sampling equivalent to the plot sampling analysis.

Recall from the description in Section 12.3.6, these sampling locations were selected using SRS from each stand's areal sampling frame.

Our aim is to step through the point sampling calculations to estimate per-acre basal area, number of trees, and volume, along with corresponding stand totals. A highly efficient workflow for these and additional estimates is developed in Section 12.6 and applied to data from both stands.

We chose to use an English BAF 20 to select measurement trees at each of the six sampling locations. These sampling locations and their associated measurement trees are shown in Figure 12.10.

We limit our analysis to overstory tree data in Stand 1. Table 12.8 provides overstory tree measurements for the three Stand 1 points. Notice in Figure 12.10 there are no overstory trees on Point 2 in Stand 1. As mentioned before, it's critical that absence of trees at a sampling location is included in subsequent estimates—meaning a sampling location with no trees is part of the sample, reflects a characteristic of the population, and needs to be included as a zero when computing population parameter estimates. We include a line for Point 2 in Table 12.8 with zero DBH and volume values to remind us to include these values in subsequent computations.

Let's begin by assuming this was a continuous tally cruise, which, recall from Section 12.4.2, means the forester only kept track of the total number of measurement trees, m, across

TABLE 12.8: Stand 1 overstory tree measurements for inventory data shown in Figure 12.10. Zero values indicate plots where no overstory trees were measured.

Point	DBH (in)	Volume (ft^3)
1	10.7	17.9
1	9.8	14.5
1	11.3	17.8
2	0.0	0.0
3	13.1	28.9
3	14.8	33.6
3	15.4	36.6

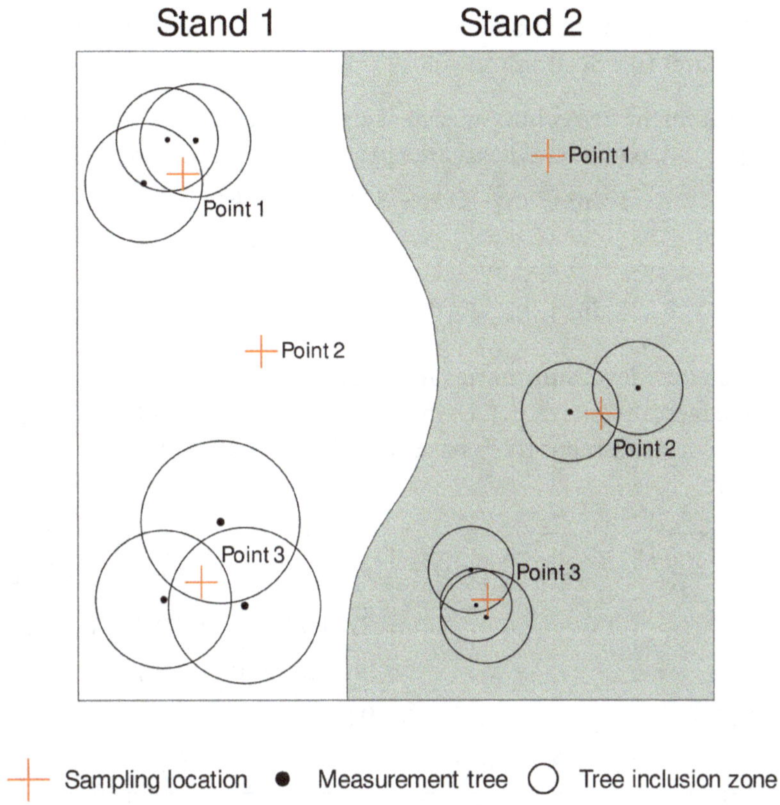

FIGURE 12.10: Toy forest inventory dataset comprising two stands and three sampling locations randomly located within each stand. An English BAF 20 was used to identify measurement trees around each sampling location. Each tree's inclusion zone is added for illustration.

the entire cruise. Looking at Figure 12.10, or by counting the non-zero rows in Table 12.8, we can see there were $m = 6$ measurement trees across the $n = 3$ sampling locations in Stand 1. Then, following (12.24), the estimate for basal area per acre is

$$\overline{\text{BA}} = \frac{m \cdot \text{BAF}}{n} = \frac{6 \cdot 20}{3} = 40 \left(\text{ft}^2/\text{acre}\right).$$

Given Stand 1's area is $A = 0.64$ acres, and following (11.25), with the slight modification of replacing N with A, the continuous tally estimate for total basal area for Stand 1 is $A \cdot \overline{\text{BA}} = A(m \cdot \text{BAF})/n = 0.64 \cdot 40 = 26\,(\text{ft}^2)$.

If you want a standard error to accompany the mean or total basal area estimates, then keep track of how many measurement trees were counted at each sampling location and compute the basal area per acre point-summaries as illustrated below.

Just like estimation using plot sampling data illustrated in Section 12.3.6, we divide the estimation process into two steps. First, compute point-level summaries for each variable (see the end of Section 12.4.1). The point-level summaries comprise each point's expanded and summarized tree measurements expressed on a per-unit-area basis (these are the sample observations). Second, use the point-level summaries to compute the desired population parameter estimates (see Section 12.4.2).

Using the measurement tree counts at each sampling location, $m_1 = 3$, $m_2 = 0$, and $m_3 = 3$, and following (12.23), point-level basal area per acre summaries are

$$\text{Point 1: } m_1 \cdot \text{BAF} = 3 \cdot 20 = 60 \left(\text{ft}^2/\text{ac}\right),$$
$$\text{Point 2: } m_2 \cdot \text{BAF} = 0 \cdot 20 = 0 \left(\text{ft}^2/\text{ac}\right),$$
$$\text{Point 3: } m_3 \cdot \text{BAF} = 3 \cdot 20 = 60 \left(\text{ft}^2/\text{ac}\right).$$

To compute the point-level summaries for all other variables, we include each measurement tree's tree factor from (12.22). Using (12.19) and (12.22), and referring to Table 12.8 to compute each measurement tree's basal area from its DBH, the point-level estimate of trees per acre are

$$\begin{aligned}
\text{Point 1: } \sum_{j=1}^{3} \text{TF}_j = \sum_{j=1}^{3} \frac{\text{BAF}}{c \cdot \text{DBH}_j^2} &= \frac{20}{c \cdot 10.7^2} + \frac{20}{c \cdot 9.8^2} + \frac{20}{c \cdot 11.3^2} \\
&= 32.02929 + 38.18235 + 28.71825 \\
&= 98.92989 \left(\text{trees}/\text{ac}\right), \\
\text{Point 2: } \qquad &= 0 \left(\text{trees}/\text{ac}\right), \\
\text{Point 3: } \sum_{j=1}^{3} \text{TF}_j = \sum_{j=1}^{3} \frac{\text{BAF}}{c \cdot \text{DBH}_j^2} &= \frac{20}{c \cdot 13.1^2} + \frac{20}{c \cdot 14.8^2} + \frac{20}{c \cdot 15.4^2} \\
&= 21.36841 + 16.74139 + 15.46228 \\
&= 53.57208 \left(\text{trees}/\text{ac}\right),
\end{aligned}$$

where $c = 0.005454$.[25]

Using (12.19) and the volume measurements in Table 12.8 along with the tree factors computed above, point-level volume per acre are

$$\begin{aligned}
\text{Point 1: } \sum_{j=1}^{3} v_j \cdot \text{TF}_j &= 17.9 \cdot 32.02929 + 14.5 \cdot 38.18235 + 17.8 \cdot 28.71825 \\
&= 573.32428 + 553.64415 + 511.18485 \\
&= 1638.15329 \left(\text{ft}^3/\text{ac}\right), \\
\text{Point 2: } \qquad &= 0 \left(\text{ft}^3/\text{ac}\right), \\
\text{Point 3: } \sum_{j=1}^{3} v_j \cdot \text{TF}_j &= 28.9 \cdot 21.36841 + 33.6 \cdot 16.74139 + 36.6 \cdot 15.46228 \\
&= 617.54714 + 562.5106 + 565.9193 \\
&= 1745.97704 \left(\text{ft}^3/\text{ac}\right),
\end{aligned}$$

where v_j is the j-th tree's volume.

---

[25] Calculations for the constant c are given in Section 5.3.1.

For reference, we collected the point-level summaries of basal area, number of trees, and volume per acre computed above into Table 12.9. Given the $n = 3$ sample observations in Table 12.9, you're back in familiar territory for computing SRS means, standard deviations, standard errors, and confidence intervals covered in Sections 11.3 and 12.4.2.

TABLE 12.9: Stand 1 overstory point-level summary for inventory data shown in Figure 12.10.

Point	Basal area (ft^2/ac)	Trees/ac	Volume (ft^3/ac)
1	60	98.9299	1638.1533
2	0	0	0
3	60	53.5721	1745.977

The code below computes each variable's mean (11.18), standard error of the mean (11.24), and 90% confidence interval. Again, as discussed in Section 12.2.1, when we select sampling locations using an areal sampling frame, we don't apply the FPC when calculating the standard error of the mean.

```
n <- 3 # Sample size.

t <- qt(p = 1 - 0.1 / 2, df = n - 1) # t-value for 90% CI.

# Basal area per acre.
ba_i <- c(60, 0, 60)
ba_bar <- mean(ba_i)
s_ba_bar <- sd(ba_i) / sqrt(n)
ci_ba <- c(ba_bar - t*s_ba_bar,
           ba_bar + t * s_ba_bar)

ba_bar #Mean basal area per acre.

#> [1] 40

ci_ba #Confidence interval for basal area per acre.

#> [1] -18.4  98.4

# Trees per acre.
trees_i <- c(98.9299, 0, 53.5721)
trees_bar <- mean(trees_i)
s_trees_bar <- sd(trees_i) / sqrt(n)
ci_trees <- c(trees_bar - t * s_trees_bar,
              trees_bar + t * s_trees_bar)
```

```
trees_bar #Mean number of trees per acre.
```

```
#> [1] 50.834
```

```
ci_trees #Confidence interval for mean number of trees per acre.
```

```
#> [1] -32.652 134.320
```

```
# Volume per acre.
vol_i <- c(1638.1533, 0, 1745.977)
vol_bar <- mean(vol_i)
s_vol_bar <- sd(vol_i) / sqrt(n)
ci_vol <- c(vol_bar - t * s_vol_bar,
            vol_bar + t * s_vol_bar)

vol_bar #Mean volume per acre.
```

```
#> [1] 1128
```

```
ci_vol #Confidence interval for volume per acre.
```

```
#> [1] -521.4 2777.5
```

The mean and associated confidence interval computed in the code above are collected into Table 12.10.

Stand 1 total and associated confidence interval for each variable can be computed using (11.25) and (11.28). These estimators for the total are modified by replacing population size N with stand area in acres $A = 0.64$. Total estimates are given in Table 12.11. Importantly, notice that obtaining totals and associated confidence intervals is as easy as scaling mean per-acre estimates given in Table 12.10 by A.

Take a moment to compare point and plot sampling estimates in Tables 12.4 and 12.5 versus Tables 12.10 and 12.11. We expect some differences because different rules are used to select measurement trees, and tree expansion factors are computed differently. In practice, point

TABLE 12.10: Stand 1 point sampling overstory mean estimates along with associated confidence interval (CI) for inventory data shown in Figure 12.7 and recorded in Table 12.8.

	$\bar{y}$	90% CI
Trees/ac	50.834	(-32.652, 134.32)
Basal area (ft^2/ac)	40.000	(-18.4, 98.4)
Volume (ft^3/ac)	1128.043	(-521.398, 2777.485)

TABLE 12.11: Stand 1 point sampling overstory total estimates along with associated confidence interval (CI) for inventory data shown in Figure 12.7 and recorded in Table 12.8.

	$\hat{t}$	90% CI
Trees	32.534	(-20.898, 85.965)
Basal area (ft^2)	25.600	(-11.776, 62.976)
Volume (ft^3)	721.948	(-333.695, 1777.59)

and plot sampling will yield different estimates for any given single sample. However, both are unbiased methods, and hence, over repeated sampling, we'd expect their estimates to agree on average.

As in the plot sampling example, the confidence intervals include negative values for strictly positive parameters. This occurs here due to small sample size and large variance, and reflects the limitations of symmetric approximation methods. See Section 12.5 for further discussion.

12.4.6 Per-unit-area estimates without measuring DBH

A key advantage to horizontal point sampling is that basal area per unit area can be estimated without measuring DBH. However, as we saw in the preceding sections, the tree factor (and hence DBH measurement) is needed to obtain estimates for other parameters. There is an approach to obtaining per-unit-area estimates using variables that scale linearly with basal area across some measure of tree height, without measuring DBH. In such cases, a tree height measurement replaces the DBH measurement. This approach is practical when a measure of tree height can be obtained more easily than DBH.

Depending on the variable of interest, the tree height measurement can take a variety of forms, e.g., total height, height to a given stem diameter, or stem segment length such as a 16 ft sawlog or 2 m bolt. The approach described in this section is often referred to as *tally by height*.

Variables associated with tree volume and weight often scale linearly with basal area across height. In some cases, such relationships are well described by the *constant form factor* equation

$$y_j = \alpha \text{DBH}_j^2 h_j, \tag{12.34}$$

where, for the j-th tree, y_j is the variable of interest, α is a parameter to be estimated, and h_j is some measure of height.

Dividing both sides of (12.34) by basal area yields the following

$$\begin{aligned}\frac{y_j}{\text{BA}_j} &= \frac{\alpha \text{DBH}_j^2 h_j}{\text{BA}_j}\\ &= \frac{\alpha\, \cancel{\text{DBH}_j^2} h_j}{c \cancel{\text{DBH}_j^2}}\\ &= \beta h_j,\end{aligned} \tag{12.35}$$

where c is the basal area constant and $\beta = \alpha/c$. Importantly, notice in (12.35) the y_j to BA_j ratio (left side) is a function of height βh_j (right side). This means that if we have a tree's height measurement (h_j) and a suitable value for β, then we can estimate the y_j to BA_j ratio. In the forestry literature, if y_j is a measure of volume, then the volume-to-basal-area ratio (i.e., y_j/BA_j) is called VBAR_j (pronounced "v bar"). Similarly, if y_j is a measure of weight, the ratio is called WBAR_j (pronounced "w bar").

We typically want a per-unit-area estimate for y_j (i.e., y_j value expanded to per-acre or per-hectare basis), not the y_j to BA_j ratio per-unit-area. To get the desired per-unit-area estimate, multiplying both sides of (12.35) by the BAF yields the following

$$\frac{\text{BAF} \cdot y_j}{\text{BA}_j} = \text{BAF}\beta h_j, \tag{12.36}$$

which can be expressed as

$$y_j \cdot \text{TF}_j = \text{BAF}\beta h_j, \tag{12.37}$$

because the tree expansion factor $\text{TF}_j = \text{BAF}/\text{BA}_j$ (12.22). Hence, (12.37) shows you can obtain the desired per-unit-area estimate for y_j (i.e., $y_j \cdot \text{TF}_j$) given your chosen BAF, a value for β, and a tree height measurement (via $\text{BAF}\beta h_j$).

For example, say you're interested in estimating lumber volume in board-feet (bd ft) using the International 1/4-inch log rule per acre.[26] Based on prior information, you choose a constant form factor equation that measures trees in 16 ft logs. Following (12.37), the j-th measurement tree's volume per acre is

$$\begin{aligned} v_j \cdot \text{TF}_j \left(\frac{\text{bd ft}}{\text{acre}}\right) &= \text{BAF}\left(\frac{\text{ft}^2}{\text{acre}}\right) \cdot \beta \left(\frac{\frac{\text{bd ft}}{\text{ft}^2}}{\text{logs}}\right) \cdot h_j\,(\text{logs}) \\ &= \text{BAF}\left(\frac{\text{ft}^2}{\text{acre}}\right) \cdot \beta h_j \left(\frac{\text{bd ft}}{\text{ft}^2}\right) \\ &= \text{BAF}\beta h_j \end{aligned} \tag{12.38}$$

Say you're given a β value of 55.41 and you conduct your cruise using an English 20 BAF prism. If the j-th measurement tree held $h_j = 2$ (16 ft logs), then following (12.38), that tree represents $\text{BAF}\beta h_j = 20 \cdot 55.41 \cdot 2 = 2216.4$ bd ft per acre.

As described toward the end of Section 12.4.1, the area-expanded estimate $y_j \cdot \text{TF}_j$ is combined with those from other measurement trees at a given sampling location to arrive at the point-level summary.

Continuing our example, say there are m_i measurement trees at the i-th sampling location and the total number of logs for those trees was $h_i = \sum_{j=1}^{m_i} h_{i,j}$. The point-level summary for the i-th location equals $v_i \cdot \text{TF}_i = 20 \cdot 55.41 \cdot h_i$. Point-level summaries from n sampling locations are used to estimate population parameters following steps in Section 12.4.2.

[26] International 1/4-inch log rule estimates board feet from a log, accounting for stem taper, wood removed from the outside of the log to square it off, and saw kerf (or the loss of wood as sawdust); see, e.g., USDA (2006).

The tally by height approach produces reasonable estimates only if the following conditions are met.

1. The linear relationship described in (12.35) holds across the range of anticipated height values.
2. Variability in the variable (e.g., volume or weight) to basal area ratio is small for a given height increment.

The linear relationship, described by the slope parameter β, is highly dependent on species, local environment, and stand characteristics, e.g., age, density, disturbance history. As a result, a β estimated for one species and stand might not be appropriate for a different species or stand.

Let's take a closer look at how one might estimate β for a given variable, species, and geographic region. Table 12.12 holds balsam fir (*Abies balsamea*) volume values published by Young (1957). Table entries, broken out by DBH and height class, summarize volume (ft^3) measurements taken on 270 trees in central Maine.

TABLE 12.12: Volume in cubic feet to a 4-inch top inside bark for balsam fir (*Abies balsamea*) in central Maine (Young 1957). Cells are colored by volume value.

	Total height (ft)				
DBH (in)	30	40	50	60	70
5	1.40	1.91	2.45		
6	2.14	2.83	3.60		
7	3.07	4.09	5.00	6.26	
8	4.20	5.60	6.82	8.42	9.85
9	5.56	7.18	8.77	10.75	12.53
10		8.90	10.90	13.33	15.56
11		10.64	13.22	16.05	18.77
12			15.53	18.78	21.83
13			18.12	21.75	25.27
14			20.87	24.76	28.66
15				28.18	32.20

Table 12.13 holds volume-to-basal-area ratios (VBARs) computed using volume and DBH from Table 12.12. For example, the VBAR in Table 12.13 for a 5 inch DBH and 30 foot tree is 10.27 (ft^3/ft^2), which, working from values in Table 12.12, is 1.40 (ft^3) divided by $c \cdot \text{DBH}^2 = 0.005454 \cdot 5^2 = 0.13635$ (ft^2).[27]

[27]Check your understanding by computing other VBARs in Table 12.13 using values in Table 12.12.

Table 12.13 cell color makes apparent the small variability among VBARs within a given 10 ft height class. Recall this is one data characteristic needed for the tally by height approach to produce reasonable estimates. The other data characteristic is a linear relationship between VBARs across the range of anticipated height values. Figure 12.11 is a scatter plot created from values in Table 12.13. Imposed on the scatter plot is a "best fit" line with intercept fixed at zero and slope β.[28] Visual inspection of the point scatter and line suggests a strong linear relationship exists across the height classes. The line's slope $\beta = 0.393$ is used in (12.37) and would yield reasonable tally by height volume per-acre estimates for balsam fir in central Maine.

12.5 Negative confidence interval bounds and defensive coding

In the illustrations for plot and point sampling (Sections 12.3.6 and 12.4.5, respectively) we pointed out that the 90% confidence intervals for some estimates include negative lower bounds. This can happen when using normal approximations to construct confidence intervals for forest parameters like stem density, basal area, or volume—quantities that are strictly positive. When sample sizes are small or variances are large, as in the toy example here ($n = 3$), confidence intervals may extend into the negative range, even though such values are not meaningful for parameters that must be zero or greater.

While this issue may arise in small-sample or high-variance settings, it's uncommon in operational forest inventory, where sample sizes are typically larger and variability lower. Still, when it does occur, it's worth recognizing and addressing. A simple option is to truncate the lower bound at zero, though this can distort the stated confidence level. More generally, if the goal is to better handle non-normal sampling distributions or enforce constraints that reflect the parameter's support, you may need to look beyond design-based inference and consider other approaches—such as model-assisted or model-based methods—where distributional assumptions are built directly into the estimator (see, e.g., Särndal et al. (1992) and Rao and Molina (2015)).

Encounters with unexpected results—like a confidence interval suggesting an impossible value—are not just statistical curiosities. They're reminders to pause and think carefully about your workflow. This kind of deliberate checking falls under the broader practice of *defensive programming*—a mindset that assumes things can go wrong and encourages coding habits that prevent silent failures. In statistical computing, this means checking inputs, intermediate results, and final outputs for consistency with both the math and the context. For example, a negative lower confidence bound for a strictly positive parameter isn't just a numerical oddity—it's a cue to revisit your data, assumptions, estimators, and code to make sure everything aligns. Thoughtful debugging isn't an afterthought; it's an essential part of reproducible and trustworthy data science.

[28] The line's slope is found using the least-squares regression method with a zero intercept constraint, implemented using the `lm()` function; see Section 16.1 for more details.

TABLE 12.13: Volume-to-basal-area ratios (VBARs) (ft^3/ft^2) for balsam fir (*Abies balsamea*) in central Maine computed using Table 12.12. Cells are colored by VBAR value.

	Total height (ft)				
DBH (in)	30	40	50	60	70
5	10.27	14.01	17.97		
6	10.90	14.41	18.34		
7	11.49	15.30	18.71	23.42	
8	12.03	16.04	19.54	24.12	28.22
9	12.59	16.25	19.85	24.33	28.36
10		16.32	19.99	24.44	28.53
11		16.12	20.03	24.32	28.44
12			19.77	23.91	27.80
13			19.66	23.60	27.42
14			19.52	23.16	26.81
15				22.96	26.24

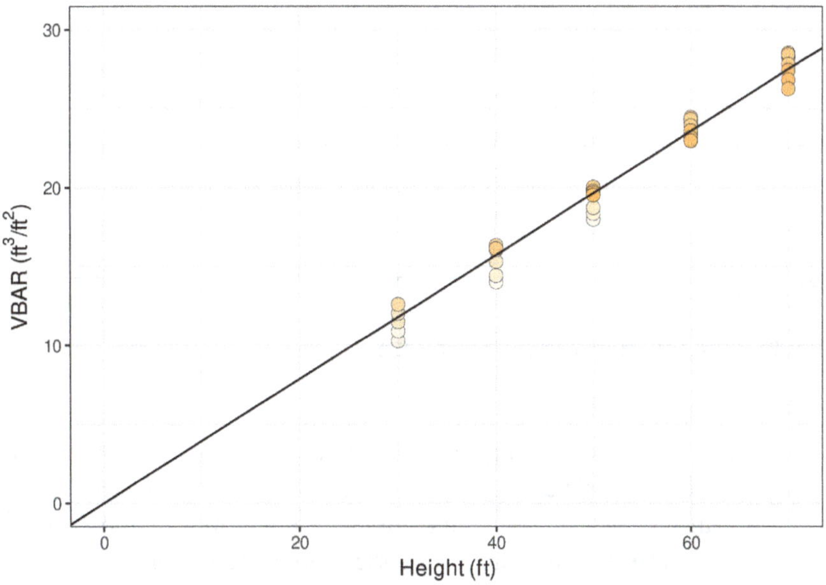

FIGURE 12.11: Scatter plot of Table 12.13 volume-to-basal-area ratios (VBARs) for balsam fir (*Abies balsamea*) in central Maine. Point colors correspond to VBAR values in Table 12.13.

Brian W. Kernighan, a pioneering computer scientist best known for co-authoring *The C Programming Language* and for his work on Unix at Bell Labs, put it succinctly:

> Everyone knows that debugging is twice as hard as writing a program in the first place. So if you're as clever as you can be when you write it, how will you ever debug it?
>
> — Brian W. Kernighan

Debugging is indeed difficult—especially in analysis workflows where silent failures can lead to plausible but incorrect results. One of the most effective strategies is to pair deliberate reasoning with simple tools that make your workflow more transparent. Again, Kernighan's advice remains timeless:

> The most effective debugging tool is still careful thought, coupled with judiciously placed print statements.
>
> — Brian W. Kernighan

12.6 Workflows for forest inventory data

12.6.1 Plot sampling illustration

Here we use `tidyverse` functions to repeat and expand on the analysis presented in Section 12.3.6. Recall that the analysis used a toy dataset to illustrate plot sampling estimation methods. As described in Section 12.3.6 and illustrated in Figure 12.7, the dataset comprises two stands, each with three overstory and nested regeneration plots. The overstory and regeneration plots are circular fixed areas with 24 ft and 6.8 ft radii, respectively. Plot sampling locations were selected using SRS from each stand's areal sampling frame and serve as the overstory plot centers.

The code below reads in the data collected on the inventory plots and prints its structure.

```
stands <- read_csv("datasets/two_stands_plot_data.csv")
stands %>% glimpse()
```

```
#> Rows: 30
```

```
#> Columns: 8
#> $ stand_id        <dbl> 1, 1, 1, 1, 1, 1, 1, 1, 1, 1, 1, 1~
#> $ plot_id         <dbl> 1, 1, 1, 1, 1, 2, 2, 2, 3, 3, 3, 3~
#> $ tree_type       <chr> "Overstory", "Overstory", "Oversto~
#> $ tree_count      <dbl> 1, 1, 1, 1, 1, 0, 1, 1, 1, 1, 1, 2~
#> $ scientific_name <chr> "Abies balsamea", "Pinus strobus",~
#> $ DBH_in          <dbl> 11.3, 9.8, 10.7, NA, NA, 0.0, NA, ~
#> $ height_ft       <dbl> 60.2, 61.1, 63.6, NA, NA, 0.0, NA,~
#> $ vol_cu_ft       <dbl> 17.8, 14.5, 17.9, NA, NA, 0.0, NA,~
```

Column definitions for `stands` are as follows.

1. `stand_id` stand number within which the plot falls, takes values 1 or 2.
2. `plot_id` plot number within the stand on which the measurements are recorded.
3. `tree_type` type of plot on which the tree was measured, takes values `Overstory` or `Regen`.
4. `tree_count` number of trees measured on the plot that have the same values for the next four columns, or zero if no trees occurred on the plot.
5. `species` measurement tree species.
6. `DBH_in` measurement tree DBH (in).
7. `height_ft` measurement tree height (ft).
8. `vol_cu_ft` measurement tree volume (ft^3).

There are a few things to notice about these data (run `stands %>% print(n = nrow(.))` to see all the data).

1. Plot numbers are unique within the stand, meaning both stand and plot must be specified to identify a plot's set of trees.
2. Because we're only interested in regeneration density by species for this survey, trees on regeneration plots have `NA` values for their DBH, height, and volume.
3. As you might have noticed in Figure 12.7, there are no overstory trees on Plot 2 in Stand 1. It's critical that the absence of trees at a sampling location be included in subsequent estimates—meaning a plot with no trees is part of the sample, reflects a characteristic of the population, and needs to be included as a zero when computing population parameter estimates. Here, we chose to record plots with no trees using a `tree_count` value of zero.
4. In addition to tracking plots with zero measurement trees, the `tree_count` column records the number of trees on a given plot that share the same species and measurements on other variables. This allows each row in the data file to represent more than one tree. It's common in timber cruises to record DBH and number of logs to the nearest whole number, in which case there are often several overstory trees on a plot with counts greater than one. Also, including a tree count variable allows us to accommodate double-count boundary trees identified using the mirage or walkthrough method described in Section 12.3.3, minimize the number of data rows needed to record the inventory data, and build an efficient analysis workflow.

To compare results with those presented in Section 12.3.6, we derive estimates for overstory per-acre number of trees, basal area, and volume, along with corresponding stand totals.

Also, following Section 12.3.6, the estimation process is done in two steps. Step 1, compute plot-level summaries for each variable. The plot-level summaries comprise each plot's expanded and summarized tree measurements expressed on a per-unit-area basis. Step 2, use the plot-level summaries from Step 1 to compute the desired population parameter estimates.

The first bit of code below makes a new tibble (`o_stands`) that includes only rows corresponding to overstory tree measurements and a column (`TF`) that holds their tree factors. As you'll see in the subsequent piped workflow, it's convenient to have a tree factor column pre-computed.[29] The tree factor equation used in `mutate()` follows (12.1).

The second bit of code below creates the plot-level per-acre summary for each variable. Importantly, notice we group by `stand_id` and `plot_id` so the subsequent `summarize()` works on trees in each plot—yielding plot-level summaries. The equations used to compute these plot-level summaries within `summarize()` follow (12.2) and parallel steps in Section 12.3.6. The only difference here is we scale by `tree_count` to accommodate how the data were recorded (i.e., `tree_count` records the number of trees with common species, DBH, height, and volume, or the absence of trees on a plot).

```
# Select overstory trees and add a TF trees/ac column.
o_stands <- stands %>%
  filter(tree_type == "Overstory") %>%
  mutate(TF = 43560 / (pi * 24^2))

# Step 1, compute per-unit-area plot-level summaries.
c <- pi / (144 * 4) # Basal area constant.

plot_summary <- o_stands %>%
  group_by(stand_id, plot_id) %>%
  summarize(trees_i = sum(tree_count * TF),
            ba_i = sum(tree_count * TF * c * DBH_in^2),
            vol_i = sum(tree_count * TF * vol_cu_ft),
            .groups = "drop_last")

plot_summary
```

```
#> # A tibble: 6 x 5
#> # Groups:   stand_id [2]
#>   stand_id plot_id trees_i  ba_i vol_i
#>      <dbl>   <dbl>   <dbl> <dbl> <dbl>
#> 1        1       1    72.2  44.4 1208.
#> 2        1       2     0     0      0 
#> 3        1       3    72.2  82.4 2386.
```

[29]This is especially true if trees are measured on different sized plots or under point sampling where the tree factor depends on tree-specific characteristics like DBH.

```
#> 4         2         1      72.2   22.2   479.
#> 5         2         2      72.2   28.9   669.
#> 6         2         3      96.3   33.2   749.
```

The `plot_summary` values for `stand_id` 1 match those computed in Section 12.3.6 and displayed in Table 12.3.

Importantly, the `summarize()` in Step 1 above uses the default `.groups = "drop_last"` rule, so the resulting `plot_summary` is grouped by `stand_id` and is ready to be summarized by stand in Step 2 below.

The code below uses the plot-level summaries (`plot_summary`) to compute estimates for the population parameters.[30] In the `summarize()`, we first compute stand-specific sample size `n` and t-value for a 90% confidence interval to parallel the analysis in Section 12.3.6, then mean, standard error, and confidence interval bounds for the three variables (`trees_i`, `ba_i`, and `vol_i`). The resulting per-acre estimates and associated confidence intervals for Stand 1 and 2 are held in `stand_estimates` row 1 and 2, respectively. Below, we pass `stand_estimates` to `glimpse()` so the output fits on the page. From this printed output, you can see `stand_id` 1 column values match those computed in Section 12.3.6 and presented in Table 12.4. Values in the second column printed below correspond to estimates for `stand_id` 2.

```
# Step 2, compute per-unit-area estimates for each stand.
stand_estimates <- plot_summary %>%
  summarize(n = n(), # Sample size.
            t = qt(p = 1 - 0.1 / 2, df = n - 1), # t-value for 90% CI.
            trees_bar = mean(trees_i),
            s_trees_bar = sd(trees_i) / sqrt(n),
            ci_lower_trees = trees_bar - t * s_trees_bar,
            ci_upper_trees = trees_bar + t * s_trees_bar,
            ba_bar = mean(ba_i),
            s_ba_bar = sd(ba_i) / sqrt(n),
            ci_lower_ba = ba_bar - t * s_ba_bar,
            ci_upper_ba = ba_bar + t * s_ba_bar,
            vol_bar = mean(vol_i),
            s_vol_bar = sd(vol_i) / sqrt(n),
            ci_lower_vol = vol_bar - t * s_vol_bar,
            ci_upper_vol = vol_bar + t * s_vol_bar)

stand_estimates %>% glimpse()
```

```
#> Rows: 2
#> Columns: 15
#> $ stand_id         <dbl> 1, 2
```

[30] Notice again that `plot_summary` is already grouped by `stand_id`, so the subsequent call to `summarize()` produces stand-level summaries, as desired.

```
#> $ n             <int> 3, 3
#> $ t             <dbl> 2.92, 2.92
#> $ trees_bar     <dbl> 48.144, 80.241
#> $ s_trees_bar   <dbl> 24.0722, 8.0241
#> $ ci_lower_trees <dbl> -22.146, 56.810
#> $ ci_upper_trees <dbl> 118.43, 103.67
#> $ ba_bar        <dbl> 42.278, 28.110
#> $ s_ba_bar      <dbl> 23.8185, 3.2139
#> $ ci_lower_ba   <dbl> -27.272, 18.725
#> $ ci_upper_ba   <dbl> 111.827, 37.494
#> $ vol_bar       <dbl> 1198.0, 632.3
#> $ s_vol_bar     <dbl> 688.670, 79.987
#> $ ci_lower_vol  <dbl> -812.91, 398.73
#> $ ci_upper_vol  <dbl> 3208.90, 865.86
```

The workflow above is good, but let's make it less verbose and more flexible. Notice in the `summarize()` above, the same operations are applied to each variable. Specifically, say `trees_i`, `ba_i`, or `vol_i` is generically represented with `y`, then the operations to compute the mean and its standard error are `mean(y)` and `sd(y)/sqrt(n)`, respectively. As illustrated in the code below, application of these common operations to each variable allows us to use `across()` (Section 7.11) to simplify the code.

In the `summarize()` below, the call to `across()` computes the mean, standard error of the mean, and confidence bounds for each variable. To simplify the code, we use a custom function `se()` to compute the standard error of the mean for each variable (writing custom functions is covered in Chapter 5). The call to `across()` also uses the anonymous function lambda syntax introduced in Section 7.11, that is, the tilde `~` and `.x` syntax.

The `.names` argument in `across()` controls how the new column names are formed. Within the string, `{.col}` is replaced by the name of the input column and `{.fn}` is replaced by the name of the applied function. This lets you define exactly how the output variable names should look. In this example, we use a small string substitution `sub('_i$', '', .col)` to remove the trailing `_i` from each input variable name before appending the function name. As a result, column names like `trees_i`, `ba_i`, and `vol_i` become `trees`, `ba`, and `vol`, producing cleaner and more readable outputs such as `y_bar_trees`, `s_y_bar_trees`, and `ci_upper_trees`.

```
# Function to compute the standard error of the mean.
se <- function(y, n){
  sd(y) / sqrt(n)
}

# Alternative Step 2, compute per-unit-area estimates for each stand.
plot_summary %>%
 summarize(
   n = n(), # Sample size.
   t = qt(p = 1 - 0.1 / 2, df = n - 1), # t-value for 90% CI.
```

```
  across(.cols = c("trees_i", "ba_i", "vol_i"),
         .fns = list(y_bar = mean,
                     s_y_bar = ~ se(.x, n),
                     ci_lower = ~ mean(.x) - t * se(.x, n),
                     ci_upper = ~ mean(.x) + t * se(.x, n)
                     ),
         .names = "{.fn}_{sub('_i$', '', .col)}")) %>%
    glimpse()
```

```
#> Rows: 2
#> Columns: 15
#> $ stand_id       <dbl> 1, 2
#> $ n              <int> 3, 3
#> $ t              <dbl> 2.92, 2.92
#> $ y_bar_trees    <dbl> 48.144, 80.241
#> $ s_y_bar_trees  <dbl> 24.0722, 8.0241
#> $ ci_lower_trees <dbl> -22.146, 56.810
#> $ ci_upper_trees <dbl> 118.43, 103.67
#> $ y_bar_ba       <dbl> 42.278, 28.110
#> $ s_y_bar_ba     <dbl> 23.8185, 3.2139
#> $ ci_lower_ba    <dbl> -27.272, 18.725
#> $ ci_upper_ba    <dbl> 111.827, 37.494
#> $ y_bar_vol      <dbl> 1198.0, 632.3
#> $ s_y_bar_vol    <dbl> 688.670, 79.987
#> $ ci_lower_vol   <dbl> -812.91, 398.73
#> $ ci_upper_vol   <dbl> 3208.90, 865.86
```

We finish this section by extending our workflow to estimate parameters for both overstory and regeneration populations. Importantly, the only thing we need to do is include those trees measured on the smaller regeneration plots (i.e., 6.8 ft radius plots) and their associated tree factors; the rest of the workflow stays pretty much the same.

Recall the `tree_type` column indicates if a tree was measured on the overstory or regeneration plot using `Overstory` and `Regen.` values, respectively. Using (12.1), the code below illustrates `if_else()` and `case_when()` within `mutate()` to add `tree_type` specific tree factors (`TF`) to `stands`. Both approaches result in the same `TF`; we include both simply for demonstration. The last line of code below is included to double check that tree factors are correctly assigned to each tree type (again, following from Section 12.3.1, one overstory tree represents approximately 24.1 trees per acre and one regeneration tree represents 300 trees per acre).

```
# Add TF trees/ac column for each tree type (plot size).
stands <- stands %>% mutate(TF = if_else(tree_type == "Overstory",
                                         43560 / (pi * 24^2),
```

```
                              43560 / (pi * 6.8^2))
                    )

# Or equivalently.
stands <- stands %>%
 mutate(TF = case_when(tree_type == "Overstory" ~ 43560 / (pi * 24^2),
                       tree_type == "Regen." ~ 43560 / (pi * 6.8^2))
         )

# Double check TF is correct for each tree type (plot size).
stands %>% group_by(tree_type) %>% distinct(TF)
```

```
#> # A tibble: 2 x 2
#> # Groups:   tree_type [2]
#>   tree_type    TF
#>   <chr>     <dbl>
#> 1 Overstory  24.1
#> 2 Regen.    300.
```

Now, given our `tree_type`-specific tree factor in `TF`, we can move to the familiar two-step estimation process. Step 1, create plot-level summaries. Step 2, summarize at the stand level using the plot-level summary. The code below is identical to what we used for the overstory-only analysis above, except we add `tree_type` as the first grouping column to produce separate overstory and regeneration plot summaries.

```
# Step 1, compute per-unit-area plot-level summaries.
c <- pi / (144 * 4) # Basal area constant.

plot_summary <- stands %>%
  group_by(tree_type, stand_id, plot_id) %>%
  summarize(trees_i = sum(tree_count * TF),
            ba_i = sum(tree_count * TF * c * DBH_in^2),
            vol_i = sum(tree_count * TF * vol_cu_ft),
            .groups = "drop_last") # Keep groups tree_type and stand_id.

plot_summary
```

```
#> # A tibble: 12 x 6
#> # Groups:   tree_type, stand_id [4]
#>    tree_type stand_id plot_id trees_i  ba_i vol_i
#>    <chr>        <dbl>   <dbl>   <dbl> <dbl> <dbl>
#>  1 Overstory        1       1    72.2  44.4 1208.
#>  2 Overstory        1       2     0     0      0
#>  3 Overstory        1       3    72.2  82.4 2386.
```

```
#>  4 Overstory        2         1     72.2   22.2  479.
#>  5 Overstory        2         2     72.2   28.9  669.
#>  6 Overstory        2         3     96.3   33.2  749.
#>  7 Regen.           1         1    600.    NA     NA
#>  8 Regen.           1         2    600.    NA     NA
#>  9 Regen.           1         3   1199.    NA     NA
#> 10 Regen.           2         1    900.    NA     NA
#> 11 Regen.           2         2    900.    NA     NA
#> 12 Regen.           2         3   1199.    NA     NA
```

Recall DBH and volume were not recorded for trees measured on regeneration plots, hence the `NA` values in the regeneration rows for `ba_i` and `vol_i`.

Stand-level estimates can now be generated by passing `plot_summary` to `summarize()`. We could use the previous overstory analysis code unchanged; however, for brevity, below we opt to compute only stand density estimates using `trees_i` for the regeneration and overstory populations.

```
# Step 2, compute per-unit-area estimates.
stand_estimates <- plot_summary %>%
  summarize(n = n(), # Sample size.
            t = qt(p = 1 - 0.1 / 2, df = n - 1), # t-value for 90% CI.
            trees_bar = mean(trees_i),
            s_trees_bar = sd(trees_i) / sqrt(n),
            ci_lower = trees_bar - t * s_trees_bar,
            ci_upper = trees_bar + t * s_trees_bar,
            .groups = "drop") # Grouping is no longer needed.

# Print per-acre estimates we care about (trees/acre).
stand_estimates %>%
  select(stand_id, tree_type, trees_bar, ci_lower, ci_upper)
```

```
#> # A tibble: 4 x 5
#>   stand_id tree_type trees_bar ci_lower ci_upper
#>      <dbl> <chr>         <dbl>    <dbl>    <dbl>
#> 1        1 Overstory      48.1    -22.1     118.
#> 2        2 Overstory      80.2     56.8     104.
#> 3        1 Regen.        800.     216.     1383.
#> 4        2 Regen.       1000.     708.     1291.
```

Notice that the output above includes estimates for all four populations: Stand 1 overstory, Stand 2 overstory, Stand 1 regeneration, and Stand 2 regeneration.

So far, we've focused on per-unit-area estimates; however, we're often equally interested in totals and their associated confidence intervals. Totals are computed using (11.25) and (11.28). In our current setting, these estimators are adapted by replacing the population size N with the stand area in acres, A.

Following the steps in Section 12.3.6, we'll scale our per-acre estimates by their respective stand areas to arrive at total estimates.

Begin by adding stand-specific areas to `stand_estimates` using the `case_when()` function introduced in Section 7.6.[31]

```
# Add a stand area column.
stand_estimates <- stand_estimates %>%
  mutate(A_ac = case_when(stand_id == 1 ~ 0.64,
                          stand_id == 2 ~ 0.68))

# Make sure A_ac is added as desired.
stand_estimates %>%
  select(stand_id, A_ac)
```

```
#> # A tibble: 4 x 2
#>   stand_id  A_ac
#>      <dbl> <dbl>
#> 1        1  0.64
#> 2        2  0.68
#> 3        1  0.64
#> 4        2  0.68
```

Next, within a call to `mutate()`, multiply `A_ac` by the mean and associated confidence interval estimates. We append `_hat` to the new columns to remind ourselves that these are totals—corresponding to the hat on $\hat{t}$ in (11.25).

```
# Compute total estimates.
stand_estimates <- stand_estimates %>%
  mutate(trees_hat = A_ac * trees_bar,
         ci_lower_hat = A_ac * ci_lower,
         ci_upper_hat = A_ac * ci_upper)

# Print total estimates of interest (total trees).
stand_estimates %>%
  select(tree_type, stand_id, trees_hat, ci_lower_hat, ci_upper_hat)
```

```
#> # A tibble: 4 x 5
#>   tree_type stand_id trees_hat ci_lower_hat ci_upper_hat
#>   <chr>        <dbl>     <dbl>        <dbl>        <dbl>
#> 1 Overstory        1      30.8        -14.2         75.8
```

[31]With only two stands, it's simple enough to use `case_when()` within `mutate()` to add stand-specific information like area. For larger datasets, a more flexible and scalable approach is to use a mutating join (Section 7.14.1) to merge stand attributes with plot- or stand-level summaries. For example, see how `left_join()` is used for this purpose in Section 15.7.

```
#> 2 Overstory          2       54.6         38.6         70.5
#> 3 Regen.             1      512.         138.         885.
#> 4 Regen.             2      680.         481.         878.
```

Notice that the Stand 1 overstory estimate for total trees matches $\hat{t}$ in Table 12.5.

As illustrated in this section, we're almost always interested in estimates for multiple parameters—and often for multiple populations. In practice, measurements on several variables (e.g., volume, basal area, and density) are commonly collected using the same sampling design, which means the same estimators can be applied to each variable with just a few lines of well-written code.

12.6.2 Point sampling illustration

Here we repeat the point sampling illustration presented in 12.4.5 using a `tidyverse` workflow. The way in which the tree factor is computed is the only major change from the plot sampling workflow covered in Section 12.6.1.

Recall from Section 12.4.5 that an English BAF of 20 was used to select overstory measurement trees at three sampling locations within each stand. Those measurements are read into `o_stands` below, and the `BAF` and `TF` columns are added. When adding the tree factor column, an `if_else()` statement is used to avoid assigning a `TF` value of `Inf` to points with no measurement trees (i.e., in our data, when `tree_count == 0`, `DBH_in` is zero, which would lead to `BAF/0 = Inf`).[32] Given a measurement tree's DBH, its `TF` is computed using (12.22).

```
BAF <- 20

c <- pi / (144 * 4) # Basal area constant.

o_stands <- read_csv("datasets/two_stands_point_data.csv")

o_stands <- o_stands %>%
  mutate(
    BAF = BAF,
    TF = if_else(tree_count == 0,
      0,
      BAF / (c * DBH_in^2)
    )
  )

# Take a look at the new BAF and TF columns.
o_stands %>%
```

[32] Because the same BAF was used for both stands, we could have simply referenced the `BAF` variable defined outside `o_stands`. However, it's common to have stand-specific BAF values, in which case you'd want to include a `BAF` column in your dataset. You could use `case_when()` to assign stand-specific values.

```
  select(stand_id, point_id, tree_count, BAF, DBH_in, TF)
```

```
#> # A tibble: 13 x 6
#>    stand_id point_id tree_count   BAF DBH_in    TF
#>       <dbl>    <dbl>      <dbl> <dbl>  <dbl> <dbl>
#>  1        1        1          1    20   10.7  32.0
#>  2        1        1          1    20    9.8  38.2
#>  3        1        1          1    20   11.3  28.7
#>  4        1        2          0    20    0     0
#>  5        1        3          1    20   13.1  21.4
#>  6        1        3          1    20   14.8  16.7
#>  7        1        3          1    20   15.4  15.5
#>  8        2        1          0    20    0     0
#>  9        2        2          1    20    8.8  47.4
#> 10        2        2          1    20    9.3  42.4
#> 11        2        3          1    20    8.8  47.4
#> 12        2        3          1    20    6.9  77.0
#> 13        2        3          1    20    8.2  54.5
```

In the code below, equations used to compute these point- and stand-level summaries parallel steps in Section 12.4.5. The only difference here is we scale by `tree_count` to accommodate how the data were recorded (i.e., `tree_count` records the number of trees with common species, DBH, height, volume, or the absence of trees on a plot). As demonstrated toward the end of Section 12.6.1, we use `across()` to simplify the stand-level summary.

```
# Step 1, compute per unit area point-level summaries.
point_summary <- o_stands %>%
  group_by(stand_id, point_id) %>%
  summarize(ba_i = sum(tree_count * BAF),
            trees_i = sum(tree_count * TF),
            vol_i = sum(tree_count * TF * vol_cu_ft),
            .groups = "drop_last")

point_summary
```

```
#> # A tibble: 6 x 5
#> # Groups:   stand_id [2]
#>   stand_id point_id  ba_i trees_i vol_i
#>      <dbl>    <dbl> <dbl>   <dbl> <dbl>
#> 1        1        1    60    98.9 1638.
#> 2        1        2     0     0      0
#> 3        1        3    60    53.6 1746.
#> 4        2        1     0     0      0
#> 5        2        2    40    89.7  948.
```

```
#> 6         2         3    60    179.  1345.
```

```
# Step 2, compute per-unit-area estimates for each stand.
point_summary %>%
 summarize(
   n = n(), # Sample size.
   t = qt(p = 1 - 0.1 / 2, df = n - 1), # t-value for 90% CI.
   across(.cols = c("ba_i", "trees_i", "vol_i"),
          .fns = list(y_bar = mean,
                      s_y_bar = ~ sd(.x)/sqrt(n),
                      ci_lower = ~ mean(.x) - t * sd(.x)/sqrt(n),
                      ci_upper = ~ mean(.x) + t * sd(.x)/sqrt(n)
                      ),
          .names = "{.fn}_{sub('_i$', '', .col)}")) %>%
  glimpse
```

```
#> Rows: 2
#> Columns: 15
#> $ stand_id       <dbl> 1, 2
#> $ n              <int> 3, 3
#> $ t              <dbl> 2.92, 2.92
#> $ y_bar_ba       <dbl> 40.000, 33.333
#> $ s_y_bar_ba     <dbl> 20.000, 17.638
#> $ ci_lower_ba    <dbl> -18.40, -18.17
#> $ ci_upper_ba    <dbl> 98.400, 84.837
#> $ y_bar_trees    <dbl> 50.833, 89.552
#> $ s_y_bar_trees  <dbl> 28.591, 51.646
#> $ ci_lower_trees <dbl> -32.652, -61.254
#> $ ci_upper_trees <dbl> 134.32, 240.36
#> $ y_bar_vol      <dbl> 1128.0, 764.3
#> $ s_y_bar_vol    <dbl> 564.86, 398.98
#> $ ci_lower_vol   <dbl> -521.38, -400.71
#> $ ci_upper_vol   <dbl> 2777.4, 1929.3
```

Point-level summaries (`point_summary`) for Stand 1 match those shown in Table 12.9, and stand-level summaries for Stand 1 match the estimates in Table 12.10. From here, you could compute stand-level totals following the approach illustrated at the end of Section 12.6.1.

12.6.3 Stand and stock table estimates

Stand and stock tables were introduced in Section 8.3 using pre-computed stand-level estimates. This section is a prequel to Section 8.3. Here, we show how to generate those estimates directly from individual tree measurements collected via plot or point sampling.

Recall from Section 8.3 that the stand table summarizes a *quantitative discrete* variable (e.g., stem count) grouped by one or more *categorical* variables (e.g., size class and species).

Similarly, the stock table summarizes a *quantitative continuous* variable (e.g., volume or biomass) grouped by one or more *categorical* variables.

In forestry applications, as detailed and illustrated in Section 8.3, stand tables typically summarize the average number of trees per unit area (e.g., acre or hectare) within an area of interest (e.g., stand, management unit, or forest) by DBH class and species. Similarly, stock tables typically summarize the average wood volume or weight per unit area within an area of interest by DBH class and species. DBH classes are DBH intervals with widths chosen to communicate situation-specific information.

The key to generating estimates for stand and stock tables is recognizing that each row and column combination defines a unique population. For example, in each of Tables 8.2 and 8.3, there are nine separate populations defined by species and DBH class combinations. Our task is to generate an estimate for each population. These estimates come from the estimators defined in Chapter 11 and illustrated for plot and point sampling in Sections 12.3.6 and 12.4.5, respectively. The only tricky part is partitioning the inventory data into these populations—a task made simple using `tidyverse` functions.

Let's build a stand and stock table using the toy dataset introduced in Section 12.3.6. We'll focus on Stand 1 overstory trees and define our populations by species and DBH class, as in Section 8.3.

We begin by reading in the data, filtering for Stand 1 overstory trees, and selecting the relevant columns.

```
stands <- read_csv("datasets/two_stands_plot_data.csv")

o_stand <- stands %>%
  filter(stand_id == 1, tree_type == "Overstory") %>%
  select(plot_id, tree_count, scientific_name, DBH_in, vol_cu_ft)

o_stand
```

```
#> # A tibble: 7 x 5
#>   plot_id tree_count scientific_name   DBH_in vol_cu_ft
#>     <dbl>      <dbl> <chr>              <dbl>     <dbl>
#> 1       1          1 Abies balsamea      11.3      17.8
#> 2       1          1 Pinus strobus        9.8      14.5
#> 3       1          1 Pinus strobus       10.7      17.9
#> 4       2          0 <NA>                 0         0
#> 5       3          1 Betula papyrifera   14.8      33.6
#> 6       3          1 Betula papyrifera   15.4      36.6
#> 7       3          1 Pinus strobus       13.1      28.9
```

Recall there were no trees on Plot 2 in Stand 1. We preserve that zero count using the `tree_count` variable, as shown in the output above. In the next step, we convert `plot_id` to a factor to ensure all plots, including those with no trees, are retained as factor levels (see

Section 4.3 for more on factor data types). We then remove the row with `tree_count == 0`, but don't worry—those zero-count plot indexes are still preserved as factor levels, and we'll add their explicit zeros back in later using `complete()`.[33]

```
# Make plot_id a factor, then remove plots with zero trees.
o_stand <- o_stand %>%
  mutate(plot_id = factor(plot_id)) %>%
  filter(tree_count != 0)
```

Next, we add zeros where species and DBH class combinations (which define populations) were not observed on a plot. As noted earlier, a plot with no trees is still part of the sample, reflects a characteristic of the population, and must be included when computing estimates.

We first add a DBH class column using `cut_width()`, which returns a factor. We also convert `scientific_name` to a factor to ensure that all levels are preserved in the call to `complete()`. This allows us to insert unobserved combinations with explicit zeros—making implicit zeros explicit, as described in Section 8.3.

```
# Add the DBH class column and ensure species is a factor.
o_stand <- o_stand %>%
  mutate(DBH_4in = cut_width(DBH_in, width = 4),
         scientific_name = factor(scientific_name))

# Make implicit zeros explicit for species and DBH class
# combinations not observed on plots.
o_stand <- o_stand %>%
  complete(plot_id, scientific_name, DBH_4in,
           fill = list(tree_count = 0, vol_cu_ft = 0))

o_stand
```

```
#> # A tibble: 28 x 6
#>    plot_id scientific_name  DBH_4in tree_count DBH_in vol_cu_ft
#>    <fct>   <fct>            <fct>        <dbl>  <dbl>     <dbl>
#>  1 1       Abies balsamea   [6,10]           0   NA         0
#>  2 1       Abies balsamea   (10,14]          1   11.3      17.8
#>  3 1       Abies balsamea   (14,18]          0   NA         0
#>  4 1       Betula papyrifera [6,10]          0   NA         0
#>  5 1       Betula papyrifera (10,14]         0   NA         0
#>  6 1       Betula papyrifera (14,18]         0   NA         0
#>  7 1       Pinus strobus    [6,10]           1    9.8      14.5
#>  8 1       Pinus strobus    (10,14]          1   10.7      17.9
```

[33]After running the code below, you can confirm that Plot 2 is still present as a factor level by calling `levels(o_stand$plot_id)`.

```
#>  9 1       Pinus strobus     (14,18]          0 NA             0
#> 10 2       Abies balsamea    [6,10]           0 NA             0
#> # i 18 more rows
```

Now `o_stand` is ready for the workflow that generates stand-level estimates for trees per acre and volume per acre. Importantly, the steps below follow the same logic as in Section 12.6.1, except here we group by species and DBH class to partition the data into the nine populations.

```
# Add a TF trees/ac column.
o_stand <- o_stand %>%
  mutate(TF = 43560 / (pi * 24^2))

# Step 1, Compute per-unit-area plot-level summaries.
plot_summary <- o_stand %>%
  group_by(scientific_name, DBH_4in, plot_id) %>%
  summarize(
    trees_i = sum(tree_count * TF),
    vol_i = sum(tree_count * TF * vol_cu_ft),
    .groups = "drop_last"
  )
```

There is a useful check here before we move on to computing the stand estimates. The data have 3 species, 3 DBH classes, and 3 plots. The product of these numbers (i.e., 27) is the expected number of rows in `plot_summary` (print `plot_summary` to confirm for yourself).

```
# Step 2, Compute per-unit-area estimates for each stand.
stand_estimates <- plot_summary %>%
  summarize(
    trees_bar = mean(trees_i),
    vol_bar = mean(vol_i)
  )

stand_estimates
```

```
#> # A tibble: 9 x 4
#> # Groups:   scientific_name [3]
#>   scientific_name   DBH_4in trees_bar vol_bar
#>   <fct>             <fct>       <dbl>   <dbl>
#> 1 Abies balsamea    [6,10]       0          0
#> 2 Abies balsamea    (10,14]      8.02     143.
#> 3 Abies balsamea    (14,18]      0          0
#> 4 Betula papyrifera [6,10]       0          0
#> 5 Betula papyrifera (10,14]      0          0
```

```
#> 6 Betula papyrifera (14,18]      16.0      563.
#> 7 Pinus strobus     [6,10]        8.02     116.
#> 8 Pinus strobus     (10,14]      16.0      376.
#> 9 Pinus strobus     (14,18]       0          0
```

`stand_estimates` hold the stand table (trees per acre in `trees_bar`) and stock table (volume per acre in `vol_bar`) values. From here, and following Section 8.3, two calls to `pivot_wider()` reshape the output into the desired wide (i.e., messy) table formats. Code to add row and column totals is provided near the end of Section 8.3.

```
# Stand table.
stand_estimates %>%
  pivot_wider(
    id_cols = scientific_name,
    names_from = DBH_4in,
    values_from = trees_bar
  )
```

```
#> # A tibble: 3 x 4
#> # Groups:   scientific_name [3]
#>   scientific_name   `[6,10]` `(10,14]` `(14,18]`
#>   <fct>                <dbl>     <dbl>     <dbl>
#> 1 Abies balsamea        0         8.02       0
#> 2 Betula papyrifera     0         0         16.0
#> 3 Pinus strobus         8.02     16.0        0
```

```
# Stock table.
stand_estimates %>%
  pivot_wider(
    id_cols = scientific_name,
    names_from = DBH_4in,
    values_from = vol_bar
  )
```

```
#> # A tibble: 3 x 4
#> # Groups:   scientific_name [3]
#>   scientific_name   `[6,10]` `(10,14]` `(14,18]`
#>   <fct>                <dbl>     <dbl>     <dbl>
#> 1 Abies balsamea          0       143.         0
#> 2 Betula papyrifera       0         0        563.
#> 3 Pinus strobus         116.      376.         0
```

12.7 Summary

This chapter connected general statistical survey sampling topics covered in Chapter 11 with forestry applications. Key to this connection is the areal sampling frame used to select sampling locations and rules—plot sampling and point sampling—used to select measurement trees at each sampling location. We introduced the tree-centered perspective as a means to better understand tree inclusion zones, the measurement tree selection process, probability of selection, and tree factors. The notion of a tree inclusion zone was also helpful for understanding boundary slopover bias and the need for slope correction. The mirage and walkthrough methods were introduced for correcting boundary slopover bias. Several different approaches for correcting slope effects were also presented.

Much of our focus was on calculating and applying tree factors for plot- and point-level summaries and how these summaries are used to estimate population parameters. We presented all topics in this chapter assuming samples were collected using SRS. Building on this foundation, the remaining chapters develop several more practical and efficient sampling designs and estimators.

Section 12.6 stepped through how to build efficient `tidyverse` workflows for estimation using samples collected via plot and point sampling. Once mastered, `dplyr`, `tidyr`, and related packages provide tools for computation on groups (e.g., via `group_by()`), and enable the creation of simple, elegant, and reusable code for even the most complex estimation tasks.

12.8 Exercises

Exercise 12.1. What is the definition of a tree factor?

Exercise 12.2. For a 1/20-th acre circular inventory plot, answer the following questions:

a. What is the area of the plot in square feet?

b. What is the radius of the plot, in feet?

c. What is the tree factor for a tree measured on this plot?

Exercise 12.3. For a circular inventory plot with an area of 300 m^2, answer the following questions:

a. What is the radius of the plot, in meters?

b. What fraction of a hectare does this plot cover?

c. What is the tree factor for a tree measured on this plot?

Exercise 12.4. What is the relationship between BAF and a tree's inclusion zone area? How does this relationship affect the number of "in" (i.e., measurement) trees at a given point? No math or code needed to answer this question.

Exercise 12.5. What is the horizontal distance multiplier (HDM) and how is it used?

Exercise 12.6. Given an English 20 BAF and using the HDMs provided in Table 12.6, determine whether the following trees are "in" or "out"—that is, whether they should be measured or not.

a. 10-inch DBH tree, 20 feet from the point.

b. 10-inch DBH tree, 19 feet from the point.

c. 60-inch DBH tree, 110 feet from the point.

d. 20-inch DBH tree, 40 feet from the point.

Exercise 12.7. Given a metric 4 BAF and using the HDMs provided in Table 12.7, determine whether the following trees are "in" or "out"—that is, whether they should be measured or not.

a. 25-cm DBH tree, 6 meters from the point.

b. 25-cm DBH tree, 7 meters from the point.

c. 150-cm DBH tree, 35 meters from the point.

d. 50-cm DBH tree, 84 meters from the point.

Exercise 12.8. Using the cruise data given in Table 12.14, answer the following questions using the three measurement trees on Plot 1. Include the units as a comment for each answer.

a. What is each measurement tree's tree factor?

b. Compute the plot-level trees per hectare.

c. Compute each tree's BA.

d. Compute each tree's BA per hectare.

TABLE 12.14: Cruise data for $n = 2$ plots. Measurements were taken on fixed-area circles plots with a 9.77 (m) radius.

Plot	Tree	DBH (cm)	Volume (m^3)
1	1	41	4
1	2	58	10
1	3	30	2
2	1	46	6
2	2	71	15

e. Compute the plot-level basal area per hectare.

f. Compute each tree's volume per hectare.

g. Compute the plot-level volume per hectare.

Exercise 12.9. Recognize the three plot-level summaries computed in Exercise 12.8(b,e,g) (i.e., per-hectare number of trees, basal area, and volume) represent one set of observations from one plot. Such plot-level summaries must be computed for each of n plots. The n plot-level summaries are the sample from which you compute the mean, standard error of the mean, and confidence interval forest-level estimates.

Using the cruise data given in Table 12.14, answer the following questions. Include the units as a comment for each answer.

a. Compute the plot-level summary for trees per hectare, basal area per hectare, and volume per hectare (hint, you already did this for Plot 1 in Exercise 12.8(b,e,g)). You'll end up with six values (i.e., number of trees, basal area, and volume for each of the two plots).

b. Use your plot-level summaries computed in Exercise 12.9(a) to compute a forest-level estimate for per-hectare number of trees, basal area, and volume. Your forest-level estimate for each is the mean ($\bar{y}$) over your plot-level summaries.

c. Use your plot-level summaries computed in Exercise 12.9(a) to compute the standard error of the mean ($s_{\bar{y}}$) for your forest-level estimate for per-hectare number of trees, basal area, and volume computed in Exercise 12.9(b). These standard errors can be used to compute a confidence interval for each forest-level estimate.

Exercise 12.10. Using the cruise data given in Table 12.15, answer the following questions using the three measurement trees at Point 1. Include the units as a comment for each answer.

a. What is the basal area per acre that each tree represents? No calculations or code needed here.

TABLE 12.15: Point sampling cruise data for $n = 2$ points with measurement trees idenfied using an English 20 BAF.

Point	Tree	DBH (in)	Volume (ft^3)
1	1	16	46
1	2	23	103
1	3	12	23
2	1	18	60
2	2	28	159

b. Compute the point-level basal area per acre.

c. Compute each tree's BA.

d. Using BA from 12.10(c) and equation (12.22), compute each tree's tree factor.

e. Compute the point-level trees per acre.

f. Compute each tree's volume per acre.

g. Compute the point-level volume per acre.

Exercise 12.11. Recognize the three point-level summaries computed in Exercise 12.10(b,e,g) (i.e., per acre basal area, number of trees, and volume) represent one set of observations from one point. Such point-level summaries must be computed for each of n points. The n point-level summaries are the sample from which you compute the mean, standard error of the mean, and confidence interval forest-level estimates.

Using the cruise data given in Table 12.15, answer the following questions. Include the units as a comment.

a. Compute the point-level summary for basal area per acre, trees per acre, and volume per acre (hint, you already did this for Point 1 in Exercise 12.10(b,e,g)). You'll end up with six values (i.e., basal area, number of trees, and volume for each of the two points).

b. Use your point-level summaries computed in Exercise 12.11(a) to compute a forest-level estimate for per acre basal area, number of trees, and volume. Your forest-level estimate for each is the mean ($\bar{y}$) over your point-level summaries.

c. Use your point-level summaries computed in Exercise 12.11(a) to compute the standard error of the mean ($s_{\bar{y}}$) for your forest-level estimate for per acre basal area, number of trees, and volume computed in Exercise 12.11(b). These standard errors can be used to compute a confidence interval for each forest-level estimate.

Exercise 12.12. Toward the end of Section 12.3.4, we outlined the *radius adjustment* approach to slope correction for fixed-area circular plots. Table 12.16 holds slope measured

TABLE 12.16: Slope and scaling factor r^* used to implement radius adjustment approach to slope correction for fixed-area circular plots.

Slope (%)	Slope (°)	r*	Slope (%)	Slope (°)	r*
0	0	1	80	38.7	1.132
10	5.71	1.002	90	42.0	1.160
20	11.3	1.010	100	45	1.189
30	16.7	1.022	110	47.7	1.219
40	21.8	1.038	120	50.2	1.250
50	26.6	1.057	130	52.4	1.281
60	31.0	1.080	140	54.5	1.312
70	35.0	1.105	150	56.3	1.343

in percent and degrees, and the associated radius scaling factor r^*. As described in Section 12.3.4, the scaling factor r^* is multiplied by the horizontal plot radius R (i.e., plot radius used in the cruise) to give the adjusted plot radius R^* (i.e., plot radius used on the slope).

Your task is to write code to reproduce r^* values in Table 12.16. Recall R's trigonometric functions assume angles are in radians, so we also provide several utility functions for converting among slope degrees, percents, and radians. Given the sequence of slopes (either percentages or degrees), your code to compute r^* should be a simple R math expression.

Include your code and the resulting r^* values as output.

```
# Some useful angle conversion functions.
r2d <- function(r) {r * 180 / pi}          # Radian to degree.
d2r <- function(d) {d * pi / 180}          # Degree to radian.
p2r <- function(p) {atan(p / 100)}         # Percent to radian.
p2d <- function(p) {r2d(atan(p / 100))} # Percent to degree.

# To get you started, we've set up the vectors of slopes below.
p_slope <- seq(from = 0, to = 150, by = 10) # slope percent.
d_slope <- p2d(p_slope) # slope degrees.
r_slope <- p2r(p_slope) # slope radians.
```

Exercise 12.13. Say you conducted a cruise using 1/20-th acre circular plots. Table 12.17 holds information about the slope on a few plots and distances from plot center to possible measurement trees. Your task is to use the radius adjustment slope correction approach to fill in missing information in Table 12.17. Specifically, given the slope, determine the scale factor (r^*) from Table 12.16. Then compute the slope adjusted plot radius (R^*) and make a determination about tree measurement status, i.e., is the tree "in" or "out" of the plot.

TABLE 12.17: Trees measured on different plot slopes. Distance from the plot center to the tree's center is provided and status is whether the tree is in or out of the 1/20-th acre plot.

Tree	Slope (%)	r^*	R^*	Distance (ft)	Status
1	30	1.022	26.91	26.5	In
2	60	1.080	28.44	29.0	Out
3	110			30.0	
4	40			27.0	
5	90			31.0	

Hint, you'll first need to compute the plot radius for the 1/20-th acre plot. The first two rows of Table 12.17 are filled in for you.

12.8.1 Penobscot Experimental Forest inventory data

The next set of exercises use the Penobscot Experimental Forest (PEF) inventory data introduced in Section 1.2.4. These data were collected as part of several long-term USDA Forest Service silvicultural studies described in Kenefic et al. (2025). The data comprise repeated measurements on fixed-area permanent sample plots (PSPs) located within management units (MUs), see Section 1.4 and Figure 1.5. The code below reads the PEF tree measurements into `pef_trees` and then takes a glimpse at the data.

```
pef_trees <- read_csv("datasets/PEF/PEF_trees.csv")
pef_trees %>% glimpse()
```

```
#> Rows: 316,837
#> Columns: 10
#> $ MU              <chr> "12", "12", "12", "12", "12"~
#> $ plot            <dbl> 11, 11, 11, 11, 11, 11, 11, ~
#> $ inv             <dbl> 17, 17, 17, 17, 17, 17, 17, ~
#> $ year            <dbl> 2014, 2014, 2014, 2014, 2014~
#> $ month           <dbl> 6, 6, 6, 6, 6, 6, 6, 6, 6, 6~
#> $ PEF_species_code <dbl> 4, 4, 7, 15, 6, 15, 16, 1, 1~
#> $ DBH_in          <dbl> 6.9, 6.4, 6.0, 6.2, 16.8, 5.~
#> $ AGB_lbs         <dbl> 219.693, 184.341, 135.754, 2~
#> $ exp_ac          <dbl> 5, 5, 5, 5, 5, 5, 5, 5, 5, 5~
#> $ study           <chr> "CMS", "CMS", "CMS", "CMS", ~
```

Rows in `pef_trees` correspond to measurement trees and columns are defined as follows.

1. `MU` management unit within which the tree was sampled. Management unit boundaries are delineated using orange polygons in Figure 1.5. MUs receive different silvicultural treatments and belong to one of two studies (see the `study` column definition below).

2. `plot` permanent sample plot number. Plot number is unique within MU, but not across MUs (i.e., the same plot number could occur in multiple MUs).
3. `inv` sequential inventory number. For a given `MU` and `plot` combination, the smallest and largest `inv` values correspond to the earliest and most current measurement, respectively. An inventory is a complete measurement of all plots within the given MU.
4. `year` year of measurement.
5. `month` month of measurement.
6. `PEF_species_code` species code used on the PEF. Common and scientific names for each code are given in `"datasets/PEF/PEF_species_codes.csv"`.
7. `DBH_in` tree DBH in inches.
8. `AGB_lbs` tree aboveground biomass (AGB) in pounds computed using DBH and species specific equations defined in Jenkins et al. (2002).
9. `exp_ac` tree expansion factor (TF) in trees per acre. As described in Section 12.3.1, `exp_ac` is the inverse of the plot area upon which the tree was measured.
10. `study` the silvicultural study to which the tree belongs. MUs, and hence trees, belong to either the Compartment Management Study (`CMS`) or Management Intensity Demonstration (`MID`).

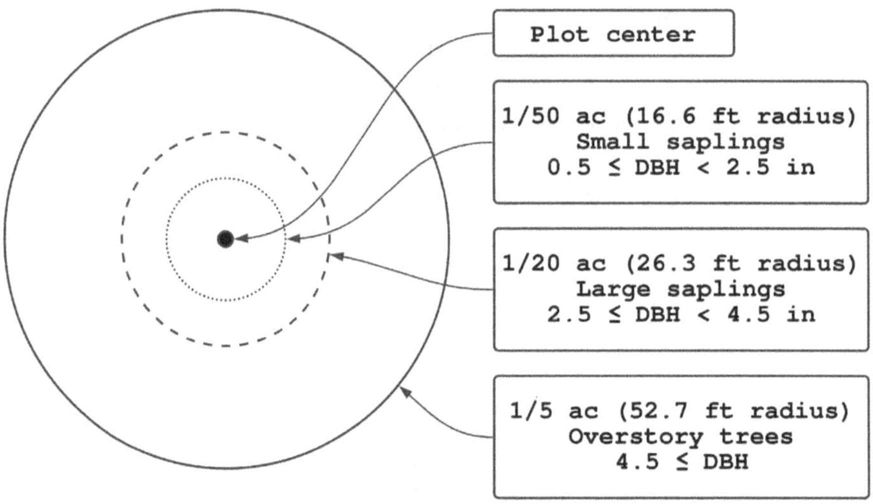

FIGURE 12.12: Penobscot Experimental Forest permanent sample plot (PSP) nested design and tree DBH measurement protocol.

Each plot depicted in Figure 1.5 was measured using the design shown in Figure 12.12.[34] As noted in Section 12.3, and illustrated using the toy inventory dataset in Section 12.6, a nested plot design that uses smaller plot size to sample smaller diameter trees reduces the time needed to conduct an inventory.

[34]This design, detailed in Waskiewicz et al. (2015), was only consistently applied to recent PEF data. If you dig into the `pef_trees` data, you'll find that prior to 2000 many tree measurements don't follow the DBH cutoffs given in Figure 12.12.

The `pef_trees` subset created in Exercises 12.14 and 12.15 is used in Exercises 12.16 through 12.18. For several exercises, we provide the `dplyr` workflow output and it's your job to write code to reproduce the output. As always, use the pipe operator when possible.

Exercise 12.14. Consider the code below used to subset `pef_trees` to form `pef_current_trees`, which includes only the most current measurement in each plot, then answer the following questions. Hint, review the `MU`, `plot`, and `inv` column definitions.

```
pef_trees_current <- pef_trees %>%
  group_by(MU, plot) %>%
  filter(inv == max(inv)) %>%
  ungroup()
```

a. Why are `MU` and `plot` in `group_by()`?

b. How does the grouped `filter(inv == max(inv))` result in each MU's most current inventory?

c. How is the tibble output from `filter()` grouped (i.e., before it's piped to `ungroup()`)?

d. What is the effect of `ungroup()` at the end of the piped workflow?

e. Why might it be a good idea to remove the grouping structure on `pef_current_trees` (i.e., why did we feel the need to add `ungroup()` to the workflow)?

Exercise 12.15. Subset `pef_trees_current`, created in Exercise 12.14, to include only trees in the MID study and name your resulting tibble `mid_trees`. Hint, following the column definitions, you'll use `study == "MID"` in your `filter()` call. Your `mid_trees` should match the output below.

```
mid_trees %>% print()
```

```
#> # A tibble: 2,496 x 10
#>    MU     plot   inv  year month PEF_species_code
#>    <chr> <dbl> <dbl> <dbl> <dbl>            <dbl>
#>  1 90       11    25  2011     5                4
#>  2 90       11    25  2011     5                4
#>  3 90       11    25  2011     5                4
#>  4 90       11    25  2011     5                6
#>  5 90       11    25  2011     5                6
#>  6 90       11    25  2011     5                4
#>  7 90       11    25  2011     5                4
#>  8 90       11    25  2011     5                6
```

```
#>  9 90        11    25  2011     5               6
#> 10 90        11    25  2011     5               6
#> # i 2,486 more rows
#> # i 4 more variables: DBH_in <dbl>, AGB_lbs <dbl>,
#> #   exp_ac <dbl>, study <chr>
```

To provide some context for subsequent exercises, we used the tree measurements held in `mid_trees` to create Figure 12.13. The subsequent exercises explore these data and generate plot- and MU-level biomass estimates.

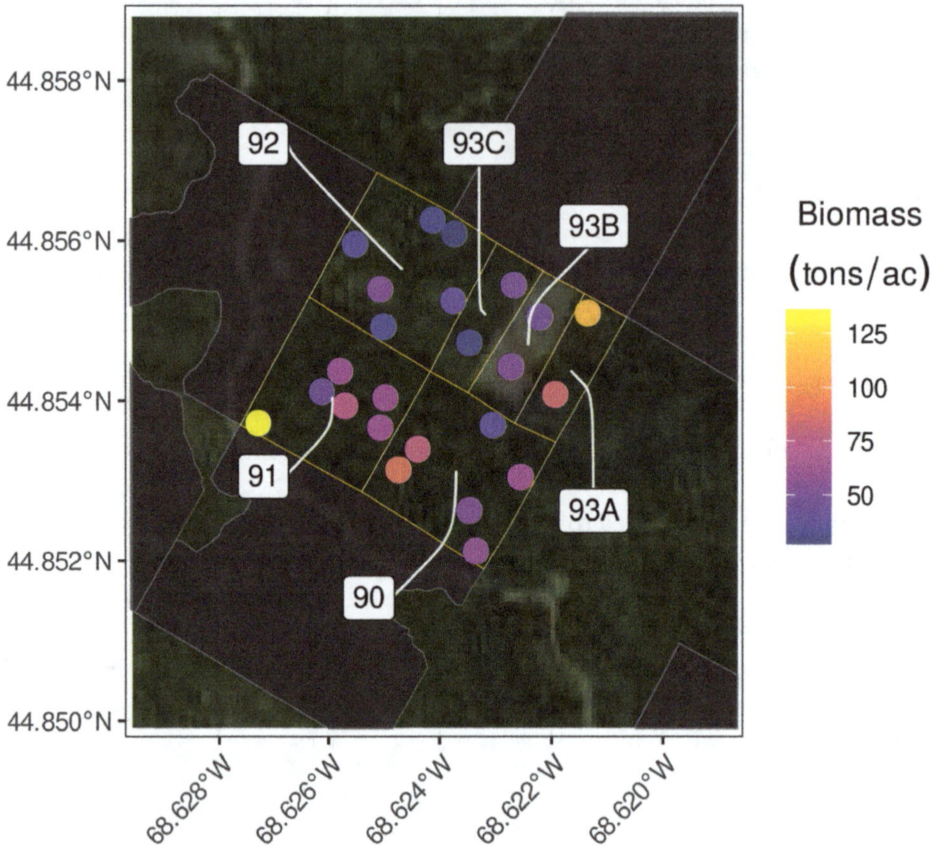

FIGURE 12.13: Penobscot Experimental Forest (PEF) management units (MUs) that are part of the Management Intensity Demonstration (MID) study are delineated in orange and labeled. Other PEF MUs are shaded gray. MID MUs are positioned near the middle of the PEF extent shown in Figure 1.5. MID permanent sample plot locations are colored by the most recent inventory's total biomass per-acre estimate for overstory trees.

Exercise 12.16. Using `mid_trees`, write code to reproduce the answer to each question below.

a. How many trees were measured across all plots? Hint, we used `nrow()`.

```
#> [1] 2496
```

b. What are the distinct MU values? Hint, consider `distinct()`.

```
#> # A tibble: 6 x 1
#>   MU
#>   <chr>
#> 1 90
#> 2 91
#> 3 92
#> 4 93A
#> 5 93B
#> 6 93C
```

c. What year was each MU inventoried? Hint, our code uses `group_by()`, `summarize()`, and `first()` to create the `inventory_year` output column.

```
#> # A tibble: 6 x 2
#>   MU    inventory_year
#>   <chr>          <dbl>
#> 1 90              2011
#> 2 91              2011
#> 3 92              2011
#> 4 93A             2010
#> 5 93B             2010
#> 6 93C             2010
```

d. How many plots in each MU? Hint, our code uses `group_by()`, `summarize()`, and `n_distinct()` to create the `n_plots` output column.

```
#> # A tibble: 6 x 2
#>   MU    n_plots
#>   <chr>   <int>
#> 1 90          6
#> 2 91          6
#> 3 92          6
#> 4 93A         2
#> 5 93B         2
#> 6 93C         2
```

e. How many unique plots were measured across all MUs? Hint, our code uses `summarize()` and `n_distinct()` to create the `n_plot` output column.

```
#> # A tibble: 1 x 1
#>   n_plots
#>     <int>
#> 1      24
```

f. Which three plots have the most trees measured? Hint, our code uses `group_by()`, `summarize()` with the argument `.groups = "drop"`, and `slice_max()` with the argument `n = 3`.

```
#> # A tibble: 3 x 3
```

```
#>    MU      plot n_trees
#>    <chr> <dbl>   <int>
#> 1 93C       12     214
#> 2 93B       12     199
#> 3 92        11     160
```

Exercise 12.17. As noted in the column definitions, `exp_ac` is the tree expansion factor (i.e., tree factor) as defined in Section 12.3.1. Use `mid_trees` to answer the following questions about the expansion factor.

a. Print the distinct values of `exp_ac`. Are they consistent with the plot areas given in Figure 12.12?

b. What is the interpretation of the `exp_ac` values? What determines a tree's expansion factor?

c. For each `exp_ac` value, what are the minimum and maximum DBH values? Write the code to reproduce our output below. Hint, we used `group_by()` and `summarize()` to create the `min_DBH_in` and `max_DBH_in` output columns. Are the DBH ranges for each expansion factor consistent with the sampling protocol given in Figure 12.12?

```
#> # A tibble: 3 x 3
#>    exp_ac min_DBH_in max_DBH_in
#>     <dbl>      <dbl>      <dbl>
#> 1       5        4.5       31.1
#> 2      20        2.5        4.4
#> 3      50        0.5        2.4
```

Exercise 12.18. This next series of questions results in plot-level summaries then MU-level estimates for per-acre number of trees, basal area, and aboveground biomass (AGB). We'll limit our estimates to overstory trees in `mid_trees` (i.e., trees measured on the 1/5-th ac plot depicted in Figure 12.12).

After question (a) below, that filters only overstory trees, this exercise follows the estimation steps laid out in Section 12.6; however, it breaks Step 1 into two parts. The first part, question (b), expands each tree's variable value to the per-unit-area basis. The second part, question (c), sums each tree's values computed in question (b) to arrive at the plot-level summaries. In practice, we can combine these two parts, as we did in Section 12.6.

a. Filter `mid_trees` to include only those trees measured on the 1/5-th ac plots and call the resulting tibble `mid_overstory_trees`. Your `mid_overstory_trees` should match the output below. Use `mid_overstory_trees` for questions b-e.

```
mid_overstory_trees %>% glimpse()
```

```
#> Rows: 1,366
#> Columns: 10
```

```
#> $ MU               <chr> "90", "90", "90", "90", "90"~
#> $ plot             <dbl> 11, 11, 11, 11, 11, 11, 11, ~
#> $ inv              <dbl> 25, 25, 25, 25, 25, 25, 25, ~
#> $ year             <dbl> 2011, 2011, 2011, 2011, 2011~
#> $ month            <dbl> 5, 5, 5, 5, 5, 5, 5, 5, 5, 5~
#> $ PEF_species_code <dbl> 6, 6, 4, 6, 6, 4, 6, 4, 4, 4~
#> $ DBH_in           <dbl> 13.9, 9.3, 16.1, 7.1, 17.6, ~
#> $ AGB_lbs          <dbl> 1207.12, 445.32, 1585.07, 22~
#> $ exp_ac           <dbl> 5, 5, 5, 5, 5, 5, 5, 5, 5, 5~
#> $ study            <chr> "MID", "MID", "MID", "MID", ~
```

b. Add the following new columns to `mid_overstory_trees`: 1) `trees_per_ac` is the number of trees per acre each tree represents; 2) `BA_sqft_per_ac` is the basal area (ft^2/ac) each tree represents; 3) `AGB_tons_per_ac` is the aboveground biomass (tons/ac) each tree represents. Note you'll need to scale `AGB_lbs` by 1/2000 to convert from pounds to tons. Hint, our code uses `mutate()`.

```
# Take a glimpse of only a few columns so they fit on the page.
mid_overstory_trees %>%
  select(MU, plot, trees_per_ac,
         BA_sqft_per_ac, AGB_tons_per_ac) %>%
  glimpse()
```

```
#> Rows: 1,366
#> Columns: 5
#> $ MU              <chr> "90", "90", "90", "90", "90",~
#> $ plot            <dbl> 11, 11, 11, 11, 11, 11, 11, 1~
#> $ trees_per_ac    <dbl> 5, 5, 5, 5, 5, 5, 5, 5, 5, 5,~
#> $ BA_sqft_per_ac  <dbl> 5.2688, 2.3586, 7.0687, 1.374~
#> $ AGB_tons_per_ac <dbl> 3.01781, 1.11329, 3.96267, 0.~
```

c. Compute the plot-level summaries by summing each plot's values for `trees_per_ac`, `BA_sqft_per_ac`, and `AGB_tons_per_ac`. Call the resulting tibble `plot_summary`. We append `_i` to the resulting variables to indicate that they correspond to the i-th plot. Hint, our code uses `group_by()` and `summarize()`.[35]

```
plot_summary %>% glimpse()
```

```
#> Rows: 24
#> Columns: 5
#> Groups: MU [6]
#> $ MU      <chr> "90", "90", "90", "90", "90", "90", "~
#> $ plot    <dbl> 11, 12, 21, 31, 51, 52, 11, 21, 31, 4~
```

[35]As an aside, plot color in Figure 12.13 corresponds to the `AGB_i` value in your resulting `plot_summary` tibble.

```
#> $ trees_i <dbl> 150, 155, 160, 250, 225, 175, 235, 28~
#> $ BA_i    <dbl> 109.483, 114.765, 98.615, 78.641, 160~
#> $ AGB_i   <dbl> 58.337, 62.491, 53.709, 37.326, 89.59~
```

d. Use the plot-level summaries for MUs 90, 91, and 92 in `plot_summary` from question (c) to estimate MU-level mean, standard error of the mean, and 95% confidence intervals for number of trees, basal area, and AGB per acre. This is Step 2 in the estimation process described in Section 12.6.
In your calculations, use the standard error given in (11.24). Call your resulting tibble `MU_estimates`. Hint, our code uses only `filter()` and `summarize()`.

```
MU_estimates %>% glimpse()
```

```
#> Rows: 3
#> Columns: 15
#> $ MU             <chr> "90", "91", "92"
#> $ n              <int> 6, 6, 6
#> $ t              <dbl> 2.5706, 2.5706, 2.5706
#> $ trees_bar      <dbl> 185.83, 263.33, 247.50
#> $ s_trees_bar    <dbl> 17.001, 36.370, 22.721
#> $ ci_lower_trees <dbl> 142.13, 169.84, 189.09
#> $ ci_upper_trees <dbl> 229.54, 356.83, 305.91
#> $ y_bar_BA       <dbl> 117.412, 138.905, 78.939
#> $ s_BA_bar       <dbl> 12.1158, 22.9370, 7.3206
#> $ ci_lower_BA    <dbl> 86.267, 79.943, 60.121
#> $ ci_upper_BA    <dbl> 148.556, 197.866, 97.757
#> $ AGB_bar        <dbl> 63.425, 73.671, 36.802
#> $ s_AGB_bar      <dbl> 7.5995, 13.0970, 3.7803
#> $ ci_lower_AGB   <dbl> 43.889, 40.004, 27.084
#> $ ci_upper_AGB   <dbl> 82.960, 107.338, 46.519
```

e. What was the sample size used to compute your MU-level estimates in question (d)? Do you think these sample sizes are large enough to provide robust estimates?

f. Use the MU-level estimates in `MU_estimates` from question (d) to compute MU-level total with associated 95% confidence intervals for AGB.[36] MU 90, 91, and 92 have an area of 9.92, 9.76, and 9.92 acres, respectively. Break your workflow into three steps. First, make a tibble called `mu_acres` with columns `MU` and `A_ac`, where `A_ac` holds the acres for each MU. Second, join `mu_acres` to `MU_estimates`.[37] Third, use `mutate()` or `transmute()` on `MU_estimates` to create a tibble that holds AGB estimated total and associated confidence

[36]We could also compute the totals for number of trees and basal area; however, in practice we typically talk about these variables on a per-unit-area basis.

[37]We used `case_when()` to accomplish these first two steps in Section 12.6.1. We're using the join approach here for demonstration—it's a better approach in practice when you're dealing with many units that need values assigned.

intervals. The output from these three steps is given below.
Step 1: Create the `mu_acres` tibble.

```
mu_acres %>% print()
```

```
#> # A tibble: 3 x 2
#>   MU    A_ac
#>   <chr> <dbl>
#> 1 90     9.92
#> 2 91     9.76
#> 3 92     9.92
```

Step 2: Join `mu_acres` to `MU_estimates`.
Step 3: Use `MU_estimates` and `transmute()` to estimate AGB total with associated confidence interval. Name the resulting tibble `MU_tot_estimates`. Hint, scale the per-unit-area estimates by their respective MU acres.

```
MU_tot_estimates %>% glimpse()
```

```
#> Rows: 3
#> Columns: 4
#> $ MU           <chr> "90", "91", "92"
#> $ AGB_hat      <dbl> 629.17, 719.03, 365.07
#> $ ci_lower_AGB <dbl> 435.38, 390.44, 268.68
#> $ ci_upper_AGB <dbl> 822.96, 1047.62, 461.47
```

13

Sampling designs and estimation in forest inventory

In Chapters 11 and 12, we focused on inference using sample data collected under a SRS design. In subsequent chapters, we step beyond SRS and introduce several additional *probability sampling* designs commonly used in forestry. Recall that a sampling design is a set of rules for selecting units from a population and for using the resulting data to generate statistically valid estimates for parameters of interest. Our focus is on four main designs that are treated in detail in the chapters that follow: 1) systematic, 2) stratified, 3) cluster sampling, and 4) multistage sampling. These designs differ in how they configure and select observations and in the estimators they prescribe. In most cases, a design is chosen to improve efficiency from a statistical, logistical, or combined standpoint.

In addition to these sampling designs, we also introduce estimation methods that incorporate auxiliary information. These methods—such as ratio and regression estimation—are not sampling designs themselves. Instead, they provide ways to improve precision by combining the sample data with information from additional variables available for some or all units in the population. These methods appear repeatedly in later chapters and are used in conjunction with multiple designs.

As discussed in Section 12.1, forest inventory typically relies on an areal sampling frame, where sampling locations are identified by placing points within the frame's spatial extent. Under plot sampling, a location might mark the center of a circular plot, a corner of a rectangular plot, or an anchor point for more complex plot shapes. Under point sampling, a location might serve as the point from which a forester projects the discerning angle to identify measurement trees. Unless stated otherwise, the sampling designs and estimators developed in subsequent chapters can be used with either plot or point sampling—that is, the design governs the selection of sampling locations and associated estimators, while a separate set of rules determines how trees are selected around each location.

As in other mensuration texts (Kershaw et al. 2016; Burkhart et al. 2018), we present the estimators associated with each design. The text by Iles (2003) is especially valuable for readers seeking practical guidance: it expands on these methods, discusses their implementation in the field, and provides real-world perspectives and advancements. Our added contribution is to show how these estimators can be implemented using `tidyverse` tools and other R packages, while also building intuition for their application. Emphasis is placed on developing computing workflows that support efficient and repeatable analysis, which are carried forward throughout the design-specific chapters that follow.

DOI: 10.1201/9781003222699-13

13.1 A gentle introduction to `sf`

Because sampling in forestry is inherently spatial, it's helpful to work with data structures that explicitly represent geographic location and geometry. In R, the primary tool for handling spatial vector data is the `sf` package.

If you've used GIS software before, many of the underlying concepts will feel familiar. Spatial vector data come in three main types: *points* (e.g., sampling locations), *lines* (e.g., transects or travel paths), and *polygons* (e.g., stands, management units, or forest extents). The `sf` package represents each of these as a special kind of data frame with an attached geometry column that stores the spatial information.

You can work with `sf` objects much like regular data frames, using familiar `tidyverse` tools such as `filter()`, `select()`, `mutate()`, and `summarize()`. Throughout the chapters that follow, we'll introduce key `sf` data structures and functions as needed. For additional background, see the `sf` documentation at https://r-spatial.github.io/sf or the package vignette via `vignette("sf1")`.

14

Systematic sampling

Systematic sampling (SYS) is the most widely used probability sampling design in forest inventory. In SYS, sampling locations are arranged in a fixed pattern across the population. To introduce randomness and ensure unbiased estimation of means and totals, the pattern—typically a rectangular grid—is positioned using a randomly chosen starting point. As illustrated in Figure 1.2, sampling locations are usually aligned along straight lines. The distance between locations on a line, known as the *sampling interval*, and the distance between lines can differ, creating a rectangular grid. This is commonly referred to as *line-plot sampling*, which is our focus here.

SYS designs are popular in forestry for several reasons. Compared to SRS, their advantages include the following.

1. Sampling locations are evenly distributed across the population, improving spatial coverage and typically yielding more precise (lower-variance) mean and total estimates when the population shows spatial structure or broad trends. When population values show little or no spatial structure, the two designs tend to provide similar precision.

2. Sampling locations are easier and faster to find, reducing cost. The cruiser can follow straight lines between locations, minimizing travel time and simplifying relocation during remeasurement.

The grid is typically positioned using GIS or spatial software (e.g., R), and the resulting points are uploaded to a GPS device for navigation. Alternatively, the cruiser may locate the first sampling point in the field and lay out the remainder using the sampling interval and grid orientation.

Key considerations when implementing a SYS design include:

1. the pattern and orientation of the grid;
2. the method used to randomly position the grid.

These points are covered in the next few sections.

DOI: 10.1201/9781003222699-14

14.1 Grid pattern, number of grid points, and orientation

In practice, either a rectangular or square grid is used. A rectangular grid defines the spacing between sampling locations along a line, denoted D_p (with the subscript p reminding us this spacing is for "plots" or "points"), and the spacing between lines, D_l. A square grid is a special case where $D_{\text{sq}} = D_p = D_l$. Unless there's a compelling reason to use unequal spacing, a square grid is a reasonable and efficient default.

Given a sample size n and total area A, the spacing between locations in a square grid is given by

$$D_{\text{sq}} = \sqrt{\frac{cA}{n}}, \tag{14.1}$$

where c is a conversion factor to express area in square distance units. For example, use $c = 10{,}000\ \text{m}^2/\text{ha}$ if A is in hectares and D_{sq} is in meters, or $c = 43{,}560\ \text{ft}^2/\text{ac}$ if A is in acres and D_{sq} is in feet.

As introduced in Section 11.3.6, sample size n should be based on the level of precision needed to meet the survey's objectives. However, in practice, it's common for experienced foresters to adopt a sampling intensity that works well for their forest type and inferential objectives. For example, the forester who conducted the Elk County cruise described in Section 1.2.2 typically uses a seven-by-seven chain SYS grid.

If you do use a sample size calculation, as recommended in Section 11.3.6, it's likely that the number of grid points won't exactly match the computed n due to how the grid falls within the boundary. In such cases, we agree with the following sage words.

> My advice is simply to take the number of points that are indicated [by the grid], and live with the slight variability. Sample sizes that are "needed" ... are just estimated anyway.
>
> — Kim Iles

If you really must have a specific sample size, then a good approach is to specify a denser grid than needed and randomly or systematically thin points until the target sample size is reached (Iles 2003).

The grid orientation can be adjusted to improve field logistics. For example, aligning lines with a forest boundary may simplify navigation. This was the approach used in the Elk County cruise (Figure 1.2).

If a known gradient exists in the population—for example, productivity change with elevation—aligning the grid to cross the gradient usually helps ensure the sample spans the range of values, which often improves precision. This benefit arises because the grid forces broad spatial patterns to be represented in the sample, patterns that might be missed when locations are placed at random.

Although rare, it's possible for a population to exhibit a systematic pattern that aligns with the grid spacing and orientation. For example, a grid aligned with plantation rows might consistently place sampling locations between rows, giving an incomplete view of the population. Rotating the grid to avoid alignment with such periodic features can prevent this problem.

14.2 Positioning the grid

To avoid bias, the grid must be randomly placed. This is typically done by selecting a *random point start* within the areal sampling frame and extending the grid in each direction from that point. If rotation is applied, it should be performed about this random point start.

The random point start can be one of the sampling locations or used as an anchor point from which to extend the grid. In the development that follows, we treat the starting point as a location on the sampling grid.

In the next section, we introduce a function that generates systematic sampling grids in R. This function allows users to specify grid spacing, orientation, and how sampling locations are ordered. The code and examples also help highlight some practical considerations when laying out systematic designs.

14.3 Creating a systematic sampling grid with `sys_grid()`

This section is presented in the style of a condensed R help page to document the `sys_grid()` function. For a deeper understanding, you're encouraged to read through the function source code in `"scripts/utils.R"` (see Section 5.4). The code comments walk through each step, and many of the elements should already be familiar.

Description

Creates a systematic grid of points within a polygonal boundary using a random start. The grid is defined by a fixed spacing between lines (`D_l`) and between points along each line (`D_p`). The grid can be rotated and the points ordered from a specified origin to facilitate field navigation.

Usage

```
sys_grid(boundary, D_l, D_p, angle = 0, origin = "se")
```

Arguments

- `boundary`: An `sf` polygon object defining the sampling boundary. Must be projected with linear units.
- `D_l`: Distance between sampling lines (x-direction) in the same units as `boundary`.
- `D_p`: Distance between points along a line (y-direction) in the same units as `boundary`.
- `angle`: Optional rotation angle (in degrees) applied to the grid. Default is `0`.
- `origin`: Optional character string indicating the origin corner used to order points. One of `"se"`, `"sw"`, `"ne"`, or `"nw"` (default is `"se"`).

Details

The `sys_grid()` function creates a rectangular grid of points with spacing determined by `D_l` and `D_p`, starting from a randomly placed point within the boundary polygon. Here's how it works.

1. A random point start is placed within the `boundary`.
2. A full grid is generated around that point, extended enough to accommodate possible rotation.
3. Points are ordered in a snaking pattern from the specified `origin`.
4. If specified, the grid is rotated by `angle` degrees about the random point start.
5. The grid is clipped to the polygon, and the resulting points are labeled and returned.

Ordering by `origin` can help plan field navigation, especially when following lines. The choice of origin affects how `line_id`, `point_id_in_line`, and `point_id` are assigned, but not the design itself.

Value

Returns a list with the following components.[1]

- `grid`: An `sf` point object with columns:
  - `point_id`: unique ID, ordered from the origin;
  - `line_id`: identifier for each line;
  - `point_id_in_line`: position along each line.
  - All columns in `boundary`.
- `path`: An `sf` `LINESTRING` connecting the points in sampling order.
- `random_point_start`: The initial random point start (an `sf` point object).
- `boundary`: The input polygon(s) used to clip the grid.

---

[1]Any of these objects can be saved to a file using `st_write()` from the `sf` package to be read by GIS or GPS software.

- type: A string used internally to verify the object when plotting.

Use plot_sys_grid() to visualize the resulting grid and path.

14.4 Systematic sampling grid illustration with sys_grid()

Let's apply sys_grid() to set up a SYS design for the Elk County cruise described in Section 1.2.2. In that cruise, the forester used a square grid with sampling locations spaced every seven chains (recall that one chain equals 66 feet).

The sys_grid() function expects the boundary polygon to be in a projection that uses the same distance units as D_l and D_p. The Elk County property is stored in the "datasets/PA/Elk_bounds" ESRI shapefile (ESRI 1998), which uses a geographic coordinate system (i.e., decimal degrees of latitude and longitude)—which isn't an intuitive measurement system for field work. Instead, we want the boundary in a projected coordinate system that allows us to specify D_l and D_p in feet or meters.

After reading the shapefile using st_read(), the code below reprojects the boundary to a projection appropriate for northern Pennsylvania and consistent with the cruise protocol. This is done using st_transform().[2]

```
# Read in the bounding polygon and reproject it.
bounds <- st_read("datasets/PA", "Elk_bounds", quiet = TRUE)
bounds <- st_transform(bounds, crs = st_crs(2271))
```

Once reprojected, bounds has units in feet, so we specify D_l and D_p in feet when calling sys_grid(). Because we don't specify values for angle and origin, the defaults of 0 and "se", respectively, are used. To ensure the same random point start is generated each time the code is run, we set the random seed prior to calling the function.

```
# Set random seed.
set.seed(1)

sys <- sys_grid(bounds, D_l = 7 * 66, D_p = 7 * 66)
```

We can visualize the resulting SYS design by passing the sys object to the plot_sys_grid() function, defined in the "scripts/utils.R" script. The resulting map is shown in Figure 14.1. We encourage you to explore the plot_sys_grid() function to see how it builds the graphic using ggplot(). Compared to the simpler examples in Chapter 9, this plot is a bit more involved, with a customized legend and multiple spatial layers.[3]

[2]The specific projection is NAD83 / Pennsylvania South (ft), which corresponds to EPSG:2271.

[3]In plot_sys_grid(), notice how the geom_sf() layers automatically render each sf object based on its geom-

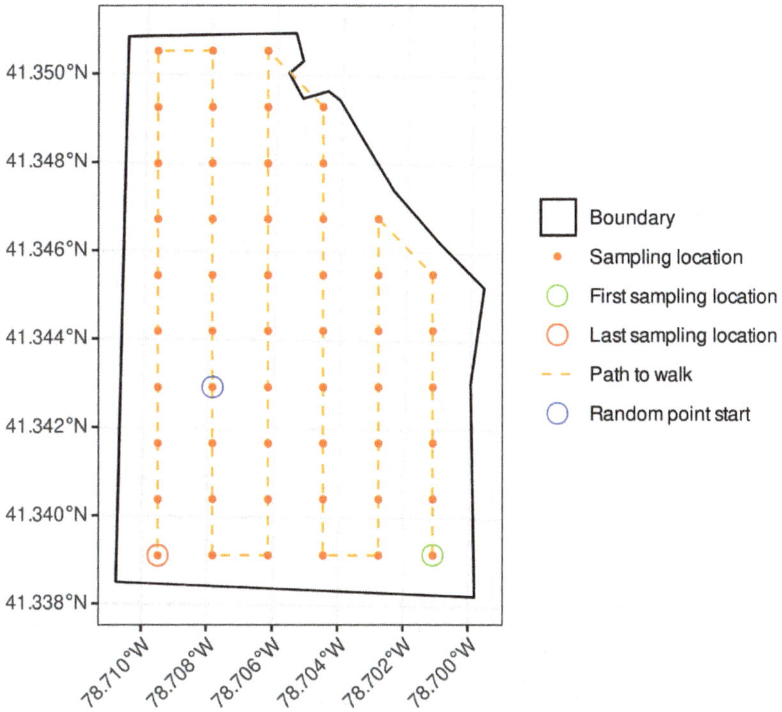

FIGURE 14.1: Systematic sampling design for an Elk County property cruise.

Let's create a few variations on the design shown in Figure 14.1 by rotating the grid and choosing a different origin—while keeping the same random point start. This isolates the effect of orientation and ordering. To reduce code clutter, we don't show the calls to `plot_sys_grid()` or the `patchwork` functions used to produce the composite graphic shown in Figure 14.2.

```
# (a) Rotate grid 45 degrees.
set.seed(1)
sys_a <- sys_grid(bounds,
                  D_l = 7 * 66, D_p = 7 * 66, angle = 45)

# (b) Northwest corner start.
set.seed(1)
sys_b <- sys_grid(bounds,
                  D_l = 7 * 66, D_p = 7 * 66, origin = "nw")

# (c) Rotation and northwest corner start.
set.seed(1)
```

etry type—points, lines, or polygons—making it easy to build clean and consistent maps. You can modify the function to adjust colors, symbols, or other plot features to suit your needs.

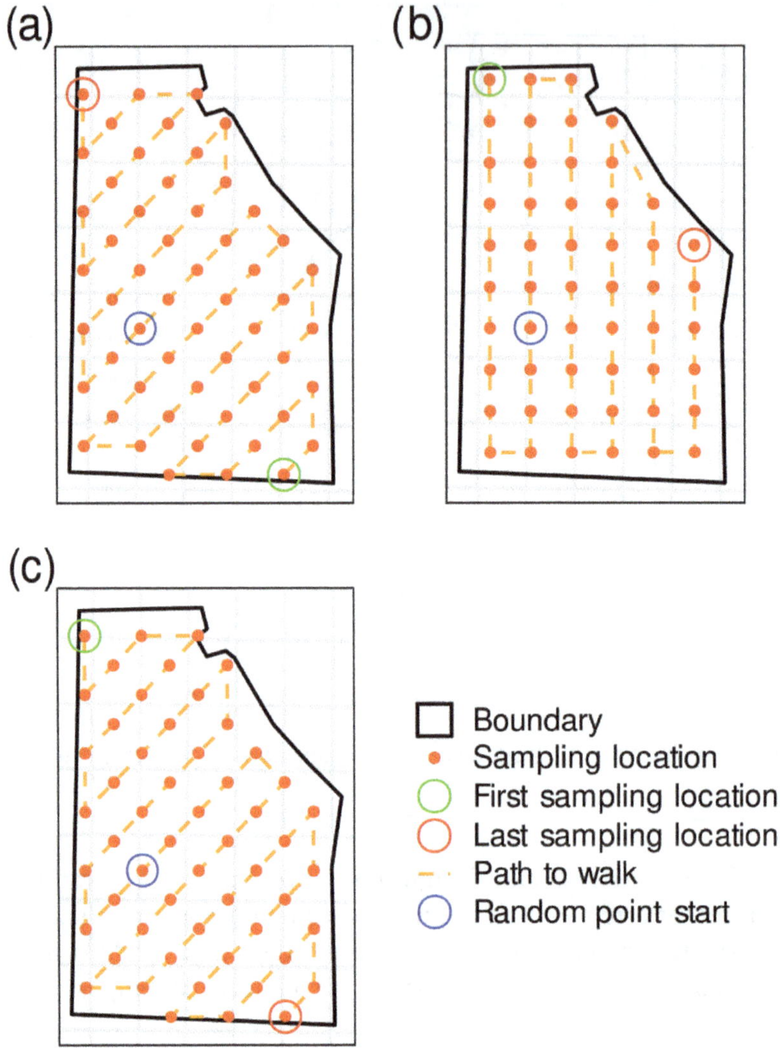

FIGURE 14.2: Systematic sampling design for an Elk County cruise under different grid rotations and starting origins.

```
sys_c <- sys_grid(bounds,
                  D_l = 7 * 66, D_p = 7 * 66,
                  angle = 45,  origin = "nw")
```

Notice in Figures 14.1 and 14.2 that the only random component of systematic sampling is the random point start. Once that point is chosen, all other sampling locations are fixed by the grid's spacing and rotation. Thus, the layout looks highly structured, but it still retains a probabilistic foundation because the sample depends on a randomly chosen start. This single source of randomness is what allows SYS to be used for statistical inference, subject to the assumptions and caveats outlined in the next section.

With the sampling grid in place, we now turn to estimation using data collected under systematic sampling.

14.5 Estimation using SYS

The SYS estimators for the population mean μ (11.1) and total τ (11.6) are the same unbiased sample mean $\bar{y}$ (11.18) and total $\hat{t}$ (11.25) used under SRS (Section 11.3). Both are unbiased under SYS.

Forest variables almost always show positive spatial dependence or broad-scale trends, where nearby locations tend to have more similar values than locations farther apart. In these settings, systematic sampling produces more stable mean estimates than SRS because the grid spreads observations evenly across the population. As a result, the sampling distribution of the SYS mean is typically narrower than that produced under SRS at the same sampling intensity. This behavior is well documented in operational inventory studies and helps explain why SYS is widely viewed as an efficient and practical design when populations exhibit spatial structure.

There is, however, no valid design-based estimator for the sample variance (and thus no estimator for the standard error of the mean) under SYS. The common workaround is to apply the SRS estimators for variance (11.20) and standard error (11.24).

The reason SYS lacks a valid variance estimator is fairly intuitive. Recall from Section 11.3 that the SRS sample variance estimator assumes the n observations are drawn *independently and at random* from the population. This independence justifies the use of $n - 1$ in the denominator of (11.20), reflecting that one degree of freedom is consumed in estimating the sample mean.

By contrast, under SYS a single *random start point* determines the entire sample. Once the start is fixed, all subsequent locations are determined by the grid layout, so the observations are not independent. From this perspective, SYS behaves more like a sample of size one than a sample of size n. Plugging $n = 1$ into the SRS variance estimator (11.20) leads to division by zero, which highlights the core issue: SYS doesn't yield a set of independent observations, and the usual SRS variance logic doesn't apply.

Despite this limitation, the SRS variance estimator is widely used as a practical approximation under SYS. It's considered safe because it tends to be conservative, producing wider than necessary confidence intervals. However, this conservatism can be wasteful, since it doesn't reflect the improved precision of the SYS mean that results from spreading the sample evenly across the population.

A brief overview of alternative SYS-specific variance estimators and spatially balanced sampling designs is offered in Section 14.7, which may be helpful for inventories conducted over larger spatial extents, including NFI-style applications.

14.6 Advantages and disadvantages of systematic sampling

Systematic sampling offers several practical benefits and is widely used in forestry, but like all designs, it involves trade-offs. Here we summarize its main strengths and limitations, particularly in comparison to SRS.

14.6.1 Advantages

- Efficient field layout: SYS designs are easy to implement in the field. Sampling locations fall on a predictable grid, allowing cruisers to follow straight lines, reducing travel time and simplifying navigation.
- Improved spatial coverage: Grid-based sampling ensures sampling locations are evenly distributed across the population, helping to capture spatial heterogeneity and reduce the risk of missing important features—a common concern with SRS.
- Simplifies remeasurement: Because sampling locations follow a fixed, repeatable pattern, they're easier to relocate during future inventory cycles.
- Cost-effective: The combination of reduced travel time and simplified logistics typically results in lower per-unit sampling cost compared to SRS.
- Increased precision: When spatial dependence or broad-scale trends are present, SYS will yield more precise estimates of the mean and total than SRS for the same sample size.

14.6.2 Disadvantages

- No valid variance estimator: SYS lacks a design-based estimator for the sample variance and standard error. While the SRS estimator is commonly used as an approximation, it can overestimate uncertainty, depending on the population structure. A variety of alternative variance estimators exist (See Section 14.7), but none fully resolve the lack of a general design-based solution.

Despite these drawbacks, SYS remains a go-to design in operational forestry due to its simplicity, spatial coverage, and low implementation cost. When used thoughtfully—with attention to population structure and proper randomization of the starting point—SYS can yield highly effective inventory data.

14.7 SYS at larger scales: Variance and design alternatives

SYS is a practical and reliable choice for stand-scale and medium-scale forest inventories. Its field efficiency, repeatability, and spatial coverage make it hard to beat when navigating sampling locations during a cruise. At larger scales—such as regional monitoring programs and NFIs—the limitations of SYS become more consequential, especially when the objective is to realize statistical efficiency gains from designs implemented at substantial cost across broad spatial and temporal extents.

Two related issues motivate the development of alternatives. First, given an existing SYS design, how can variance be estimated in a way that better reflects the sampling distribution of the mean? Second, can alternative sampling designs be constructed that retain the space-covering qualities of SYS while providing stronger design-based support for inference over large areas?

The first issue is addressed through the development of SYS-specific variance estimators. On the second issue, spatially balanced probability sampling designs extend the core idea behind SYS by deliberately spreading sampling locations across the population so that nearby locations are unlikely to be selected together. By enforcing spatial spread at the design stage, these methods help samples capture broad spatial patterns more efficiently.

On the first point, applying the SRS variance estimator to SYS samples is known to produce standard errors that are too large when broad spatial trends are present. These inflated standard errors can obscure much of the efficiency that SYS provides in such settings. This behavior has been documented in several forestry applications; see, for example, Magnussen et al. (2020) and Räty et al. (2020). To address this, SYS-specific variance estimators have been proposed. Magnussen et al. (2020) provide an overview, historical perspective, and comparison of these methods. Variants of these estimators, including the local-difference estimators originally proposed by Matérn (1947), have been applied in the Swedish and Finnish NFIs (Ranneby et al. (1987); Heikkinen (2006)). While these estimators often outperform the SRS estimator and reveal more of the precision inherent in SYS, their performance still depends on how the systematic grid interacts with spatial patterns in the population.

Alongside these developments, a number of spatially balanced probability sampling designs have been proposed for large-area inventories. A leading example is the local pivotal method (LPM; Grafström et al. (2012); Grafström et al. (2017)), which enforces spatial spread directly in the sample selection process. When spatially explicit auxiliary variables correlated with the parameters of interest are available—for example, remote-sensing-based productivity layers—LPM can also spread the sample in auxiliary-variable space. LPM and related designs have been applied in recent NFI-scale studies (e.g., Grafström et al. (2017) and Heikkinen et al. (2025)). Implementations of LPM are available in R via the `BalancedSampling` package (Grafström and Prentius 2024), with related spatially balanced designs available in `spsurvey` (Dumelle et al. 2023). See also Brus (2022) for instructional treatments of these methods and their implementation in R.

For most stand- and property-scale inventories, SYS coupled with the SRS variance estimator remains a practical—though potentially overly conservative—choice. At regional or national scales, where accurate variance estimation in the presence of broad spatial structure is essential, SYS paired with specialized variance estimators or spatially balanced designs such as LPM provides useful alternatives.

This brief section highlights options that become more relevant as inventory scale increases. Keep these alternative variance estimators and spatially balanced designs in mind when SYS is applied in later chapters.

14.8 Exercises

14.8.1 Elk County timber cruise

The Elk County, Pennsylvania cruise and resulting data, introduced in Sections 1.2.2 and 8.5, were collected on a 271-acre forested property. The cruise included 54 sampling points where the forester recorded tree species, DBH, height, and sawlog potential. Sampling points were laid out on a 7-by-7 chain grid (1 chain = 66 ft), giving roughly one point per 5 acres, using a systematic sampling design. At each point, tally trees were selected with an English 10 BAF prism.

Figure 14.3 shows the sampling layout, with points colored by the number of 16-ft hardwood sawlogs per acre. The dotted orange line indicates the transect, or path, followed by the forester between points.

Over the next several exercises, you'll use the Elk County cruise data to generate property-level estimates for hardwood trees classified as acceptable growing stock (AGS) sawtimber. Specifically, you'll estimate mean basal area per acre (ft^2), mean trees per acre (density), and mean number of 16-ft sawlogs per acre, and report 95 percent confidence intervals for each estimate.

All estimates will be based on the tree measurements collected at the $n = 54$ sampling points described above. These data are provided in `"Elk_hardwood_sawtimber_trees_BAF10.csv"` found in `"datasets/PA"`.

Use the following steps to complete your task.

1. Read in the tree measurement data and examine how they're organized. Each row represents a measurement tree ("in" tree) tallied using point sampling, except at points with no trees. The `tree_count` column shows how many trees the row represents—here, values are either 1 or 0. Points with no trees appear as a single row with zeros for numeric variables and `"No trees"` in the `species` column. For this cruise, `tree_count` never exceeds 1; however, we still recommend carrying `tree_count` through your workflow (see Section 12.6.2) so your code stays reusable for other datasets where a row may represent more than one tree.

2. Generate point-level summaries for basal area per acre, trees per acre, and logs per acre. Following Section 12.4, recall that under horizontal point sampling:
    a. Basal area per acre at a point equals the number of measurement trees at that point multiplied by the BAF;

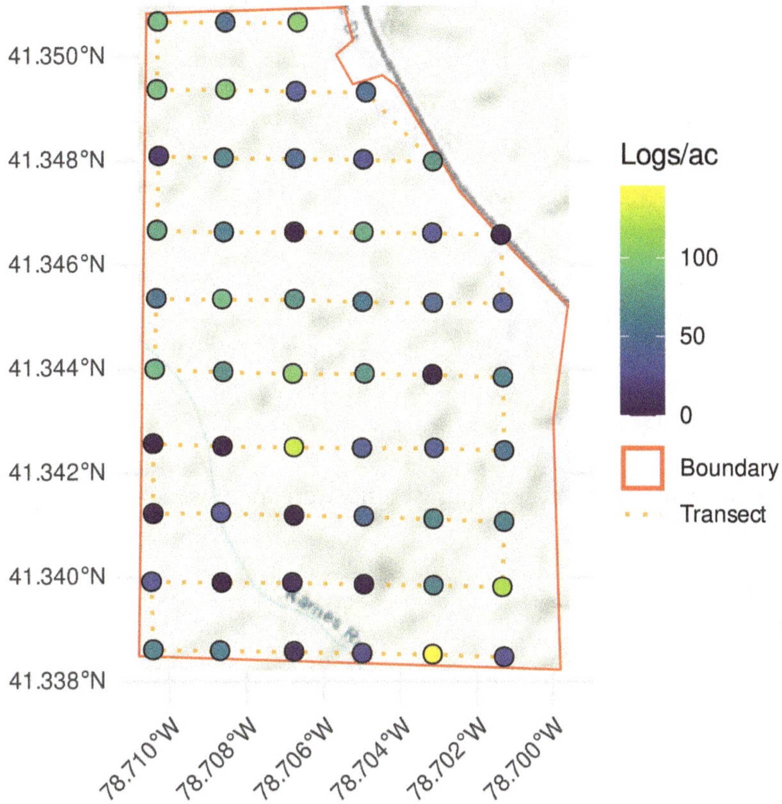

FIGURE 14.3: Systematic sample layout for the Elk County timber cruise in northwestern Pennsylvania. Points are colored by the number of 16-ft hardwood sawlogs per acre. The dotted orange line shows the path followed by the forester between points. An English 10 BAF prism was used to select tally trees at each point.

  b. Trees per acre at a point equals the sum of the tree factors (TF) for all measurement trees at that point, where each tree's TF is defined in (12.22);
  c. Logs per acre at a point equals the sum of each measurement tree's number of logs multiplied by that tree's TF.

3. Using the n point-level summaries from Step 2, estimate the property's mean per-acre basal area, density, and number of logs, and report 95 percent confidence intervals for each. Compute sample means and standard errors using the SRS estimators in (11.18) and (11.24), then form confidence intervals using (11.35). Finally, scale the per-acre means by the total property area to obtain total estimates, and report the corresponding 95 percent confidence intervals for totals.

To get you started, we read in the data below. Use these data for all subsequent exercises in this section. Your job is to write the `tidyverse` code needed to reproduce the output we provide.

```
elk_trees <-
  read_csv("datasets/PA/Elk_hardwood_sawtimber_trees_BAF10.csv")

elk_trees %>% glimpse()
```

```
#> Rows: 251
#> Columns: 5
#> $ point_id   <dbl> 1, 1, 1, 1, 1, 1, 1, 1, 1, 2, 2, 2~
#> $ tree_count <dbl> 1, 1, 1, 1, 1, 1, 1, 1, 1, 1, 1, 1~
#> $ species    <chr> "Black Cherry", "Black Cherry", "S~
#> $ DBH_in     <dbl> 12, 16, 16, 18, 16, 16, 16, 14, 30~
#> $ logs_16ft  <dbl> 1.0, 1.0, 2.0, 2.0, 1.5, 2.0, 2.0,~
```

Exercise 14.1. How many unique sampling points were measured on the property? Confirm your answer using Figure 14.3.

```
#> # A tibble: 1 x 1
#>   n_points
#>      <int>
#> 1       54
```

Exercise 14.2. How many points have no measurement trees?

Hint, `tree_count` is zero when no trees were tallied at that point.

```
#> # A tibble: 1 x 1
#>   n_points_w_zero_trees
#>                   <int>
#> 1                     9
```

Exercise 14.3. Excluding points with no trees, what are the minimum and maximum numbers of tallied trees across points?

Hint, use `filter()`, `group_by()`, and two calls to `summarize()`.

```
#> # A tibble: 1 x 2
#>   m_trees_min m_trees_max
#>         <dbl>       <dbl>
#> 1           1          12
```

Exercise 14.4. Now start the workflow for property-level estimates.

Add a tree factor column so that each tree has its own tree factor. Recall from Section 12.4 that, for point sampling, the tree factor is a function of the BAF and the tree's DBH. Following the methods in Section 12.4 and the computing steps in Section 12.6.2, add the `TF` column to `elk_trees`. When adding `TF`, use `if_else()` inside `mutate()` so that `TF` is zero when `tree_count` is zero.

For readability, we placed `TF` immediately after `tree_count`. Part of the resulting tibble is shown below.

```
elk_trees
```

```
#> # A tibble: 251 x 6
#>    point_id tree_count    TF species   DBH_in logs_16ft
#>       <dbl>      <dbl> <dbl> <chr>      <dbl>     <dbl>
#>  1        1          1 12.7  Black Ch~     12       1
#>  2        1          1  7.16 Black Ch~     16       1
#>  3        1          1  7.16 Sugar Ma~     16       2
#>  4        1          1  5.66 Sugar Ma~     18       2
#>  5        1          1  7.16 Sugar Ma~     16       1.5
#>  6        1          1  7.16 Black Ch~     16       2
#>  7        1          1  7.16 Black Ch~     16       2
#>  8        1          1  9.35 Black Ch~     14       1.5
#>  9        1          1  2.04 Sugar Ma~     30       2
#> 10        2          1  7.16 Red Maple     16       1.5
#> # i 241 more rows
```

Exercise 14.5. Compute the per-acre point-level summaries for basal area, density, and logs. Assign your result to `point_summary`.

We append `_i` to each variable to indicate that it corresponds to the i-th point and to match the notation used in Chapter 12. Our result is printed below.

```
point_summary
```

```
#> # A tibble: 54 x 4
#>    point_id  ba_i trees_i logs_i
#>       <dbl> <dbl>   <dbl>  <dbl>
#>  1        1    90    65.6  103.
#>  2        2    50    26.6   52.4
#>  3        3    70    56.4   90.3
#>  4        4    80    58.3   91.7
#>  5        5    80    56.4   98.2
#>  6        6    40    25.7   36.1
#>  7        7    50    32.4   51.5
#>  8        8    80    56.4   73.2
#>  9        9    30    16.3   29.1
#> 10       10    40    31.5   47.2
#> # i 44 more rows
```

Exercise 14.6. Using your `point_summary` from Exercise 14.5, compute property-level estimates for the per-acre sample mean ($\bar{y}$) and standard error of the mean ($s_{\bar{y}}$), along with 95% confidence intervals for basal area, density, and logs per acre.

Means are calculated using (11.18), standard errors using (11.24), and confidence intervals using (11.35).

Assign your result to `property_summary`. Our result is printed below and includes the intermediate values of sample size n and t-value for reference.

```
property_summary %>% glimpse()
```

```
#> Rows: 1
#> Columns: 14
#> $ n               <int> 54
#> $ t               <dbl> 2.0057
#> $ ba_bar          <dbl> 44.815
#> $ s_ba_bar        <dbl> 4.1776
#> $ ba_ci_lower     <dbl> 36.436
#> $ ba_ci_upper     <dbl> 53.194
#> $ trees_bar       <dbl> 34.199
#> $ s_trees_bar     <dbl> 3.2352
#> $ trees_ci_lower <dbl> 27.71
#> $ trees_ci_upper <dbl> 40.688
#> $ logs_bar        <dbl> 51.372
#> $ s_logs_bar      <dbl> 4.9521
#> $ logs_ci_lower   <dbl> 41.44
#> $ logs_ci_upper   <dbl> 61.305
```

Exercise 14.7. Using your `property_summary` from Exercise 14.6, compute property-level totals for basal area, density, and logs. You can obtain these by scaling the per-acre estimates and their confidence limits by the property area of 271 acres. Assign your result to `property_summary_totals`. Our result is printed below.

Hint, there's no need to recalculate anything—use a single `mutate()` or `transmute()` on `property_summary` to multiply each mean and confidence interval bound by the property area. We used `transmute()` to retain only the desired total estimates in `property_summary_totals`.

```
property_summary_totals %>% glimpse()
```

```
#> Rows: 1
#> Columns: 10
#> $ A_ac             <dbl> 271
#> $ ba_total         <dbl> 12145
#> $ ba_total_lower   <dbl> 9874.1
```

```
#> $ ba_total_upper    <dbl> 14416
#> $ trees_total       <dbl> 9268
#> $ trees_total_lower <dbl> 7509.5
#> $ trees_total_upper <dbl> 11027
#> $ logs_total        <dbl> 13922
#> $ logs_total_lower  <dbl> 11230
#> $ logs_total_upper  <dbl> 16614
```

14.8.1.1 Cruise stand and stock tables

Next, you'll apply what you learned about stand and stock tables in Sections 8.3 and 12.6.3 to the Elk County cruise data.

Your stand table will summarize trees per acre by species and DBH class, and your stock table will summarize logs per acre by species and DBH class. Both tables will use the same `elk_trees` dataset you worked with in the previous exercises to generate property-level estimates.

Let's begin with a fresh copy of `elk_trees`, ensure that point IDs are stored as factors, and remove rows where `tree_count == 0`.

```
elk_trees <-
  read_csv("datasets/PA/Elk_hardwood_sawtimber_trees_BAF10.csv")

elk_trees <- elk_trees %>%
  mutate(point_id = factor(point_id)) %>%
  filter(tree_count != 0)

elk_trees %>% glimpse()
```

```
#> Rows: 242
#> Columns: 5
#> $ point_id   <fct> 1, 1, 1, 1, 1, 1, 1, 1, 1, 2, 2, 2~
#> $ tree_count <dbl> 1, 1, 1, 1, 1, 1, 1, 1, 1, 1, 1, 1~
#> $ species    <chr> "Black Cherry", "Black Cherry", "S~
#> $ DBH_in     <dbl> 12, 16, 16, 18, 16, 16, 16, 14, 30~
#> $ logs_16ft  <dbl> 1.0, 1.0, 2.0, 2.0, 1.5, 2.0, 2.0,~
```

Exercise 14.8. Add a column of 4-inch DBH classes to `elk_trees` for summarizing trees and logs per acre.

Your result should match ours printed below.

```
elk_trees
```

```
#> # A tibble: 242 x 6
```

```
#>    point_id tree_count species       DBH_in logs_16ft DBH_4in
#>    <fct>         <dbl> <chr>          <dbl>     <dbl> <fct>
#>  1 1                 1 Black Cherry      12       1   [10,14]
#>  2 1                 1 Black Cherry      16       1   (14,18]
#>  3 1                 1 Sugar Maple       16       2   (14,18]
#>  4 1                 1 Sugar Maple       18       2   (14,18]
#>  5 1                 1 Sugar Maple       16       1.5 (14,18]
#>  6 1                 1 Black Cherry      16       2   (14,18]
#>  7 1                 1 Black Cherry      16       2   (14,18]
#>  8 1                 1 Black Cherry      14       1.5 [10,14]
#>  9 1                 1 Sugar Maple       30       2   (26,30]
#> 10 2                 1 Red Maple         16       1.5 (14,18]
#> # i 232 more rows
```

Exercise 14.9. Use `complete()` to add unobserved species and DBH class combinations for each point, filling missing `tree_count` and `logs_16ft` values with zeros. Assign your output to `elk_trees_complete`.

We printed our result below. Make sure yours matches ours. Check the number of rows—if your count is smaller, confirm that your `complete()` call includes `point_id`, `species`, and `DBH_4in`.

As discussed in Section 12.6.3, including these zeros ensures that every species and DBH class combination contributes correctly to the property-level estimates.

```
elk_trees_complete
```

```
#> # A tibble: 1,989 x 6
#>    point_id species      DBH_4in tree_count DBH_in logs_16ft
#>    <fct>    <chr>        <fct>        <dbl>  <dbl>     <dbl>
#>  1 1        Beech        [10,14]          0     NA         0
#>  2 1        Beech        (14,18]          0     NA         0
#>  3 1        Beech        (18,22]          0     NA         0
#>  4 1        Beech        (22,26]          0     NA         0
#>  5 1        Beech        (26,30]          0     NA         0
#>  6 1        Black Cherry [10,14]          1     12         1
#>  7 1        Black Cherry [10,14]          1     14         1.5
#>  8 1        Black Cherry (14,18]          1     16         1
#>  9 1        Black Cherry (14,18]          1     16         2
#> 10 1        Black Cherry (14,18]          1     16         2
#> # i 1,979 more rows
```

Exercise 14.10. Add a tree factor (`TF`) column to `elk_trees_complete`. Use the same `if_else()` approach as in Exercise 14.4, setting `TF = 0` when `tree_count == 0`. Your result should match ours printed below.

```
elk_trees_complete %>% glimpse()
```

```
#> Rows: 1,989
#> Columns: 7
#> $ point_id   <fct> 1, 1, 1, 1, 1, 1, 1, 1, 1, 1, 1, 1~
#> $ species    <chr> "Beech", "Beech", "Beech", "Beech"~
#> $ DBH_4in    <fct> "[10,14]", "(14,18]", "(18,22]", "~
#> $ tree_count <dbl> 0, 0, 0, 0, 0, 1, 1, 1, 1, 1, 0, 0~
#> $ DBH_in     <dbl> NA, NA, NA, NA, NA, 12, 14, 16, 16~
#> $ logs_16ft  <dbl> 0.0, 0.0, 0.0, 0.0, 0.0, 1.0, 1.5,~
#> $ TF         <dbl> 0.0000, 0.0000, 0.0000, 0.0000, 0.~
```

Exercise 14.11. Compute point-level summaries for trees per acre and logs per acre by species, DBH class, and point.

As shown in Section 12.6.3, group by `species`, `DBH_4in`, and `point_id`, then summarize totals. Assign your result to `point_summary`.

Your output should match ours printed below. There is a useful check here. Note that there are 7 species, 5 DBH classes, and 54 points. The product of these numbers (i.e., 1890) is the expected number of rows in `point_summary` below.

```
point_summary
```

```
#> # A tibble: 1,890 x 5
#>    species DBH_4in point_id trees_i logs_i
#>    <chr>   <fct>   <fct>      <dbl>  <dbl>
#>  1 Beech   [10,14] 1           0       0
#>  2 Beech   [10,14] 2           0       0
#>  3 Beech   [10,14] 3          12.7    12.7
#>  4 Beech   [10,14] 4           9.35   14.0
#>  5 Beech   [10,14] 5           0       0
#>  6 Beech   [10,14] 6           0       0
#>  7 Beech   [10,14] 7           0       0
#>  8 Beech   [10,14] 8           0       0
#>  9 Beech   [10,14] 9           0       0
#> 10 Beech   [10,14] 10          0       0
#> # i 1,880 more rows
```

Exercise 14.12. Using your `point_summary`, compute property-level means for trees per acre and logs per acre by species and DBH class. These means ($\bar{y}$) represent the per-acre estimates for each species and DBH combination.

You'll group by `species` and `DBH_4in`, and take the mean of `trees_i` and `logs_i`.

Assign your result to `property_summary`. Your output should match ours printed below.

```
property_summary
```

```
#> # A tibble: 35 x 4
#>    species       DBH_4in trees_hat logs_hat
#>    <chr>         <fct>       <dbl>    <dbl>
#>  1 Beech         [10,14]    1.70      1.79
#>  2 Beech         (14,18]    0.398     0.663
#>  3 Beech         (18,22]    0.0849    0.127
#>  4 Beech         (22,26]    0         0
#>  5 Beech         (26,30]    0         0
#>  6 Black Cherry  [10,14]    1.75      2.09
#>  7 Black Cherry  (14,18]    1.79      3.59
#>  8 Black Cherry  (18,22]    0.480     1.10
#>  9 Black Cherry  (22,26]    0         0
#> 10 Black Cherry  (26,30]    0         0
#> # i 25 more rows
```

Exercise 14.13. Finally, use `pivot_wider()` on your `property_summary` to create the stand and stock tables.

Your stand table should match ours printed below and display trees per acre by species and DBH class.

```
#> # A tibble: 7 x 6
#>   species        `[10,14]` `(14,18]` `(18,22]` `(22,26]` `(26,30]`
#>   <chr>              <dbl>     <dbl>     <dbl>     <dbl>     <dbl>
#> 1 Beech              1.70      0.398    0.0849    0         0
#> 2 Black Cherry       1.75      1.79     0.480     0         0
#> 3 Cucumber Tree      0.173     0.342    0.155     0.0589    0
#> 4 Red Maple         13.7       7.13     2.68      0.118     0.0433
#> 5 Sugar Maple        1.51      0.712    0         0         0.0377
#> 6 Yellow Birch       1.04      0.237    0         0         0
#> 7 Yellow Poplar      0         0        0         0.0502    0
```

Then, repeat the process to create the stock table showing logs per acre by species and DBH class. Your result should match ours printed below.

```
#> # A tibble: 7 x 6
#>   species        `[10,14]` `(14,18]` `(18,22]` `(22,26]` `(26,30]`
#>   <chr>              <dbl>     <dbl>     <dbl>     <dbl>     <dbl>
#> 1 Beech              1.79      0.663     0.127     0         0
#> 2 Black Cherry       2.09      3.59      1.10      0         0
#> 3 Cucumber Tree      0.260     0.960     0.465     0.177     0
#> 4 Red Maple         16.5      13.1       5.43      0.236     0.108
```

```
#> 5 Sugar Maple      1.68     1.29     0     0        0.0755
#> 6 Yellow Birch     1.13     0.409    0     0        0
#> 7 Yellow Poplar    0        0        0     0.151    0
```

Typically, we'd finish off these stand and stock tables by adding row and column totals, the code for which is provided near the end of Section 8.3.

15

Stratified sampling

In forest inventory, it's common to estimate population parameters not only for the entire area of interest but also for smaller subareas, or *strata*. *Stratification* refers to dividing the population into non-overlapping subareas, either to improve statistical efficiency or to ensure estimates are available for specific subareas of interest. Ideally, strata are defined using characteristics expected to influence the parameter being estimated, making them internally homogeneous but meaningfully different from one another. Common stratification variables include species, age, productivity, and management unit.

Stratified sampling involves drawing a probability sample independently within each stratum. While sampling within strata is most often done using SRS or SYS, other probability sampling methods can be used. The main benefits of stratification are potential gains in statistical efficiency, i.e., reduced standard errors, and the ability to produce both overall and stratum-specific estimates.

Stratification is implemented in one of two ways.

- *Pre-stratification*: Strata are delineated before sampling begins, letting us allocate stratum-specific sample sizes based on stratum size, expected variability, and sampling costs. This gives full control over the number of observations placed in each stratum. In this case, the area of each stratum must be known ahead of time.
- *Post-stratification*: Here, sampling proceeds without regard to strata, and strata are assigned after the fact based on information collected during or after sampling. This still supports stratum-specific estimation, but the per-stratum sample sizes aren't under our control. The area of each stratum may be determined after sampling.

The main advantage of pre-stratification is the ability to apply *optimum allocation*—assigning more observations to strata that are large or exhibit greater variability, and fewer to those that are small, more homogeneous, or costly to sample. This improves statistical efficiency while controlling sampling costs.

There are substantial statistical and cost-saving benefits to stratification. Even a poor choice of strata yields unbiased estimates and, in almost all practical cases, performs no worse than unstratified sampling. At worst, you'll end up with about the same standard error as you'd get without stratification; at best, you can dramatically improve statistical and cost efficiency.

Our focus for the remainder of this section is on pre-stratification, where strata are defined in advance and sample sizes can be controlled by design.

DOI: 10.1201/9781003222699-15

Stratified sampling proceeds in two steps.

1. *Assignment*: Before any measurements are taken, areas of the population—such as stands or management units—are assigned to strata using available information.
2. *Measurement*: A sample is then drawn within each stratum. At least two observations per stratum are needed to compute within-stratum variance.

15.1 Choosing stratum membership

Stratification begins by subdividing the population into distinct strata based on available information. Assignment can be guided by prior inventory data (e.g., species, age, height, volume, or biomass), expert judgment, or ancillary information like topography, canopy structure, or remote sensing-based spectral or structural measurements. Strata may also align with management units or areas where a separate estimate is desired.

The key requirement is that strata don't overlap. For example, if strata are defined using polygons—where one or more polygons comprise a stratum—those polygons must be spatially distinct, i.e., their boundaries don't overlap. Each location in the population must belong to one and only one stratum.

Useful strata group together areas that are similar to each other and different from the rest. Ideally, variance within a stratum is low, and variance between strata is large. These characteristics improve statistical efficiency.

A stratum can consist of spatially contiguous or disjoint areas. For example, a stratum may comprise spatially adjacent stands or stands scattered across the forest. It's the internal consistency of each stratum that counts—not whether the areas are next to each other on the map.

Stratum assignment doesn't introduce bias in the estimate, even if the grouping doesn't reflect real differences in the population. As long as every part of the population belongs to one and only one stratum, and sampling within each follows the intended design, the overall estimate remains unbiased. A poor choice of strata mainly affects estimate precision rather than accuracy—at worst giving a standard error about the same as you'd get without stratification, and at best providing a clear gain in efficiency.

In the next section, we introduce the estimators used in stratified sampling.

15.2 Estimation using stratified sampling

Throughout this section, and in the subsequent illustration, we assume that sampling within each stratum is done using SRS or SYS, and that SRS estimators are used for the stratum means and variances. A key feature of stratified sampling is that these samples are drawn *independently within each stratum*. A separate probability sample is selected in each stratum, and the resulting stratum-specific estimators are treated as independent by design. Other within-stratum designs are possible but may require different estimators; see, e.g., Section 17.6.

We introduce the following notation:

- H number of strata in the population;
- N_h population size in the h-th stratum;
- $N = \sum_{h=1}^{H} N_h$ total population size across all strata;
- n_h sample size in the h-th stratum;
- $n = \sum_{h=1}^{H} n_h$ total sample size across all strata.

This notation and the estimators that follow assume a finite population. Section 15.5 describes the adjustments needed when working from an areal sampling frame. We present estimators both with and without the FPC, which lets us introduce the finite-population results in a logical order and later makes it easier to see how these same estimators change under an areal (effectively infinite) sampling frame.

The stratum population mean μ_h is estimated using the sample mean

$$\bar{y}_h = \frac{\sum_{i=1}^{n_h} y_{hi}}{n_h}, \tag{15.1}$$

where y_{hi} is the i-th observation in the h-th stratum.

The overall population mean μ (i.e., across all strata) is estimated using

$$\bar{y} = \frac{\sum_{h=1}^{H} N_h \bar{y}_h}{N}. \tag{15.2}$$

This is a weighted sum of the stratum-specific sample means, with weights given by each stratum's proportion of the total population size, i.e., N_h/N.

The population total τ is estimated using

$$\hat{t} = \sum_{h=1}^{H} N_h \bar{y}_h = N\bar{y}. \tag{15.3}$$

The sample variance within each stratum, s_h^2, is estimated using the SRS variance estimator (11.20). If standard errors and confidence intervals are desired for individual strata, use (11.23) or (11.24) and (11.35), respectively.

The standard error of the overall mean $\bar{y}$ is

$$s_{\bar{y}} = \sqrt{\frac{1}{N^2} \sum_{h=1}^{H} \left[\frac{N_h^2 s_h^2}{n_h} \left(1 - \frac{n_h}{N_h}\right) \right]}, \tag{15.4}$$

and, if the FPC is omitted,

$$s_{\bar{y}} = \sqrt{\frac{1}{N^2} \sum_{h=1}^{H} \frac{N_h^2 s_h^2}{n_h}}. \tag{15.5}$$

The standard error for the total estimate $\hat{t}$ is

$$s_{\hat{t}} = N s_{\bar{y}}. \tag{15.6}$$

Confidence intervals for the mean follow the usual form

$$\bar{y} \pm \mathrm{t}_{\mathrm{df}, 1-\frac{\alpha}{2}}\, s_{\bar{y}}, \tag{15.7}$$

with degrees of freedom computed as

$$\mathrm{df} = \sum_{h=1}^{H} (n_h - 1) = n - H. \tag{15.8}$$

Confidence intervals for $\hat{t}$ follow the same form using $s_{\hat{t}}$.

Later sections introduce estimators that work directly with totals rather than means. If the total is the quantity of interest, it's often convenient to express the stratified estimators in terms of the stratum totals. Nothing new is happening here—the total estimator simply uses the same stratum weights already introduced for the mean estimator, but applied to the stratum-level totals. In particular,

$$\hat{t} = \sum_{h=1}^{H} \hat{t}_h, \tag{15.9}$$

where $\hat{t}_h$ is the estimated total in stratum h. Because the strata are treated independently in both design and estimation, we can compute the standard error of the total using a direct sum of their estimated sampling variances

$$s_{\hat{t}} = \sqrt{\sum_{h=1}^{H} s^2_{\hat{t}_h}}, \tag{15.10}$$

where $s_{\hat{t}_h}$ is the estimated standard error of the total in stratum h.

15.3 Sample size determination

After identifying strata and selecting a sampling design, the next step is to determine the total sample size n. This sample size can either be fixed in advance based on available resources or determined based on a desired level of precision for estimating a population mean or total.

Stratified sampling supports two common allocation strategies: *proportional allocation* and *optimum allocation*. Proportional allocation assigns sample size to each stratum based solely on its relative area. Optimum allocation also considers variability and, if cost is taken into account, balances that against the relative cost of sampling. The goal is to allocate effort where it provides the greatest reduction in the variance of the estimated mean (and hence its standard error) per unit cost.

When the total sample size n is unknown, it can be estimated using equations similar to those for SRS (see Section 11.3.6). These equations have the same circularity issue as before:

n appears on both sides of the equation. As with SRS, this can be resolved using the iterative algorithm described in Section 11.3.6.

Importantly, these sample size estimates are intended to meet allowable error targets for the overall population, not for individual strata.

15.3.1 Total sample size with proportional allocation

As introduced in Section 11.3.6, we can estimate sample size given an allowable error E that is in the same units as $\bar{y}$ and an estimate for each stratum's sample variance s^2_h. Because we haven't yet taken the sample, the values for s^2_h must come from somewhere else, e.g., from prior experience or inventory data from a similar population.

The total sample size required under proportional allocation is given by

$$n = \frac{N \sum_{h=1}^{H} N_h s^2_h}{\frac{N^2 E^2}{\text{t}^2} + \sum_{h=1}^{H} N_h s^2_h}, \tag{15.11}$$

where t is the t-value corresponding to the desired confidence level $(1 - \alpha)$.

The sample size (15.11) without the FPC is

$$n = \frac{\text{t}^2 \sum_{h=1}^{H} N_h s^2_h}{N E^2}. \tag{15.12}$$

15.3.2 Total sample size with optimum allocation

Optimum allocation (often called Neyman allocation) minimizes the variance of the stratified mean by accounting for both stratum size and variability. The total sample size is

$$n = \frac{\left(\sum_{h=1}^{H} N_h s_h\right)^2}{\frac{N^2 E^2}{\text{t}^2} + \sum_{h=1}^{H} N_h s^2_h}. \tag{15.13}$$

The sample size (15.13) without the FPC is

$$n = \frac{\text{t}^2 \left(\sum_{h=1}^{H} N_h s_h\right)^2}{N^2 E^2}. \tag{15.14}$$

15.4 Allocation of samples to strata

Given the total sample size n, the next step is to allocate observations among the strata using equations specific to proportional or optimum allocation. Whichever approach was used to determine n should also be used to allocate samples to the strata.

15.4.1 Proportional allocation

In proportional allocation, each stratum receives a share of the total sample based solely on its size. The sample size in stratum h is given by

$$n_h = n\frac{N_h}{N}. \tag{15.15}$$

This approach is easy to apply but ignores differences in variability across strata.

15.4.2 Optimum allocation

In optimum allocation, sample sizes are chosen to minimize the variance of the stratified mean. The number of samples in stratum h is computed as

$$n_h = n\frac{N_h s_h}{\sum_{h=1}^{H} N_h s_h}. \tag{15.16}$$

This method improves efficiency by assigning more samples to strata that are larger or more variable.

15.4.3 Optimum allocation with cost constraints

If sampling costs differ between strata, we can modify the optimum allocation formula to incorporate cost. Let c_h denote the per-unit cost of sampling a single observation in stratum h. The cost-adjusted optimum allocation is

$$n_h = n\frac{N_h s_h/\sqrt{c_h}}{\sum_{h=1}^{H} N_h s_h/\sqrt{c_h}}. \tag{15.17}$$

This formula reduces sample sizes in strata that are more expensive to survey, while still accounting for size and variability. It's particularly useful in operational forest inventories where travel cost, terrain, or accessibility varies across the landscape.

15.5 Stratified sampling under an areal sampling frame

In most forest inventory applications, stratified sampling is implemented under an areal sampling frame. The population has a geographic *population extent* of known total area A, partitioned into H non-overlapping strata with known areas A_h, such that $A = \sum_{h=1}^{H} A_h$. Within each stratum, sampling locations are selected from the continuous set of point locations contained in that stratum, rather than from a finite list of population elements (see Section 12.2).

Under an areal sampling frame, the interpretation changes but the estimator structure doesn't. Observations y_{hi} are expressed on a per-unit-area basis (e.g., tons/ha), so $\bar{y}_h$ estimates the stratum mean per unit area. Stratified estimators therefore target a population mean per unit area, with the corresponding population total obtained by multiplying by the total area A.

Because the population within each stratum is treated as effectively infinite, the finite-population quantities N_h and N are replaced by A_h and A. In particular, the stratified mean estimator in (15.2) applies directly after making this substitution.

Variance estimation likewise proceeds using the stratified estimators without the FPC. The standard error of the stratified mean follows (15.5), and the standard error of the total follows (15.6) after replacing N with A.

The same logic applies when working in terms of stratum totals. The total estimator in (15.9) and its standard error in (15.10) carry over after interpreting $\hat{t}_h$ as the contribution from stratum h based on its area A_h and per-unit-area estimate $\bar{y}_h$. As in the finite-population case, stratum-level contributions combine because strata are treated independently by design.

Sample size determination under an areal sampling frame also follows directly from earlier results. In practice, this means using the stratified sample size formulas without the FPC—namely (15.12) or (15.14)—with stratum areas A_h replacing N_h and total area A replacing N. As before, these expressions are intended to meet allowable error targets for the overall population mean, not for individual strata.

Overall, the differences relative to the finite-population setting are largely interpretive: population sizes are replaced by areas, the FPC is omitted, and all quantities are understood on a per-unit-area basis.

With this areal interpretation in mind, we now turn to the practical problem of laying out systematic sampling grids within strata.

15.5.1 Sampling in strata and systematic designs

In the development above, we assumed that the n_h observations within each stratum are drawn according to an independent probability sampling design, such as SRS or SYS. In practice, stratified sampling is often implemented using SYS within strata rather than SRS (see Section 14).

A common approach is to place a separate systematic grid within each stratum, using a random point start for each grid, as described in Section 14.2. Grid spacing or orientation may differ across strata, reflecting differences in stratum size, shape, or field logistics. This approach preserves the space-filling advantages of SYS while retaining the basic stratified structure.

It's also common to use a *single* systematic grid that spans all strata. This is often attractive operationally, since cruisers can follow continuous straight sampling lines that cross stratum boundaries without interruption. Under this design, only one random point start is used to position the grid over the entire population, and stratum membership is assigned based on each plot's location.

When a single systematic grid is used across all strata, sampling is no longer independent across strata in the strict design-based sense, since all strata share a common random point start. In practice, however, this dependence is typically weak and is rarely, if ever, of practical concern. As with systematic sampling more generally, variance estimation proceeds using standard SRS-based approximations, and the stratified estimators presented above remain the standard working estimators.

In both cases, the inferential considerations associated with SYS apply directly. As discussed in Chapter 14, systematic samples are generated from a small number of random start points (one per stratum in the first case, one overall in the second), and variance estimation typically relies on SRS-based approximations. These issues aren't unique to stratified designs, and we don't revisit them here. Instead, we proceed under the same working assumptions used elsewhere with SYS, focusing on how stratification affects estimation and allocation rather than on the finer details of SYS variance behavior.

15.6 Creating a stratified systematic sampling grid

Once strata have been defined and sample sizes allocated (either proportionally or optimally), the next step is to lay out a systematic sampling grid within each stratum. This can be done using functions that extend the `sys_grid()` approach to stratified designs.

One option is to build stratum-specific grids with the `sys_grid()` function described in Section 14.3. This gives full flexibility in grid spacing, orientation, and pattern, and can be useful when strata are large or widely separated.

In many applications, however, a separate grid for each stratum complicates field logistics. When strata are small or adjacent, it's often easier for field crews to work from a single systematic grid that covers the entire population. The two functions below support this approach by generating a unified grid across all strata under either proportional or optimum allocation. As in Section 14.3, the documentation is presented in the style of condensed R help pages.

15.6.1 Grid generation for proportional allocation `strat_prop_sys_grid()`

Description

Generates a systematic sampling grid using fixed spacing between lines (`D_l`) and points (`D_p`) across the entire area. Points are assigned to strata based on spatial intersection with the input boundary. This is designed for proportional allocation, where grid density is uniform across strata. Built on `sys_grid()` with added logic to track stratum membership and within-stratum ordering.

Usage

```
strat_prop_sys_grid(boundary, D_l, D_p, angle = 0, origin = "se",
                    strat_id_col)
```

Arguments

- `boundary`: An `sf` polygon object defining the stratum boundaries. Must be projected with linear units.
- `D_l`: Distance between sampling lines (x-direction) in the same units as `boundary`.
- `D_p`: Distance between points along a line (y-direction) in the same units as `boundary`.
- `angle`: Optional rotation angle (in degrees). Default is `0`.
- `origin`: Corner from which to order lines and points. One of `"se"`, `"sw"`, `"ne"`, or `"nw"`. Default is `"se"`.
- `strat_id_col`: Name of the column in `boundary` that identifies strata.

Details

This function builds a systematic grid across the union of stratum polygons using `sys_grid()`, then assigns each point to a stratum via spatial intersection. Each sampling location is tagged with its unique point ID as well as its position within each line and stratum. This approach simplifies logistics when a single, uniform grid is acceptable for all strata.

Value

Returns a list with the following components.

- `grid`: An `sf` object of all points, with columns for unique `point_id`, `stratum_id`, and other useful indexing IDs. Columns in `boundary` are appended.
- `path`: An `sf` `LINESTRING` connecting grid points in order.
- `random_point_start`: The initial random point start.
- `boundary`: The input polygon with stratum attributes.
- `type`: A string used internally for validation when plotting.

Use `plot_sys_grid()` to visualize the resulting grid and path.

15.6.2 Grid generation for optimum allocation `strat_opt_sys_grid()`

Description

Creates a stratified systematic sampling grid that supports optimum allocation, assigning different numbers of grid points to each stratum. Builds a dense grid using `sys_grid()`, then intersects it with stratum polygons and thins the points in each stratum to approximate the target sample sizes. The final number of points per stratum may not match the targets exactly, but the function tries to get as close as possible. Supports both random and sequential thinning.

Usage

```
strat_opt_sys_grid(boundary, D_l, angle = 0, origin = "se",
                   n_col, strat_id_col, thinning_method = "random",
                   D_p = NULL)
```

Arguments

- `boundary`: An `sf` polygon object defining the stratum boundaries. Must be projected with linear units.
- `D_l`: Distance between sampling lines (x-direction) in the same units as `boundary`.
- `angle`: Optional rotation angle (in degrees). Default is `0`.
- `origin`: Corner from which to order lines and points. One of `"se"`, `"sw"`, `"ne"`, or `"nw"`. Default is `"se"`.
- `n_col`: Name of the column in `boundary` containing target sample sizes per stratum.
- `strat_id_col`: Name of the column in `boundary` that identifies strata.
- `thinning_method`: Method used to thin grid points within each stratum. Either `"random"` or `"sequential"`. Default is `"random"`.
- `D_p`: Optional distance between points along a line. If not supplied, it's computed automatically from the stratum with the densest sampling, using the same units as `boundary`.

Details

This function is designed for optimum allocation, where sample size varies across strata. It uses `sys_grid()` to build an initial grid, assigns each point to a stratum, then selects a subset of points from each stratum using the specified thinning method. Additional columns in the output identify both the full grid and the subset of points selected for measurement.

Value

Returns a list with the following components.

- `grid`: An `sf` object of all points, with logical `measure` column and columns for unique `point_id`, `stratum_id`, and other useful indexing IDs. Columns in `boundary` are appended.
- `measure_path`: A `LINESTRING` showing the path through the selected (`measure = TRUE`) points.
- `path`: The full path through all grid points.
- `random_point_start`: The initial random point start.
- `boundary`: The original input polygon.
- `type`: A string used internally for validation when plotting.

Use `plot_strat_opt_sys_grid()` to visualize the sampled grid and path.

15.7 Illustrative analysis

We use the Harvard Forest dataset introduced in Section 1.2.5 to walk through an example of stratified estimation. The dataset is small and a bit toy-like, but a complete census of trees

FIGURE 15.1: Harvard Forest population extent partitioned into four strata.

is available, which lets us compare our stratified estimates directly to the true population values.

As discussed in Section 15.5, this illustration adopts an areal sampling frame interpretation. Observations are expressed on a per-unit-area basis, stratum areas replace finite population sizes, and inference follows the stratified estimators without the FPC.

Figure 15.1 shows the extent of the 35 ha forest, partitioned into four strata. Strata 1 and 4 are mixed conifer-hardwood forest, Stratum 2 is conifer-dominated, and Stratum 3 is a sparsely forested wetland.

Our goal is to estimate average AGB per hectare and total AGB for each stratum and for the entire forest. The population of interest includes all stems 12.7 cm DBH and larger. We'll also break out estimates by species groups of interest.

The cruise will use fixed-area circular plots with a radius of 9 meters. At each selected sampling location, all trees meeting the DBH threshold will be measured. See Section 12.3 for additional details on plot-based sampling.

We'd like our AGB estimates to be within $\pm$ 35 Mg/ha of the population value with 95% confidence. This allowable error corresponds to roughly 10% of the expected AGB.

TABLE 15.1: Information used for the Harvard Forest example. The stratum-specific aboveground biomass (AGB) standard deviation s_h is an estimate used in sample size calculations.

Stratum ID	A_h (ha)	s_h (Mg/ha)
1	1.24	100
2	8.93	100
3	2.25	75
4	22.58	125

The cruise will use a stratified systematic sampling design, with sample sizes determined by optimum allocation as described in the preceding sections.

We begin by computing the total sample size using the stratum-specific information in Table 15.1 and the optimum allocation formula (15.14). As discussed in Section 11.3.6, this formula involves a circular dependency: we're solving for n on the left, but also need it to determine the appropriate t-value on the right, which depends on the degrees of freedom.

To resolve this circular dependency, we use the iterative algorithm described in Section 11.3.6. The code below applies that method using a `for` loop, as introduced in Section 5.2.3, to iteratively solve for n until it converges.[1]

```
H <- 4 # Number of strata.
A_h <- c(1.24, 8.93, 2.25, 22.58) # Stratum area (ha).
A <- sum(A_h) # Total area (ha).
E <- 35 # Allowable error for AGB (Mg/ha).
alpha <- 0.05 # For 95% confidence level.
s_h <- c(100, 100, 75, 125) # Stratum standard deviation (Mg/ha).
n_start <- 25 # Initial guess for total sample size.

n_prev <- n_start
for (i in 1:3) {
  t <- qt(p = 1 - alpha / 2, df = n_prev - H) # df = n - H
  n_next <- ceiling((t / (A * E))^2 * sum(A_h * s_h)^2)
  cat("Iteration", i, ": n =", n_next, "\n")
  n_prev <- n_next
}
```

```
#> Iteration 1 : n = 47
#> Iteration 2 : n = 44
#> Iteration 3 : n = 44
```

[1]We use a `for` loop here to save space. Compare with the hardcoded approach in Section 11.3.6.1.

```
n <- n_prev # Final sample size.
```

The algorithm converged quickly to $n = 44$. In the code below, we allocate the total sample size among strata using the optimum allocation formula (15.16).

```
n_h <- n * (A_h * s_h) / sum(A_h * s_h)
n_h <- ceiling(n_h) # Round up to the nearest whole number.
n_h
```

```
#> [1]  2 10  2 31
```

Note that due to rounding, we end up with $n = 45$ plots in the final design.

Now, let's build our sampling grid using `strat_opt_sys_grid()`. As described in Section 15.6, this function requires an `sf` polygon object that delineates the stratum boundaries. As briefly noted in Section 13.1, `sf` objects are data frames with a spatial geometry column and can be manipulated using `tidyverse` functions.

The code below reads in the Harvard Forest stratum boundaries as an `sf` polygon object, adds a new column for the target stratum-specific sample sizes, and prints the data frame portion of the object so we can inspect its columns. `st_drop_geometry()` removes the spatial geometry column to simplify printed output.

```
bounds <- st_read("datasets/HF", "HF_stratified", quiet = TRUE)

bounds <- bounds %>% mutate(n_h = n_h, .after = stratum_id)

bounds %>% st_drop_geometry()
```

```
#>   stratum_id n_h      ha
#> 1          1   2  1.2361
#> 2          2  10  8.9315
#> 3          3   2  2.2494
#> 4          4  31 22.5830
```

With a unique identifier column for each stratum (`stratum_id`) and the target sample sizes (`n_h`), we're ready to pass `bounds` to `strat_opt_sys_grid()`. This and other sampling functions are defined in `"scripts/utils.R"`.

Before calling `strat_opt_sys_grid()` in the code below, we set a random seed to ensure reproducible results.

```
source("scripts/utils.R")

set.seed(1)
```

```
strat_sys <- strat_opt_sys_grid(bounds, D_l = 50, n_col = "n_h",
                                 strat_id_col = "stratum_id",
                                 thinning_method = "random")
```

The call to `strat_opt_sys_grid()` above sets the distance between lines to 50 meters and allows the function to compute the sampling interval along lines (`D_p`) automatically.[2]

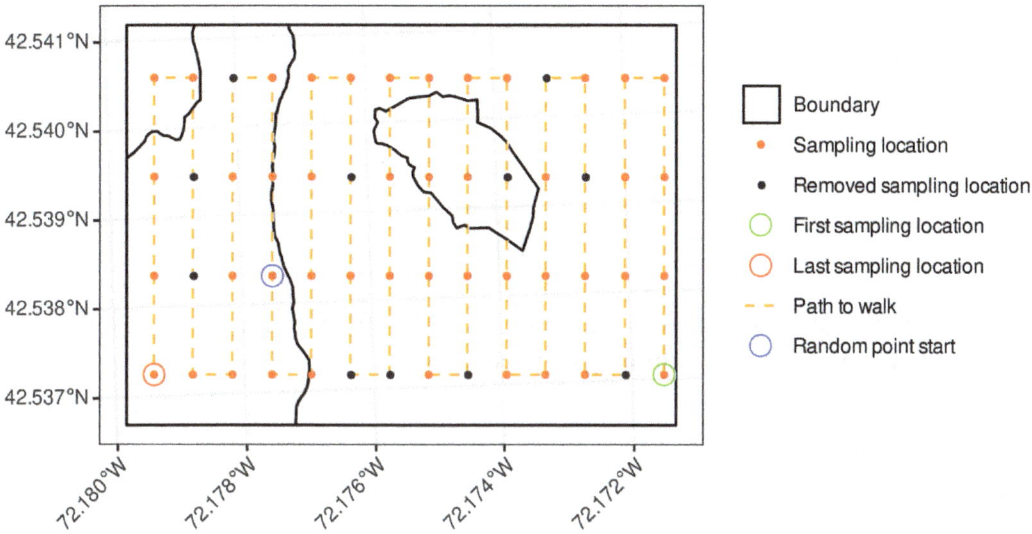

FIGURE 15.2: Systematic sampling with stratification for the Harvard Forest example. Illustrates random removal of sampling locations to approximate the target optimum sample size.

Figure 15.2, created using `plot_strat_opt_sys_grid(strat_sys, path = "lines")`, shows the full systematic grid, the selected sampling locations, and those removed from the grid to meet the target sample sizes (labeled as "Removed sampling location" in the figure legend).

If you only want to visualize the final sampling locations, set the `path` argument in `plot_strat_opt_sys_grid()` to `"points"` as shown below in Figure 15.3.

We can confirm the final sample size by stratum using the code below, which shows the target sample sizes were achieved. This won't always be the case, depending on how the sampling grid intersects with each stratum.

```
strat_sys$grid %>%
  filter(measure) %>%
  group_by(stratum_id) %>%
  summarize(realized_n_h = n()) %>%
  st_drop_geometry()
```

[2]The `bounds` object is projected in meters. You can check an `sf` object's distance units using `st_crs(bounds)$units` or `units::deparse_unit(st_distance(bounds))`.

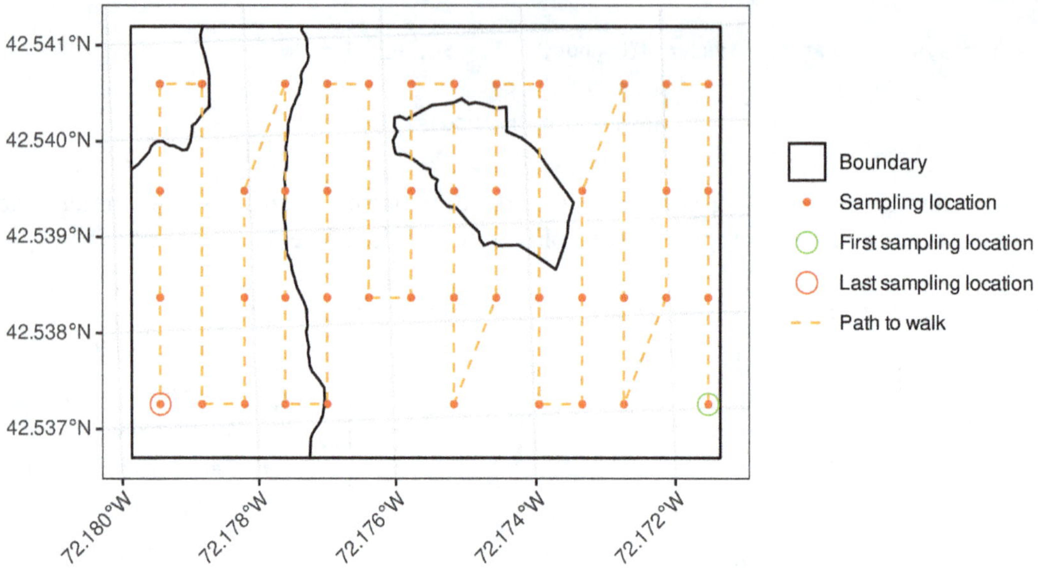

FIGURE 15.3: Systematic sampling with stratification for the Harvard Forest example. Illustrates the path between sampling locations.

```
#> # A tibble: 4 x 2
#>   stratum_id realized_n_h
#> *      <int>        <int>
#> 1          1            2
#> 2          2           10
#> 3          3            2
#> 4          4           31
```

Given the sampling locations, we're ready to head to the field (virtually, in this case). For our virtual cruise, we'll use a 9 meter radius plot centered on the sampling locations. These plot areas are shown as gray circles in Figure 15.4.

As always, if any portion of a plot extends beyond a boundary, then a boundary correction method must be applied—this includes stratum boundaries. Figure 15.4 highlights the three plots with boundary slopover and applies the mirage method described in Section 12.3.3.

The cruise tree data are stored in `"datasets/HF/HF_strat_sys_n45.csv"` and are read into `plot_trees` below. Each row represents one or more trees. The dataset includes a `plot_id` column that uniquely identifies plots across all strata, and a `tree_id` column that identifies individual trees within each plot. The `stratum_id` column indicates the stratum in which the plot falls. Tree counts are recorded in the `tree_count` column, which equals 2 when a tree is double-counted using mirage correction. Additional columns include `agb_Mg` (aboveground biomass in metric tons), `dbh_cm` (diameter at breast height in centimeters), and `species` (tree species).

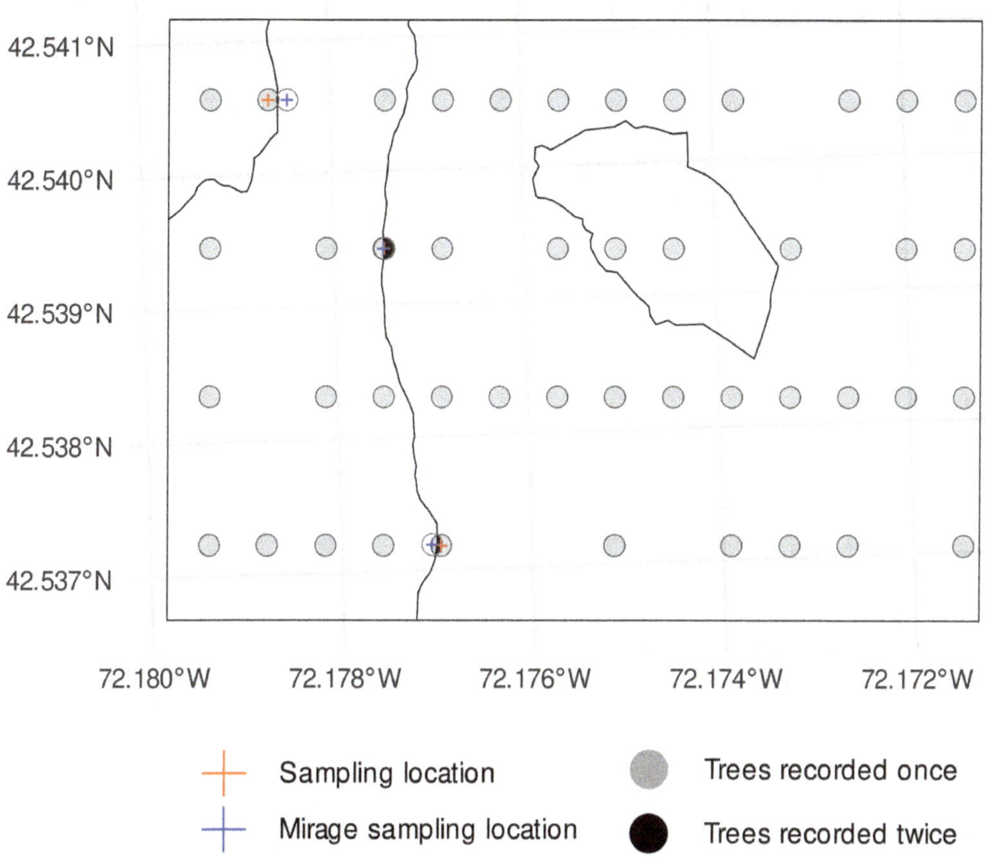

FIGURE 15.4: Systematic sampling with stratification for the Harvard Forest example. Fixed-area plots are shown at each sampling location, including those that extend beyond stratum boundaries and require boundary slopover correction. The mirage method is used for these plots; see Section 12.3.3 for details.

```
plot_trees <- read_csv("datasets/HF/HF_strat_sys_n45.csv")

plot_trees %>% glimpse()
```

```
#> Rows: 797
#> Columns: 7
#> $ plot_id    <dbl> 1, 1, 1, 1, 1, 1, 1, 1, 1, 1, 2, 2~
#> $ tree_id    <dbl> 1, 2, 3, 4, 5, 6, 7, 8, 9, 10, 1, ~
#> $ stratum_id <dbl> 4, 4, 4, 4, 4, 4, 4, 4, 4, 4, 4, 4~
#> $ tree_count <dbl> 1, 1, 1, 1, 1, 1, 1, 1, 1, 1, 1, 1~
#> $ agb_Mg     <dbl> 0.086166, 1.853968, 2.227097, 1.27~
#> $ dbh_cm     <dbl> 14.4, 44.6, 48.0, 38.4, 27.0, 41.8~
#> $ species    <chr> "Acer rubrum", "Quercus rubra", "Q~
```

Given these tree data, and following the steps laid out in Section 12.6.1, our first task is to expand tree-level agb_Mg measurements to a per-hectare basis using the appropriate tree factor, then summarize these values at the plot level. These plot-level values correspond to the y_{hi} terms in (15.1), i.e., the i-th plot measurement in the h-th stratum.

To compute these plot-level AGB values, we group by plot_id and use summarize() to calculate y_hi, the plot-level AGB in Mg/ha. We also retain each plot's stratum_id using first(stratum_id).[3]

```
# Add tree factor for the given plot area.
r <- 9 # Plot radius (m).

plot_trees <- plot_trees %>%
  mutate(TF = 10000 / (pi * r^2))

# Compute plot-level AGB estimates.
plot_summary <- plot_trees %>%
  group_by(plot_id) %>%
  summarize(y_hi = sum(TF * tree_count * agb_Mg),
            stratum_id = first(stratum_id))

plot_summary
```

```
#> # A tibble: 45 x 3
#>    plot_id  y_hi stratum_id
#>      <dbl> <dbl>      <dbl>
#>  1       1  468.          4
#>  2       2  197.          4
#>  3       3  253.          4
#>  4       4  419.          4
#>  5       5  174.          4
#>  6       6  388.          4
#>  7       7  579.          4
#>  8       8  346.          4
#>  9       9  364.          4
#> 10      10  158.          4
#> # i 35 more rows
```

Next, we compute the stratum-level mean $\bar{y}_h$ and standard deviation s_h by grouping plot_summary by stratum_id. We also compute the number of plots within each stratum for convenient reference in the next step.

[3]This works because plot_id is unique across strata. If that's not the case, for example if plot IDs are reused within each stratum, we'd need to include stratum_id in the grouping to avoid mixing trees from different strata that share the same plot ID.

```
stratum_summary <- plot_summary %>%
  group_by(stratum_id) %>%
  summarize(
    n_h = n(),
    y_bar_h = mean(y_hi),
    s_h = sd(y_hi)
  )

stratum_summary
```

```
#> # A tibble: 4 x 4
#>   stratum_id   n_h y_bar_h   s_h
#>        <dbl> <int>   <dbl> <dbl>
#> 1          1     2   311.   38.3
#> 2          2    10   474.  140.
#> 3          3     2    63.0  51.0
#> 4          4    31   348.  108.
```

These stratum-level estimates are useful when reporting means, totals, and associated uncertainty for individual strata. For now, however, we focus on estimating the overall population across strata.

The estimators for the stratified population mean and its standard error are given in (15.2) and (15.5), respectively. Both require n_h, A_h, and the stratum-level summaries. To streamline the workflow, we first join the stratum areas to `stratum_summary` and then apply the estimators to compute 95% confidence intervals.

```
# Join stratum areas.
stratum_area <- tibble(
  stratum_id = 1:4,
  A_h = c(1.24, 8.93, 2.25, 22.58)
)

stratum_summary <- stratum_summary %>%
  left_join(
    stratum_area,
    by = "stratum_id"
  )

# Estimate population mean and 95% CI.
pop_summary <- stratum_summary %>%
  summarize(
    A = sum(A_h),
    y_bar = sum(A_h * y_bar_h) / A,
```

```
    s_y_bar = sqrt(1 / A^2 * sum((A_h^2 * s_h^2) / n_h)),
    t = qt(df = sum(n_h - 1), p = 1 - 0.05 / 2),
    y_ci_lower = y_bar - t * s_y_bar,
    y_ci_upper = y_bar + t * s_y_bar
  )

pop_summary %>%
  select(y_bar, y_ci_lower, y_ci_upper)
```

```
#> # A tibble: 1 x 3
#>   y_bar y_ci_lower y_ci_upper
#>   <dbl>      <dbl>      <dbl>
#> 1  360.       326.       395.
```

To estimate the population total and its confidence interval, we can simply scale the per-hectare estimates in `pop_summary` by the population's total area of 35 ha.

```
pop_summary %>%
  transmute(
    t_hat = A * y_bar,
    t_ci_lower = A * y_ci_lower,
    t_ci_upper = A * y_ci_upper
  )
```

```
#> # A tibble: 1 x 3
#>    t_hat t_ci_lower t_ci_upper
#>    <dbl>      <dbl>      <dbl>
#> 1 12615.     11411.     13819.
```

Alternatively, and mathematically equivalently, we can build the population total and its confidence interval directly from stratum-level total estimates. In the code below, the `mutate()` computes $\hat{t}_h$ and $s_{\hat{t}_h}$, and then, following (15.9) and (15.10), the `summarize()` computes the population total estimate $\hat{t}$, its standard error $s_{\hat{t}}$, and the corresponding confidence interval.

```
stratum_summary %>%
  mutate(
    t_hat_h = A_h * y_bar_h,
    s_t_hat_h = A_h * s_h / sqrt(n_h)
  ) %>%
  summarize(
    H = n(),
    n = sum(n_h),
    t_hat = sum(t_hat_h),
```

```
    s_t_hat = sqrt(sum(s_t_hat_h^2)),
    t = qt(df = n - H, p = 1 - 0.05 / 2),
    t_ci_lower = t_hat - t * s_t_hat,
    t_ci_upper = t_hat + t * s_t_hat
  ) %>%
  select(t_hat, t_ci_lower, t_ci_upper)
```

```
#> # A tibble: 1 x 3
#>    t_hat t_ci_lower t_ci_upper
#>    <dbl>      <dbl>      <dbl>
#> 1 12615.     11411.     13819.
```

15.7.1 Checking coverage and allowable error

How well did our estimates perform in terms of uncertainty quantification? The census-based AGB for the Harvard Forest is 341.17 Mg/ha, with a forest-wide total of 11,941 Mg. Our estimated mean and total, with 95% confidence intervals, were 360.44 (326.03, 394.84) Mg/ha and 12,615 (11,411, 13,819) Mg, respectively. Based on this sample, both confidence intervals successfully captured the true values.

If the theory holds, we'd expect about 95% of confidence intervals from repeated samples to cover the true mean. Because we have a census, we can perform repeated sampling to empirically check that the coverage rate matches this expectation. We can also evaluate whether our optimum allocation strategy meets the target allowable error of $E = 35$ Mg/ha with 95% confidence.

Using the estimated n and n_h, we drew 1,000 samples from the population using stratified SYS. For each, we computed the mean AGB and its 95% confidence interval. These 1,000 replicate estimates are stored in `"datasets/HF/strat_sys_agb_ests_1000reps.csv"`. The code below reads in these estimates and computes two key empirical rates: the proportion of intervals that covered the true mean, and the proportion with absolute error less than the allowable error $E = 35$ Mg/ha. Both rates are very close to 95%, which is strong evidence that the design-based theory is holding up in this example!

```
agb_ests_reps <- read_csv("datasets/HF/strat_sys_agb_ests_1000reps.csv")

# Compute the true mean AGB (Mg/ha) from the census.
y_true <- read_csv("datasets/HF/HF_trees.csv") %>%
  filter(dbh_cm > 12.7) %>%
  summarize(y_true = sum(agb_Mg) / A) %>%
  pull()

# Percent of confidence intervals that cover the true mean.
mean(y_true > agb_ests_reps$ci_lower_95 &
```

```
    y_true < agb_ests_reps$ci_upper_95) * 100
```

```
#> [1] 98.1
```

```
# Percent of estimates within the allowable error E
mean(abs(agb_ests_reps$y_bar - y_true) < E) * 100
```

```
#> [1] 97.3
```

15.7.2 Efficiency gains from stratification

Stratified sampling is especially effective when variation within strata is smaller than variation between them, allowing for more precise estimates than an unstratified sample of the same size. In our Harvard Forest example, we defined four strata. With the exception of the small Stratum 3, the others were fairly similar, so we might not expect dramatic efficiency gains. As a reminder, *efficiency* here means achieving greater precision—that is, a smaller standard error—for the same number of plots compared to an unstratified design.

To compare estimator performance (i.e., unstratified vs. stratified), we repeated the procedure from Section 15.7.1, drawing 1,000 independent samples from the same population. For each sample, we computed the mean AGB and its 95% confidence interval using both unstratified and stratified estimators.

The unstratified replicate estimates are stored in `"datasets/HF/sys_agb_ests_1000reps.csv"`. The stratified replicates are the same as those used in Section 15.7.1. For each estimator, we computed the average 95% confidence interval width across replicates. These were 84 and 73 (Mg/ha) for the unstratified and stratified cases, respectively.

To quantify the gain in precision from stratification, we applied the same logic used in (11.44), replacing variances with average confidence interval widths. This substitution is reasonable because confidence interval width is directly proportional to the standard error and provides an intuitive, interpretable measure of estimator precision. The resulting percent gain in precision was 13%, reflecting the reduction in average interval width achieved by incorporating strata.

This confirms that even modest stratification can yield measurable gains in precision, making it a worthwhile method when stratifying variables are available.

15.7.3 Group-level estimates from stratified data

So far, we've focused on estimating AGB within strata and for the overall population. In many applications, however, we're also interested in estimates for *subgroups* of trees within each stratum and across the entire forest. Common examples include estimates by species, age class, forest type, product type, or functional group.

Let's walk through one such example.

Eastern hemlock (*Tsuga canadensis*) is under threat throughout the eastern US due to the spread of hemlock woolly adelgid (HWA), an invasive insect that has caused widespread mortality. Because HWA is present in the region of Massachusetts where our data were collected, it's reasonable to ask how much AGB would be lost if all eastern hemlock trees in the forest were killed.[4]

We begin by creating a new column in `plot_trees` called `species_groups`, which assigns each tree to one of two groups of interest:

- `"Eastern hemlock"` (`species == "Tsuga canadensis"`);
- `"Other"` (`species != "Tsuga canadensis"`).

Both `species_groups` and `plot_id` are converted to factors. This prepares the data for use with `complete()`, which relies on factor levels to insert rows for combinations of plot and group that may not appear in the observed data.

```
plot_trees <- plot_trees %>%
  mutate(
    species_groups = factor(if_else(species == "Tsuga canadensis",
                                    "Eastern hemlock",
                                    "Other")),
    plot_id = factor(plot_id)
  )
```

As noted in Chapter 12, our convention is to retain plots with no trees by recording a `tree_count` of zero. This ensures that all sampled plots—including those with zero trees of a given group—are preserved in summaries and downstream calculations. The key point is that all sampled plots contribute an observation for each species group, even if that observation is zero. We used a similar approach when building stand and stock tables in Section 12.6.3.

Next, we compute plot-level AGB for each species group.

```
plot_summary <- plot_trees %>%
  group_by(plot_id, species_groups) %>%
  summarize(
    y_hi = sum(TF * tree_count * agb_Mg),
    stratum_id = first(stratum_id),
    .groups = "drop"
  )
```

This yields one row per plot per group when trees from that group are present. If no trees from a group were observed on a given plot, that combination doesn't appear yet.

[4]At the time of writing, much of the hemlock at Harvard Forest and across the broader region had already succumbed to HWA (Orwig et al. 2012).

The code below uses `complete()` to add the missing plot-group combinations, inserting explicit zeros for plots where a species group wasn't observed. We then use `tidyr::fill()` to restore `stratum_id` for the newly created rows. `fill()` restores (i.e., fills) missing values in selected columns using the next or previous entry, which is specified by the `.direction` argument. Grouping by `plot_id` before calling `fill()` ensures values are only propagated within plots and not across strata.

```
plot_summary_filled <- plot_summary %>%
  complete(plot_id, species_groups, fill = list(y_hi = 0)) %>%
  group_by(plot_id) %>%
  fill(stratum_id, .direction = "downup") %>%
  ungroup()
```

At this point, the dataset contains one observation per sampled plot per species group. Plots with zero biomass for a given group are represented by valid zero-valued observations, which still count toward the stratum-specific sample size and subsequent estimates. This distinction matters: zeros are data, not missing values, and they reflect true absence of a species on a plot.

We now summarize to the stratum level, computing the mean AGB per hectare, $\bar{y}_h$, along with standard errors and 95% confidence intervals for each species group.

```
stratum_summary <- plot_summary_filled %>%
  group_by(stratum_id, species_groups) %>%
  summarize(
    n_h = n(),
    y_bar_h = mean(y_hi),
    s_h = sd(y_hi),
    s_y_bar_h = s_h / sqrt(n_h),
    t = qt(df = n_h - 1, p = 1 - 0.05 / 2),
    y_ci_lower_h = y_bar_h - t * s_y_bar_h,
    y_ci_upper_h = y_bar_h + t * s_y_bar_h,
    .groups = "drop"
  )

stratum_summary %>%
  select(stratum_id, species_groups, y_bar_h, y_ci_lower_h, y_ci_upper_h)
```

```
#> # A tibble: 8 x 5
#>   stratum_id species_groups  y_bar_h y_ci_lower_h y_ci_upper_h
#>        <dbl> <fct>             <dbl>        <dbl>        <dbl>
#> 1          1 Eastern hemlock    89.8        -678.         857.
#> 2          1 Other             221.         -891.        1333.
#> 3          2 Eastern hemlock   255.          101.         410.
#> 4          2 Other             219.          110.         327.
```

```
#> 5          3 Eastern hemlock     0          0           0
#> 6          3 Other              63.0      -396.        522.
#> 7          4 Eastern hemlock    48.3        24.3        72.2
#> 8          4 Other             300.        258.        341.
```

These results show that Stratum 2 carries about 250 Mg/ha of eastern hemlock, which is substantially more than any of the other strata. Not surprisingly, with small sample sizes of $n_h = 2$, Strata 1 and 3 have low precision, resulting in wide confidence intervals (and negative lower bounds; see Section 12.5 for interpretation and discussion). No eastern hemlock was observed on the plots in Stratum 3, so its estimated mean is zero. To help visualize these stratum-level differences, Figure 15.5 maps the estimated mean eastern hemlock AGB per hectare in each stratum.

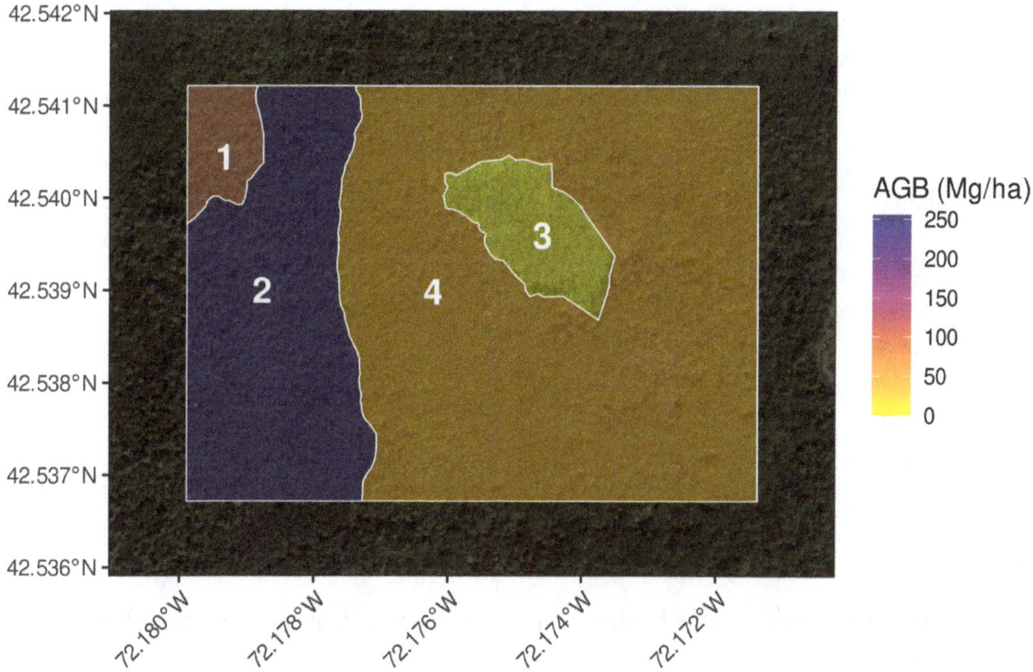

FIGURE 15.5: Estimated mean aboveground biomass (AGB) of eastern hemlock (*Tsuga canadensis*) per hectare in each stratum. Estimates are based on plot-level measurements and stratified sampling.

To estimate forest-wide AGB by species group, we follow the same steps used for the combined-species estimates. As before, we can either scale the forest-wide mean AGB per hectare by the total forest area or work directly from stratum-level total estimates. The estimator logic is unchanged; the only substantive modification to the code is the addition of `group_by(species_groups)`. This illustrates an important advantage of the workflow: once the estimators are written correctly, extending them to subgroups requires no new statistical machinery—only a change in how the data are grouped.

We compute species-group-specific forest-wide mean AGB per hectare and its standard error using the stratified mean estimators in (15.2) and (15.5), respectively. Again, the only change from before is the addition of `group_by(species_groups)`.

```
# Join the stratum areas.
stratum_summary <-  stratum_summary %>%
  left_join(stratum_area, by = "stratum_id")

# Estimate population mean and 95% CI.
pop_summary <- stratum_summary %>%
  group_by(species_groups) %>%
  summarize(
    A = sum(A_h),
    y_bar = sum(A_h * y_bar_h) / A,
    s_y_bar = sqrt(1 / A^2 * sum((A_h^2 * s_h^2) / n_h)),
    t = qt(df = sum(n_h - 1), p = 1 - 0.05 / 2),
    y_ci_lower = y_bar - t * s_y_bar,
    y_ci_upper = y_bar + t * s_y_bar
  )

pop_summary %>%
  select(y_bar, y_ci_lower, y_ci_upper)
```

```
#> # A tibble: 2 x 3
#>   y_bar y_ci_lower y_ci_upper
#>   <dbl>      <dbl>      <dbl>
#> 1  99.5       61.0       138.
#> 2 261.       224.        298.
```

Scaling these mean estimates by the total forest area yields equivalent forest-wide total estimates.

```
pop_summary %>%
  transmute(
    t_hat = A * y_bar,
    t_ci_lower = A * y_ci_lower,
    t_ci_upper = A * y_ci_upper
  )
```

```
#> # A tibble: 2 x 3
#>   t_hat t_ci_lower t_ci_upper
#>   <dbl>      <dbl>      <dbl>
#> 1 3482.      2134.      4831.
#> 2 9133.      7838.     10428.
```

Alternatively, we can work the population totals up directly from stratum-level totals. In the code below, `mutate()` computes $\hat{t}_h$ and $s_{\hat{t}_h}$, and then, following (15.9) and (15.10), `summarize()` computes the population total estimate $\hat{t}$, its standard error $s_{\hat{t}}$, and the corresponding confidence interval. Again, the only difference from the combined-species estimates is the addition of `group_by(species_groups)`.

```
stratum_summary %>%
  mutate(
    t_hat_h = A_h * y_bar_h,
    s_t_hat_h = A_h * s_h / sqrt(n_h)
  ) %>%
  group_by(species_groups) %>%
  summarize(
    H = n(),
    n = sum(n_h),
    t_hat = sum(t_hat_h),
    s_t_hat = sqrt(sum(s_t_hat_h^2)),
    t = qt(df = n - H, p = 1 - 0.05 / 2),
    t_ci_lower = t_hat - t * s_t_hat,
    t_ci_upper = t_hat + t * s_t_hat
  ) %>%
  select(t_hat, t_ci_lower, t_ci_upper)
```

```
#> # A tibble: 2 x 3
#>   t_hat t_ci_lower t_ci_upper
#>   <dbl>      <dbl>      <dbl>
#> 1 3482.      2134.      4831.
#> 2 9133.      7838.     10428.
```

So, to answer our initial question, *how much AGB would be lost if all eastern hemlock trees in the forest were killed by HWA?* We estimate a total loss of 3,482 Mg, with a 95% confidence interval of (2,134, 4,831) Mg. The majority of this biomass loss comes from Stratum 2.

15.8 Advantages and disadvantages of stratified sampling

Stratified sampling is widely used in forest inventory, especially when auxiliary information is available to define meaningful strata. Below, we summarize some key advantages and disadvantages of this design.

15.8.1 Advantages

- Improved precision: Stratification improves precision when units are relatively homogeneous within strata (low within-stratum variance) and heterogeneous across strata. By reducing within-stratum variability (and with sensible allocation), stratified sampling reduces the overall variance of the estimator relative to unstratified designs.

- Supports subregion estimation: Stratification enables valid estimation not just for the entire population, but also for each individual stratum. This is useful when separate summaries are needed for forest types, disturbance regimes, management units, ownership classes, or other subgroups.

- Enables efficient allocation: With *optimum allocation*, more samples can be directed to strata that are larger or more variable, improving efficiency without increasing total cost.

- Compatible with other designs: Different sampling strategies—such as SRS, SYS, or others—can be used within strata, depending on operational needs or design goals. Design-based stratified estimators remain valid provided the within-stratum estimators are valid for the chosen design and sampling is independent across strata.

- Robust to poor strata choices: Even with poor strata assignment, stratified sampling often performs comparably to unstratified sampling and is unlikely to perform much worse, particularly when allocation is roughly proportional or near-optimal.

- Supports post-stratification: If stratum membership isn't known at the time of sampling, post-stratification can still be used. When post-strata explain variation in the parameter of interest and stratum weights are known, post-stratification can improve precision and supports domain-level summaries.

15.8.2 Disadvantages

- Area requirements for pre-stratification: A key requirement of pre-stratification is that the size (or area) of each stratum must be known to support design-based estimation. This can be a limitation if strata are based on incomplete or evolving information.

- Limited allocation flexibility in post-stratification: Post-stratified designs don't allow proportional or optimum allocation, since sample sizes per stratum aren't controlled in advance. They may also result in too few (or zero) samples in some strata to estimate variance reliably.

- Added bookkeeping: Stratified sampling requires tracking stratum membership, managing stratum-specific calculations, and combining results with appropriate weights.

15.9 Exercises

15.9.1 Mountain Creek tract analysis

These data come from managed lands in the southern Oregon Coast Range owned by the Cow Creek Band of Umpqua Tribe of Indians. The project area is the Mountain Creek tract, a one-square-mile section of the Public Land Survey System (see Figure 15.6). The aerial image is the Exhibit B map for Mountain Creek, the standard map used in project documents to show unit boundaries and other features.

The Mountain Creek tract is divided into discrete management units, which serve as the strata for sampling. Some units are intended for overstory harvest, while others are designated for regeneration or other management purposes. Estimates were needed for each unit as well as for the entire tract, including overstory summaries and regeneration counts to support future silvicultural planning.

There are 13 units, ranging in area from 3.16 acres to 91.41 acres. The total area across all units is 412.58 acres. The inventory followed a systematic (SYS) design, with $n = 73$ sampling locations—about one point every 6 acres.

At each location, overstory trees (trees greater than 5 inches DBH) were tallied using a 20 BAF (English units). In addition, a 1/100-th acre fixed-area subplot was established to measure regeneration (trees less than or equal to 5 inches DBH).

While the intent was to implement a standard systematic design, a few minor deviations appear to have occurred during field layout. In operational inventory settings, it's not uncommon to encounter such departures from the intended sampling plan. Although these modifications represent a shift from the original design and should be made with care, we'll assume for this exercise that any changes were implemented in a way consistent with systematic sampling and remain valid for estimation. This lets us proceed with design-based inference under the assumption that the overall sampling plan remains approximately systematic. Situations like this reflect the realities of working with field data and provide a useful opportunity to think critically about how design assumptions align with the data we actually receive as analysts.

Figure 15.6 shows the unit boundaries and sampling locations for the Mountain Creek tract. A copy of the inventory map is also available as a PDF in `"datasets/CCBUTI/CCBUTI_inventory_map.pdf"`.

The data are provided in two separate files located in the `"datasets/CCBUTI"` directory: `"CCBUTI_trees.csv"` contains tree-level overstory and regeneration measurements, and `"CCBUTI_units.csv"` contains unit information, including unit identifiers and areas. The code below reads each file and takes a quick look at their structure.

```
ccbuti_trees <- read_csv("datasets/CCBUTI/CCBUTI_trees.csv")
ccbuti_trees %>% glimpse()
```

FIGURE 15.6: Management unit boundaries and systematic sample layout for timber cruise and regeneration survey conducted on lands owned by the Cow Creek Band of Umpqua Tribe of Indians in the southern Oregon Coast Range. Overstory trees were tallied using a 20 BAF point sample, and regeneration was measured on a 1/100-th acre fixed-area subplot at each location.

```
#> Rows: 288
#> Columns: 8
#> $ unit_id    <dbl> 2, 2, 2, 2, 2, 2, 2, 2, 2, 2, 3, 3~
#> $ point_id   <dbl> 6, 6, 6, 6, 6, 7, 7, 8, 8, 8, 20, ~
#> $ type       <chr> "Overstory", "Regeneration", "Rege~
#> $ tree_count <dbl> 0, 6, 1, 1, 1, 0, 1, 1, 1, 1, 0, 1~
#> $ spp_code   <chr> NA, "df", "df", "df", "df", NA, "d~
#> $ DBH_in     <dbl> NA, 2.0, 3.0, 4.0, 1.0, NA, 2.0, 6~
#> $ height_ft  <dbl> NA, 15, 17, 20, 15, NA, 15, 29, 16~
#> $ vol_cuft   <dbl> NA, NA, NA, NA, NA, NA, NA, 1.3461~
```

```
ccbuti_units <- read_csv("datasets/CCBUTI/CCBUTI_units.csv")
ccbuti_units %>% glimpse()
```

```
#> Rows: 13
#> Columns: 2
#> $ unit_id <dbl> 2, 3, 4, 5, 6, 7, 8, 9, 10, 11, 12, 1~
#> $ acres   <dbl> 20.73, 56.33, 33.30, 91.41, 3.16, 35.~
```

Rows in `ccbuti_trees` correspond to measured trees (both overstory and regeneration), or placeholders for sampling locations without trees. Sampling locations (`point_id`) are labeled as shown in Figure 15.6; note that `point_id` is not nested within `unit_id`. Columns in `ccbuti_trees` are defined as follows.

1. `unit_id` unit identifier in which the sampling point falls.
2. `point_id` unique identifier for each sampling point (not nested within `unit_id`).
3. `type` tree type, either `"Overstory"` (BAF 20 at each point) or `"Regeneration"` (1/100-th acre fixed-area subplot co-located at each point).
4. `tree_count` number of trees a given row represents. A value of zero is a placeholder and means no trees of that type were observed. When count is zero, all subsequent columns have an `NA` value.
5. `spp_code` species code matching the Code column in Table 15.2.
6. `DBH_in` tree diameter at breast height (DBH) in inches. Overstory trees are those with DBH > 5 inches, and regeneration trees are those with DBH ≤ 5 inches.
7. `height_ft` tree height in feet.
8. `vol_cuft` overstory tree merchantable volume (ft^3) to a 4-in top. Regeneration trees have `NA`. Volume estimates are based on species, DBH, and height using the US National Scale Volume and Biomass Estimators (NSVB) (Westfall et al. 2024).

Rows in `ccbuti_units` correspond to harvest units and columns are defined as follows.

1. `unit_id` unit identifier (common to both the tree and unit files).
2. `acres` unit area in acres.

In the following exercises, you'll use these datasets to compute overstory tree basal area, density (trees per unit area), and merchantable volume estimates, as well as regeneration

TABLE 15.2: Species codes and associated species names recorded on the Mountain Creek tract in the southern Oregon Coast Range.

Species list	
Code	Species name
wp	western white pine
wh	western hemlock
df	Douglas fir

density summaries at both the unit- and tract-level.

For most exercises, we provide the desired output—your task is to write the `tidyverse` code to reproduce it.

Exercise 15.1. What is the total number of points across all units? Confirm your answer using Figure 15.6 or the inventory map `"datasets/CCBUTI/CCBUTI_inventory_map.pdf"`.

As you check the map, notice that points are not labeled consecutively. This is not an issue, just something to be aware of—we only care that each point has a unique ID value.[5]

```
#> # A tibble: 1 x 1
#>         n
#>     <int>
#> 1      73
```

Exercise 15.2. What are the distinct `unit_id` values and how many units are there? Confirm your answer using Figure 15.6 or the inventory map `"datasets/CCBUTI/CCBUTI_inventory_map.pdf"`.

Note that there is no unit ID 1 in these data, but the units are consecutively numbered starting from ID 2.

```
#> # A tibble: 13 x 1
#>    unit_id
#>      <dbl>
#>  1       2
#>  2       3
#>  3       4
#>  4       5
#>  5       6
```

[5]Run this code to see the distinct point IDs. You can confirm this by running `ccbuti_trees %>% distinct(point_id) %>% arrange(point_id) %>% print(n = Inf)`, and this code if you'd like to see which points are missing in what would otherwise be a consecutive ordering `setdiff(min(ccbuti_trees$point_id):max(ccbuti_trees$point_id), point_ids)`.

```
#>  6         7
#>  7         8
#>  8         9
#>  9        10
#> 10        11
#> 11        12
#> 12        13
#> 13        14
```

Exercise 15.3. What is the total number of points in each unit? Arrange the results in ascending order by `unit_id`. Confirm your answer using Figure 15.6.

When we apply the stratified estimator using units as strata, remember that we need at least two observations in each stratum to compute its sample variance. Do all units meet this requirement?

Hint, start with a pipe from `ccbuti_trees` and use a sequence of functions like `group_by()`, `summarize()`, and `arrange()`. You will also need a function in `summarize()` that counts the number of distinct points within each unit.

```
#> # A tibble: 13 x 2
#>    unit_id n_points
#>      <dbl>    <int>
#>  1       2        3
#>  2       3       10
#>  3       4        7
#>  4       5       15
#>  5       6        3
#>  6       7        5
#>  7       8        3
#>  8       9       10
#>  9      10        3
#> 10      11        3
#> 11      12        3
#> 12      13        4
#> 13      14        4
```

Exercise 15.4. Of the total number of sampling locations, how many overstory points and regeneration plots have no trees (i.e., `tree_count` of zero)?

Hint, one approach is to sum the occurrences of `tree_count == 0` by `type`, since a `tree_count` of zero indicates no trees were measured at that location.

```
#> # A tibble: 2 x 2
#>   type         zero_trees
#>   <chr>             <int>
#> 1 Overstory            30
#> 2 Regeneration         22
```

Exercise 15.5. Using `ccbuti_trees`, compute *overstory* species abundance as a percent of total overstory trees, arranged in descending order. What is the most abundant overstory species?

Species percent abundance is defined as

$$p_j = 100\frac{m_j}{\sum_{j=1}^{J} m_j}, \tag{15.18}$$

where m_j is the overstory stem count for species j (the sum of `tree_count` for overstory trees only) and J is the number of overstory species.

Getting the right answer is a bit of a process. We suggest the following steps.

1. Filter for overstory trees only.
2. Filter out rows with missing species codes (these are placeholders for points with no trees).
3. Group the data by species code.
4. Compute `m_j = sum(tree_count)`, via a `summarize()`.
5. Compute the total number of trees `M = sum(m_j)` and `p_j = 100 * m_j / M`, via `mutate()`.
6. Arrange species in descending order of `p_j`.
7. Print a small table with species code, `m_j`, and `p_j`. As an internal check, make sure the `p_j` values sum to 100.

We suggest starting with a pipe from `ccbuti_trees` and using `filter()`, `group_by()`, `summarize()`, `mutate()`, and `arrange()`. Alternatively, `count(spp_code, wt = tree_count, name = "m_j")` can replace steps 3-4.[6]

You should get the following output.

```
#> # A tibble: 3 x 4
#>   spp_code   m_j     M    p_j
#>   <chr>    <dbl> <dbl>  <dbl>
#> 1 df         104   113 92.0
#> 2 wh           8   113  7.08
#> 3 wp           1   113  0.885
```

Note, in the output above `M` is an intermediate value and is the denominator of (15.18), i.e., the total number of trees across all species.

Exercise 15.6. Modify your code from Exercise 15.5 to compute *regeneration* species abundance as a percent of total regeneration trees, arranged in descending order. What is the most abundant regeneration species?

You should get the following output.

[6]Take a look at the `count()` manual page to understand the suggested argument values.

```
#> # A tibble: 2 x 4
#>   spp_code   m_j     M   p_j
#>   <chr>    <dbl> <dbl> <dbl>
#> 1 df         134   144 93.1
#> 2 wh          10   144  6.94
```

Exercise 15.7. The next several exercises work toward stratified estimates for unit- and tract-level overstory basal area, density, and volume.

Begin by creating a subset of overstory trees from `ccbuti_trees`. Then add a tree factor column so that each tree has its own tree factor. Recall from Section 12.4 that, for point sampling, the tree factor is a function of the BAF and the tree's DBH.

Following the methods in Section 12.4 and the specific computing steps in Section 12.6.2, we called our overstory subset `o_ccbuti_trees` and our tree factor column `TF`. When adding the `TF` values, use an `ifelse()` within your `mutate()` call to ensure a `TF` value of zero when `tree_count` is zero.

For visualization, we placed the new `TF` column immediately after `tree_count`. Part of the resulting tibble is shown below.

```
o_ccbuti_trees
```

```
#> # A tibble: 141 x 9
#>    unit_id point_id type      tree_count    TF spp_code
#>      <dbl>    <dbl> <chr>          <dbl> <dbl> <chr>
#>  1       2        6 Overstory          0   0   <NA>
#>  2       2        7 Overstory          0   0   <NA>
#>  3       2        8 Overstory          1 102.  df
#>  4       3       20 Overstory          0   0   <NA>
#>  5       3       21 Overstory          0   0   <NA>
#>  6       3       22 Overstory          0   0   <NA>
#>  7       3       23 Overstory          1  89.5 df
#>  8       3       24 Overstory          0   0   <NA>
#>  9       3       25 Overstory          0   0   <NA>
#> 10       3       26 Overstory          0   0   <NA>
#> # i 131 more rows
#> # i 3 more variables: DBH_in <dbl>, height_ft <dbl>,
#> #   vol_cuft <dbl>
```

Exercise 15.8. Using `o_ccbuti_trees` from Exercise 15.7, compute point-level summaries for per-acre basal area, tree density, and volume. Call the resulting tibble `point_summary`.

In our output below, `BA_hi`, `trees_hi`, and `vol_hi` represent per-acre basal area, density, and volume, respectively. The `_hi` suffix indicates that these are values for the i-th point within the h-th unit (stratum). We grouped by both `unit_id` and `point_id`. While it would have been sufficient to group by `point_id` alone (because point IDs are unique across all units), we included `unit_id` so it's available later when computing unit-level estimates in Exercise 15.9.

Hint, the general steps for generating these point-level summaries are illustrated in Section 12.6.2. Some care is needed when computing `vol_hi` because `vol_cuft` values associated with points that have no trees (i.e., `tree_count == 0`) are `NA`. We'd suggest using an `ifelse()` within `mutate()` in your piped sequence to change those `vol_cuft` `NA` values to 0; otherwise, the computation `vol_hi = sum(TF * tree_count * vol_cuft)` will include `NA` values and the result will be `NA`.

```
point_summary
```

```
#> # A tibble: 73 x 5
#> # Groups:   unit_id [13]
#>    unit_id point_id BA_hi trees_hi vol_hi
#>      <dbl>    <dbl> <dbl>    <dbl>  <dbl>
#>  1       2        6     0        0      0
#>  2       2        7     0        0      0
#>  3       2        8    20      102.   137.
#>  4       3       20     0        0      0
#>  5       3       21     0        0      0
#>  6       3       22     0        0      0
#>  7       3       23    20      89.5   175.
#>  8       3       24     0        0      0
#>  9       3       25     0        0      0
#> 10       3       26     0        0      0
#> # i 63 more rows
```

Exercise 15.9. Using `point_summary` from Exercise 15.8, compute unit-level estimates for the per-acre sample mean ($\bar{y}_h$) and standard deviation (s_h) for basal area, density, and volume. Also compute the sample size within each unit (number of points n_h).

We'll call these unit-level summaries `unit_summary` and add the `_h` suffix to indicate that these are the unit estimates. In our output, the unit-level sample means are `BA_bar_h`, `tree_bar_h`, and `vol_bar_h`, and the sample standard deviations are `s_BA_h`, `s_tree_h`, and `s_vol_h`.

Note, the estimation methods are laid out in Sections 15.2 and 15.5, and the illustration in Section 15.7 can be adapted to this setting; for example, the `stratum_summary` in Section 15.7 is analogous to the `unit_summary` you're computing here.

```
unit_summary %>% glimpse()
```

```
#> Rows: 13
#> Columns: 8
```

```
#> $ unit_id      <dbl> 2, 3, 4, 5, 6, 7, 8, 9, 10, 11, 1~
#> $ n_h          <int> 3, 10, 7, 15, 3, 5, 3, 10, 3, 3, ~
#> $ BA_bar_h     <dbl> 6.6667, 8.0000, 0.0000, 20.0000, ~
#> $ s_BA_h       <dbl> 11.547, 13.984, 0.000, 21.381, 0.~
#> $ trees_bar_h <dbl> 33.953, 41.612, 0.000, 94.422, 0.~
#> $ s_trees_h    <dbl> 58.808, 71.630, 0.000, 100.517, 0~
#> $ vol_bar_h    <dbl> 45.705, 74.946, 0.000, 184.072, 0~
#> $ s_vol_h      <dbl> 79.163, 133.023, 0.000, 231.089, ~
```

Exercise 15.10. Given the unit-level sample estimates in `unit_summary` from Exercise 15.9, you can compute unit-specific standard errors and confidence intervals for both means and totals. This involves applying the standard error estimators for the mean and total, then calculating their corresponding confidence intervals as detailed in Section 11.3. We'll return to that in Exercise 15.12. For now, let's pause and build some intuition about this tract based on the unit-level estimates in `unit_summary`.

In addition to basal area, density, and volume estimates, compute the quadratic mean diameter (QMD) for each unit. In forestry, when we talk about mean or average tree diameter, we're almost always referring to the quadratic mean, not the arithmetic mean.[7] Because it weights larger trees more heavily, QMD is preferred over the arithmetic mean diameter for describing stand structure and development. QMD represents the DBH of a tree with the stand's average basal area.

Unlike basal area, density, or volume, which can be expressed on a per-unit-area basis, QMD is a tree-level mean summarizing the average size of individual trees in a stand.

Foresters use QMD, together with tree density, to understand stand structure—how large the trees are, how many occupy a given area, and where a stand might be in its development using stand management diagrams. These measures help evaluate growing space and competition, describe stand conditions, and guide management actions such as thinning.

Following from the quadratic mean in (11.15), we define the QMD as

$$\sqrt{\frac{\sum_{i=1}^{m} DBH_i^2}{m}}, \tag{15.19}$$

where m is the number of trees.

If we have basal area and tree density expressed on a per-unit-area basis, then an equivalent expression to (15.19) is

$$\sqrt{\frac{\text{BA}}{(c \cdot m)}}, \tag{15.20}$$

where BA is basal area, m is tree density, and c is a unit conversion constant (if BA is in square feet then $c = \pi/(144 \cdot 4) \approx 0.005454$, and if BA is in square meters then $c = \pi/(10000 \cdot 4) \approx 0.0000785$). See Curtis and Marshall (2000) and Ducey and Kershaw

[7]While quadratic mean diameter is commonly abbreviated QMD in the US, it's often written as D_g elsewhere, where D stands for diameter and the g subscript comes from Grundfläche, German for basal area.

(2023) for details on QMD's long history of use in forestry, its relationship with stand parameters like volume, and alternative formulations.

For this exercise, compute each unit's QMD using the basal area and density estimates from Exercise 15.9. So following (15.20) we're computing

$$\text{QMD}_h = \sqrt{\frac{\overline{\text{BA}}_h}{c \cdot \overline{\text{trees}}_h}}, \tag{15.21}$$

where $\overline{\text{BA}}_h$ and $\overline{\text{trees}}_h$ are the per-acre mean basal area (ft^2/ac) and number of trees within the h-th unit, and c is defined in (15.20). In `unit_summary`, $\overline{\text{BA}}_h$ and $\overline{\text{trees}}_h$ correspond to `BA_bar_h` and `trees_bar_h`, respectively.

Write a piped sequence to:

- compute `QMD_h` using (15.21) (use `mutate()` and wrap the formula in `ifelse()` to avoid dividing by zero when `trees_bar_h == 0`);
- select only `unit_id`, `BA_bar_h`, `trees_bar_h`, `vol_bar_h`, and the computed `QMD_h`;
- print the results arranged in descending order of `QMD_h`.

Do not add `QMD_h` back into `unit_summary`—we only want to inspect it for this exercise. After running your code, compare your results to the output provided to check your work.

Based on density, volume, and QMD, what general patterns of stand development do you observe across the units?

```
#> # A tibble: 13 x 5
#>    unit_id BA_bar_h trees_bar_h vol_bar_h QMD_h
#>      <dbl>    <dbl>       <dbl>     <dbl> <dbl>
#>  1      11   127.        127.      3523.  13.5
#>  2      10    46.7        58.3      932.  12.1
#>  3      13    55          83.3     1092.  11.0
#>  4      14    65         123.      1771.   9.84
#>  5       9    66         125.      1313.   9.83
#>  6      12    40         103.       643.   8.44
#>  7       7    16          64.9      186.   6.72
#>  8       5    20          94.4      184.   6.23
#>  9       2     6.67       34.0       45.7  6
#> 10       3     8          41.6       74.9  5.94
#> 11       4     0           0          0    0
#> 12       6     0           0          0    0
#> 13       8     0           0          0    0
```

Exercise 15.11. Let's push on with computing the tract-level estimates using `unit_summary` from Exercise 15.9. Join the unit area stored in `ccbuti_units` to `unit_summary`. Then, to stay consistent with our notation, rename the area column `acres` to `A_h`.

Here's the result after the join and rename (we also relocated `A_h` to immediately follow `unit_id`, just for visualization).

```
unit_summary %>% glimpse()
```

```
#> Rows: 13
#> Columns: 9
#> $ unit_id     <dbl> 2, 3, 4, 5, 6, 7, 8, 9, 10, 11, 1~
#> $ A_h         <dbl> 20.73, 56.33, 33.30, 91.41, 3.16,~
#> $ n_h         <int> 3, 10, 7, 15, 3, 5, 3, 10, 3, 3, ~
#> $ BA_bar_h    <dbl> 6.6667, 8.0000, 0.0000, 20.0000, ~
#> $ s_BA_h      <dbl> 11.547, 13.984, 0.000, 21.381, 0.~
#> $ trees_bar_h <dbl> 33.953, 41.612, 0.000, 94.422, 0.~
#> $ s_trees_h   <dbl> 58.808, 71.630, 0.000, 100.517, 0~
#> $ vol_bar_h   <dbl> 45.705, 74.946, 0.000, 184.072, 0~
#> $ s_vol_h     <dbl> 79.163, 133.023, 0.000, 231.089, ~
```

Exercise 15.12. Following Exercise 15.11, `unit_summary` has everything needed to apply the stratified estimators for means, standard errors, and confidence intervals for basal area, density, and volume.

To keep things simple and stay focused as we work toward the tract-level estimates, we'll narrow our attention to volume.

From `unit_summary`, select the columns `unit_id`, `A_h`, `n_h`, `vol_bar_h`, and `s_vol_h`, and call the resulting subset `unit_summary_vol`.

```
unit_summary_vol
```

```
#> # A tibble: 13 x 5
#>    unit_id   A_h   n_h vol_bar_h s_vol_h
#>      <dbl> <dbl> <int>     <dbl>   <dbl>
#>  1       2 20.7      3      45.7    79.2
#>  2       3 56.3     10      74.9   133.
#>  3       4 33.3      7       0       0
#>  4       5 91.4     15     184.    231.
#>  5       6  3.16     3       0       0
#>  6       7 35.9      5     186.    219.
#>  7       8 16.4      3       0       0
#>  8       9 58.1     10    1313.    809.
#>  9      10 21.0      3     932.    463.
#> 10      11 13.9      3    3523.   3294.
#> 11      12 18.6      3     643.    423.
#> 12      13 13.2      4    1092.    718.
#> 13      14 30.6      4    1771.   1516.
```

Exercise 15.13. Using `unit_summary_vol` from Exercise 15.12, apply the stratified mean and standard error estimators (15.2) and (15.5), respectively, and compute 95% confidence intervals for mean volume. These are per-acre estimates (ft^3/ac).

In the output below, we left the total tract area (`A`), standard error, and t-value so you can check your steps. The estimates show a mean volume of 615.72 ft^3/ac with a 95% confidence interval of (422.98, 808.46) ft^3/ac. Given that many units have little or no overstory, such a low tract-level estimate should not be surprising.

```
#> Rows: 1
#> Columns: 6
#> $ A           <dbl> 412.58
#> $ vol_bar     <dbl> 615.72
#> $ s_vol_bar   <dbl> 96.355
#> $ t           <dbl> 2.0003
#> $ vol_ci_lower <dbl> 422.98
#> $ vol_ci_upper <dbl> 808.46
```

Exercise 15.14. Make a few modifications to your code from Exercise 15.13 to apply the stratified total and standard error estimators in (15.3) and (15.6), respectively, and compute 95% confidence intervals for total volume. These are tract-level totals (ft^3).

```
#> Rows: 1
#> Columns: 6
#> $ A           <dbl> 412.58
#> $ vol_hat     <dbl> 254033
#> $ s_vol_hat   <dbl> 39754
#> $ t           <dbl> 2.0003
#> $ vol_ci_lower <dbl> 174514
#> $ vol_ci_upper <dbl> 333553
```

Exercise 15.15. For an internal check, assume there is no stratification and compute the tract-level estimate of mean volume with its standard error. Starting from `o_ccbuti_trees`, apply the point-sampling estimators from Section 12.4, using the computing steps in Section 12.6.2. Your output should match the output below. Recall, these are per-acre estimates (ft^3/ac).

How does this mean and its standard error compare with the stratified results from Exercise 15.13? In particular, how do the standard errors (i.e., `s_vol_bar`) compare, and why are there differences? Does stratification improve statistical efficiency here?

```
#> Rows: 1
#> Columns: 6
#> $ n           <int> 73
#> $ vol_bar     <dbl> 608.94
#> $ s_vol_bar   <dbl> 128.84
#> $ t           <dbl> 1.9935
```

```
#> $ vol_ci_lower <dbl> 352.11
#> $ vol_ci_upper <dbl> 865.78
```

Exercise 15.16. Let's turn to the regeneration data. Recall, this survey was conducted on fixed-area 1/100-acre plots that were co-located with the overstory sampling locations. Our focus here is on regeneration within individual units, not on tract-level estimates, so we'll stop after generating unit-level summaries and will not apply tract-level stratified estimators.

Begin by creating a subset of regeneration trees from `ccbuti_trees`. Then add a tree factor column so that each tree has its own tree factor. Recall from Section 12.3 that under plot sampling the tree factor is only a function of the plot area, and therefore will be the same for all trees.

Following methods in Section 12.3 and the computing steps in Section 12.6.1, we called our regeneration subset `r_ccbuti_trees` and our tree factor column `TF`. For visualization, we placed the new `TF` column immediately after `tree_count`. Part of the resulting tibble is shown below.

```
r_ccbuti_trees %>% glimpse()
```

```
#> Rows: 147
#> Columns: 9
#> $ unit_id    <dbl> 2, 2, 2, 2, 2, 2, 2, 3, 3, 3, 3, 3~
#> $ point_id   <dbl> 6, 6, 6, 6, 7, 8, 8, 20, 20, 21, 2~
#> $ type       <chr> "Regeneration", "Regeneration", "R~
#> $ tree_count <dbl> 6, 1, 1, 1, 1, 1, 1, 1, 1, 1, 1, 1~
#> $ TF         <dbl> 100, 100, 100, 100, 100, 100, 100,~
#> $ spp_code   <chr> "df", "df", "df", "df", "df", "df"~
#> $ DBH_in     <dbl> 2.0, 3.0, 4.0, 1.0, 2.0, 1.0, 5.0,~
#> $ height_ft  <dbl> 15, 17, 20, 15, 15, 16, 26, 26, 26~
#> $ vol_cuft   <dbl> NA, NA, NA, NA, NA, NA, NA, NA, NA~
```

Exercise 15.17. Using `r_ccbuti_trees` from Exercise 15.16, compute plot-level summaries for per-acre tree density. Call the resulting tibble `plot_summary`.

In our output below, `trees_hi` represents per-acre tree density. The `_hi` suffix indicates that these are values for the i-th plot within the h-th unit. We grouped by both `unit_id` and `point_id`. While grouping by `point_id` alone would have been sufficient (because point IDs are unique across all units), we included `unit_id` so it will be available later when computing unit-level estimates in Exercise 15.18.

```
plot_summary
```

```
#> # A tibble: 73 x 3
#> # Groups:   unit_id [13]
#>    unit_id point_id trees_hi
#>      <dbl>    <dbl>    <dbl>
#>  1       2        6      900
#>  2       2        7      100
#>  3       2        8      200
#>  4       3       20      200
#>  5       3       21      400
#>  6       3       22      200
#>  7       3       23      100
#>  8       3       24      200
#>  9       3       25      300
#> 10       3       26      200
#> # i 63 more rows
```

Exercise 15.18. Using `plot_summary` from Exercise 15.17, compute the unit-level sample size and mean number of regeneration trees per acre, along with the associated standard error. Assign your output to `unit_summary`. It should match our output below (which we arranged in decreasing order of `trees_bar_h`).

You can perform some simple checks on your output here. For example, there is a single regeneration tree measured in Unit 13 across 4 plots, hence the estimated 25 trees per acre (i.e., 100/4). Similarly, there are no regeneration trees in Units 10 and 11, which is why those units have estimates of zero.

While the survey data do provide an estimate of regeneration within each unit, the sample sizes are quite small, and the resulting standard errors can be large. If more precise estimates are needed for management decisions, one might consider ways to increase the sample size without incurring substantial additional cost. For example, you could combine similar units to increase sample size. Alternatively, cluster designs like those discussed in Section 17 could be used to place more regeneration subplots at each sampling location.

```
unit_summary %>% arrange(desc(trees_bar_h))
```

```
#> # A tibble: 13 x 4
#>    unit_id   n_h trees_bar_h s_trees_bar_h
#>      <dbl> <int>       <dbl>         <dbl>
#>  1       7     5        620          193. 
#>  2       8     3        567.          88.2
#>  3       2     3        400          252. 
#>  4      12     3        267.          66.7
#>  5       4     7        229.          60.6
#>  6       3    10        210           27.7
#>  7       6     3        200            0  
```

```
#>  8         5    15         173.          38.4
#>  9        14     4         125           62.9
#> 10        13     4          25           25
#> 11         9    10          10           10
#> 12        10     3           0            0
#> 13        11     3           0            0
```

Exercise 15.19. It's interesting to examine the relationship between density and mean DBH for these data. We generally expect density to decrease as mean DBH increases, because stands in early successional stages lose individual trees to growth-based crowding and other sources of mortality.

The code below calculates the quadratic mean DBH for each unit using the individual tree measurements. You'll note the code uses the quadratic mean DBH formula (15.19). We use `weighted.mean()` to appropriately weight each DBH measurement; the weight effectively repeats the given value according to the tree's count and tree factor.[8]

After computing these means, we join them to `unit_summary` from Exercise 15.18 and arrange by `trees_bar_h`. Your job is to follow the code and decide whether there is any apparent relationship between mean DBH and density.

```
tree_level <- r_ccbuti_trees %>%
  group_by(unit_id) %>%
  summarize(
    QMD_h = sqrt(weighted.mean(DBH_in^2, w = tree_count * TF))
    )

unit_summary %>%
  left_join(tree_level) %>%
  arrange(desc(trees_bar_h))
```

```
#> Joining with `by = join_by(unit_id)`
```

```
#> # A tibble: 13 x 5
#>     unit_id   n_h trees_bar_h s_trees_bar_h  QMD_h
#>       <dbl> <int>       <dbl>         <dbl>  <dbl>
#>  1        7     5         620          193.   2.69
#>  2        8     3         567.          88.2  1.61
#>  3        2     3         400          252.   2.58
#>  4       12     3         267.          66.7  2.69
#>  5        4     7         229.          60.6  1.97
#>  6        3    10         210           27.7  3.65
#>  7        6     3         200            0    3.13
```

[8]This demonstrates an alternative approach to the one used in Exercise 15.10, which started from stand-level estimates. Both approaches will produce the same QMD.

```
#>  8        5    15         173.          38.4   3.13
#>  9       14     4         125           62.9   3.20
#> 10       13     4          25           25     0
#> 11        9    10          10           10     2
#> 12       10     3           0            0   NaN
#> 13       11     3           0            0   NaN
```

16

Estimation using covariates

Up to this point, our estimators have relied primarily on direct measurements of the variable of interest, y, collected from sampled units. In many settings, however, we also have access to auxiliary information—additional variables that are already available or inexpensive to obtain. When such auxiliary data are related to y, they can be used to improve estimation and increase efficiency. A covariate x is one form of auxiliary information: it's a variable that is related to, but distinct from, y. When x is related to y, incorporating it into estimation can reduce standard errors and, in some settings, lower field measurement costs.

We have already seen covariates at work indirectly. For example, in stratified sampling (Chapter 15), a stratifying variable improved estimation by grouping similar units together, which reduced within-stratum variability and led to more precise estimates. Here, we treat x as an explicit part of the estimator.

In this chapter, we'll consider several different estimators. The choice of estimator depends on the nature of the relationship between y and x.

- *Regression estimation* is preferred when the relationship is approximately linear with a nonzero intercept and the variability of y remains roughly constant across values of x (Figure 16.1(a)).
- *Ratio estimation* is preferred when the relationship is approximately proportional—passing through (or near) the origin—and the variability of y increases with x (Figure 16.1(b,c)). We'll detail two common ratio estimators, referred to as the *ratio-of-means* and *mean-of-ratios*. The ratio-of-means works well when variability in y increases roughly in proportion to x (Figure 16.1(b)), while the mean-of-ratios is useful when variability in y increases with x^2 (Figure 16.1(c)) or when unit-level ratios are of direct interest.

In Sections 16.1 and 16.2, we detail regression and ratio estimators for settings where the population mean of x (i.e., μ_x) is known and used to improve estimation of the population mean of y (i.e., μ_y). Then, in Section 16.3, we relax the assumption that μ_x is known and instead rely on its estimate $\bar{x}$ to improve estimation of μ_y.[1]

[1]Throughout this chapter, we follow the classical forestry inventory treatment of regression and ratio methods, drawing largely from Cochran (1977) and Freese (1962).

DOI: 10.1201/9781003222699-16

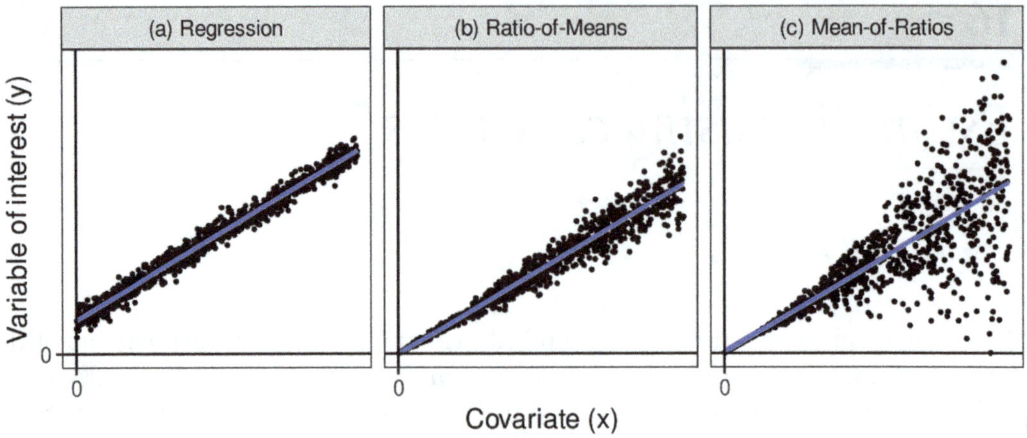

FIGURE 16.1: Three straight-line relationship patterns. (a) Nonzero intercept with roughly constant variance. (b) Passes through the origin with variance increasing in proportion to x. (c) Passes through the origin with variance increasing in proportion to x^2.

16.1 Regression estimation

A common way to describe the relationship between two variables is with a straight line. Suppose we have a variable of interest y and a covariate x. The general equation for a line is

$$y = b_0 + b_1 x, \tag{16.1}$$

where b_0 is the point where the line crosses the y-axis (the intercept) and b_1 is the slope, which tells us how much y changes for a one-unit increase in x.

Suppose we have a population of paired values (x_i, y_i) for $i = 1, \ldots, N$. From this population we select a sample of n pairs, which could produce a scatter plot like Figure 16.1(a). With these data we can fit a straight line using *ordinary least squares regression*. This method finds the values of b_0 and b_1 that minimize the squared vertical distances between the observed points and the line. In R, this is done with `lm()`, which returns estimates for b_0 and b_1. This is simple linear regression in a nutshell.

Our goal here isn't to build a predictive regression model or test hypotheses about regression coefficients. Instead, we're after a practical estimator for μ_y that borrows strength from a linear relationship with x. We treat the regression equation as a stepping stone: by fitting a line through the sample data, we derive an estimator that incorporates known information about the population mean of x. Let's walk through how this works.

To see how regression leads to an estimator, note that in ordinary least squares the fitted regression line always passes through the point formed by the sample means, $(\bar{x}, \bar{y})$. Substituting these values into (16.1) gives

$$\bar{y} = b_0 + b_1 \bar{x}. \tag{16.2}$$

This identity can be rearranged to express the intercept in terms of sample quantities

$$b_0 = \bar{y} - b_1\bar{x}. \tag{16.3}$$

Now suppose the population mean of the covariate, μ_x, is known. If the fitted line is a reasonable approximation to the population relationship between y and x, then evaluating the line at $x = \mu_x$ provides a natural estimator for the population mean of y. We denote this regression estimator by $\bar{y}_{\text{R}}$, to distinguish it from the usual sample mean $\bar{y}$, and write

$$\bar{y}_{\text{R}} = b_0 + b_1\mu_x. \tag{16.4}$$

Substituting b_0 from (16.3) into (16.4) gives

$$\bar{y}_{\text{R}} = (\bar{y} - b_1\bar{x}) + b_1\mu_x, \tag{16.5}$$

which simplifies to the regression estimator

$$\bar{y}_{\text{R}} = \bar{y} + b_1(\mu_x - \bar{x}). \tag{16.6}$$

From (16.6) we see the logic of regression estimation. The usual sample mean $\bar{y}$ serves as the baseline. The adjustment term $b_1(\mu_x - \bar{x})$ shifts this baseline according to how far the sample mean of x is from its population mean, scaled by the slope. If $\bar{x}$ is less than μ_x, the estimate is adjusted upward; if greater, it's adjusted downward. In this way, the regression estimator incorporates auxiliary information from x to improve our estimate of μ_y. The extent of improvement depends on how strong and linear the relationship is between x and y.

The slope in (16.6) is computed from the sample as

$$b_1 = \frac{S_{xy}}{S_{xx}}, \tag{16.7}$$

where S denotes a sum of squares (or cross-products). In particular,

$$S_{xy} = \sum_{i=1}^{n}(x_i - \bar{x})(y_i - \bar{y}) \tag{16.8}$$

and

$$S_{xx} = \sum_{i=1}^{n}(x_i - \bar{x})^2. \tag{16.9}$$

The standard error of $\bar{y}_{\text{R}}$ can be estimated as

$$s_{\bar{y}_{\text{R}}} = \sqrt{s^2_{y|x}\left(\frac{1}{n} + \frac{(\mu_x - \bar{x})^2}{S_{xx}}\right)\left(1 - \frac{n}{N}\right)}, \tag{16.10}$$

where $s^2_{y|x}$ is the residual variance of y about the fitted regression line,

$$s^2_{y|x} = \frac{S_{yy} - \frac{S^2_{xy}}{S_{xx}}}{n - 2} \tag{16.11}$$

and

$$S_{yy} = \sum_{i=1}^{n}(y_i - \bar{y})^2. \tag{16.12}$$

The FPC term should be removed from (16.10) when appropriate, see Section 12.2.1.

To estimate the population total we scale by N

$$\hat{t}_{\mathrm{R}} = N\bar{y}_{\mathrm{R}} \tag{16.13}$$

and its standard error is

$$s_{\hat{t}_{\mathrm{R}}} = N s_{\bar{y}_{\mathrm{R}}}. \tag{16.14}$$

Confidence intervals for the mean and total follow the usual form

$$\bar{y}_{\mathrm{R}} \pm \mathrm{t}_{\mathrm{df},1-\frac{\alpha}{2}}\, s_{\bar{y}_{\mathrm{R}}},$$
$$\hat{t}_{\mathrm{R}} \pm \mathrm{t}_{\mathrm{df},1-\frac{\alpha}{2}}\, s_{\hat{t}_{\mathrm{R}}},$$

with degrees of freedom $\mathrm{df} = n - 2$.

Like regression itself, regression estimation works best when two assumptions hold: (1) the mean of y varies approximately linearly with x, and (2) the variability of y around the fitted line is approximately constant across the range of x. If the first assumption fails, a transformation of x may help. If the second fails and variance increases with x, a variance-stabilizing transformation may be considered, or ratio estimation (Section 16.2) might be more appropriate.

16.2 Ratio estimation

Ratio estimation assumes a proportional relationship between y and x, so that $y \approx Rx$ for some constant multiplier R (where $\approx$ means "approximately equal"). This implies the mean relationship between y and x passes through (or near) the origin. We estimate R from the sample as $\hat{R}$ and then use it to estimate the population mean as $\bar{y}_{\text{ratio}} = \hat{R}\mu_x$, and the population total as $\hat{t}_{\text{ratio}} = \hat{R}\tau_x$. Ratio estimators are generally biased, though this bias often decreases with larger sample sizes.

There are two common ways to define $\hat{R}$, leading to two distinct ratio estimators. To keep things clear, we use the subscript RM for the *ratio-of-means* estimator and MR for the *mean-of-ratios* estimator in the expressions that follow.

16.2.1 Ratio-of-means

The estimator of the population mean is

$$\bar{y}_{\mathrm{RM}} = \hat{R}\,\mu_x, \tag{16.15}$$

where

$$\hat{R} = \frac{\bar{y}}{\bar{x}}. \tag{16.16}$$

Equivalently, this ratio can be written

$$\hat{R} = \frac{\sum_{i=1}^{n} y_i}{\sum_{i=1}^{n} x_i}, \tag{16.17}$$

since the factors of n in the sample means cancel out.[2]

[2] Looking at (16.17), the estimator might arguably be better described as a *ratio-of-totals*, since it's equivalent to the ratio of summed values.

The standard error of $\bar{y}_{\text{RM}}$ can be estimated as

$$s_{\bar{y}_{\text{RM}}} = \sqrt{\left(\frac{\mu_x}{\bar{x}}\right)^2 \left(\frac{s_y^2 + \hat{R}^2 s_x^2 - 2\hat{R}s_{xy}}{n}\right)\left(1 - \frac{n}{N}\right)}, \tag{16.18}$$

where s_y^2 and s_x^2 are the sample variances of y and x, and

$$s_{xy} = \frac{1}{n-1}\sum_{i=1}^{n}(x_i - \bar{x})(y_i - \bar{y}) \tag{16.19}$$

is the sample covariance between x and y.[3] The FPC term should be removed from (16.18) when appropriate, see Section 12.2.1.

The population total is estimated using

$$\hat{t}_{\text{RM}} = N\bar{y}_{\text{RM}} \tag{16.20}$$

and its standard error is

$$s_{\hat{t}_{\text{RM}}} = N s_{\bar{y}_{\text{RM}}}. \tag{16.21}$$

Confidence intervals for the mean and total follow the usual form

$$\bar{y}_{\text{RM}} \pm \text{t}_{\text{df},1-\frac{\alpha}{2}}\, s_{\bar{y}_{\text{RM}}},$$

$$\hat{t}_{\text{RM}} \pm \text{t}_{\text{df},1-\frac{\alpha}{2}}\, s_{\hat{t}_{\text{RM}}},$$

with degrees of freedom $\text{df} = n - 1$.

16.2.2 Mean-of-ratios

The estimator of the population mean is

$$\bar{y}_{\text{MR}} = \hat{R}\mu_x, \tag{16.22}$$

where

$$\hat{R} = \frac{1}{n}\sum_{i=1}^{n} r_i, \tag{16.23}$$

and $r_i = y_i/x_i$ is the unit ratio.

The standard error of $\bar{y}_{\text{MR}}$ can be estimated as

$$s_{\bar{y}_{\text{MR}}} = \mu_x \sqrt{\frac{s_r^2}{n}\left(1 - \frac{n}{N}\right)}, \tag{16.24}$$

where

$$s_r^2 = \frac{1}{n-1}\sum_{i=1}^{n}\left(r_i - \hat{R}\right)^2. \tag{16.25}$$

The FPC term should be removed from (16.24) when appropriate, see Section 12.2.1.

The population total is estimated using

$$\hat{t}_{\text{MR}} = N\bar{y}_{\text{MR}} \tag{16.26}$$

[3] In R, the sample variance and covariance can be computed using `var()` and `cov()`, respectively.

and its standard error is

$$s_{\hat{t}_{\text{MR}}} = N s_{\bar{y}_{\text{MR}}}. \tag{16.27}$$

Confidence intervals for the mean and total follow the usual form

$$\bar{y}_{\text{MR}} \pm \text{t}_{\text{df},1-\frac{\alpha}{2}}\, s_{\bar{y}_{\text{MR}}},$$
$$\hat{t}_{\text{MR}} \pm \text{t}_{\text{df},1-\frac{\alpha}{2}}\, s_{\hat{t}_{\text{MR}}},$$

with degrees of freedom $\text{df} = n - 1$.

16.3 Double sampling (Two-phase)

Double sampling is useful when the covariate x is inexpensive to measure but its population mean μ_x is unknown. The design proceeds in two phases. In phase 1, we draw a large sample of size n_1 and measure only x. In phase 2, we draw a nested subsample of size n_2 from the phase-1 units and measure y. As a result, we have a large sample of x values and a smaller set of paired observations (x, y).

The two phases play complementary roles. Phase 1 provides a precise estimate of the population mean of x, which anchors the estimation. Phase 2 concentrates effort on measuring y, allowing us to estimate how y varies with x—for example, through a fitted slope b_1 or a ratio $\hat{R}$.

As in the regression and ratio estimators introduced in the preceding sections, the idea is to start with the mean of y from the smaller sample and then adjust it based on the relationship between x and y and how far the phase-2 sample mean $\bar{x}_2$ is from the phase-1 mean $\bar{x}_1$. In essence, phase 1 tells us where x is centered, and phase 2 tells us how changes in x translate into changes in y. When n_1 is large, most of the uncertainty in the estimator arises from residual variation around the fitted relationship, rather than from estimating $\bar{x}_1$.

16.3.1 Regression estimation

The regression estimator of the population mean under double sampling is

$$\bar{y}_{\text{Rd}} = \bar{y}_2 + b_1\left(\bar{x}_1 - \bar{x}_2\right), \tag{16.28}$$

where the slope b_1 is defined in (16.7) and is computed here using only the n_2 phase-2 observations.

The standard error of $\bar{y}_{\text{Rd}}$ can be approximated as

$$s_{\bar{y}_{\text{Rd}}} = \sqrt{s^2_{y|x}\left(\frac{1}{n_2} + \frac{(\bar{x}_1 - \bar{x}_2)^2}{S_{xx}}\right)\left(1 - \frac{n_2}{n_1}\right) + \frac{s^2_y}{n_1}\left(1 - \frac{n_1}{N}\right)}, \tag{16.29}$$

where $s^2_{y|x}$ is the residual variance defined in (16.11), again computed using the phase-2 data.

Note that (16.29) includes two FPC terms: one for the phase-2 sample nested within phase 1, and one for the phase-1 sample drawn from the population. As before, the second FPC is removed under an areal frame (see Section 12.2.1), but we always retain the first FPC since the phase-1 sample is finite.

The population total is estimated as

$$\hat{t}_{\text{Rd}} = N\bar{y}_{\text{Rd}}, \tag{16.30}$$

with standard error

$$s_{\hat{t}_{\text{Rd}}} = Ns_{\bar{y}_{\text{Rd}}}. \tag{16.31}$$

Confidence intervals follow the usual form

$$\bar{y}_{\text{Rd}} \pm \text{t}_{\text{df},1-\frac{\alpha}{2}}\, s_{\bar{y}_{\text{Rd}}},$$

$$\hat{t}_{\text{Rd}} \pm \text{t}_{\text{df},1-\frac{\alpha}{2}}\, s_{\hat{t}_{\text{Rd}}},$$

with degrees of freedom $\text{df} = n_2 - 2$.

16.3.2 Ratio-of-means

The estimator of the population mean is

$$\bar{y}_{\text{RMd}} = \hat{R}\,\bar{x}_1, \tag{16.32}$$

where

$$\hat{R} = \frac{\bar{y}_2}{\bar{x}_2}. \tag{16.33}$$

The standard error of $\bar{y}_{\text{RMd}}$ can be estimated as

$$s_{\bar{y}_{\text{RMd}}} = \sqrt{\left(\frac{\bar{x}_1}{\bar{x}_2}\right)^2 \left(\frac{s_y^2 + \hat{R}^2 s_x^2 - 2\hat{R}s_{xy}}{n_2}\right)\left(1 - \frac{n_2}{n_1}\right) + \frac{s_y^2}{n_1}\left(1 - \frac{n_1}{N}\right)}. \tag{16.34}$$

If you're using an areal sampling frame, it's appropriate to drop the second FPC term (see Section 12.2.1 for discussion).

The population total is estimated using

$$\hat{t}_{\text{RMd}} = N\bar{y}_{\text{RMd}} \tag{16.35}$$

and its standard error is

$$s_{\hat{t}_{\text{RMd}}} = Ns_{\bar{y}_{\text{RMd}}}. \tag{16.36}$$

Confidence intervals for the mean and total follow the usual form

$$\bar{y}_{\text{RMd}} \pm \text{t}_{\text{df},1-\frac{\alpha}{2}}\, s_{\bar{y}_{\text{RMd}}},$$

$$\hat{t}_{\text{RMd}} \pm \text{t}_{\text{df},1-\frac{\alpha}{2}}\, s_{\hat{t}_{\text{RMd}}},$$

with degrees of freedom $\text{df} = n_2 - 1$.

16.3.3 Mean-of-ratios

The estimator of the population mean is

$$\bar{y}_{\text{MRd}} = \hat{R}\,\bar{x}_1, \tag{16.37}$$

where

$$\hat{R} = \frac{1}{n_2}\sum_{i=1}^{n_2} r_i, \tag{16.38}$$

and $r_i = y_i/x_i$ is the unit ratio.

The standard error of $\bar{y}_{\text{MRd}}$ can be estimated as

$$s_{\bar{y}_{\text{MRd}}} = \sqrt{\bar{x}_1^2\left(\frac{s_r^2}{n_2}\right)\left(1-\frac{n_2}{n_1}\right) + \frac{s_y^2}{n_1}\left(1-\frac{n_1}{N}\right)}, \tag{16.39}$$

where s_r^2 was defined in (16.25) and computed using the phase-2 data. If you're using an areal sampling frame, it's appropriate to drop the second FPC term (see Section 12.2.1 for discussion).

The population total is estimated using

$$\hat{t}_{\text{MRd}} = N\bar{y}_{\text{MRd}} \tag{16.40}$$

and its standard error is

$$s_{\hat{t}_{\text{MRd}}} = N s_{\bar{y}_{\text{MRd}}}. \tag{16.41}$$

Confidence intervals for the mean and total follow the usual form

$$\bar{y}_{\text{MRd}} \pm t_{\text{df},1-\frac{\alpha}{2}}\; s_{\bar{y}_{\text{MRd}}},$$

$$\hat{t}_{\text{MRd}} \pm t_{\text{df},1-\frac{\alpha}{2}}\; s_{\hat{t}_{\text{MRd}}},$$

with degrees of freedom $\text{df} = n_2 - 1$.

16.4 Illustrative analysis and extensions

The next few sections illustrate how the estimators we've introduced are applied in forest inventory settings. Each example highlights how auxiliary information in the form of a covariate can improve efficiency, and how the choice of estimator depends on available data, the relationship between variables, sampling designs, and field constraints.

We begin in Section 16.4.1 with a timber sale cruise, where ratio estimation offers a simple and efficient approach to estimating total volume. Section 16.4.2 turns to the challenge of quantifying insect damage using remotely sensed data, where double sampling proves useful. Finally, Sections 16.5 and 16.6 explore more advanced inventory designs built around ratio and regression estimators.

16.4.1 Using ratio estimators to estimate timber volume

This illustration is adapted from a short essay by Ken Desmarais, who at the time was the Assistant District Ranger in the White Mountain National Forest, New Hampshire. The piece, entitled *Marcy and Ted Take a Crack at Ratio Estimators*, is both fun and informative, and we encourage you to read the original (included in the `datasets/WMNF` directory). Here we present the key elements, focusing on the methods and analysis.

16.4.1.1 Setting

In this example, the crew needed to mark every merchantable tree and estimate its volume for a spruce-fir timber sale in the White Mountain National Forest. This is a demanding task if DBH and height must be measured for every stem and then entered into a volume equation to calculate volume.

Rather than measure every tree, Marcy, the heroine of the story, suggests using ratio estimation. Her idea is to guess DBH and height for each tree, yielding quick but coarse volume estimates, and then measure DBH and height for a random subset of trees to get accurate volumes. The guesses were made by ocular assessment only—no instruments—just a quick guess at the tree's DBH class and number of 16-foot logs (height), which were then run through a volume table to produce coarse volume estimates. The ratio estimator is used to correct these guesses, producing an accurate and precise estimate of the total sale volume.

Mapping this back to our regression sampling methods, the coarse (guessed) tree volumes play the role of x. Since every tree is guessed, the population mean μ_x is known. The y values come from trees whose DBH and height were measured and then run through a volume equation. The parameter of interest is μ_y, the true mean tree volume, which can then be scaled by N to estimate the total sale volume $\tau_y = N\mu_y$.

16.4.1.2 Sampling design and implementation

1. Each tree in the sale was marked and assigned an ocular estimate of 2-inch DBH class and number of 16-foot logs. These coarse values were run through a balsam-fir volume table to yield a guessed volume (in board feet).

2. A subsample of trees was selected for measurement. This subsample had to be chosen using a random mechanism (e.g., SRS or SYS). To simplify implementation in the field, a SYS design was used. After a random start, every 200-th tree received a DBH and height measurement. The crew figured they needed about 20 observations to estimate the relationship between guessed and estimated volume, i.e., $n = 20$ (x, y) pairs should give a good estimate of R. They estimated about 4,000 trees in the sale, so a SYS sampling interval of 200 should yield the desired sample size n.

3. To start the systematic sample, a random start between 1 and 200 was chosen; the draw produced 97. Starting from the first tree, each was assigned a guessed volume. When the crew reached the 97-th tree, they guessed as usual but also took a measurement. The same was done for the 297-th tree, the 497-th, and so on, until all trees were marked.

TABLE 16.1: Paired guessed and measured tree volumes for the subsample of 20 spruce-fir trees in the White Mountain National Forest.

		Guessed			Measured		
Tree	Species	DBH (in)	Logs (16-ft)	Vol. (bd ft)	DBH (in)	Height (ft)	Vol. (bd ft)
1	rs	10	1.5	48	10.4	27	57.9
2	bf	8	1.0	24	8.6	18	27.4
3	rs	12	1.5	74	11.8	28	79.9
4	rs	10	1.5	48	10.2	25	52.4
5	rs	12	2.0	92	12.4	36	107.0
6	rs	12	2.0	92	12.1	38	104.9
7	bf	10	1.5	48	10.7	24	57.0
8	rs	14	2.5	153	14.8	43	182.7
9	rs	12	2.0	92	11.5	31	80.8
10	rs	10	1.5	48	10.7	28	63.6
11	rs	8	1.0	24	7.7	18	20.6
12	bf	8	1.0	24	8.6	19	28.5
13	rs	8	1.0	24	8.2	16	22.2
14	rs	10	1.0	36	10.8	18	47.6
15	rs	8	1.0	24	8.1	14	19.5
16	bf	12	1.5	74	12.2	27	84.2
17	rs	14	1.5	105	14.8	28	134.6
18	bf	12	2.0	92	11.7	30	82.3
19	rs	10	1.0	36	10.3	18	42.5
20	rs	10	1.5	48	9.7	25	46.5

16.4.1.3 Data and analysis

In the end, $N = 4{,}181$ trees were marked and included in the sale. Because the total population size is known and finite (4,181 marked trees) we will apply estimators for a finite population. Table 16.1 shows the $n = 20$ trees that received both a guess and a measurement. Guessed volumes are coarse values from a volume table; estimated volumes are based on DBH and height measurements entered into a volume equation, providing more accurate estimates.[4] The mean guessed volume across all N trees was $\mu_x = 62.1$ board feet per tree.

The paired data in Table 16.1 are available in `datasets/WMNF/sf_xy.csv`. We read them into `sf_xy` and use a scatter plot (Figure 16.2) to characterize the relationship between guessed and measured volumes, helping inform our choice of estimator.

[4]Volume for *guessed* trees is based on Girard Form Class 78, International 1/4-inch rule in Table 21 of Mesavage and Girard (1956) (8-inch DBH volumes interpolated from the published 10-inch values). For example, looking at the first row in Table 16.1, a 10-inch DBH tree with 1.5 16-foot logs has a volume of 48 board feet according to Table 21. Volumes for *measured* trees use the corresponding Form Class 78, International 1/4-inch formula published in Wiant (1986). This formula was used to create R functions in Exercise 5.3. For instance, to calculate the *measured* volume of a 10.4-inch DBH tree with 27 feet of merchantable height (i.e., 27/16=1.6875 16-foot logs), use `vol_I(10.4, 27/16)`, which returns 57.9 board feet after rounding.

```
sf_xy <- read_csv("datasets/WMNF/sf_xy.csv")
sf_xy %>% glimpse()
```

```
#> Rows: 20
#> Columns: 8
#> $ tree_id      <dbl> 1, 2, 3, 4, 5, 6, 7, 8, 9, 10, 1~
#> $ species      <chr> "rs", "bf", "rs", "rs", "rs", "r~
#> $ dbh_guess    <dbl> 10, 8, 12, 10, 12, 12, 10, 14, 1~
#> $ logs_guess   <dbl> 1.5, 1.0, 1.5, 1.5, 2.0, 2.0, 1.~
#> $ vol_guess    <dbl> 48, 24, 74, 48, 92, 92, 48, 153,~
#> $ dbh_measure  <dbl> 10.4, 8.6, 11.8, 10.2, 12.4, 12.~
#> $ ht_measure   <dbl> 27, 18, 28, 25, 36, 38, 24, 43, ~
#> $ vol_measured <dbl> 57.9, 27.4, 79.9, 52.4, 107.0, 1~
```

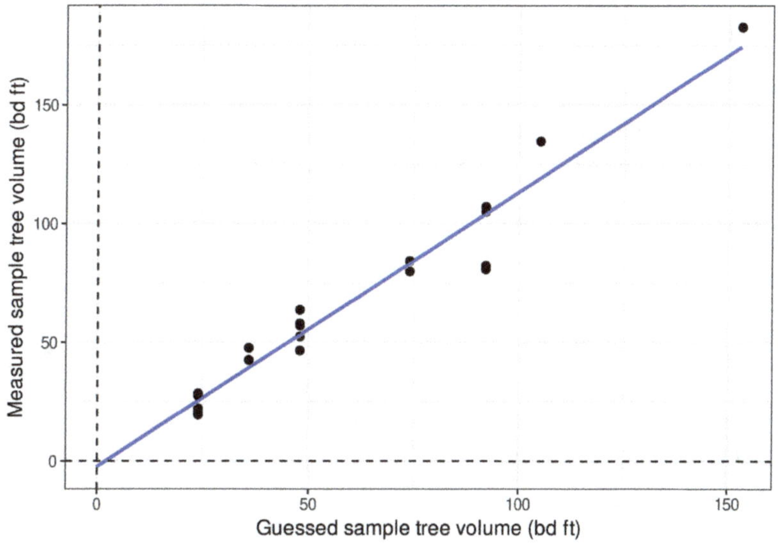

FIGURE 16.2: Guessed (x) versus measured (y) tree volume for the paired trees listed in Table 16.1. A simple linear regression line is drawn for exploratory analysis.

The scatter in Figure 16.2 shows a strong positive relationship between guessed (x) and measured (y) volumes, and the fitted line crosses the y-axis near the origin. With only 20 pairs it's hard to assess variance patterns, but the spread looks roughly constant. This supports using the ratio-of-means estimator defined in Section 16.2.1.

The code below computes the ratio-of-means estimates for mean volume (16.15) and total volume (16.20), along with their respective standard errors (16.18) and (16.21), and a 95% confidence interval for the total.

```
# Some known values.
N <- 4181 # Trees.
```

```
mu_x <- 62.1 # (bd ft/tree).
n <- nrow(sf_xy)

# Pull out variables.
x <- sf_xy %>% pull(vol_guess)
y <- sf_xy %>% pull(vol_measured)

# Ratio estimate.
R_hat <- sum(y) / sum(x)

# Mean estimate (bd ft/tree).
y_bar <- R_hat * mu_x

# Total estimate (bd ft).
t_hat <- N * y_bar

# Standard error of the mean.
s_y_bar <- sqrt((mu_x / mean(x))^2 *
  (var(y) + R_hat^2 * var(x) - 2 * R_hat * cov(x, y)) / n *
  (1 - n / N))

# Standard error of the total.
s_t_hat <- N * s_y_bar

# 95% CI
t <- qt(df = n - 1, p = 1 - 0.05 / 2)
ci_lower <- t_hat - t * s_t_hat
ci_upper <- t_hat + t * s_t_hat
```

From the code above, after rounding, the computed $\hat{R}$ is 1.11. This multiplier converts the population mean of x into an estimate of the population mean of y. The estimated total sale volume is 288,941 board feet, with a 95% confidence interval of (269,717, 308,165).

These results come from just 20 measured trees. Guessing takes time, but it's a fraction of what it would take to measure DBH and height on every stem. That's the payoff of a good design: many quick, inexpensive values adjusted by a small set of careful measurements. We'll return to variations on this theme in the sections that follow and in the chapter exercises.

16.4.2 Double sampling for dead-tree totals

In the previous example, the auxiliary variable x was observed for every tree in the sale. In practice, we often can't afford to collect x for the entire population. As described in Section 16.3, a common workaround is *double sampling*, where a relatively inexpensive measurement is taken on a large phase-1 sample, and a more accurate but costly measurement is taken on a smaller, nested phase-2 sample. Remote sensing is often used in phase 1 to provide broad, inexpensive coverage, while phase 2 relies on targeted field data collection for more detailed and accurate measurements.

This example involves a forest manager who wants to estimate the total number of dead trees in a 400-acre area under heavy insect infestation. The area is divided into $N = 200$ equal-area tiles. Aerial imagery-based mortality counts are collected for a phase-1 sample of $n_1 = 100$ tiles. From those, a nested phase-2 subsample of $n_2 = 10$ tiles is visited for a field-based mortality survey. The aerial counts (x) are fairly accurate, but the manager is concerned about missing subcanopy mortality and other potential biases due to image interpretation. The field-based counts (y) are more accurate.

Our goal is to estimate total mortality across the full area using a double sampling estimator. For comparison, we'll also estimate total mortality using only the phase-2 field measurements, ignoring the phase-1 data.

16.4.2.1 Data and analysis

Here, the phase-1 and phase-2 data are stored in a single file, with one row per sampled tile.

Column names are:

- `tile_id` unique identifier for each sampled tile (n_1 rows);
- `x` aerial image count (available for all n_1 tiles);
- `y` field-based mortality count (available for n_2 tiles; `NA` otherwise).

We begin by reading the data into `mort`. From this, we extract three vectors: x_1 contains all phase-1 counts; x_2 is the subset of aerial counts paired with field measurements; and y_2 holds the phase-2 field measurements. These align with the notation used in our estimators.

```
mort <- read_csv("datasets/insects/mortality_two_phase.csv")

# Extract phase-1 sample.
x_1 <- mort %>% pull(x)

# Extract phase-2 sample.
x_2 <- mort %>% filter(!is.na(y)) %>% pull(x)
y_2 <- mort %>% filter(!is.na(y)) %>% pull(y)
```

Figure 16.3 shows the paired (x_2, y_2) values for the ten tiles with both aerial and field-based counts. The plot affirms the manager's concerns: assuming the field survey is more accurate, the image-based survey tends to undercount smaller values and overcount larger ones. But that's okay—the important thing is the strong linear relationship between y and x. The fitted line has a y-intercept well above the origin, the relationship appears roughly proportional, and the scatter of points about that line has fairly constant variance across values of x. Given these observations, the regression estimator is more appropriate than the ratio estimator. However, for comparison, we'll apply both the regression and

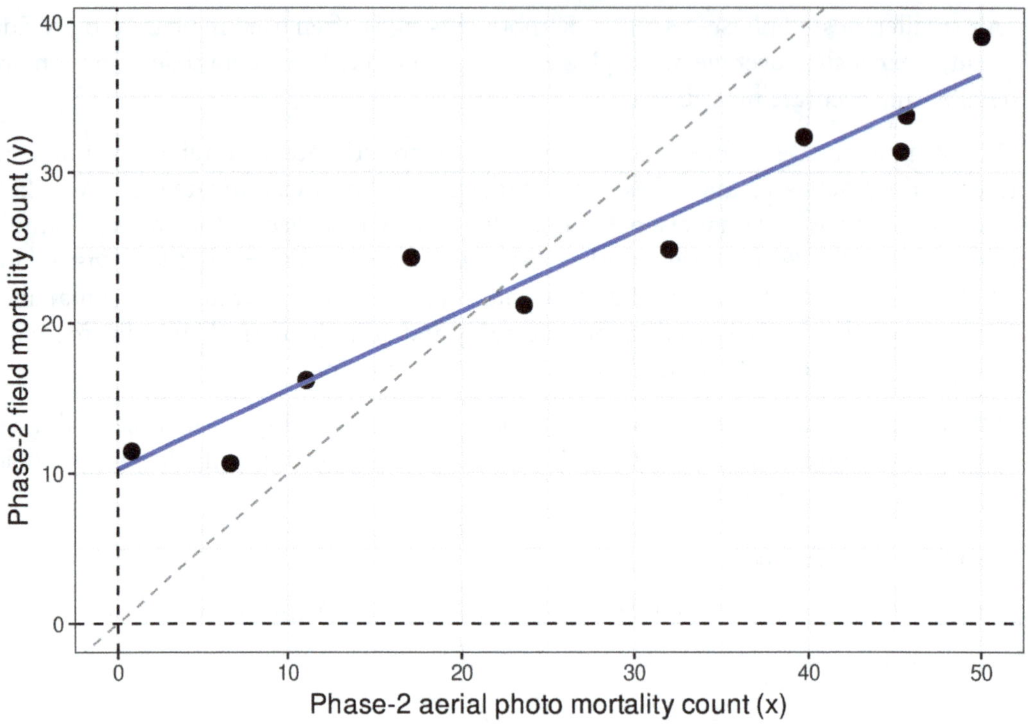

FIGURE 16.3: Paired aerial (x_2) and field (y_2) mortality counts for ten tiles. The fitted regression line is used only to characterize the relationship and help choose an appropriate estimator. The gray dashed line marks the 1:1 reference.

ratio-of-means estimators, and compare their results to a baseline SRS estimate based on the n_2 field measurements alone.

As noted in Section 11.3.7, the *relative standard error* (RSE) is a useful way to express estimation uncertainty as a percentage of the estimate itself. It shares the same mathematical form as the general relative variation measure in (11.43), with the standard error of the estimator in the numerator and the estimate itself in the denominator.

In this case, since we're estimating the total number of dead trees, the numerator is the standard error of the total, and the denominator is the estimated total. All else equal, estimators with lower RSE are preferred because they reflect greater precision relative to the size of the estimate.

The code below applies the three estimators and collects their results into a summary table.

```
# Known values.
N <- 200
n_1 <- length(x_1)
n_2 <- length(x_2)
```

```
# Means, variances, and covariance.
x_bar_1 <- mean(x_1)
x_bar_2 <- mean(x_2)
y_bar_2 <- mean(y_2)
s_sq_x_2 <- var(x_2)
s_sq_y_2 <- var(y_2)
s_xy <- cov(x_2, y_2)

# Ratio-of-means (double sampling).
R_hat_RMd <- y_bar_2 / x_bar_2
y_bar_RMd <- R_hat_RMd * x_bar_1
s_y_bar_RMd <- sqrt((1 - n_2 / n_1) *
  ((x_bar_1 / x_bar_2)^2 *
    ((s_sq_y_2 +
      R_hat_RMd^2 * s_sq_x_2 -
      2 * R_hat_RMd * s_xy) / n_2)) +
  (s_sq_y_2 / n_1) * (1 - n_1 / N))

t_hat_RMd <- N * y_bar_RMd
s_t_hat_RMd <- N * s_y_bar_RMd

RSE_t_hat_RMd <- 100 * s_t_hat_RMd / t_hat_RMd

# Regression (double sampling).
S_xx <- sum((x_2 - x_bar_2)^2)
S_yy <- sum((y_2 - y_bar_2)^2)
S_xy <- sum((x_2 - x_bar_2) * (y_2 - y_bar_2))
b1 <- S_xy / S_xx
s_sq_y_given_x <- (S_yy - S_xy^2 / S_xx) / (n_2 - 2)

y_bar_Rd <- y_bar_2 + b1 * (x_bar_1 - x_bar_2)

s_y_bar_Rd <- sqrt(s_sq_y_given_x *
  (1 / n_2 + (x_bar_1 - x_bar_2)^2 / S_xx) *
  (1 - n_2 / n_1) +
  (s_sq_y_2 / n_1) *
    (1 - n_1 / N))

t_hat_Rd <- N * y_bar_Rd
s_t_hat_Rd <- N * s_y_bar_Rd

RSE_t_hat_Rd <- 100 * s_t_hat_Rd / t_hat_Rd
```

```
# Phase-2 only, SRS.
y_bar_SRS <- y_bar_2
s_y_bar_SRS <- sqrt((1 - n_2 / N) * (s_sq_y_2 / n_2))

t_hat_SRS <- N * y_bar_SRS
s_t_hat_SRS <- N * s_y_bar_SRS

RSE_t_hat_SRS <- 100 * s_t_hat_SRS / t_hat_SRS

# Output summary.
tibble(
  method = c("SRS (Phase-2)", "Ratio (RMd)", "Regression (Rd)"),
  y_bar = c(y_bar_SRS, y_bar_RMd, y_bar_Rd),
  s_y_bar = c(s_y_bar_SRS, s_y_bar_RMd, s_y_bar_Rd),
  t_hat = c(t_hat_SRS, t_hat_RMd, t_hat_Rd),
  s_t_hat = c(s_t_hat_SRS, s_t_hat_RMd, s_t_hat_Rd),
  RSE_t_hat = c(RSE_t_hat_SRS, RSE_t_hat_RMd, RSE_t_hat_Rd)
)
```

```
#> # A tibble: 3 x 6
#>   method          y_bar s_y_bar t_hat s_t_hat RSE_t_hat
#>   <chr>           <dbl>   <dbl> <dbl>   <dbl>     <dbl>
#> 1 SRS (Phase-2)    24.5    2.99 4904.    598.     12.2
#> 2 Ratio (RMd)      23.9    2.21 4788.    442.      9.24
#> 3 Regression (Rd)  24.2    1.06 4836.    212.      4.39
```

The results in the table above highlight the benefit of incorporating auxiliary information from the phase-1 aerial counts. Both the regression and ratio estimators yield substantially smaller RSEs compared with the phase-2-only SRS estimator.

Because these are simulated data, we know the true population values: $\mu_y = 24.75$ dead trees per tile and $\tau_y = 4,951$ dead trees across the full 400 acres.

As suggested by the exploratory plot in Figure 16.3, the regression estimator is more appropriate than the ratio estimator due to the nonzero intercept. Ratio estimation assumes a proportional relationship between y and x that passes through the origin. When that assumption is violated—in particular, when the relationship has a nonzero intercept—the ratio estimator can be biased. The relatively low ratio estimate hints at this bias, but a full repeated sampling analysis—like the one in Section 15.7.1—would be needed to quantify it. Still, based on the evidence from Figure 16.3 and the markedly lower RSE, it's clear that the regression estimator is both more appropriate and more efficient for these data.

This example underscores the importance of EDA when working with auxiliary variables. A simple scatterplot with a fitted line can reveal whether the relationship between x and y departs from the zero-intercept assumption underlying ratio estimation. In this case, visual inspection of the (x, y) relationship provided an early and reliable indication that regression was the better choice.

16.5 Big BAF sampling

This section continues our focus on combining inexpensive measurements with a targeted subset of more costly measurements to improve parameter estimates in terms of both statistical and operational efficiency. Big BAF sampling fits squarely within this theme. It builds on the principles of horizontal point sampling (Section 12.4) and introduces an elegant way to obtain per-unit-area and total estimates by splitting the estimation process into two parts: a count-based estimate of basal area (BA) per unit area, and a tree-based estimate of volume-to-basal-area ratio (VBAR). While the method was originally devised to estimate volume, it can be applied more broadly to other parameters of interest. As initially introduced in Section 12.4.6, think of the "V" in VBAR more generically as the *variable* of interest.

The beauty of the Big BAF method lies in its simplicity. The basic idea is straightforward: at each sampling location, conduct a tree count using a standard BAF (e.g., one that yields about 4 to 10 "in" trees). These tree counts provide an estimate of BA per unit area using point-level summaries, as described in Section 12.4. Then, using a second sweep at the same sampling location, apply a larger BAF to identify a subset of trees to measure. These measurement trees are used to calculate tree-level VBAR values by measuring DBH (used to compute BA) and height, which together are used to estimate volume (or another variable of interest). The average of these tree-level VBAR values is computed across all measurement trees—that is, pooled across sampling locations. This average VBAR is then multiplied by the BA estimate from the tree counts to yield an estimate of volume per unit area.

Counting trees with a standard-size BAF is fast and inexpensive, which makes it feasible to visit many sample locations. Measuring DBH and other tree variables to calculate VBAR, by contrast, is more time-consuming and costly. The Big BAF method balances this tradeoff by using a big BAF to identify a smaller subset of trees to measure at each location. For example, if a standard-size BAF yields an average of 8 count trees per point, a big BAF that is four times larger would result in about 2 measurement trees on average—just one-fourth as many.

Studies and personal observations have shown that BA per-unit-area estimates are virtually always more variable than VBAR estimates, so it's generally more efficient to invest in a larger number of sampling locations (for BA counts) and fewer tree measurements (for VBAR). The Big BAF design supports this strategy naturally and allows us to concentrate effort where it's most needed.

Kim Iles conceived the Big BAF method, building on earlier related ideas and methods developed by John F. Bell and others, as well as foundational work by Lewis R. Grosenbaugh in the 1940s and 1950s. A thorough treatment of the topic—including historical background, estimator behavior, variations, and sample size versus cost considerations—can be found in Iles (2003), Marshall et al. (2004), Gove et al. (2020), and references therein. These sources provide practical guidance on selecting both the number of sampling locations and the number of measurement trees, along with deeper insights into estimator performance and design tradeoffs.

Big BAF is closely related to double sampling and ratio estimation, although it's not like the double sampling introduced in Section 16.3, where phase-1 and phase-2 are a sample and subsample of sampling units (e.g., plots or points). Under Big BAF, the phases occur at the tree level: phase-1 can be seen as all count trees identified using the standard-size BAF and used to estimate BA per unit area through point-level summaries; phase-2 consists of the subset of trees measured at each sampling location using the big BAF and used to estimate VBAR.[5] As we'll see in the next section, the estimator follows the same spirit as other ratio estimators—correcting a coarse, inexpensive estimate using a more accurate, but costly, measurement.

16.5.1 Estimation using Big BAF

As described above, the Big BAF estimator comprises two components: (1) the mean basal area per unit area ($\overline{\text{BA}}$), based on point-level estimates using a standard-size BAF, and (2) the mean volume-to-basal-area ratio ($\overline{\text{VBAR}}$), based on tree-level measurements using a big BAF. Throughout, we'll use superscripts s and b to distinguish quantities associated with the small (standard-size) and big BAFs, respectively.

Following the methods in Section 12.4, the mean basal area per unit area is estimated using

$$\overline{\text{BA}} = \frac{1}{n}\sum_{i=1}^{n} \text{BA}_i, \tag{16.42}$$

where n is the number of sampling locations and $\text{BA}_i = m_i^s \cdot \text{BAF}^s$ is the observed basal area per unit area at the i-th location, with m_i^s being the number of count trees selected using the small BAF (BAF^s).

We define VBAR for the j-th measurement tree at the i-th sampling location as

$$\text{VBAR}_{ij} = \frac{v_{ij}}{BA_{ij}} \tag{16.43}$$

and compute the mean across all measurement trees as

$$\overline{\text{VBAR}} = \frac{1}{m^b}\sum_{i=1}^{n}\sum_{j=1}^{m_i^b} \text{VBAR}_{ij}, \tag{16.44}$$

where m_i^b is the number of measurement trees selected using the big BAF (BAF^b) at the i-th sampling location, n is the number of sampling locations, and $m^b = \sum_{i=1}^{n} m_i^b$ is the total number of measurement trees across all locations. Note that, unlike the point-

[5]Technically, the VBAR estimate could come from a separate set of sampling locations or even a different sampling design; it's just convenient to take spatially paired measurements.

level summary of BA, which includes locations with zero count trees, the mean $\overline{\text{VBAR}}$ is defined only over measurement trees, so only locations with at least one measurement tree contribute to the sum (i.e., the i-th sampling location could have $m_i^b = 0$ and hence doesn't contribute to the sum).

The Big BAF estimator for volume per unit area is then given by

$$\hat{v} = \overline{\text{BA}} \cdot \overline{\text{VBAR}}. \tag{16.45}$$

It's useful to verify how the units work out when applying (16.45). For example, suppose volume is measured in cubic meters and basal area is reported in square meters per hectare. Then (16.45) with units becomes

$$\begin{aligned} \hat{v}\left(\frac{\text{m}^3}{\text{ha}}\right) &= \overline{\text{BA}}\left(\frac{\text{m}^2}{\text{ha}}\right) \cdot \overline{\text{VBAR}}\left(\frac{\text{m}^3}{\text{m}^2}\right) \\ &= \overline{\text{BA}}\left(\frac{\cancel{\text{m}^2}}{\text{ha}}\right) \cdot \overline{\text{VBAR}}\left(\frac{\text{m}^3}{\cancel{\text{m}^2}}\right). \end{aligned} \tag{16.46}$$

The m^2 units cancel, yielding the desired volume per unit area.

Because $\hat{v}$ is the product of two sample means—each estimated from a different sample—its standard error must account for the variability in both components. A common and practical approach is to use *Bruce's formula*, which assumes that $\overline{\text{BA}}$ and $\overline{\text{VBAR}}$ are uncorrelated and combines their uncertainty in quadrature

$$s_{\hat{v}} \approx \sqrt{s^2_{\overline{\text{BA}}} + s^2_{\overline{\text{VBAR}}}}. \tag{16.47}$$

This approach is well established in forestry sampling (Bell and Alexander 1957; Bruce 1961) and is explained in the context of Big BAF estimation by Gove et al. (2020).

In practice, Bruce's formula is implemented by combining uncertainty on a common relative scale, rather than by directly summing standard errors expressed in different units.

Each standard error is estimated from its respective sample data using (11.24).

The standard error for $\overline{\text{BA}}$ is

$$s_{\overline{\text{BA}}} = \frac{s_{\text{BA}}}{\sqrt{n}}, \tag{16.48}$$

where

$$s_{\text{BA}} = \sqrt{\frac{1}{n-1}\sum_{i=1}^{n}\left(\text{BA}_i - \overline{\text{BA}}\right)^2}. \tag{16.49}$$

The standard error for $\overline{\text{VBAR}}$ is

$$s_{\overline{\text{VBAR}}} = \frac{s_{\text{VBAR}}}{\sqrt{m^b}}, \tag{16.50}$$

where

$$s_{\text{VBAR}} = \sqrt{\frac{1}{m^b-1}\sum_{i=1}^{n}\sum_{j=1}^{m_i^b}\left(\text{VBAR}_{ij} - \overline{\text{VBAR}}\right)^2}. \tag{16.51}$$

Because $\overline{\text{BA}}$ and $\overline{\text{VBAR}}$ are on wholly different scales (e.g., m^2/ha and m^3/m^2, respectively), we apply (16.47) using *percent standard errors*. This avoids scale incompatibility and offers a clearer sense of relative precision. Rewriting (16.47) using percent standard errors gives

$$s_{\hat{v}\%} \approx \sqrt{s^2_{\overline{\text{BA}}\%} + s^2_{\overline{\text{VBAR}}\%}}, \tag{16.52}$$

where the percent standard errors for $\overline{\text{BA}}$ and $\overline{\text{VBAR}}$ are defined as

$$s_{\overline{\text{BA}}\%} = 100 \cdot \frac{s_{\overline{\text{BA}}}}{\overline{\text{BA}}} \tag{16.53}$$

and

$$s_{\overline{\text{VBAR}}\%} = 100 \cdot \frac{s_{\overline{\text{VBAR}}}}{\overline{\text{VBAR}}}. \tag{16.54}$$

In addition to aligning the two error components on a common scale, expressing the standard errors as percents in (16.52) highlights how each component contributes to the overall uncertainty in $\hat{v}$. It also provides a clear framework for evaluating tradeoffs between count and measurement effort; see, e.g., Iles (2003) and Marshall et al. (2004).

To return the standard error to the original units of $\hat{v}$, simply multiply the percent standard error by the estimate and divide by 100

$$s_{\hat{v}} = \hat{v} \cdot \frac{s_{\hat{v}\%}}{100}. \tag{16.55}$$

If a confidence interval is desired, a t-value can be computed using the Welch–Satterthwaite approximation for degrees of freedom (Welch 1947). This approach is appropriate when combining uncertainty from two independent sample means with unequal variances, and it aligns with our application of Bruce's formula to estimate the standard error. The approximate degrees of freedom are given by

$$\text{df} \approx \frac{s^2_{\overline{\text{BA}}\%} + s^2_{\overline{\text{VBAR}}\%}}{\frac{s^4_{\overline{\text{BA}}\%}}{n-1} + \frac{s^4_{\overline{\text{VBAR}}\%}}{m^b-1}}. \tag{16.56}$$

With this degrees of freedom approximation, the confidence interval for $\hat{v}$ is

$$\hat{v} \pm \text{t}_{\text{df},1-\frac{\alpha}{2}}\, s_{\hat{v}}. \tag{16.57}$$

To estimate totals, simply multiply the mean estimate $\hat{v}$ and its standard error $s_{\hat{v}}$ by the population area.

Some general observations about Big BAF. As mentioned earlier, the estimates for BA and VBAR need not come from the same locations or sampling designs; it's simply convenient to pair them. When sampling locations follow a SYS design, combining the two sampling efforts at the same points helps ensure good spatial coverage and representation of the population.

The method offers great flexibility in how BA and VBAR are sampled, and at what intensity. Working with percent standard errors makes *optimal* allocation of sampling resources a straightforward exercise, assuming costs can be assigned to sampling locations (i.e., cost per unit of n), tree counting (for BA), and tree measurement (for VBAR), and that reasonable percent standard error estimates are available (e.g., from prior inventories).

The standard error estimators presented here for $\overline{\text{VBAR}}$ and $\hat{v}$ rely on two distinct approximations. First, the standard error for $\overline{\text{VBAR}}$ treats the sampled tree-level ratios $\text{VBAR}_{ij} = v_{ij}/\text{BA}_{ij}$ as the observed values and applies the usual sample mean and variance formulas to those ratios. Although each VBAR_{ij} is computed from two measurements taken on the same tree, the ratio itself is the unit of analysis for estimation. As noted by Gregoire and Valentine (2008) (p. 259), a more exact variance estimator could be constructed that explicitly accounts for the ratio structure, but simulation results in Gove et al. (2020) show that the simpler approach yields nearly identical inferences in practice.

Second, Bruce's formula for the standard error of $\hat{v}$ assumes that the two summary estimators being combined—the mean basal area per unit area, $\overline{\text{BA}}$, and the mean volume-to-basal-area ratio, $\overline{\text{VBAR}}$—have negligible covariance. This assumption concerns the relationship between the estimators themselves, not the relationship between volume and basal area at the tree level, which are necessarily related by construction.

While this negligible-covariance assumption may not hold exactly in all settings, it's often a reasonable approximation in forest inventory applications, particularly when basal area is estimated from point-level counts and VBAR is estimated from a separate set of measurement trees. Even when some correlation is present, empirical studies suggest that omitting the covariance term typically has little practical impact on inference (see, e.g., Gove et al. (2020)). For these reasons, Bruce's formula remains a widely used and operationally effective approach for estimating uncertainty under Big BAF sampling.

As noted by Marshall et al. (2004), a useful feature of Big BAF is that it provides an internal check on basal area estimates. Because the count and measurement data rely on different BAFs, they yield separate estimates of BA. If those estimates differ substantially, it can indicate potential issues with field protocols that warrant closer review.

If estimates are to be broken down by species, product type, strata, or any other category requiring separate VBAR estimates, care must be taken to ensure enough measurement trees are sampled to support reliable estimation within each group. The more categories considered, the more measurement trees are needed to inform the VBAR estimates—which has cost implications.

16.5.2 Applying Big BAF

To illustrate Big BAF, we return once more to the Harvard Forest dataset. As in earlier examples, this is a simplified setting, but it allows us to walk through the full workflow using real data and compare sample-based estimates to known population values.

The objective is to estimate mean AGB per hectare for all stems with DBH $\geq$ 12.7 cm across the 35 ha forest.

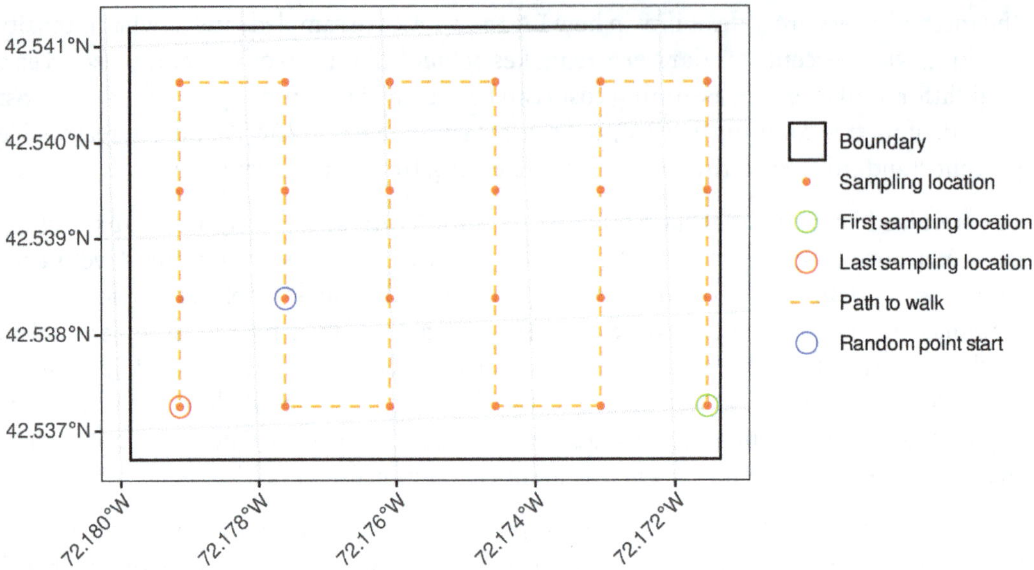

FIGURE 16.4: Systematic sample locations for the Harvard Forest Big BAF example.

We again use the `sys_grid()` function, introduced in Chapter 14, to generate a systematic grid of sampling locations across the forest. For this example, the spacing is 125-by-125 meters, yielding $n = 24$ sampling locations (Figure 16.4).

We'll use metric $\text{BAF}^s = 5$ for count trees and $\text{BAF}^b = 20.25$ for measurement trees.[6] This setup implies about one measurement tree for every four count trees (since $5/20.25 \approx 0.25$).

For both count and measurement trees, the walkthrough method was used to address boundary slopover bias. Trees with walkthrough points falling outside the boundary received a count of two; see Section 12.3.3 for details. As in previous sections, sampling locations with no trees were assigned a tree count of zero.

In the analysis that follows, we'll first apply the Big BAF mean and standard error estimators outlined in the previous section. Then, for comparison, we'll repeat the analysis assuming measurements were taken on all count trees (i.e., all "in" trees under BAF^s), representing a more traditional cruise.

The data are located in the `"datasets/BigBAF"` directory. `"HF_BAF_s_count.csv"` contains the count data collected using BAF^s and `"HF_BAF_b_measure.csv"` contains the measurement data collected using BAF^b. We begin by reading these files and inspecting their structure.

```
count_trees <- read_csv("datasets/BigBAF/HF_BAF_s_count.csv")
measurement_trees <- read_csv("datasets/BigBAF/HF_BAF_b_measure.csv")
```

[6]The wide-scale relascope includes larger BAF increments, such as 20.25 m^2/ha, suitable for Big BAF sampling. To convert from metric to English units (ft^2/ac), multiply by 4.356; the chosen BAFs correspond approximately to English BAFs of 21.8 (small) and 88.2 (big).

```r
glimpse(count_trees)
```

```
#> Rows: 194
#> Columns: 2
#> $ point_id   <dbl> 1, 1, 1, 1, 1, 1, 1, 1, 2, 2, 2, 3~
#> $ tree_count <dbl> 1, 1, 1, 1, 1, 1, 1, 1, 1, 1, 2, 1~
```

```r
glimpse(measurement_trees)
```

```
#> Rows: 54
#> Columns: 4
#> $ point_id   <dbl> 1, 1, 2, 3, 4, 4, 4, 5, 6, 6, 6, 7~
#> $ dbh_cm     <dbl> 48.0, 48.6, 22.6, 0.0, 15.9, 33.6,~
#> $ agb_Mg     <dbl> 2.22710, 2.29694, 0.44771, 0.00000~
#> $ tree_count <dbl> 1, 1, 1, 0, 1, 1, 1, 0, 1, 1, 1, 1~
```

Let's get a sense of the number of count (m_i^s) and measurement trees (m_i^b), based on their respective BAFs, across sampling locations (which are uniquely identified using `point_id`).

```r
count_trees %>%
  group_by(point_id) %>%
  summarize(m_s_i = sum(tree_count)) %>%
  summarize(min = min(m_s_i), max = max(m_s_i),
            mean = mean(m_s_i), total = sum(m_s_i))
```

```
#> # A tibble: 1 x 4
#>     min   max  mean total
#>   <dbl> <dbl> <dbl> <dbl>
#> 1     1    18  8.12   195
```

```r
measurement_trees %>%
  group_by(point_id) %>%
  summarize(m_b_i = sum(tree_count)) %>%
  summarize(min = min(m_b_i), max = max(m_b_i),
            mean = mean(m_b_i), total = sum(m_b_i),
            n_zero = sum(m_b_i == 0))
```

```
#> # A tibble: 1 x 5
#>     min   max  mean total n_zero
#>   <dbl> <dbl> <dbl> <dbl>  <int>
#> 1     0     5  2.08    50      4
```

Given these summaries, as expected, there are roughly four times as many count trees as measurement trees (195 vs. 50 total). The number of count trees per sampling location ranged from 1 to 18, with an average of about 8. This wide range in the number of "in" trees reflects the spatial variability in basal area across the forest. Measurement trees ranged from 0 to 5 per location, with an average of about 2. As indicated by `n_zero`, 4 sampling locations had no measurement trees.

Next, using the count trees, we estimate mean basal area (BA, in m^2/ha) using (16.42), standard error of the mean using (16.48), and standard error expressed as a percent using (16.53).

```
BA <- small_BAF_count %>%
    group_by(point_id) %>%
    summarize(m = sum(tree_count), # (trees per point).
              BA = m * BAF_small) %>% # (m^2/ha per point).
    summarize(n = n(), # Number of points.
              mean = mean(BA), # (m^2/ha).
              cv = 100 * sd(BA)/mean, # Coefficient of variation.
              se = sd(BA) / sqrt(n), # (m^2/ha).
              se_percent = 100 * se / mean) # Std. err. as % of mean.

BA
```

```
#> # A tibble: 1 x 5
#>       n  mean    cv    se se_percent
#>   <int> <dbl> <dbl> <dbl>      <dbl>
#> 1    24  40.6  47.1  3.90       9.61
```

Now we estimate mean VBAR (Mg/m^2) using (16.44), standard error using (16.50), and standard error expressed as a percent of the mean using (16.54).

Recall that in our cruise data, we included a `tree_count` of zero to track which sampling locations had no measurement trees. However, as noted below (16.44), these zeros should not be included in the calculation of $\overline{\text{VBAR}}$, so we filter them out using `filter(tree_count > 0)` in the subsequent piped sequence below.

After removing those rows, `tree_count` takes on values of 1 or 2. A value of 2 indicates the tree was double-counted based on the walkthrough method. Because we're computing the mean across trees (not across sampling points), the cleanest way to handle this is to duplicate the row for any tree with `tree_count == 2`. We do this using `uncount(tree_count)` from the `tidyr` package, which replicates rows according to a count column.[7]

```
# Basal area conversion factor.
c <- pi / (10000 * 4)

VBAR <- big_BAF_measure %>%
```

[7]We didn't cover the `tidyr::count()` and `uncount()` functions, but do check them out—they're very handy.

```
  filter(tree_count > 0) %>%
    uncount(tree_count) %>%
    mutate(BA = c * dbh_cm^2, # (m^2).
           VBAR = agb_Mg / BA) %>% # (Mg/m^2 per tree).
    summarize(m = n(), # Number of trees.
              mean = mean(VBAR), # (Mg/m^2).
              cv = 100 * sd(VBAR) / mean, # Coefficient of variation.
              se = sd(VBAR) / sqrt(m), # (Mg/m^2).
              se_percent = 100 * se / mean) # Std. err. as % of mean.

VBAR
```

```
#> # A tibble: 1 x 5
#>       m  mean    cv    se se_percent
#>   <int> <dbl> <dbl> <dbl>      <dbl>
#> 1    50  8.93  40.3 0.508       5.69
```

Lastly, given BA and VBAR, we compute the Big BAF estimate of AGB per unit area. The code below implements (16.45) and (16.52), and also converts the percent standard error into the standard error in AGB units using (16.55).

```
# Mean estimate (Mg/ha).
agb_Mg_ha <- VBAR$mean * BA$mean
agb_Mg_ha
```

```
#> [1] 362.69
```

```
# Standard error as percent of biomass.
se_agb_Mg_percent <- sqrt(BA$se_percent^2 + VBAR$se_percent^2)

se_agb_Mg_percent
```

```
#> [1] 11.172
```

```
# Standard error (Mg/ha).
se_agb_Mg <- agb_Mg_ha * se_agb_Mg_percent/100

se_agb_Mg
```

```
#> [1] 40.519
```

As discussed earlier, it's expected for VBAR to be less variable than BA. This pattern holds for these data as well, with $\text{CV}_{\text{VBAR}} = 40.27\%$ and $\text{CV}_{\text{BA}} = 47.09\%$ reflecting the relative variability observed in the sample data, and $s_{\overline{\text{VBAR}}\%} = 5.69$ and $s_{\overline{\text{BA}}\%} = 9.61$ reflecting the relative variability of the sampling estimates. This suggests that greater precision could be achieved by adding more sample locations to improve the estimate of BA, rather than by selecting a smaller BAF^b to increase the number of measurement trees.

We can then construct a confidence interval for the mean AGB estimate using the approximate degrees of freedom formula in (16.56) and the interval expression in (16.57).

```
BA_df <- BA$n - 1
VBAR_df <- VBAR$m - 1

df <- (BA$se_percent^2 + VBAR$se_percent^2)^2 /
  (BA$se_percent^4 / BA_df + VBAR$se_percent^4 / VBAR_df)

t <- qt(p = 1 - 0.05 / 2, df = df)

c(agb_Mg_ha - t * se_agb_Mg, agb_Mg_ha + t * se_agb_Mg)
```

```
#> [1] 280.78 444.60
```

We see that Big BAF yields a mean of 362.69 (Mg/ha) with a 95% confidence interval from 280.78 to 444.6, which covers the true biomass of 341.17.

The code below shows what we would get using a traditional cruise that measures every “in” tree using BAF^s. These data are held in `"datasets/BigBAF/HF_BAF_s_measure.csv"`.

```
measurement_trees <- read_csv("datasets/BigBAF/HF_BAF_s_measure.csv")

measure_all <- measurement_trees %>%
  mutate(TF = ifelse(tree_count != 0, BAF_small / (c * dbh_cm^2), 0)) %>%
  group_by(point_id) %>%
  summarize(agb_Mg = sum(TF * tree_count * agb_Mg)) %>% # (Mg/ha per point).
  summarize(
    n = n(), # Number of points.
    mean = mean(agb_Mg), # (Mg/ha).
    se = sd(agb_Mg) / sqrt(n), # (Mg/ha).
    se_percent = 100 * se / mean
  ) # Std. err. as % of mean.

agb_Mg_ha <- measure_all$mean
se_agb_Mg <- measure_all$se
se_agb_Mg_percent <- measure_all$se_percent

agb_Mg_ha
```

```
#> [1] 349.77
```

```
se_agb_Mg
```

```
#> [1] 29.091
```

```
se_agb_Mg_percent
```

```
#> [1] 8.3171
```

The Big BAF approach offers a clear tradeoff: by measuring 145 fewer trees (the difference between the number of trees counted with BAF^s and BAF^b), we reduce field effort and associated costs. In exchange, we accept a modest increase in uncertainty—with percent standard error rising from 8.32 to 11.17, or standard error in AGB units from 29.09 to 40.52. If you don't want that increased uncertainty, you can simply add more count trees to improve the BA estimate at very low cost—reinvesting part of the time saved by not measuring those 145 trees (see Iles (2003) for a detailed discussion of Big BAF cruise planning and associated software).

16.6 3P sampling

Probability proportional to prediction (3P) sampling directs detailed measurements toward the most influential units in a population. It begins by making a quick, inexpensive prediction for every unit—something easy to obtain, like an ocular estimate or a simple field measurement. Units with larger predictions are more likely to be selected for full measurement, focusing effort on those expected to have the greatest influence on population parameter estimates. Predictions and measurements are combined using a mean-of-ratios estimator, which is appropriate for the tree-level selection mechanism in 3P sampling and accounts for unequal selection probabilities. The resulting design focuses measurement effort on influential units while still producing accurate, stable estimates under reasonable conditions.

Classical 3P sampling was introduced by Lewis R. Grosenbaugh in the 1960s (see, e.g., Grosenbaugh (1965)) and remains the most widely applied version of the method. While originally developed for estimating timber volume, 3P is useful in any setting where the variable of interest is costly to measure but can be predicted with reasonable accuracy. We'll use total volume to motivate the discussion, but the same logic applies to other forest parameters. Classical 3P sampling involves the following steps.

1. Predict. Visit every tree and record a quick prediction x that scales with the variable of interest y. Here, x is the covariate. This might be an ocular estimate of volume or a simple dimensional measure like DBH or height. The goal isn't precision, but consistency—x should be roughly proportional (or transformable to be proportional) to the measurement y.

2. Select and measure. Use the predictions to assign each tree a selection probability proportional to x. Select a subsample accordingly, and for each selected tree, collect the measurement y.

3. Estimate. Combine the paired x and y values using a mean-of-ratios estimator. This yields an estimate of the population-level ratio between the measurement and the prediction. Scaling this ratio by the known mean or total of x (recorded for all units) gives an estimate of the mean or total of y. The estimator naturally accounts for the unequal selection probabilities.

Classical 3P is most effective when the variable of interest is expensive to measure, predictions are available for all trees, and the measurement-to-prediction ratio is relatively stable—even if the variable itself is highly variable.

The design concentrates measurement effort on the most influential trees while still visiting every tree, making it especially well suited for small areas or situations where all trees must be assessed (e.g., for marking).

In these settings, 3P offers several advantages:

- straightforward implementation with minimal training;
- every tree is visited, avoiding boundary complications;
- high efficiency, often yielding small standard errors relative to effort;
- detailed data that support stand and stock tables by species and diameter class;
- better estimation for rare species;
- no need to survey tract area, which can save considerable time in some settings.

Although our focus is on classical 3P, several useful variants exist. One example is point-3P sampling, which extends the idea of classical 3P to point sampling. At each point, trees determined to be "in" using the usual angle method (e.g., prism, angle gauge, or relascope) each receive a prediction x (e.g., DBH or height), which then determines the probability they're measured for y. It has all the advantages of classical 3P, but doesn't require every tree in the population to receive a prediction. An overview of point-3P and related methods is provided by West (2011).

For a deeper treatment of 3P sampling, Iles (2003) provides a comprehensive coverage, including historical context, practical guidance on implementation, and common pitfalls with solutions. For a more theoretical perspective, see Gregoire and Valentine (2008), which situates 3P within the broader class of unequal-probability sampling designs.

The next section further develops the classical 3P design, detailing its implementation step-by-step.

16.6.1 Sampling design and implementation

General steps in implementing a 3P cruise are as follows.

1. Determine the sample size and selection threshold.

- The sample size n refers to the number of trees that will receive detailed measurements of y. As with SRS, n is typically chosen based on available resources or an allowable error. The SRS sample size formula from Section 11.3.6 can be used, with the variance (or CV) reflecting anticipated variability in the measurement-to-prediction ratio y/x.
- To implement the selection rule, define a threshold $\text{max}_z = \hat{t}_x/n$, where $\hat{t}_x$ is a pre-cruise estimate of the population total of x. The i-th tree is selected to measure y_i with probability x_i/max_z.[8] Choosing max_z this way yields an expected sample size of n under the 3P selection rule described below.

2. Traverse the population and apply the 3P selection rule.

- Visit each tree in turn:
    - record a quick prediction x_i;
    - draw a random number z_i from a uniform distribution $U(0, \text{max}_z)$;
    - if $x_i \geq z_i$, select the tree for measurement and record y_i; otherwise, retain only x_i.
- Continue until all trees have been visited.
- The result is a complete set of predictions $\{x_1, \ldots, x_N\}$ and a subsample of about n trees with paired values (x_i, y_i).

Estimation procedures for totals, means, and standard errors based on the 3P sample are presented in Section 16.6.2.

The actual sample size achieved after the cruise depends on your initial estimate $\hat{t}_x$. If this estimate is smaller than the true total $\tau_x = \sum_{i=1}^{N} x_i$, you'll likely end up with fewer than n sampled trees. If it's larger, you'll end up with more. A poor guess at τ_x won't bias your estimates—it only affects the final sample size.

Iles (2003) outlines a practical approach for ensuring that you hit the desired sample size. He suggests deliberately selecting a large $\hat{t}_x$ that results in too many sampled trees—for example, using a value roughly twice the anticipated τ_x. As you work through the population, use a random mechanism to measure only half of the selected trees, and flag the others (which he calls "insurance trees"). At the end of the cruise, if your actual sample size is below target, you can return to the insurance trees and randomly select enough of them for measurement to meet your target n.

As noted earlier, x should be proportional to y, meaning we assume a relationship of the form $y = Rx$, where R is a scalar to be estimated. This implies a linear relationship through the origin—when $x = 0$, then $y = 0$. For example, if y is tree volume in m^3, a natural choice for x is also volume, possibly in different units, with any mismatch absorbed into the estimate of R. More generally, x can be any quantity that scales proportionally with volume.

---

[8] Trees with $x_i > \text{max}_z$ are always selected. See the discussion of "sure-to-be-measured" trees later in this section.

A good choice for x is DBH, for several reasons. First, it's often easier to give a quick and reasonably accurate guess of a tree's DBH than its volume. Second, if DBH is recorded for every tree, we can produce detailed stand and stock tables by diameter class (and species, if recorded). This adds value beyond simple estimation of totals and means, especially in operational settings.

While DBH is a convenient and useful choice, it's not itself proportional to volume, so it must be transformed prior to use as the 3P x variable.

One option for this transformation is to use a DBH-volume table to get a volume value to use for x. For example, you might use the one in Section 16.4.1, which takes DBH and height as inputs. If height isn't practical to estimate (or even guess) for every tree, a DBH-only volume equation can be used instead. A simple form is volume $= \alpha_0 \text{DBH}^{\alpha_1}$, where α_0 and α_1 are parameters taken from the literature or estimated from prior data. The exponent α_1 ensures that x and y are proportional, and α_0 scales the ratio so that R is approximately one. While this scaling isn't necessary for valid estimation, it can make it easier to choose max_z, since we usually have better intuition about total volume than about the total of DBH raised to a power.

If you read other references on 3P, you may come across the term "sure-to-be-measured" trees. In our notation, these are trees with predictions x_i exceeding the threshold max_z. Given how we define max_z, such trees are unlikely unless your pre-cruise estimate $\hat{t}_x$ is far too low and/or the target sample size n is very large. When $x_i > \text{max}_z$, the tree must be selected for measurement and flagged in the cruise data accordingly. These trees are treated as a separate stratum in which all units are measured (i.e., a census), so the standard error for that stratum is zero.[9]

In some cases—such as for high-value trees or rare species—you might choose to measure all occurrences and treat them as their own stratum. This ensures complete accounting for critical trees while still allowing 3P to be applied to the remainder of the population.

If you anticipate wanting to summarize results by category—such as species, diameter class, or product type—be sure to record those variables during the cruise. These additional data allow estimates to be broken down by group in addition to providing overall totals.

Once all trees have been visited and predictions and measurements collected, the next step is to estimate the population mean, total, and associated standard errors, as outlined in the next section.

16.6.2 Estimation using 3P

Once the cruise data are in hand—that is, a complete set of predictions for all trees and detailed measurements for a subsample—the goal is to estimate the population mean and total of y. Because trees with larger predictions are more likely to be selected for measurement, the observed measurements don't form a simple random sample, so we can't just average them. A design-consistent estimator is therefore needed to account for the

[9] As we'll see later, the 3P estimator includes an FPC, so this census stratum's standard error will automatically be zero.

unequal selection probabilities induced by 3P sampling. Under the working assumption that y is proportional to x, the measurement-to-prediction ratio is approximately stable, motivating a mean-of-ratios estimator.

Following the notation from Section 16.2.2, we use the mean-of-ratios estimator, expressed as

$$\hat{R} = \frac{1}{n}\sum_{i=1}^{n} r_i,$$

where $r_i = y_i/x_i$ is the unit ratio.

Then, following (16.22) and (16.26), we multiply the estimated ratio $\hat{R}$ by the known population mean of predictions, $\mu_x = \tau_x/N$, to estimate the population mean and total

$$\bar{y} = \hat{R}\mu_x,$$
$$\hat{t} = N\bar{y}.$$

Here and throughout this section, the mean-of-ratios (MR) subscript on $\bar{y}$ and $\hat{t}$ is suppressed to reduce notational clutter.

Standard errors, confidence intervals, and interpretation follow directly from the general mean-of-ratios case in Section 16.2.2. Because the sample is drawn from a known, finite population of size N, the standard error formula includes the FPC. In most 3P applications, however, the sample fraction is small enough that the FPC has little practical effect on the resulting standard errors.

Notice that once we have $\hat{R}$, we could go directly to estimating the total from the total of predictions, i.e., $\hat{t} = \hat{R}\tau_x$, which is perfectly valid. However, we're following the estimators in Section 16.2.2, which work up the total from the mean μ_x instead.

16.6.3 Estimation using 3P by group

In many cases, you'll want to summarize estimates by one or more categorical variables, such as species, DBH class, and/or stand. Even when you're not specifically interested in stratified summaries, it may be that the relationship between predictions and measured values differs systematically across categories. In these cases, estimation accuracy and precision can often be improved by applying group-specific ratios.

To support such estimation, the relevant categorical variable(s) must be recorded for *all* trees. For example, if you want estimates by species, then species must be recorded for every tree in the population. The same applies for DBH class, stand, or any other grouping variable of interest.

Let the population be divided into H groups (e.g., species), with known group sizes N_h and mean predicted values μ_{x_h} (both are known after the cruise). Within each group, using the 3P rule a subsample of n_h trees is measured to obtain paired (x_{hi}, y_{hi}) values. For each group $h = 1, \dots, H$, we compute the sample mean of unit ratios

$$\hat{R}_h = \frac{1}{n_h}\sum_{i=1}^{n_h} r_{hi}, \tag{16.58}$$

with $r_{hi} = y_{hi}/x_{hi}$.

The estimated mean for group h is

$$\bar{y}_h = \hat{R}_h \mu_{x_h} \tag{16.59}$$

and the corresponding total is

$$\hat{t}_h = N_h \bar{y}_h. \tag{16.60}$$

The sample variance of unit ratios within each group is

$$s^2_{r_h} = \frac{1}{n_h - 1} \sum_{i=1}^{n_h} \left(r_{hi} - \hat{R}_h \right)^2. \tag{16.61}$$

The standard error of $\bar{y}_h$ is estimated as

$$s_{\bar{y}_h} = \mu_{x_h} \sqrt{\frac{s^2_{r_h}}{n_h} \left(1 - \frac{n_h}{N_h} \right)} \tag{16.62}$$

and the standard error for $\hat{t}_h$ is

$$s_{\hat{t}_h} = N_h s_{\bar{y}_h}. \tag{16.63}$$

To obtain overall estimates, we apply the standard stratified estimators

$$\bar{y} = \frac{1}{N} \sum_{h=1}^{H} N_h \bar{y}_h, \tag{16.64}$$

$$\hat{t} = N\bar{y} = \sum_{h=1}^{H} \hat{t}_h. \tag{16.65}$$

The standard error of the overall mean is

$$s_{\bar{y}} = \sqrt{\frac{1}{N^2} \sum_{h=1}^{H} N_h^2 s^2_{\bar{y}_h}} \tag{16.66}$$

and the standard error of the total is

$$s_{\hat{t}} = N s_{\bar{y}}. \tag{16.67}$$

This approach allows 3P sampling to accommodate known groupings in the population, deliver category-specific summaries, and potentially improve estimates of population parameters.

16.6.4 Applying 3P

We'll demonstrate 3P using a simulated dataset. Although synthetic, the tree values are derived from allometric equations and reflect the type of data encountered in practice. A key advantage of using simulated data is that the true population parameters are known, allowing us to assess the estimator's performance directly.

Our goal is to estimate the total volume of a mixed-species stand in coastal British Columbia, Canada. We also want species-specific estimates, as well as stand and stock tables by species and DBH class. Although unknown at the outset, the stand contains $N = 1{,}000$ trees and two species.

For this analysis, the 3P cruise uses volume predicted from an ocular DBH estimate as the x covariate, with y also being volume. Both volume variables are expressed in cubic meters.

We begin by fitting a local DBH–volume equation using data from a mix of the anticipated species. In practice, these data might come from a pilot cruise of the stand or from a previous cruise in a similar stand. Once the DBH–volume equation is in place, we follow the 3P steps outlined in the previous section.

Our local DBH–volume equation takes the form

$$y = \alpha_0 \, x^{\alpha_1}, \tag{16.68}$$

where y is volume (m^3), x is DBH (cm), and the parameters are estimated to be $\alpha_0 = 0.00034$ and $\alpha_1 = 2.3$.[10]

Based on similar stands and prior experience, our pre-cruise estimate of total volume is 2,000 m^3. We'll aim to collect $n = 25$ measurements to estimate R. Given this information, the maximum threshold is $z_{\text{max}} = 2000/25 = 80$.

We follow the 3P cruise steps.

1. Each tree in the population is visited, its species is recorded, and an ocular estimate of DBH (cm) is obtained. This DBH is used to compute a predicted volume using (16.68), which is also recorded.
2. A tree is selected for measurement if its predicted volume exceeds a random number drawn from $U(0, 80)$. If selected, its volume is accurately determined and recorded.

Both the predicted volume (i.e., given the DBH prediction) and random number can be generated on a tablet or smartphone as each tree is visited.

At the end of the cruise, we know N= 1,000 and have predicted DBH and volume for all trees. We also have a subsample of measured trees. These data are stored in `"datasets/BC_3P/predicted.csv"` and `"datasets/BC_3P/measured.csv"`, respectively. Shared columns in both files include `tree_id`, `species`, `dbh_cm_predicted`, and `vol_m3_predicted`. The measured file also contains `vol_m3_measured`. These files are read below, and the 3P estimator is applied under the assumption of a common R across species.

```
predicted <- read_csv("datasets/BC_3P/predicted.csv")
measured <- read_csv("datasets/BC_3P/measured.csv")

predicted %>% glimpse()
```

```
#> Rows: 1,000
```

[10]You can reproduce the estimates for α_0 and α_1 using `nls(vol_m3 ~ alpha_0 * dbh_cm^alpha_1, start = list(alpha_0 = 0.001, alpha_1 = 2.5), data = dbh_vol)` with `dbh_vol` read from `"datasets/BC_3P/dbh_vol.csv"`.

```
#> Columns: 4
#> $ tree_id          <dbl> 1, 2, 3, 4, 5, 6, 7, 8, 9, 1~
#> $ species          <chr> "A", "A", "B", "A", "A", "B"~
#> $ dbh_cm_predicted <dbl> 52.5, 37.5, 37.5, 47.5, 52.5~
#> $ vol_m3_predicted <dbl> 2.7064, 1.2609, 1.2609, 2.15~
```

```
measured %>% glimpse()
```

```
#> Rows: 34
#> Columns: 5
#> $ tree_id          <dbl> 4, 43, 54, 57, 66, 81, 93, 9~
#> $ species          <chr> "A", "B", "B", "A", "B", "A"~
#> $ dbh_cm_predicted <dbl> 47.5, 42.5, 47.5, 57.5, 52.5~
#> $ vol_m3_predicted <dbl> 2.1564, 1.6752, 2.1564, 3.32~
#> $ vol_m3_measured  <dbl> 2.1970, 2.2316, 2.7591, 3.25~
```

As you can see above, the `measured` file has 34 rows, which is our actual sample size n at the end of the cruise (just a bit off from our target of 25). This might be our first clue that the pre-cruise guess about the sum of the estimates may have been a little low. But that's okay. It doesn't bias our estimates, we just ended up measuring a few more trees than anticipated, and our reward is a slightly larger sample to inform the estimates worked up in the code below.

We first compute the population mean of predicted volume, μ_x, then, within the `summarize()`, we estimate the mean-of-ratios estimator $\hat{R}$ (16.23), mean estimated volume $\bar{y}$ (16.22), sample variance of the ratios s_r^2 (16.25), standard error of the mean $s_{\bar{y}}$ (16.24), and the associated estimates for the total volume $\hat{t}$ and its standard error $s_{\hat{t}}$ using (16.26) and (16.27), respectively.

```
# Known from cruise.
N <- nrow(predicted)

# Compute mu_x (and tau_x for fun).
mu_x <- mean(predicted$vol_m3_predicted)
tau_x <- sum(predicted$vol_m3_predicted)

# Estimates for volume.
measured %>%
  summarize(
    n = n(),
    R_hat = mean(vol_m3_measured / vol_m3_predicted),
    y_bar = R_hat * mu_x,
    s_sq_r = var(vol_m3_measured / vol_m3_predicted),
    s_y_bar = mu_x * sqrt(s_sq_r / n * (1 - n / N)),
    t_hat = N * y_bar,
```

```
    s_t_hat = N * s_y_bar,
    t = qt(df = n - 1, p = 1 - 0.05 / 2),
    t_ci_lower = t_hat - s_t_hat * t,
    t_ci_upper = t_hat + s_t_hat * t,
  ) %>% select(n, R_hat, t_hat, s_t_hat, t_ci_lower, t_ci_upper)
```

```
#> # A tibble: 1 x 6
#>       n R_hat t_hat s_t_hat t_ci_lower t_ci_upper
#>   <int> <dbl> <dbl>   <dbl>      <dbl>      <dbl>
#> 1    34  1.05 2653.    62.5      2526.      2780.
```

So, how'd we do? Our estimated total, with associated 95% confidence interval, is 2,653 (2,526, 2,780) cubic meters. With only $n = 34$ measurements, we achieved an estimated total error of 2.4%, where the percent is computed as the RSE (relative standard error; see (11.43))!

Because this is simulated data, we know the true volume is $\mu_y = 2,590$ cubic meters. For this sample, our 95% confidence interval covers the true value, and the estimated total is remarkably close to the actual volume.

As for the difference between our target and actual sample size n, we had initially guessed that τ_x was 2,000. After the cruise, we see that this was an underestimate—the actual population total of x is 2,525 (from the `tau_x` value in the code above). Again, this isn't a problem; it just means we might end up with a few more measurement trees than intended, as we did here.

Now let's work up species-specific estimates following the methods in Section 16.6.3. We simulated these data with two species, labeled A and B. Figure 16.5 shows the relationship between predicted and measured volume by species for the $n = 34$ trees used to estimate ratios. For exploratory analysis, we added best-fit regression lines by species. The plot suggests that species A and B have different ratios, which in turn suggests we might improve estimation by using species-specific ratios.[11]

We'll implement the estimator in steps, starting with species-specific summaries of the predictions, measurements, and estimates. We'll then apply the stratified estimator to combine these summaries and produce stand-level estimates of the mean and total.

```
pred_summary <- predicted %>%
  group_by(species) %>%
  summarize(
    N_h = n(),
    mu_xh = mean(vol_m3_predicted),
    tau_xh = sum(vol_m3_predicted)
  )
```

[11]Since these are simulated data, we know they were generated with slightly different ratios.

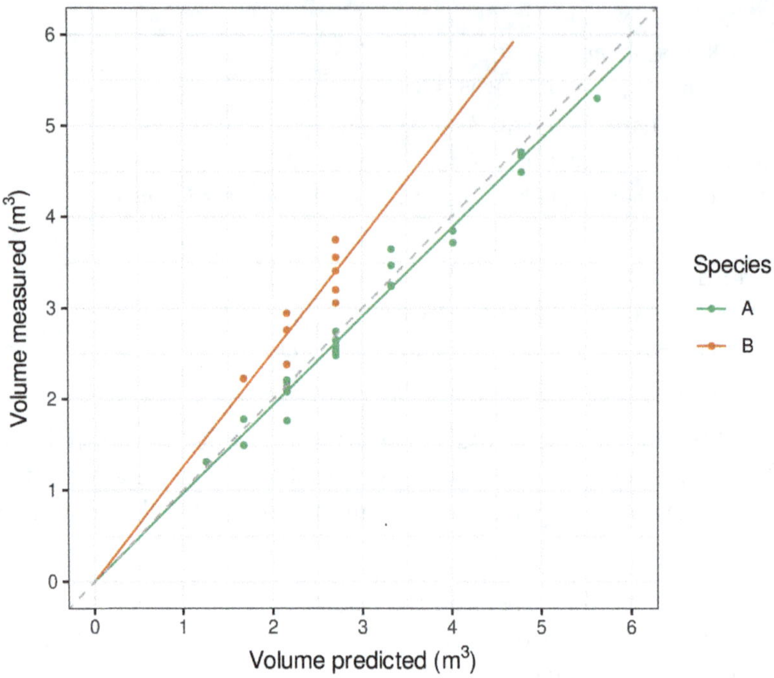

FIGURE 16.5: Predicted versus measured volume by species for the simulated 3P dataset. Lines are fit using least squares regression constrained to pass through the origin and included for visual exploration. The dashed line represents the one-to-one line.

```
pred_summary
```

```
#> # A tibble: 2 x 4
#>   species   N_h mu_xh tau_xh
#>   <chr>   <int> <dbl>  <dbl>
#> 1 A         734  2.62  1920.
#> 2 B         266  2.27   605.
```

From the prediction summary, we see that species A dominates the stand, with roughly three times as many stems (N_h) and three times the volume (τ_{xh}) as species B. We'll need N_h and μ_{x_h} in `pred_summary` for the subsequent estimation after computing species-specific ratios and variances below.

```
measure_summary <- measured %>%
  group_by(species) %>%
  summarize(
    n_h = n(),
    R_hat_h = mean(vol_m3_measured / vol_m3_predicted),
    s_sq_rh = var(vol_m3_measured / vol_m3_predicted),
```

```
)

measure_summary
```

```
#> # A tibble: 2 x 4
#>   species   n_h R_hat_h s_sq_rh
#>   <chr>   <int>   <dbl>   <dbl>
#> 1 A          25   0.975 0.00336
#> 2 B           9   1.26  0.0103
```

Estimates for species-specific $\hat{R}_h$ suggest the relationship between predicted and measured volume varies by species—our predictions tend to overestimate volume for species A and underestimate it for species B. Again, this is okay as long as the over- or underestimation is proportional across the observed pairs of x and y values—and, by assumption, across the full population.

For example, $\hat{R}$ for species B is 1.26, so if we predict volumes of 1, 2, 3, and 4 cubic meters, we expect measured values of 1.26, 2.53, 3.79, and 5.05 cubic meters, respectively. In other words, y and x are proportional. These proportional relationships are evident for species A and B in the exploratory plot shown in Figure 16.5, supporting our assumptions about the linear relationship between predicted and measured volumes.

Below are the species-specific estimates, followed by stand-level means and totals obtained by combining across strata. We begin by joining the species population information in `pred_summary` to `measure_summary`.

```
measure_summary <- left_join(measure_summary,
  pred_summary,
  by = join_by(species)
)

# Species-specific estimates.
measure_summary %>%
  mutate(
    y_bar_h = R_hat_h * mu_xh,
    s_y_bar_h = mu_xh * sqrt(s_sq_rh / n_h * (1 - n_h / N_h)),
    t_hat_h = N_h * y_bar_h,
    s_t_hat_h = N_h * s_y_bar_h,
    t_h = qt(df = n_h - 1, p = 1 - 0.05 / 2),
    t_ci_lower_h = t_hat_h - s_t_hat_h * t_h,
    t_ci_upper_h = t_hat_h + s_t_hat_h * t_h,
  ) %>%
  select(n_h, t_hat_h, s_t_hat_h, t_ci_lower_h, t_ci_upper_h)
```

```
#> # A tibble: 2 x 5
```

```
#>      n_h t_hat_h s_t_hat_h t_ci_lower_h t_ci_upper_h
#>    <int>   <dbl>     <dbl>        <dbl>        <dbl>
#> 1     25   1871.      21.9        1826.        1916.
#> 2      9    764.      20.1         717.         810.
```

Using a sample size of 25 and 9 measurements for species A and B, respectively, we achieved RSE values of 1.2% and 2.6% for the corresponding species totals. As defined in (11.43), the RSE is computed as $100 \cdot s_{\hat{t}_h}/\hat{t}_h$.

Our estimated totals, with associated 95% confidence intervals, are 1,871 (1,826, 1,916) and 764 (717, 810) cubic meters for species A and B, respectively. The true volume totals are A = 1,851 and B = 740 cubic meters.

Now we'll estimate the stand-level mean and total by combining the species-specific results using their population weights (again following estimators in Section 16.6.3).

```
measure_summary %>%
  mutate(
    y_bar_h = R_hat_h * mu_xh,
    s_y_bar_h = mu_xh * sqrt(s_sq_rh / n_h * (1 - n_h / N_h)),
  ) %>%
  summarize(
    N = sum(N_h),
    y_bar = 1 / N * sum(N_h * y_bar_h),
    s_y_bar = sqrt(1 / N^2 * sum(N_h^2 * s_y_bar_h^2)),
    t_hat = N * y_bar,
    s_t_hat = N * s_y_bar,
    t = qt(df = sum(n_h) - n(), p = 1 - 0.05 / 2),
    t_ci_lower = t_hat - s_t_hat * t,
    t_ci_upper = t_hat + s_t_hat * t
  ) %>%
  select(t_hat, s_t_hat, t_ci_lower, t_ci_upper)
```

```
#> # A tibble: 1 x 4
#>   t_hat s_t_hat t_ci_lower t_ci_upper
#>   <dbl>   <dbl>      <dbl>      <dbl>
#> 1 2635.    29.7      2575.      2696.
```

The species-specific stratified estimator produced an RSE of 1.1% for the total volume. The estimated total volume, with a 95% confidence interval, is 2,635 (2,575, 2,696) cubic meters. Recall, again, that the true volume is $\mu_y = 2,590$ cubic meters.

In the first analysis, we ignored potential species differences and estimated a single pooled ratio, which yielded an RSE of 2.4%. We then allowed for species-specific ratios and applied the stratified estimator, which substantially reduced the RSE to 1.1%—less than half the pooled-ratio value. This demonstrates the value of using category-specific ratios when there are indications of differences (e.g., as seen in exploratory analysis) and enough observations to support separate estimation.

16.7 Advantages and disadvantages of estimation with covariates

Estimation with covariates improves efficiency by incorporating auxiliary information—such as remotely sensed data, ocular estimates, or other inexpensive field measures—into the estimation process. When the covariate is informative, this approach can substantially improve precision or reduce field effort. Below, we summarize the main advantages and limitations of covariate-based estimation.

16.7.1 Advantages

- Improved precision: When the covariate is strongly related to the parameter of interest, ratio and regression estimators can yield substantially smaller standard errors than estimators that ignore auxiliary information—often achieving higher precision with the same or even smaller sample size.
- Statistical and practical efficiency: Using inexpensive covariates allows effort to be shifted toward a smaller number of careful measurements, improving cost efficiency without sacrificing accuracy.
- Flexible and widely applicable: Covariate-based estimators can be paired naturally with common probability sampling designs and are especially useful in settings where full measurement of the variable of interest is impractical.
- Supports small-area and category-specific inference: When covariates are available for the full population or a large initial sample, estimates can be extended to subregions, species groups, or other categories without requiring additional measurements.
- Multiple estimator choices: Ratio and regression estimators provide flexible options that can be matched to the form and strength of the relationship between the covariate and the parameter of interest.

16.7.2 Disadvantages

- Dependence on the strength of the relationship: Covariate-based estimators are most effective when a reasonably linear or proportional relationship exists between the covariate and the variable of interest. When this relationship is weak, gains in precision may be modest.
- Small-sample behavior: Regression and ratio estimators can exhibit a small bias when sample sizes are limited, although this typically decreases as the number of sampled units increases. Departures from the assumed relationship mainly reduce gains in precision rather than leading to systematically misleading estimates.

- Greater analytical effort: Selecting an appropriate estimator, exploring the relationship between variables, and assessing whether assumptions such as linearity or proportionality are reasonable requires more judgment and diagnostic work than simpler estimators that rely only on direct measurements.
- Choice among estimators requires expertise: Different covariate-based estimators are suited to different relationship structures and sampling implementations. Choosing among regression, ratio-of-means, and mean-of-ratios estimators requires understanding how variability in the variable of interest changes with the covariate and how the covariate is used in the sampling process. In designs such as 3P sampling, inappropriate estimator choices can lead not only to reduced efficiency but also to biased estimates.
- Risk of extrapolation: Care is needed when applying covariate-based estimators beyond the range of observed data. Estimates can become unstable and highly variable if the relationship between the covariate and the parameter of interest is poorly supported in parts of the population.

16.8 Exercises

16.8.1 Estimating forest carbon using ALS and inventory data

A manager has LiDAR-derived canopy height measurements covering a 1,000 ha forest and field-based measurements of forest carbon for a set of inventory plots collected using systematic sampling. The LiDAR data were obtained using a drone-based aerial laser scanning (ALS) system. The goal is to estimate the mean and total forest carbon (Mg C/ha) across the forest using the covariate-based estimators developed in this chapter.

In this setting, LiDAR-derived canopy height is the covariate x, and field-measured forest carbon is the variable of interest y. For each inventory plot, the manager observes both the LiDAR-derived canopy height over the plot footprint and the corresponding field-based carbon measurement.

Because canopy height is available for all locations over the forest, the population mean canopy height μ_x is known and equals 24.19 m. This makes the setting appropriate for regression or ratio estimation with a known covariate mean.

The code below reads in the paired (x, y) values for the n field plots and shows the file structure. Columns in `c_xy` are:

- `plot_id` plot identifier;
- `x` LiDAR-derived canopy height (m);
- `y` field-measured forest carbon (Mg C/ha).

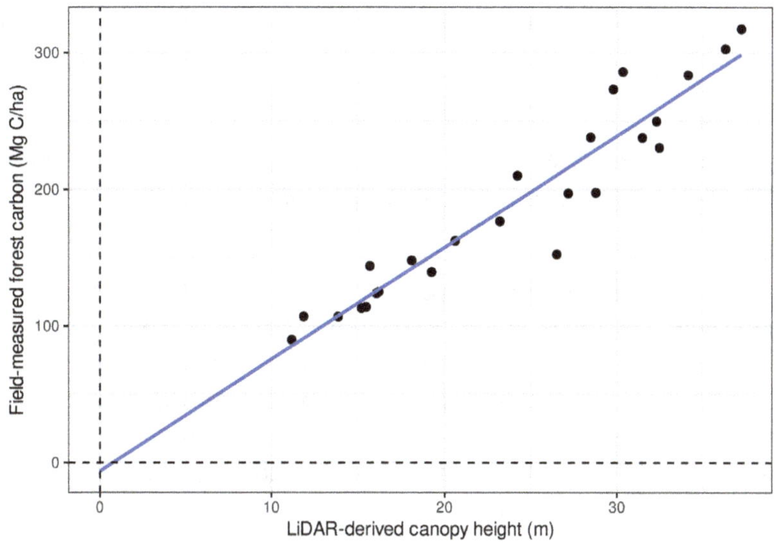

FIGURE 16.6: LiDAR-derived canopy height (x) versus field-measured forest carbon (y) for the paired plots used for Exercise 16.1.

```
c_xy <- read_csv("datasets/canopy_height_carbon.csv")

c_xy %>% glimpse()
```

```
#> Rows: 25
#> Columns: 3
#> $ plot_id <dbl> 31, 244, 63, 688, 304, 508, 311, 945,~
#> $ x       <dbl> 20.617, 11.886, 32.479, 18.130, 30.33~
#> $ y       <dbl> 162.21, 107.03, 230.45, 147.89, 285.8~
```

Next, to assess the relationship between x and y, the code below plots the paired observations along with a best-fit line.

```
ggplot(c_xy, aes(x = x, y = y)) +
  geom_point(size = 2) +
  geom_smooth(method = "lm", formula = y ~ x, se = FALSE, fullrange = TRUE) +
  geom_hline(yintercept = 0, linetype = "dashed") +
  geom_vline(xintercept = 0, linetype = "dashed") +
  theme_bw() +
  labs(x = "LiDAR-derived canopy height (m)",
       y = "Field-measured forest carbon (Mg C/ha)")
```

Exercise 16.1.

1. Based on Figure 16.6, explain in one sentence why the ratio-of-means estimator is appropriate in this setting.

2. Using the ratio-of-means estimator in Section 16.2.1, compute the population mean forest carbon $\bar{y}_{\text{RM}}$, its standard error $s_{\bar{y}_{\text{RM}}}$ (with no FPC), and a 95% confidence interval for $\bar{y}_{\text{RM}}$.

3. Using your estimate of the mean forest carbon and forest area A = 1,000 ha, estimate the total forest carbon $\hat{t}_{\text{RM}}$ and its 95% confidence interval.

4. Suppose the manager ignored the LiDAR canopy height data entirely and relied only on the n field-measured forest carbon values, treating them as a SRS from the population. Using these data, estimate the population mean forest carbon $\bar{y}$, its standard error $s_{\bar{y}}$ (with no FPC), and a 95% confidence interval for $\bar{y}$. Then, using (11.44), quantify the percent improvement in precision of the ratio-of-means estimator relative to the SRS estimator. When applying (11.44), take estimator A to be the SRS estimator and estimator B to be the ratio-of-means estimator, so that $s_A = s_{\bar{y}}$ and $s_B = s_{\bar{y}_{\text{RM}}}$. Based on this metric, what percent precision is gained by incorporating the LiDAR covariate in this example?

To check that you're on the right track, your ratio-of-means estimate of the total $\hat{t}_{\text{RM}}$ should be approximately 191,629 Mg C, with a 95% confidence interval of (182,583, 200,675) Mg C.

16.8.2 Estimating growing stock volume using PLS and inventory data

An agency is evaluating an inventory design that couples measurements collected with handheld personal laser scanning (PLS) devices with traditional field measurements. Such PLS data are increasingly used to augment or, in some cases, replace traditional tree measurement methods (see, e.g., Gollob et al. (2020) and Nothdurft et al. (2025)). The goal is to produce high-quality growing stock volume (GSV) estimates for management units while reducing field costs (i.e., crew time).

PLS-based measurements can be collected much more quickly than traditional tree measurements, offering substantial gains in efficiency and cost savings. At the same time, given the long history of field-based inventory, traditional field-based GSV estimates are often treated as a trusted benchmark. In this setting, the estimators introduced in this chapter allow these measurement approaches to be combined, using field measurements to correct or calibrate PLS-based estimates when needed.

Under the proposed design, the field crew collects PLS data on a dense systematic plot grid across a management unit, yielding a large phase-1 sample of PLS-based GSV values. From this phase-1 sample, they then select a nested random subsample of plots for traditional field measurement, producing paired observations of PLS-based and field-based GSV for use in estimation.

The two-phase data are stored in `"datasets/pls_two_phase_gsv.csv"`. This file has columns:

- `plot_id` sampling location identifier;
- `x` PLS-derived GSV (m^3/ha);
- `y` traditional field-measured GSV (m^3/ha), observed for the phase-2 subsample only (`NA` otherwise).

The code below reads the two-phase dataset and shows the file structure.

```
gsv_xy <- read_csv("datasets/pls_two_phase_gsv.csv")

gsv_xy %>% glimpse()
```

```
#> Rows: 100
#> Columns: 3
#> $ plot_id <dbl> 121, 164, 285, 292, 327, 329, 353, 68~
#> $ x       <dbl> 943.23, 767.39, 159.81, 229.08, 794.2~
#> $ y       <dbl> NA, NA, NA, NA, 793.72, 422.24, NA, N~
```

Next, to assess the relationship between the phase-2 data x_2 and y_2, the code below plots the paired observations along with a best-fit line.

Exercise 16.2.

1. Using Figure 16.7 of (x_2, y_2), assess whether the fitted line passes near the origin and whether variance appears roughly constant or increases with x. Based on this assessment, which estimator would you expect to be more appropriate: the regression estimator or the ratio-of-means estimator?

2. Using the double sampling estimators in Sections 16.3.1 and 16.3.2, compute:
   - the regression estimate of the population mean, $\bar{y}_{\text{Rd}}$, its standard error $s_{\bar{y}_{\text{Rd}}}$, and a 95% confidence interval;

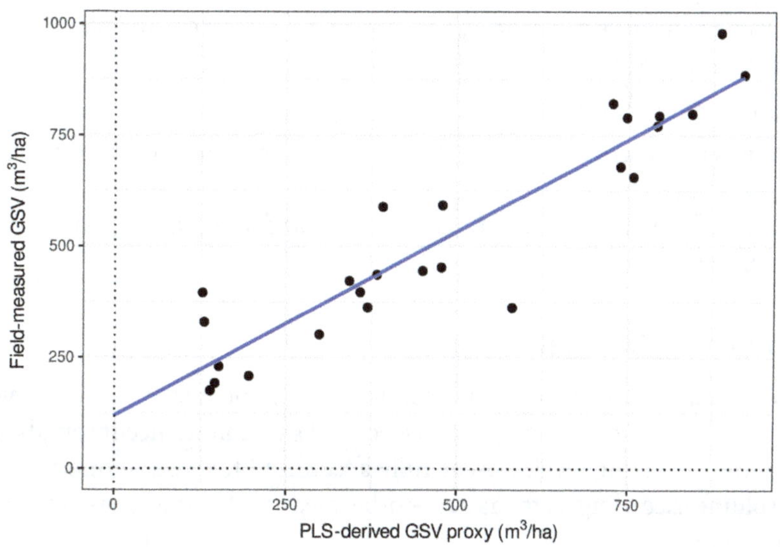

FIGURE 16.7: PLS-derived proxy (x_2) versus field-measured growing stock volume (y_2) for the paired phase-2 plots used for Exercise 16.2.

- the ratio-of-means estimate of the population mean, $\bar{y}_{\text{RMd}}$, its standard error $s_{\bar{y}_{\text{RMd}}}$, and a 95% confidence interval. Note, because an areal sampling frame is used, the second FPC term in (16.29) and (16.34) is omitted.

3. The management unit has area $A = 10{,}000$ ha. Scale each mean estimate to obtain the total GSV, i.e., $\hat{t}_{\text{Rd}}$ and $\hat{t}_{\text{RMd}}$, along with their respective 95% confidence intervals.

4. Using the percent improvement in precision metric defined in (11.44), quantify the percent improvement in precision of the regression estimator relative to the ratio-of-means estimator. When applying (11.44), take estimator A to be the ratio-of-means estimator and estimator B to be the regression estimator, so that $s_A = s_{\bar{y}_{\text{RMd}}}$ and $s_B = s_{\bar{y}_{\text{Rd}}}$. Based on this metric, which estimator is preferred in this setting? Does this result agree with your initial assessment based on Figure 16.7?

5. Suppose the agency ignored the PLS data entirely and collected only traditional field measurements of GSV, yielding a sample of size n_2 of y values. Using these data, estimate the population mean $\bar{y}$, its standard error, and a 95% confidence interval. Then, using (11.44), quantify the percent improvement in precision of the regression estimator relative to this SRS estimator (again treating the SRS estimator as A and the regression estimator as B). How much precision is gained by incorporating the PLS covariate in this example?

To check that you're on the right track, you should get something close to the following.

- Mean estimates (m^3/ha)
  - Regression estimator: $\bar{y}_{\text{Rd}} = 524.91$, $s_{\bar{y}_{\text{Rd}}} = 28.64$, and 95% confidence interval (465.67, 584.15).
  - Ratio-of-means estimator: $\bar{y}_{\text{RMd}} = 525.57$, $s_{\bar{y}_{\text{RMd}}} = 30.66$, and 95% confidence interval (462.3, 588.84).
- Total estimates over $A = 10{,}000$ ha (m^3)
  - Regression estimator: $\hat{t}_{\text{Rd}} = 5{,}249{,}112$ and 95% confidence interval (4,656,715, 5,841,510).
  - Ratio-of-means estimator: $\hat{t}_{\text{RMd}} = 5{,}255{,}726$ and 95% confidence interval (4,623,009, 5,888,442).

16.8.3 Northern California 3P cruise

Suppose you conducted a 3P cruise on a conifer stand in northern California with the goal of estimating the total volume (ft^3) with associated 95% confidence intervals. Because the landowner wanted all high-value trees fully accounted for, you defined any tree with a predicted volume exceeding 85 ft^3 as sure-to-be-measured. These trees are always selected for measurement.

Following the 3P sampling steps in Section 16.6.1, this 85 ft^3 cutoff also serves as the upper bound of the uniform distribution used to generate the random number z_i. As each tree was encountered in the field, you recorded its predicted volume x_i and drew a corresponding

$z_i \sim U(0, 85)$. A tree was selected for measurement when $x_i \geq z_i$, in which case its volume y_i was determined from tree measurements (e.g., DBH, height) and a volume table. Trees with $x_i < z_i$ were not measured and retain their prediction only.

This procedure produces three categories of trees:

- census of predicted volume x_i for trees with $x_i \leq 85$ ft^3 (these are used to compute μ_x and τ_x),
- sample of measured volume y_i with paired (x_i, y_i) values (used to estimate $\hat{R}$ and other parameters),
- census of sure-to-be-measured volume for trees with $x_i > 85$ ft^3 (added directly to the estimated total and confidence interval bounds).

As discussed in Section 16.6.2, sure-to-be-measured trees form a census stratum with zero sampling variance. Their volumes are added directly to the estimated total, and they contribute no sampling uncertainty.

Table 16.2 lists the 30 trees in the population along with their predicted values, random draws, and (when selected) measured volumes. For trees in the sure-to-be-measured stratum, only the measured volume is shown to avoid confusion with the unused prediction. As you inspect the table, note that under the 3P design, whenever $x_i \geq z_i$, the observed volume y_i is recorded.

TABLE 16.2: Field data from a 3P cruise in northern California conifers. All values are in cubic foot volume.

Tree	x_i	z_i	y_i	Sure-to-be-measured	Tree	x_i	z_i	y_i	Sure-to-be-measured
1	54	57			16	48	48	54	
2	48	48	51		17	60	51		
3				99	18	54	48		
4	48	54			19	63	57	60	
5	57	54	63		20	51	66		
6	51	60			21				96
7	60	69			22	66	57	69	
8	51	45	54		23	57	66		
9				93	24	69	75		
10	45	51			25	51	63		
11	63	72			26	57	48	60	
12	60	54	66		27	66	66	69	
13	72	75			28	51	63		
14	57	63			29				102
15				93	30	66	57	63	

We now turn to estimation. The first step is to read the three files that store the data in Table 16.2. After reading them in, run `glimpse()` on each to verify their column names and values, noting that the files use descriptive column names while the table uses the notation introduced for 3P sampling.

```
predicted <- read_csv("datasets/CA_3P/predicted.csv")
measured <- read_csv("datasets/CA_3P/measured.csv")
sure <- read_csv("datasets/CA_3P/sure_to_be_measured.csv")
```

Exercise 16.3. In this first exercise, ignore the sure-to-be-measured trees and apply the 3P estimators detailed in Section 16.6.2. Here are the steps.

1. Compute N, μ_x, and τ_x from the x_i in `predicted`.
2. Compute n, $\hat{R}$ (16.23), and s_r^2 (16.25) from the paired (x_i, y_i) in `measured`.
3. Given values and estimates from Steps 1 and 2, compute $\bar{y}$ (16.22), $s_{\bar{y}}$ (16.24), $\hat{t} = N\bar{y}$ (16.26), and $s_{\hat{t}}$ (16.27).
4. Construct a 95% confidence interval for $\hat{t}$ using $s_{\hat{t}}$ and a t-value with $n - 1$ degrees of freedom.

Your sample size should be $n = 10$. After rounding, you should get an $\hat{R}$ of 1.05, s_r^2 of 0.003, and $\hat{t}$ of 1496.54 ft^3 with a standard error of 20.19 ft^3. Your 95% confidence interval for the total should be (1450.88, 1542.21) ft^3.

Exercise 16.4. Now modify your code to add in the sure-to-be-measured tree totals using the following steps.

1. Compute t_{sure} by summing the measured volumes in `sure_to_be_measured`.
2. Compute $\hat{t}_{\text{adj}} = \hat{t} + t_{\text{sure}}$.
3. Shift both confidence interval endpoints by adding t_{sure}.

As discussed in Section 16.6.2, the standard error $s_{\hat{t}}$ doesn't change because the sure-to-be-measured trees are a census and hence contribute no sampling variance.

Your sure-to-be-measured total t_{sure} should be 483. Your adjusted total $\hat{t}_{\text{adj}}$ should be 1979.54 ft^3, with 95% confidence interval (1933.88, 2025.21) ft^3.

17

Cluster sampling

Cluster sampling is commonly used when it's impractical or costly to sample widely dispersed locations. In this design, the primary sampling unit (PSU) is a *cluster* of smaller secondary sampling units (SSUs), where measurements are taken. SSUs are nested within PSUs, introducing additional design considerations compared to the sampling designs introduced earlier.

Cluster sampling is particularly useful when:

- the cost of reaching widely dispersed sampling locations is high;
- operational constraints require grouping sampling locations for efficient data collection;
- measurement of a sampling location is inexpensive relative to travel costs;
- or the sampling frame is only available at the PSU level.

These conditions are common in forest inventory, where measurement costs are low relative to the cost of moving a field crew. Cluster designs reduce travel time by grouping nearby SSUs within the same PSU, allowing crews to measure multiple locations with minimal movement. This approach is especially useful when sampling large areas. For example, cluster designs are widely used in NFI programs, where travel costs are high and field crews must cover broad regions efficiently.

The tradeoff, however, is a potential loss of statistical efficiency. Because of spatial autocorrelation—the tendency for observations located close together to be more similar than those farther apart—SSUs in the same PSU often exhibit less variability than locations sampled independently across space, such as in the SYS designs shown in Chapter 14. As a result, cluster sampling typically produces less precise estimates than a SRS with the same number of SSUs. This is a classic cost-precision tradeoff: cluster sampling reduces field effort but can increase sampling error.

17.1 Clusters, strata, and systematic sampling

Although clusters (PSUs) and strata both divide the population into groups, their motivation differs. In *stratified sampling*, the population is partitioned into non-overlapping subgroups (*strata*) using auxiliary variables, and sampling is conducted independently within each stratum. The goal is to increase precision by concentrating sampling effort within relatively homogeneous subgroups.

DOI: 10.1201/9781003222699-17

In *cluster sampling*, the goal is not to improve statistical efficiency directly, but to simplify or reduce the cost of data collection. PSUs are often based on geographic proximity or field logistics. A subset of PSUs is selected (e.g., via SRS, SYS, or another probability sampling method), and measurements are collected from all or some SSUs within each selected PSU.

The contrast can be summarized as follows.

- *Stratification* aims to reduce within-group variability and improve precision.
- *Clustering* groups nearby SSUs into PSUs to reduce travel effort, but because those SSUs tend to be similar, precision is often lower than if the same number of SSUs were sampled independently.

Both approaches are valid and often used together—for example, in *stratified cluster sampling*, where PSUs are selected independently within each stratum.

SYS, as described in Chapter 14, can be viewed conceptually as a form of cluster sampling with a single PSU consisting of n evenly spaced SSUs. This reinforces why SYS doesn't have a design-based variance estimator (Section 14.5)—with only one randomly selected PSU, between-PSU variation cannot be estimated.

17.2 Elements of cluster sampling design

When implementing a cluster sampling design, several key elements must be considered.

1. *Cluster layout and anchoring.* PSUs may be naturally defined (e.g., administrative units or predefined sampling units) or imposed by the analyst and selected using a probability sampling method. Under an areal sampling frame, PSUs act as anchor points around which SSUs are arranged. In many applications, clustering is motivated by logistical considerations: PSUs group nearby SSUs to reduce movement or setup costs, with the understanding that this grouping can increase within-cluster similarity.

2. *Method of selecting clusters.* PSUs can be selected using any valid probability sampling method. For finite populations, SRS and SYS are common. When auxiliary information is available, stratified sampling or PPS sampling may improve precision while maintaining logistical feasibility. The choice of selection method determines inclusion probabilities and, therefore, the appropriate estimators and variance estimators.

3. *Cluster shape and SSU configuration.* If not naturally defined, SSUs are arranged in a fixed geometric pattern around an anchor point. The spacing and geometry of SSUs within each PSU strongly influence within-cluster correlation and, therefore, statistical efficiency: SSUs that are spread farther apart tend to be less redundant because spatial autocorrelation weakens with distance, while tightly clustered SSUs provide more similar information. This statistical benefit must be balanced against measurement and movement costs within the PSU. SSU size can be scaled to efficiently measure different variables, and nested configurations can be used when needed.

4. *Number of clusters and SSUs per cluster.* In a finite-population setting, each PSU contains a fixed and known set of SSUs. When clusters are imposed by design, the number, size, and arrangement of SSUs are design choices. In either case, increasing the number of PSUs generally improves precision more than increasing the number of SSUs per PSU, because PSUs tend to provide more independent information than SSUs within the same PSU.

These elements interact to determine the balance between statistical precision and operational cost. In forest inventory, the goal is often to design a layout that captures sufficient spatial variability while remaining cost-effective and logistically manageable.

17.3 Choosing the number of PSUs and SSUs

Unlike stratified sampling, where well-established formulas exist for allocating sample sizes across strata, sample size planning for clusters is more heuristic and context-specific. The main tradeoff is between the number of PSUs and the number of SSUs within each PSU. Adding SSUs improves precision only up to a point; when observations are spatially correlated, the marginal benefit of additional SSUs declines rapidly. In practice, reallocating effort toward sampling more PSUs is usually a more effective way to reduce sampling error.

In most forest inventory settings, these decisions are shaped by cost and logistics. Analytical expressions for optimal allocation exist but require estimates of within- and between-PSU variability, which are rarely available before sampling begins. Some survey efforts use simulation to explore tradeoffs under realistic spatial and operational constraints. Additional discussion can be found in Freese (1962), Lohr (2019) Section 5.5, Cochran (1977) Section 9.6, and Brus (2022) Chapter 6.

There are many variations of cluster sampling. This section focuses on *single-stage* cluster sampling, where all SSUs within each selected PSU are measured. When observing every SSU is impractical, a subset of SSUs is sampled instead—a design known as *two-stage* (or *multistage*) sampling, covered in Chapter 18.

17.4 Estimation using cluster sampling

Here, we focus on single-stage cluster sampling, in which every SSU within a selected PSU is measured, and we introduce estimators for both equal- and unequal-probability selection of PSUs.

We use the following notation:

- N number of PSUs in the population;
- m_i number of SSUs in the i-th PSU;

- $M = \sum_{i=1}^{N} m_i$ number of SSUs in the population;
- $\bar{M} = M/N$ population mean number of SSUs per PSU;
- n number of PSUs in the sample;
- $\bar{m} = \frac{1}{n}\sum_{i=1}^{n} m_i$ mean number of SSUs in the sampled PSUs;
- y_{ij} value of the j-th SSU in the i-th PSU;
- $y_i = \sum_{j=1}^{m_i} y_{ij}$ total for PSU i.

This notation and the estimators that follow assume a finite population with a fixed set of N PSUs (and their associated SSUs). Section 17.5 describes how closely related estimators are used when working from an areal sampling frame.

17.4.1 Sampling PSUs with equal probability

17.4.1.1 Unbiased estimators

An unbiased estimator for the population total τ is

$$\hat{t} = \frac{N}{n}\sum_{i=1}^{n} y_i. \tag{17.1}$$

The standard error of $\hat{t}$ is estimated as

$$s_{\hat{t}} = N\sqrt{\left(1 - \frac{n}{N}\right)\frac{s^2}{n}}, \tag{17.2}$$

where

$$s^2 = \frac{1}{n-1}\sum_{i=1}^{n}\left(y_i - \frac{\hat{t}}{N}\right)^2 \tag{17.3}$$

is the sample variance among PSU totals.

Given $\hat{t}$, the population mean μ is estimated as

$$\bar{y} = \frac{\hat{t}}{M}, \tag{17.4}$$

with standard error

$$s_{\bar{y}} = \frac{s_{\hat{t}}}{M}. \tag{17.5}$$

Confidence intervals for the population mean and total are

$$\bar{y} \pm \mathrm{t}_{\mathrm{df},1-\frac{\alpha}{2}}\, s_{\bar{y}} \tag{17.6}$$

and

$$\hat{t} \pm \mathrm{t}_{\mathrm{df},1-\frac{\alpha}{2}}\, s_{\hat{t}}. \tag{17.7}$$

Here, $\mathrm{df} = n - 1$, reflecting that PSUs—not SSUs—are the independent sampling units.

17.4.1.2 Ratio estimators

Using (17.4) requires knowledge of M (the total number of SSUs). If M is unknown, the ratio estimator below can be used. Even when M is known, the ratio estimator often provides more stable or efficient estimates than the unbiased estimator. The ratio estimator for $\bar{y}$ is not unbiased; however, the bias is typically small for moderate to large sample sizes.

The population mean μ is estimated using

$$\bar{y} = \frac{\sum_{i=1}^{n} y_i}{\sum_{i=1}^{n} m_i}. \tag{17.8}$$

The standard error of $\bar{y}$ is estimated as

$$s_{\bar{y}} = \sqrt{\left(1 - \frac{n}{N}\right) \frac{1}{n\bar{M}^2} \frac{\sum_{i=1}^{n} (y_i - \bar{y}m_i)^2}{n-1}}, \tag{17.9}$$

where $\bar{M}$ is replaced by $\bar{m}$ when $\bar{M}$ is unknown.

Given the sample mean $\bar{y}$ and its standard error $s_{\bar{y}}$, the population total and its standard error are estimated as

$$\hat{t} = M\bar{y} \tag{17.10}$$

and

$$s_{\hat{t}} = M s_{\bar{y}}. \tag{17.11}$$

Confidence intervals for the mean and total follow (17.6) and (17.7), respectively.

17.4.2 Sampling PSUs with probability proportional to size

When auxiliary information describing cluster size is available prior to sampling, efficiency can be improved by selecting PSUs with *probability proportional to size* (PPS). Here, "size" refers to any nonnegative quantity correlated with the PSU total.[1]

We assume the following design elements.

- Each PSU has a known per-draw selection probability p_i, with $p_i \propto x_i$ for some nonnegative size measure x_i (where $\propto$ means "proportional to"), and $\sum_{i=1}^{N} p_i = 1$. A common choice is $x_i = m_i$, giving $p_i = m_i/M$ where $M = \sum_{i=1}^{N} m_i$.
- Sampling is conducted with replacement, producing n independent draws. If a PSU is selected more than once, a new PSU-level measurement y_i is obtained. Because PSUs are selected with replacement, the draws are independent and no FPC is applied.

The population total τ is estimated using

$$\hat{t} = \frac{1}{n} \sum_{i=1}^{n} \frac{y_i}{p_i}. \tag{17.12}$$

The sum is taken over the n PSUs selected with replacement. Again, if the same PSU appears multiple times, its contributions are treated as independent terms.

---

[1]This is the same idea introduced in point sampling (Section 12.4), where trees were selected with probability proportional to basal area, and in 3P sampling (Section 16.6), where trees were selected with probability proportional to prediction. Here, we apply the same concept to PSUs, selecting them with probability proportional to a measure of their size.

The standard error of $\hat{t}$ is estimated as

$$s_{\hat{t}} = \sqrt{\frac{1}{n(n-1)} \sum_{i=1}^{n} \left(\frac{y_i}{p_i} - \hat{t} \right)^2}. \tag{17.13}$$

Given $\hat{t}$, the estimator for the population mean is

$$\bar{y} = \frac{\hat{t}}{M}, \tag{17.14}$$

with standard error

$$s_{\bar{y}} = \frac{s_{\hat{t}}}{M}. \tag{17.15}$$

Confidence intervals for the mean and total follow (17.6) and (17.7), respectively.

This estimator is unbiased for both the total and the mean, and the associated variance estimators are unbiased as well. PPS sampling is particularly useful when cluster sizes vary substantially and are correlated with the parameter of interest.

17.5 Cluster sampling under an areal sampling frame

Cluster sampling under an *areal sampling frame* arises most often in large-area forest inventories, particularly NFIs and other long-term monitoring programs. In these settings, PSUs are located over a continuous geographic region using probability-based spatial designs—most commonly SYS or spatially balanced variants—and each PSU anchors a fixed configuration of SSUs, such as fixed-area plots or point samples.

SSUs within a cluster follow a simple, repeatable geometric layout relative to the PSU anchor. Common configurations include squares, rectangles, lines, or L-shaped arrangements. These layouts are often chosen for field efficiency, interpretability, and consistency over time. However, in some NFIs the orientation, spacing, and overall extent of the SSU configuration are also informed by anticipated environmental gradients and spatial autocorrelation, with the goal of improving information capture while remaining operationally efficient; see, e.g., the Swedish and Finnish designs described in Tomppo et al. (2010). Once defined, the SSU configuration is treated as part of the sampling design and applied uniformly across all PSUs.

This setting differs fundamentally from the finite-population cluster framework introduced earlier. PSUs aren't selected from a finite list, and SSUs don't correspond to a fixed population of elements. Instead, inference targets population means or totals *per unit area* over a geographic region of known area A. Finite-population quantities such as N and M don't appear explicitly, and finite population corrections aren't used.

Because cluster layouts are fixed in space, SSUs inevitably interact with population and stratum boundaries. Operational NFIs address these boundary effects using a variety of design-based strategies that emphasize robustness and operational simplicity rather than exact local geometric correction. A useful conceptual contrast is between geometry-explicit inclusion-zone methods (e.g., Valentine et al. (2006) or Gregoire and Valentine (2008), Section 7.5.6) and the ratio-based estimators used in many operational inventories, including those employed by several NFIs (Tomppo et al. 2010). These approaches typically avoid plot-level boundary corrections (e.g., mirage or walkthrough methods) by handling boundary effects implicitly through design choices and estimation. An example of one such NFI approach is introduced in the next section and illustrated in Section 17.7.2.

17.5.1 Ratio estimation using SSU centers

Here, we describe a ratio estimator used in several operational NFIs (see Tomppo (2006), Section 11.3.3.3) and consistent with the ratio-estimation framework introduced in Section 16.2.

A key requirement for this estimator is that PSUs are selected over a region that extends beyond the population (or stratum) boundary by a sufficient buffer distance. This buffer must be large enough that the full SSU configuration associated with any sampled PSU can occur entirely outside the population boundary. In other words, PSU locations are generated over a spatial region larger than the population of interest, while estimation is restricted to the population itself. This design allows SSU center indicators to provide a consistent measure of sampling effort near boundaries and eliminates the need for explicit geometric boundary corrections (e.g., mirage or walkthrough methods). The role of this buffer, and its interaction with population boundaries, is illustrated in Section 17.7.2.

Each PSU contains a fixed configuration of SSUs determined by the sampling design. Tree-level expansion is defined at the SSU level, exactly as in plot and point sampling (Sections 12.3.1 and 12.4.1). Expansion factors aren't modified for proximity to population or stratum boundaries.

The population mean per unit area is estimated using the cluster ratio estimator

$$\bar{y} = \frac{\sum_{i=1}^{n} y_i}{\sum_{i=1}^{n} z_i}, \tag{17.16}$$

where n is the number of sampled PSUs. This is a *ratio-of-means* estimator in the sense of Section 16.2.1[2], with PSU-level contributions aggregated across clusters and normalized by the total number of SSU centers falling within the population or stratum boundary.

The PSU-level numerator in (17.16) is

$$y_i = \sum_{j=1}^{m_i} y_{ij}, \tag{17.17}$$

where m_i is the number of SSUs in PSU i, as defined by the cluster layout, and y_{ij} is the per-unit-area contribution from SSU j in PSU i *assigned to the population or stratum of interest*. This quantity is computed as the sum of expanded tree-level measurements for trees that are both measured on SSU j and belong to the population or stratum, based on each tree's location or condition.

---

[2]Strictly speaking, this estimator could be described as a *ratio-of-totals*, since it's equivalent to the ratio of summed numerator and denominator values rather than the ratio of two sample means.

Tree expansion factors are defined in the same way as in plot or point sampling and reflect each tree's inclusion probability under the SSU design. Trees that fall outside the population or stratum do not contribute to y_{ij}, even if they're measured on the SSU.

The PSU-level denominator in (17.16) is

$$z_i = \sum_{j=1}^{m_i} z_{ij}, \tag{17.18}$$

where z_{ij} is an SSU-level center indicator

$$z_{ij} = \begin{cases} 1, & \text{if the center of SSU } j \text{ lies within the population (or stratum)}, \\ 0, & \text{otherwise.} \end{cases}$$

Trees may contribute to the numerator even when they occur on SSUs whose centers lie outside the population or stratum, while the denominator is defined entirely by center-point inclusion.

The standard error of $\bar{y}$ is estimated as

$$s_{\bar{y}} = \sqrt{\frac{1}{n\bar{z}^2} \frac{\sum_{i=1}^{n}(y_i - \bar{y}z_i)^2}{n-1}}, \tag{17.19}$$

with $\bar{z} = \frac{1}{n}\sum_{i=1}^{n} z_i$.

The population total is estimated as

$$\hat{t} = A\bar{y}, \tag{17.20}$$

where A is the total area of the population (or stratum) of interest, with standard error

$$s_{\hat{t}} = As_{\bar{y}}. \tag{17.21}$$

The independent sampling units here are PSUs, not SSUs. All variability is assessed at the PSU level, and the ratio form of the estimator is essential whenever the number of SSU centers contributing to the denominator varies across PSUs, as commonly occurs near population or stratum boundaries.

This estimator provides a practical and elegant solution to boundary problems in large-area cluster sampling. Rather than redefining SSUs or performing explicit geometric corrections, boundary effects are handled implicitly by assigning expanded tree-level contributions to computation units and using center-point counts to represent sampling effort. Under systematic or spatially balanced sampling, over- and under-coverage near boundaries cancel in expectation, yielding unbiased estimates for large areas.

Two limitations are worth noting. First, because SSUs are not subdivided when they intersect a population or stratum boundary, SSU-level means for partially intersecting SSUs are not defined. Trees are assigned to the numerator based on whether they fall within the population or stratum, while SSUs contribute to the denominator only through the center-point indicator. As a result, it's not possible to recover SSU-level means for boundary-crossing SSUs under this formulation.

Second, for very small strata, it's possible for trees to contribute to the numerator while no SSU centers fall within the stratum, leading to an undefined mean. These situations are rarely consequential in routine NFI applications focused on large-area reporting but can matter for small-area estimation.

The Finnish NFI implements (17.16) using the SSU center indicator z_{ij}, a simple binary measure of whether each SSU center falls inside the population or stratum. This approach is easy for field crews to apply and keeps data collection efficient, but it carries the limitations noted above. The Swedish NFI uses the same ratio-estimation framework, but defines the denominator in terms of the *area of each SSU* that falls within the population or stratum (Fridman et al. 2014). While this requires additional effort to determine partial SSU areas, it avoids zero-denominator cases and permits area-weighted SSU-level means for SSUs that intersect boundaries. Together, these two approaches highlight defensible tradeoffs between operational simplicity and statistical detail.

17.6 Stratified cluster sampling

Stratified cluster sampling extends the ideas introduced in stratified sampling (Chapter 15). The population is first divided into H non-overlapping strata, and a cluster sampling design is then applied independently within each stratum.

Conceptually, the stratification step is unchanged. The key difference is that, within each stratum, estimation is based on PSU-level quantities rather than individual observations. As a result, the stratum-level inputs to the stratified estimators are cluster-level means, totals, and standard errors.

Within each stratum $h = 1, \dots, H$:

1. select n_h PSUs using an appropriate probability sampling method (e.g., SRS, SYS, or PPS);
2. within each sampled PSU, measure all SSUs and compute PSU-level quantities (e.g., y_{hi});
3. use the appropriate cluster estimator to obtain a stratum-level mean $\bar{y}_h$ and its standard error $s_{\bar{y}_h}$.

The choice of estimator in Step 3 depends on the sampling frame. For finite populations, unbiased, ratio, or PPS cluster estimators may be used (Sections 17.4.1.1, 17.4.1.2, and 17.4.2). Under an areal sampling frame, ratio estimators based on SSU center inclusion or related proportional-weighting approaches are typically used (Section 17.5).

Once stratum-level estimates are available, they are combined exactly as in standard stratified sampling. The overall mean is

$$\bar{y} = \frac{\sum_{h=1}^{H} M_h \bar{y}_h}{M}, \tag{17.22}$$

where M_h is the size of stratum h and $M = \sum_{h=1}^{H} M_h$.

In a finite-population setting, M_h and M correspond to counts of SSUs (or other appropriate size measures). Under an areal sampling frame, they correspond to stratum area A_h and total area A, respectively (see Section 15.5). In practice, this amounts to replacing counts with areas while retaining the same estimator structure.

The standard error of the stratified mean is

$$s_{\bar{y}} = \sqrt{\frac{\sum_{h=1}^{H} M_h^2 s_{\bar{y}_h}^2}{M^2}}, \tag{17.23}$$

which follows directly from combining independent stratum-level estimates.

If stratum-level totals $\hat{t}_h$ are estimated directly, the overall total and its standard error are

$$\hat{t} = \sum_{h=1}^{H} \hat{t}_h \tag{17.24}$$

and

$$s_{\hat{t}} = \sqrt{\sum_{h=1}^{H} s_{\hat{t}_h}^2}. \tag{17.25}$$

Confidence intervals for stratified cluster estimators follow the same form as in Section 15.2, with degrees of freedom

$$\text{df} = n - H, \tag{17.26}$$

where $n = \sum_{h=1}^{H} n_h$ is the total number of sampled PSUs. This reflects that PSUs—not SSUs—are the independent sampling units.

17.7 Illustrative analyses

The following two examples demonstrate how the cluster sampling estimators introduced earlier behave in practice. To facilitate performance comparisons, both examples use datasets for which a complete census is available, allowing us to assess bias, precision, and interval coverage under repeated sampling.

The first example uses a Portland, Oregon park tree inventory to estimate total tree carbon across the city's parks. This is a finite-population setting, and we examine the unbiased, ratio, and PPS estimators. The second example revisits the Harvard Forest dataset and illustrates the ratio estimator under an areal (infinite-population) sampling frame using fixed-area subplots arranged in PSU-level clusters.

These examples highlight implementation using efficient `tidyverse` workflows and show how estimator behavior differs under realistic forest inventory conditions.

17.7.1 Portland park tree inventory

The Portland Parks and Recreation Urban Forestry division undertook a major effort to collect a complete census of all trees in the city's parks and streets, producing one of the

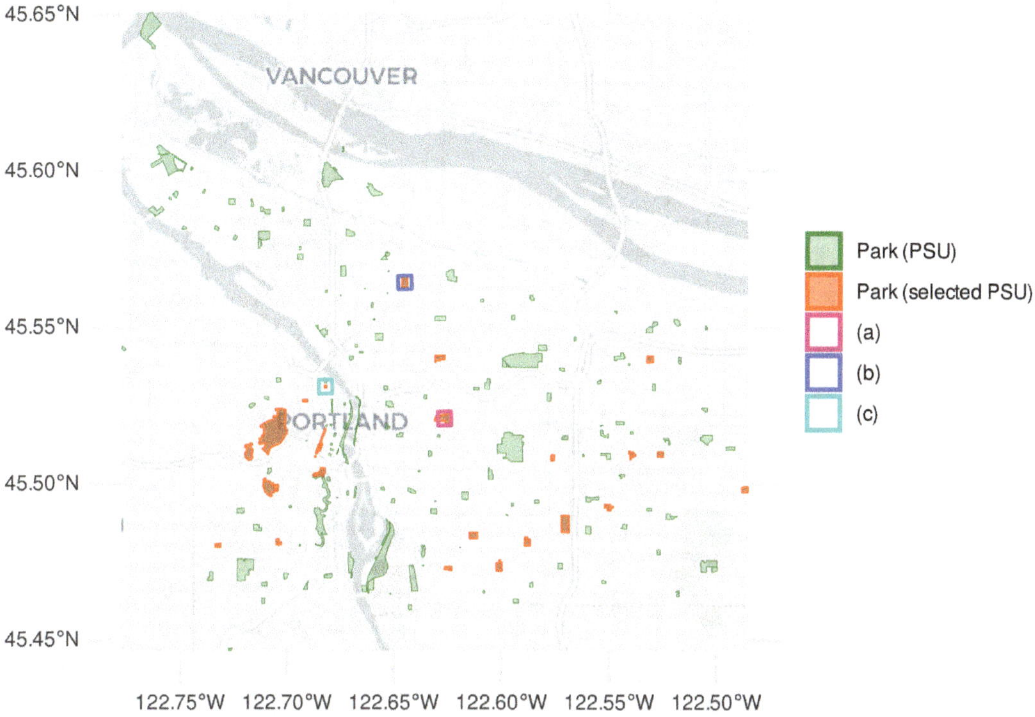

FIGURE 17.1: Portland parks cluster sampling example with parks as primary sampling units (PSUs) and trees as secondary sampling units (SSUs). PSU selection uses probability proportional to size (PPS) based on the number of trees in each park. Close-up views of the parks labeled (a), (b), and (c) are provided in Figure 17.2.

most detailed municipal inventories in the US (DiSalvo et al. 2017).[3] Here we focus on the park portion of that census. Each park was surveyed to record tree locations, species, and structural measurements. A primary objective of the census was to estimate environmental benefits for the city's trees, including carbon storage. To do so, tree-level allometric equations from the i-Tree software suite (Nowak 2021) were used to derive carbon estimates.

For our analysis, we treat this census as the underlying population from which samples could be drawn. Mapping back to the notation of Section 17.4, there are $N = 177$ parks (PSUs) containing a total of $M = 25{,}257$ trees (SSUs). The true total carbon is $\tau = 26{,}349$ tons (1 ton = 2,000 pounds). Each park i contains m_i trees, each with a carbon value y_{ij}.

Figure 17.1 shows an overview of the city's parks, with parks treated as PSUs and trees as SSUs. Close-up views of three example parks are shown in Figure 17.2, illustrating the variation in park size and tree density that motivates the sampling approaches considered below.

---

[3]An R package, `pdxTrees` (McConville and Caldwell 2020), archives these data and provides helper functions for extraction and visualization.

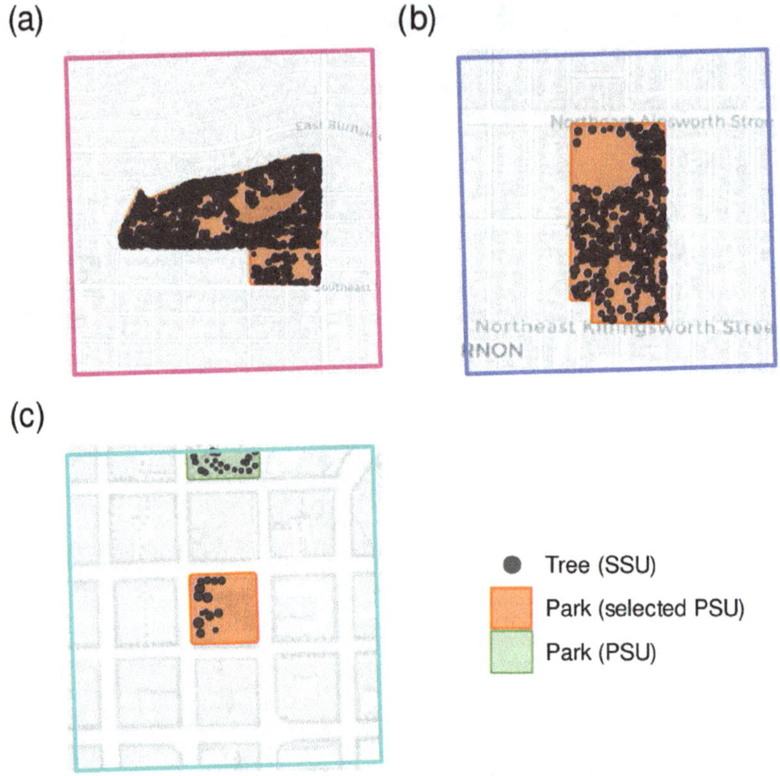

FIGURE 17.2: Close-up views of the parks (PSUs) labeled (a), (b), and (c) in Figure 17.1. These correspond to `park_id` values 12, 79, and 170, which are Alberta Park, Laurelhurst Park, and Tanner Springs Park, respectively.

To make the example more realistic, we assume we do *not* have access to the full set of tree-level measurements, but we do know the total number of trees in each park. This assumption is increasingly plausible due to advances in remote sensing and individual tree detection, where identifying crowns and estimating tree counts from imagery has become routine; see, e.g., Ventura et al. (2024) and Mehtätalo et al. (2022). Under this setup, the sampling problem reduces to selecting which parks (PSUs) to visit for detailed tree carbon measurement (SSUs).

Before proceeding, it's useful to explore the relationship between park size and total carbon. In practice we wouldn't know carbon for every park, but we would know m_i, and it's reasonable to expect that parks with more trees store more carbon. Understanding this relationship helps evaluate which estimators are appropriate and whether their variance estimates are reliable. For exploratory purposes, we temporarily use the census to examine this relationship.

Figure 17.3 shows a strong association between tree count (m_i) and total carbon (y_i). The marginal histograms emphasize the heavy right-skew in both variables and the near-linear increase in y_i with m_i.[4]

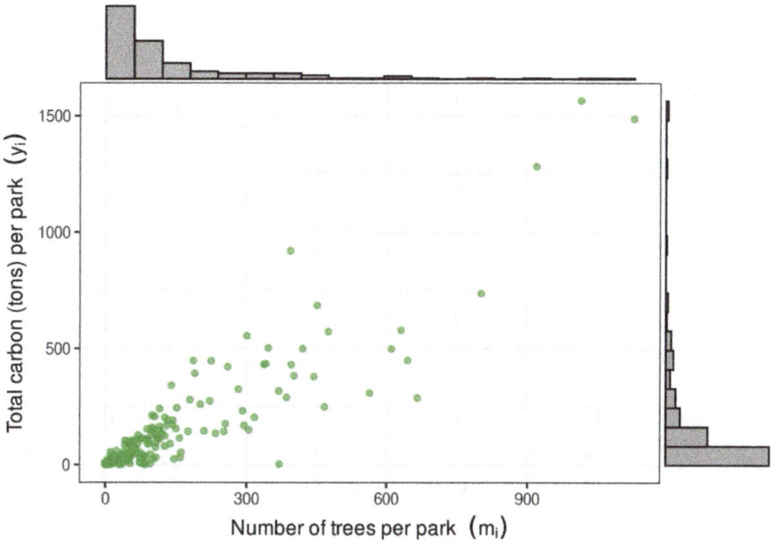

FIGURE 17.3: Relationship between the number of trees (m_i) and total carbon (y_i) across all Portland parks. Marginal histograms show the distribution of each variable.

The heavy right-tail in both park size and total carbon means that a small number of parks contribute a disproportionately large share of the population total. As discussed in Section 11.3.4, standard confidence intervals rely on the estimator's sampling distribution being roughly normal. This can happen even when the underlying data distribution is skewed. But when the population is highly skewed and the sample size is modest, the sampling distribution may remain skewed as well. In these cases, normal-based confidence intervals (like the t-based ones we use here) can miss their intended coverage.[5]

One way to help alleviate this problem is to use a stratified design (Chapter 15) that ensures both large, high-carbon parks and smaller parks are represented in the sample. By forcing the sample to cover the full range of park sizes, stratification reduces the chance of missing influential large parks and limits their impact on the overall estimator. This often stabilizes the sampling distribution and improves precision, leading to confidence intervals with coverage much closer to the intended (i.e., nominal) level.

To explore these estimators and the effect of stratification, we implement:

1. equal-probability PSU sampling with the unbiased estimator (Section 17.4.1.1);
2. equal-probability PSU sampling with the ratio estimator (Section 17.4.1.2);
3. PPS sampling with the PPS estimator (Section 17.4.2).

[4]The figure uses `ggMarginal()` from the `ggExtra` package to add marginal histograms to a scatterplot.

[5]Undercoverage is common when a few extreme units dominate the estimator's variability, though overcoverage can also occur if the variance estimator is overly conservative.

We then stratify the parks into two groups based on tree count (small versus large) and repeat these same estimators under a stratified cluster sampling design. Finally, using the full census, we carry out a repeated sampling analysis to compare bias, precision, and confidence interval coverage across all six estimators—the unstratified and stratified versions of the unbiased, ratio, and PPS estimators.

17.7.1.1 Applying the estimators without stratification

We begin by estimating total carbon using a sample of $n = 25$ PSUs (parks) selected using SRS and PPS. Three data files are used in this analysis.

- `"datasets/PDX/all_parks.csv"` contains the full set of PSUs in the population. Each row corresponds to a park, with columns for a unique park identifier (`park_id`), the number of trees in that park (m_i), and the corresponding total carbon (y_i), computed as the sum of that park's y_{ij} tree values.
- `"datasets/PDX/cluster_equal_prob_park_sample_n25.csv"` contains the 25-park sample drawn using SRS from `"all_parks.csv"`. The code below generated this sample using the tibble `all_parks` read from `"all_parks.csv"`.

```
set.seed(1)
all_parks %>%
  slice_sample(n = 25, replace = FALSE)
```

- `"datasets/PDX/cluster_pps_park_sample_n25.csv"` contains the 25-park sample drawn using PPS from `"all_parks.csv"`, with selection probabilities proportional to each park's tree count (m_i). This sample was generated using the code below and selected parks are shown in Figure 17.1.

```
set.seed(1)
all_parks %>%
  slice_sample(n = 25, weight_by = m_i, replace = TRUE)
```

  Under this PPS-with-replacement design, the per-draw selection probability for park i is $p_i = m_i/M$, consistent with the estimator in Section 17.4.2.

The code below reads and inspects each data file.

```
all_parks <- read_csv("datasets/PDX/all_parks.csv")
all_parks %>% glimpse()
```

```
#> Rows: 177
#> Columns: 3
#> $ park_id <dbl> 1, 2, 3, 4, 5, 6, 7, 8, 9, 10, 11, 12~
#> $ m_i     <dbl> 26, 253, 100, 75, 110, 60, 318, 921, ~
```

```
#> $ y_i     <dbl> 7.85925, 143.06605, 160.04115, 7.3109~
```

```
equal_prob_parks <-
  read_csv("datasets/PDX/cluster_equal_prob_park_sample_n25.csv")
equal_prob_parks %>% glimpse()
```

```
#> Rows: 25
#> Columns: 3
#> $ park_id <dbl> 68, 167, 129, 162, 43, 14, 51, 85, 21~
#> $ m_i     <dbl> 49, 108, 4, 96, 1, 90, 35, 35, 338, 9~
#> $ y_i     <dbl> 37.60115, 30.29090, 2.20830, 43.03720~
```

```
pps_parks <- read_csv("datasets/PDX/cluster_pps_park_sample_n25.csv")
pps_parks %>% glimpse()
```

```
#> Rows: 25
#> Columns: 3
#> $ park_id <dbl> 87, 94, 58, 10, 70, 109, 92, 11, 36, ~
#> $ m_i     <dbl> 565, 403, 261, 66, 647, 71, 48, 181, ~
#> $ y_i     <dbl> 308.2177, 382.0469, 421.2472, 108.513~
```

17.7.1.1.1 *Equal-probability sampling: unbiased estimator*

The unbiased estimator for the total and its standard error follow (17.1) and (17.2).

```
N <- nrow(all_parks)

# Apply the unbiased estimator.
est_unbiased <- equal_prob_parks %>%
  summarize(
    n = n(),
    t_hat = N / n * sum(y_i),
    s_sq = 1 / (n - 1) * sum((y_i - t_hat / N)^2),
    s_t_hat = N * sqrt((1 - n / N) * s_sq / n),
    t = qt(df = n - 1, p = 1 - 0.05 / 2),
    t_ci_lower = t_hat - t * s_t_hat,
    t_ci_upper = t_hat + t * s_t_hat
  )

est_unbiased %>%
  select(t_hat, t_ci_lower, t_ci_upper)
```

```
#> # A tibble: 1 x 3
#>    t_hat t_ci_lower t_ci_upper
#>    <dbl>      <dbl>      <dbl>
```

```
#> 1 32277.      10724.      53830.
```

The census-based total carbon across all parks is $\tau = 26{,}349$ tons. The unbiased estimate for the total, with corresponding 95% confidence intervals, was 32,277 (10,724, 53,830) tons. Based on this sample, the confidence interval captured the true value.

17.7.1.1.2 *Equal-probability sampling: ratio estimator*

The ratio estimator for the mean and total, and their corresponding standard errors, follow (17.8), (17.10), (17.9), and (17.11).

```
N <- nrow(all_parks)
M <- sum(all_parks$m_i)
M_bar <- M / N

# Apply the ratio estimator.
est_ratio <- equal_prob_parks %>%
  summarize(
    n = n(),
    y_bar = sum(y_i) / sum(m_i),
    s_y_bar = sqrt((1 - n / N) * (1 / (n * M_bar^2)) *
      sum((y_i - y_bar * m_i)^2) / (n - 1)),
    t = qt(df = n - 1, p = 1 - 0.05 / 2),
    y_ci_lower = y_bar - t * s_y_bar,
    y_ci_upper = y_bar + t * s_y_bar
  ) %>%
  mutate(
    t_hat = M * y_bar,
    t_ci_lower = M * y_ci_lower,
    t_ci_upper = M * y_ci_upper
  ) %>%
  select(t_hat, t_ci_lower, t_ci_upper)

est_ratio
```

```
#> # A tibble: 1 x 3
#>     t_hat t_ci_lower t_ci_upper
#>     <dbl>      <dbl>      <dbl>
#> 1 28098.      19754.      36441.
```

Again, the census-based total carbon across all parks is $\tau = 26{,}349$ tons. The ratio estimate for the total, with corresponding 95% confidence intervals, was 28,098 (19,754, 36,441) tons. Based on this sample, the confidence interval captured the true value.

17.7.1.1.3 *PPS sampling estimator*

The PPS estimator for the total and its standard error follow (17.12) and (17.13).

```
M <- sum(all_parks$m_i)

# Apply the PPS estimator.
est_pps <- pps_parks %>%
  mutate(p_i = m_i / M) %>%
  summarize(
    n = n(),
    t_hat = (1 / n) * sum(y_i / p_i),
    s_t_hat = sqrt(1 / (n * (n - 1)) * sum((y_i / p_i - t_hat)^2)),
    t = qt(df = n - 1, p = 1 - 0.05 / 2),
    t_ci_lower = t_hat - t * s_t_hat,
    t_ci_upper = t_hat + t * s_t_hat
  ) %>%
  select(t_hat, t_ci_lower, t_ci_upper)

est_pps
```

```
#> # A tibble: 1 x 3
#>    t_hat t_ci_lower t_ci_upper
#>    <dbl>      <dbl>      <dbl>
#> 1 28029.     22735.     33323.
```

Again, the census-based total carbon across all parks is τ= 26,349 tons. The PPS estimate for the total, with corresponding 95% confidence intervals, was 28,029 (22,735, 33,323) tons. Based on this sample, the confidence interval captured the true value.

As with the other designs, a single sample doesn't tell us much about estimator performance. We can, however, see the confidence interval narrowing as we move from the unbiased estimator to the ratio estimator and then to the PPS estimator. In Section 17.7.1.3, we'll check whether this pattern holds across repeated samples or if it was just a quirk of this one sample.

17.7.1.2 Applying the estimators with stratification

Following the steps in Section 15.1, we begin by defining the strata. Because park-level tree counts (m_i) are known for all parks, they offer a natural basis for stratification. Figure 17.4 shows histograms of tree count (m_i) and total carbon (y_i), with colors indicating the two proposed strata. The strata were created by splitting the parks based on the cumulative proportion of total tree counts, separating parks with relatively few trees from those with many. The cutoff is subjective—the goal was simply to isolate the handful of parks with exceptionally high tree counts.

This figure also shows that using m_i for stratification produces nearly the same cutoff for total carbon. We can confirm this here only because the full population data are available. In most applications, we'd rely on the m_i values alone and would assume the strong positive relationship with total carbon seen earlier in Figure 17.3.

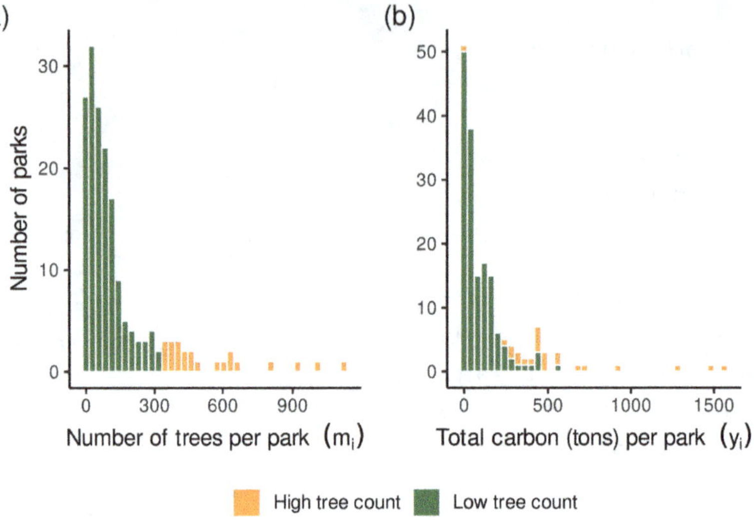

FIGURE 17.4: Histograms showing how Portland parks were divided into low- and high-tree-count strata based on the number of trees per park (m_i). The total carbon (y_i) distribution shows a nearly identical separation, suggesting that tree counts can serve as an effective stratifying variable.

The strong right-skew in both variables highlights that a small number of large parks account for a disproportionate share of the population total. Stratifying on m_i helps ensure these influential parks are represented in the sample, reducing extreme sample-to-sample variability in the resulting estimates. This approach aims to address the issues raised in Section 11.3.4, where strong skew can cause equal-probability confidence intervals to have unreliable coverage.

To keep the stratified and unstratified designs comparable, we select a total of $n = 25$ parks, allocating a higher sampling fraction (50%) to the high-tree-count stratum and a lower fraction (8%) to the low-tree-count stratum. In practice, such allocation might be driven by cost or logistics.

Sampling is carried out independently within each stratum. Under the equal-probability design, parks are selected with equal probability within their stratum. Under the PPS design, selection probabilities are proportional to each park's number of trees (m_{hi}), so that $p_{hi} = m_{hi}/M_h$.

The PSU-level sample data are provided in the following files. In addition to the columns `park_id`, `m_i`, and `y_i`, these files also include a column `stratum` indicating whether each park falls into the high- or low-tree-count stratum.

- `"datasets/PDX/strat_all_parks.csv"` is the same as `"all_parks.csv"` from Section 17.7.1.1, but with the addition of the `stratum` column.
- `"datasets/PDX/cluster_strat_equal_prob_park_sample_n25.csv"` contains the park-level data for the stratified equal-probability sample.

- `"datasets/PDX/cluster_strat_pps_park_sample_n25.csv"` contains the park-level data for the stratified PPS sample.

Following Section 17.6, we next apply the three estimators—unbiased, ratio, and PPS—within this stratified framework. Each estimator is applied within strata and then combined to obtain an overall estimate of the population total τ. As before, the results reflect a single realization of the sampling process and are intended for illustration.

17.7.1.2.1 *Equal-probability stratified sampling: unbiased estimator*

We first apply the unbiased cluster estimator within each stratum and then combine the stratum-level results to obtain the stratified estimate for the population total and its standard error, using (17.24) and (17.25), respectively.

```
# Read the stratified equal-probability sample.
strat_all_parks <-
  read_csv("datasets/PDX/strat_all_parks.csv")

strat_equal_prob_parks <-
  read_csv("datasets/PDX/cluster_strat_equal_prob_park_sample_n25.csv")

strat_size <- strat_all_parks %>%
  group_by(stratum) %>%
  summarize(N_h = n())

# Apply the unbiased estimator.
est_unbiased <- strat_equal_prob_parks %>%
  left_join(strat_size, by = "stratum") %>%
  group_by(stratum) %>%
  summarize(
    N_h = first(N_h),
    n_h = n(),
    t_hat_h = N_h / n_h * sum(y_i),
    s_sq_h = 1 / (n_h - 1) * sum((y_i - t_hat_h / N_h)^2),
    s_t_hat_h = N_h * sqrt((1 - n_h / N_h) * s_sq_h / n_h)
  ) %>% summarise(
    t_hat = sum(t_hat_h),
    s_t_hat = sqrt(sum(s_t_hat_h^2)),
    H = n_distinct(stratum),
    n = sum(n_h),
    t = qt(df = n - H, p = 1 - 0.05 / 2),
    t_ci_lower = t_hat - t * s_t_hat,
    t_ci_upper = t_hat + t * s_t_hat
  )
```

```
est_unbiased %>%
  select(t_hat, t_ci_lower, t_ci_upper)
```

```
#> # A tibble: 1 x 3
#>    t_hat t_ci_lower t_ci_upper
#>    <dbl>      <dbl>      <dbl>
#> 1 24379.     16333.     32425.
```

The census-based total carbon across all parks is τ= 26,349 tons. The stratified unbiased estimate for the total, with corresponding 95% confidence intervals, was 24,379 (16,333, 32,425) tons. Based on this sample, the confidence interval captured the true value.

17.7.1.2.2 *Equal-probability stratified sampling: ratio estimator*

We apply the ratio estimator within each stratum and then combine the stratum-level estimates to obtain the stratified ratio estimate for the population total and its standard error, following (17.24) and (17.25).

```
strat_size <- strat_all_parks %>%
  group_by(stratum) %>%
  summarize(
    N_h = n(),
    M_h = sum(m_i),
    M_bar_h = M_h / N_h
  )

# Apply the ratio estimator.
est_ratio <- strat_equal_prob_parks %>%
  left_join(strat_size, by = "stratum") %>%
  group_by(stratum) %>%
  summarize(
    N_h = first(N_h),
    M_h = first(M_h),
    M_bar_h = first(M_bar_h),
    n_h = n(),
    y_bar_h = sum(y_i) / sum(m_i),
    s_y_bar_h = sqrt((1 - n_h / N_h) * (1 / (n_h * M_bar_h^2)) *
      sum((y_i - y_bar_h * m_i)^2) / (n_h - 1)),
    t_hat_h = M_h * y_bar_h,
    s_t_hat_h = M_h * s_y_bar_h
  ) %>% summarise(
    t_hat = sum(t_hat_h),
    s_t_hat = sqrt(sum(s_t_hat_h^2)),
    H = n_distinct(stratum),
```

```
    n = sum(n_h),
    t = qt(df = n - H, p = 1 - 0.05 / 2),
    t_ci_lower = t_hat - t * s_t_hat,
    t_ci_upper = t_hat + t * s_t_hat
  )

est_ratio %>%
  select(t_hat, t_ci_lower, t_ci_upper)
```

```
#> # A tibble: 1 x 3
#>    t_hat t_ci_lower t_ci_upper
#>    <dbl>      <dbl>      <dbl>
#> 1 26818.     21117.     32519.
```

Again, the census-based total carbon across all parks is $\tau = 26{,}349$ tons. The stratified ratio estimate for the total, with corresponding 95% confidence intervals, was 26,818 (21,117, 32,519) tons. Based on this sample, the confidence interval captured the true value.

17.7.1.2.3 *PPS stratified sampling estimator*

We apply the PPS estimator within each stratum and then combine the stratum-level total estimates and their associated standard errors to obtain the stratified PPS estimate for the population total and its standard error, following (17.24) and (17.25).

```
pps_parks <-
  read_csv("datasets/PDX/cluster_strat_pps_park_sample_n25.csv")

strat_size <- strat_all_parks %>%
  group_by(stratum) %>%
  summarize(M_h = sum(m_i))

# Apply the PPS estimator.
est_pps <- pps_parks %>%
  left_join(strat_size, by = "stratum") %>%
  mutate(p_i = m_i / M_h) %>%
  group_by(stratum) %>%
  summarize(
    n_h = n(),
    t_hat_h = (1 / n_h) * sum(y_i / p_i),
    s_t_hat_h = sqrt(1 / (n_h * (n_h - 1)) *
                       sum((y_i / p_i - t_hat_h)^2)),
  ) %>% summarise(
    t_hat = sum(t_hat_h),
    s_t_hat = sqrt(sum(s_t_hat_h^2)),
```

```
    H = n_distinct(stratum),
    n = sum(n_h),
    t = qt(df = n - H, p = 1 - 0.05 / 2),
    t_ci_lower = t_hat - t * s_t_hat,
    t_ci_upper = t_hat + t * s_t_hat
  )

est_pps %>%
  select(t_hat, t_ci_lower, t_ci_upper)
```

```
#> # A tibble: 1 x 3
#>    t_hat t_ci_lower t_ci_upper
#>    <dbl>      <dbl>      <dbl>
#> 1 23794.     19759.     27829.
```

Again, the census-based total carbon across all parks is τ= 26,349 tons. The stratified PPS estimate for the total, with corresponding 95% confidence intervals, was 23,794 (19,759, 27,829) tons. Based on this sample, the confidence interval captured the true value.

Taken together, these single-sample results illustrate how each estimator behaves for one particular draw from the Portland population. They also show how stratification and PPS sampling aim to stabilize estimates when a few large parks dominate the total. To see how these patterns play out over many possible samples, and to compare the designs on equal footing, the next section presents a repeated-sampling study based on the full census.

17.7.1.3 Repeated-sampling comparison

In the previous sections, the estimators provided single-sample estimates of the population total and associated 95% confidence intervals. Those results are useful for illustration, but they reflect only one realization of the sampling process. Here, we examine how the estimators behave under repeated sampling by drawing many samples from the same finite population and summarizing their performance across repetitions. This allows us to assess whether the apparent gains from stratification and PPS sampling persist across many samples or were simply artifacts of the earlier single-sample examples.

Under repeated sampling, each sample yields an estimate of the same fixed population total. By examining how these estimates behave across many samples—how they vary from sample to sample, how often their confidence intervals contain the true total, and whether they tend to be centered near the true total—we can assess estimator performance more reliably than from a single sample.

The results in Table 17.1 summarize how the six estimators behave under repeated sampling from the Portland parks population. As seen earlier in Figure 17.3, total carbon per park is strongly right-skewed, with a small number of large parks contributing a disproportionate share of the population total. Strong skewness, coupled with modest sample sizes, can leave the sampling distribution skewed as well, which in turn affects estimator performance—especially confidence interval coverage.

TABLE 17.1: Comparison of six estimators for total park-level carbon under repeated sampling from the Portland parks population. Columns show the mean difference from the true population total (i.e., the repeated-sampling mean deviation), the corresponding mean relative difference, the relative standard error (RSE), and 95% confidence interval coverage, based on 500,000 samples each of size $n = 25$.

Design	Estimator	Mean difference (tons)	Mean relative difference (%)	RSE (%)	95% Coverage
Equal prob.	Unbiased	-8.9	-0.034	26.8	87.6
Equal prob.	Ratio	-92.9	-0.353	12.0	86.9
PPS	PPS	8.3	0.031	10.1	94.9
Strat. equal prob.	Unbiased	1.1	0.004	16.3	91.6
Strat. equal prob.	Ratio	-56.3	-0.214	9.9	89.0
Strat. PPS	PPS	2.3	0.009	10.1	94.8

In Table 17.1, the columns Mean difference (tons) and Mean relative (%) report the mean difference between the estimated total and the true population total across the repeated samples, along with the corresponding mean relative difference. These quantities indicate whether an estimator tends to overshoot or undershoot the true total on average. Under repeated sampling, this mean difference corresponds to the estimation bias defined in Section 11.2.3, but we report it here directly for descriptive clarity.

For the equal-probability unbiased estimator, we expect this mean difference to be zero under repeated sampling, and the results are consistent with that expectation. The PPS estimator shows similar expected behavior. The ratio estimator, which is not exactly unbiased, exhibits a very small mean relative difference for both the unstratified and stratified versions (i.e., -0.353% and -0.214%, respectively), consistent with its repeated-sampling properties at this sample size.

The key differences among the estimators appear in the relative standard error (RSE %) and 95% coverage columns. Under unstratified equal-probability sampling, the unbiased estimator has a large RSE of about 26.8%, reflecting substantial sample-to-sample variability when a draw happens to miss one or more of the largest parks. Additionally, the unbiased estimator of the population total is known to be inefficient when cluster sizes are unequal and cluster totals are roughly proportional to m_i (see Lohr (2019), Section 5.3). The unstratified ratio estimator improves precision considerably, reducing the RSE to about 12%. Both unstratified equal-probability estimators have lower-than-nominal coverage (87.6% and 86.9%), a consequence of applying normal-based confidence intervals to a highly skewed population.

Stratifying on tree count helps stabilize both equal-probability estimators. For the unbiased estimator, the RSE drops from 26.8% to 16.3%, and for the ratio estimator it drops from 12% to 9.9%. Average confidence interval coverage improves modestly as well (91.6% and 89%). This pattern illustrates how stratification can reduce variability when a small number of large units dominate the population total.

The PPS estimators stand out most clearly. Both the unstratified and stratified PPS designs achieve low RSE (about 10.1%) and near-nominal coverage (94.8%). Their estimates are centered close to the true total, but their precision and interval performance are markedly better. For a population with the level of skewness seen in Figure 17.3, PPS sampling uses the tree counts to reliably locate the large, influential parks in nearly every sample, making it the most stable design among those considered.

We'll return to this dataset in Section 18.6.1, where a two-stage design will be used to subsample trees for measurement within selected parks.

17.7.2 Harvard Forest biomass

To illustrate cluster sampling under an areal sampling frame, we return to the Harvard Forest dataset. We first apply the cluster sampling ratio estimator defined in Section 17.5 to a single unstratified sample, and then extend the same logic to a stratified setting using the estimators introduced in Section 17.6.

Our goal is to estimate average AGB per hectare and total AGB across the forest, considering all stems with DBH $\geq$ 12.7 cm.

We use a systematic design to place the PSUs. Each PSU is centered on a SYS grid point, and around each center we lay out three fixed-area circular SSUs (subplots). The subplots are positioned 15 meters from the PSU center at azimuths of 360°, 120°, and 240°, each with a 5-meter radius. For estimation, tree-level contributions are assigned to the population based on tree location (i.e., whether trees fall within the subplot and within the boundary), while SSU centers determine the denominator of the cluster ratio estimator, as described in Section 17.5.1.

As described in Section 14, the `sys_grid()` function generates the systematic grid of PSU centers using a randomly chosen starting point. For this example, the spacing is 100-by-100 meters, which yields $n = 42$ PSUs. Figure 17.5 shows the resulting layout.

The cluster layout shown in Figure 17.5 highlights several features that are important for estimation under an areal sampling frame. PSUs are placed using a systematic design over a *buffered* Harvard Forest boundary. This buffer ensures that the full SSU configuration associated with a PSU can occur entirely outside the forest boundary, which is required for SSU center indicators to provide a consistent measure of sampling effort near boundaries.

Each PSU anchors a fixed configuration of three circular SSUs arranged in a triangle-shaped pattern. Near the boundary, some SSUs extend partially outside the forest and are therefore *clipped* to the boundary. Trees are included in estimation only if they fall within the portion of an SSU that lies inside the forest. In contrast, whether an SSU contributes to the denominator of the estimator depends solely on whether its *center point* lies within the forest, regardless of how much of the SSU area lies inside the boundary.

The tree-level data used for estimation are stored in `"datasets/HF/HF_cluster_sys.csv"` and read into `ssu_trees` below.

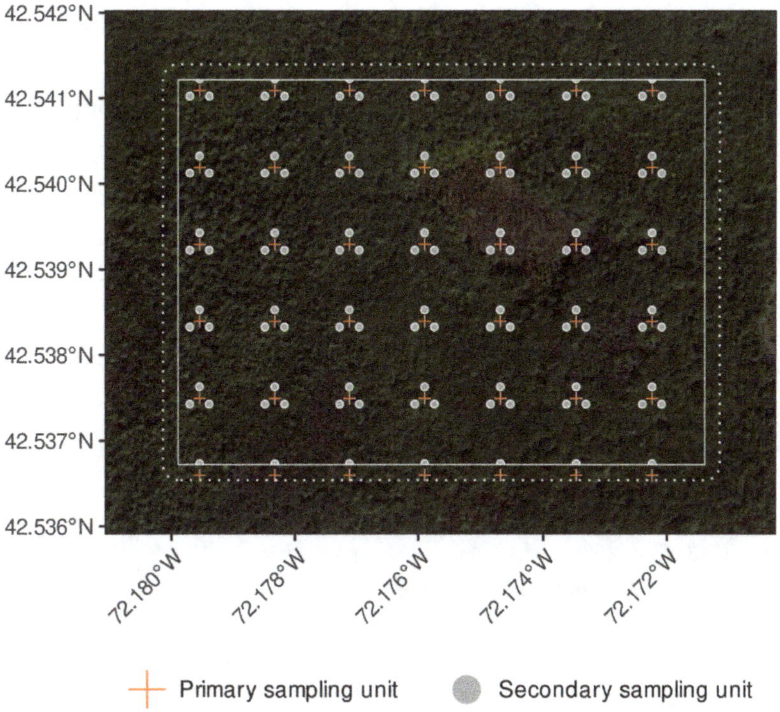

FIGURE 17.5: Cluster sampling under an areal sampling frame at Harvard Forest. Primary sampling units (PSUs) are placed using a systematic design over a buffered population extent to allow complete cluster exclusion near boundaries. Secondary sampling units (SSUs) are fixed-area circular plots arranged in a triangle-shaped configuration around each PSU center. SSUs are clipped to the population boundary for tree inclusion, while SSU centers falling inside the boundary define the denominator of the ratio estimator.

```
ssu_trees <- read_csv("datasets/HF/HF_cluster_sys.csv")
ssu_trees %>% glimpse()
```

```
#> Rows: 588
#> Columns: 4
#> $ psu_id <dbl> 1, 2, 2, 2, 2, 2, 2, 2, 2, 2, 2, 2, 2,~
#> $ ssu_id <dbl> 1, 1, 1, 1, 1, 1, 2, 2, 2, 2, 3, 3, 3,~
#> $ z_ij   <dbl> 1, 1, 1, 1, 1, 1, 1, 1, 1, 1, 1, 1, 1,~
#> $ agb_Mg <dbl> 1.083639, 1.041079, 2.192626, 1.970485~
```

Each row of `ssu_trees` corresponds to a single tree measurement collected on the portion of an SSU that lies within the forest boundary. The identifiers `psu_id` and `ssu_id` indicate the PSU and SSU to which each tree belongs, reflecting the nested structure of the cluster design.

The column `agb_Mg` gives the tree's aboveground biomass in metric tons. The indicator variable `z_ij` records whether the *center of the SSU* lies within the forest boundary (`z_ij = 1`) or outside (`z_ij = 0`). This variable plays a key role in defining the denominator of the cluster ratio estimator under an areal sampling frame.

Before proceeding with estimation, we verify the number of sampled PSUs and the number of SSUs associated with each PSU.

```
ssu_trees %>%
  summarize(n = n_distinct(psu_id))
```

```
#> # A tibble: 1 x 1
#>       n
#>   <int>
#> 1    42
```

```
ssu_trees %>%
  group_by(psu_id) %>%
  summarize(m_i = n_distinct(ssu_id)) %>%
  distinct(m_i)
```

```
#> # A tibble: 2 x 1
#>     m_i
#>   <int>
#> 1     1
#> 2     3
```

In this realization, PSUs have either one or three SSUs that fall at least partially within the forest boundary.

We begin by computing SSU-level contributions. Tree-level AGB is expanded to a per-hectare basis using the standard tree factor for 5-meter-radius circular plots. Expanded tree values are then summed within each SSU to obtain y_{ij}.

```
r <- 5 # Plot radius (m).

ssu_summary <- ssu_trees %>%
  mutate(TF = 10000 / (pi * r^2)) %>%
  group_by(psu_id, ssu_id) %>%
  summarize(
    y_ij = sum(TF * agb_Mg),
    z_ij = first(z_ij),
    .groups = "drop"
  )

ssu_summary
```

```
#> # A tibble: 112 x 4
```

```
#>    psu_id ssu_id  y_ij  z_ij
#>     <dbl>  <dbl> <dbl> <dbl>
#>  1      1      1 138.      1
#>  2      2      1 891.      1
#>  3      2      2 564.      1
#>  4      2      3 371.      1
#>  5      3      1  42.9     1
#>  6      3      2 544.      1
#>  7      3      3 403.      1
#>  8      4      1 724.      1
#>  9      4      2 202.      1
#> 10      4      3 308.      1
#> # i 102 more rows
```

Next, we aggregate to the PSU level. The PSU-level numerator contribution y_i is the sum of SSU contributions within the cluster. The corresponding denominator contribution z_i is the number of SSU centers that lie within the forest.

```
psu_summary <- ssu_summary %>%
  group_by(psu_id) %>%
  summarize(
    y_i = sum(y_ij),
    z_i = sum(z_ij)
  )
```

Using the cluster ratio estimator in (17.16) and its variance estimator in (17.19), we estimate the mean aboveground biomass per hectare. The forest-wide total is obtained by multiplying the mean by the known forest area A.

```
A <- 35 # Total area (ha).

psu_summary %>%
  summarize(
    y_bar = sum(y_i) / sum(z_i),
    n = n(),
    z_bar = mean(z_i),
    s_y_bar = sqrt(
      1 / (n * z_bar^2) *
        sum((y_i - y_bar * z_i)^2) / (n - 1)
    ),
    t = qt(1 - 0.05 / 2, df = n - 1),
    y_ci_lower = y_bar - t * s_y_bar,
    y_ci_upper = y_bar + t * s_y_bar
  ) %>%
```

```
  select(y_bar, y_ci_lower, y_ci_upper) %>%
  mutate(
    t_hat = A * y_bar,
    t_ci_lower = A * y_ci_lower,
    t_ci_upper = A * y_ci_upper
  )
```

```
#> # A tibble: 1 x 6
#>   y_bar y_ci_lower y_ci_upper  t_hat t_ci_lower t_ci_upper
#>   <dbl>      <dbl>      <dbl>  <dbl>      <dbl>      <dbl>
#> 1  355.       302.       409. 12441.     10579.     14302.
```

For reference, the census-based aboveground biomass for the Harvard Forest is 341.46 Mg/ha, with a forest-wide total of 11,951 Mg.

For this realization, the estimated mean and total, along with their 95% confidence intervals, successfully capture the census values.

17.7.2.1 Stratified cluster sampling

We now extend the same cluster design to a stratified setting using the Harvard Forest strata introduced in Section 15.7 and shown in Figure 17.6. As before, inference targets stratum-specific and overall means *per unit area* under an areal sampling frame, and estimation proceeds by applying the cluster ratio estimator separately within each stratum.[6]

The same systematic sample of PSUs used in the unstratified example is retained. For estimation within a given stratum, tree-level contributions are assigned to that stratum based on tree location. This means an SSU can contribute to a stratum's *numerator* whenever part of the SSU overlaps the stratum and trees from that overlapped portion belong to it, even if the SSU center lies outside the stratum. In contrast, the *denominator* is based only on SSU center points: an SSU contributes one unit of sampling effort to stratum h if and only if its center lies in stratum h.

As a result, PSUs near stratum boundaries may have numerator contributions coming from SSUs on either side of a boundary, while denominator contributions depend strictly on where SSU centers fall. This is handled naturally by the ratio form of the estimator described in Section 17.5.1.

The stratified cruise data are read below. As the `glimpse()` output shows, SSU IDs are nested within PSU IDs, which are in turn nested within stratum IDs. This nesting structure determines how summaries and estimators are computed.[7]

---

[6]Strictly speaking, stratified estimators assume independent sampling within strata. Here, a single systematic design is applied across all strata for simplicity.

[7]Notice there are more rows in the stratified `ssu_trees` than in the unstratified `ssu_trees`. This is because a few SSUs straddle stratum boundaries.

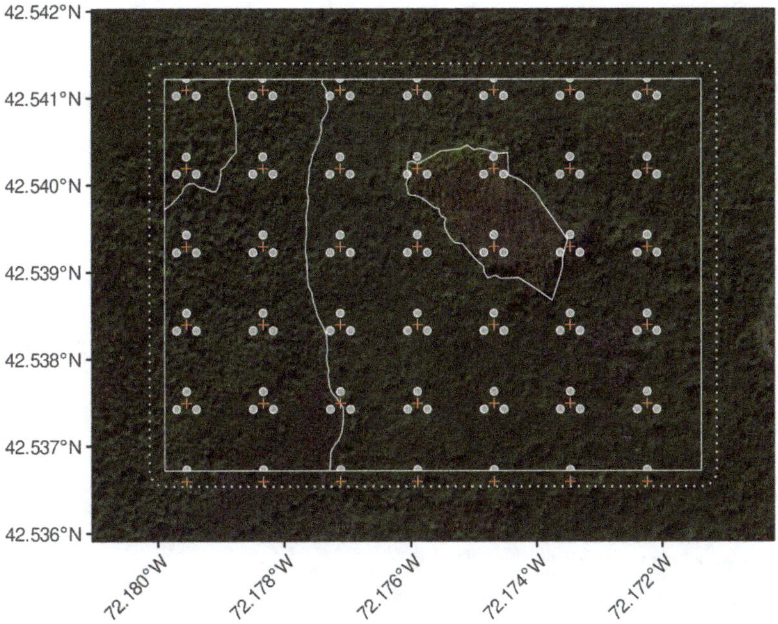

FIGURE 17.6: Stratified cluster sampling under an areal sampling frame at Harvard Forest. The same systematic PSU layout is used across all strata. SSUs are clipped to stratum boundaries for tree inclusion, while SSU centers falling inside a stratum define the denominator of the cluster ratio estimator for that stratum.

```
ssu_trees <- read_csv("datasets/HF/HF_strat_cluster_sys.csv")
ssu_trees %>% glimpse()
```

```
#> Rows: 590
#> Columns: 5
#> $ psu_id     <dbl> 41, 41, 41, 41, 41, 41, 41, 41, 41~
#> $ ssu_id     <dbl> 1, 1, 1, 1, 1, 1, 2, 2, 2, 3, 3, 3~
#> $ stratum_id <dbl> 1, 1, 1, 1, 1, 1, 1, 1, 1, 1, 1, 1~
#> $ z_ij       <dbl> 1, 1, 1, 1, 1, 1, 1, 1, 1, 1, 1, 1~
#> $ agb_Mg     <dbl> 0.071185, 0.194714, 0.263038, 0.70~
```

We now walk through the estimation workflow described in Section 17.6. As before, AGB is expanded to a per-hectare basis at the SSU level, then aggregated to the PSU and stratum levels. For this stratified example we follow the same logic as Section 17.5.1, so the SSU-level numerator is based on tree locations while the SSU-center indicator `z_ij` defines the denominator contributions.

```
r <- 5 # Plot radius (m).

ssu_summary <- ssu_trees %>%
```

```
  mutate(TF = 10000 / (pi * r^2)) %>%
  group_by(stratum_id, psu_id, ssu_id) %>%
  summarize(
    y_hij = sum(TF * agb_Mg),
    z_hij = first(z_ij)
  )
```

Next, for each stratum we compute PSU-level numerator and denominator contributions, then obtain stratum-level means and standard errors using the cluster ratio estimator in (17.16) and variance estimator in (17.19).

```
stratum_summary <- ssu_summary %>%
  group_by(stratum_id, psu_id) %>%
  summarize(
    y_hi = sum(y_hij),
    z_hi = sum(z_hij)
  ) %>%
  group_by(stratum_id) %>%
  summarize(
    y_bar_h = sum(y_hi) / sum(z_hi),
    n_h = n(),
    z_bar_h = mean(z_hi),
    s_y_bar_h = sqrt(
      1 / (n_h * z_bar_h^2) *
        sum((y_hi - y_bar_h * z_hi)^2) / (n_h - 1)
    )
  )

stratum_summary
```

```
#> # A tibble: 4 x 5
#>   stratum_id y_bar_h   n_h z_bar_h s_y_bar_h
#>        <dbl>   <dbl> <int>   <dbl>     <dbl>
#> 1          1   252.      2    2.5      89.2
#> 2          2   387.     12    2.17     36.4
#> 3          3    34.9     4    2         5.88
#> 4          4   390.     28    2.36     29.4
```

Given stratum-level estimates, we combine them using the stratified estimators (17.22) and (17.23). Under an areal sampling frame, stratum areas A_h play the same role that M_h does in the finite-population stratified estimators, so we weight stratum means by area and combine stratum variances accordingly.

```
stratum_summary <- stratum_summary %>%
  left_join(stratum_area, by = "stratum_id")

stratum_summary %>%
  summarize(
    A = sum(A_h),
    y_bar = sum(A_h * y_bar_h) / A,
    s_y_bar = sqrt(sum(A_h^2 * s_y_bar_h^2) / A^2),
    df = sum(n_h - 1),
    t = qt(1 - 0.05 / 2, df = df),
    ci_lower = y_bar - t * s_y_bar,
    ci_upper = y_bar + t * s_y_bar
  ) %>%
  select(y_bar, ci_lower, ci_upper)
```

```
#> # A tibble: 1 x 3
#>   y_bar ci_lower ci_upper
#>   <dbl>    <dbl>    <dbl>
#> 1  361.     318.     404.
```

For this realization, the stratified cluster estimate yields a slightly narrower confidence interval than the unstratified estimate. Because this comparison is based on a single systematic sample, we defer interpretation and turn next to a repeated-sampling analysis to more reliably assess gains in precision attributable to stratification.

17.7.2.2 Comparing stratified and unstratified cluster sampling

In this section, we compare the performance of unstratified and stratified cluster sampling under an areal sampling frame using *repeated systematic samples* from the Harvard Forest population. In both cases, clusters are treated as collections of fixed-area SSUs, and estimation follows the SSU-center–based cluster ratio estimator described in Section 17.5.1, consistent with the Finnish NFI.

For each design, we repeatedly generate a new systematic sample of PSUs over the buffered population extent, construct identical cluster layouts around each PSU center, and apply the same ratio estimator used in the single-sample examples in the preceding sections. Each iteration yields an estimate of mean AGB per hectare and a 95% confidence interval, from which empirical coverage and average interval width are compared across designs.

The empirical coverage rates were 97.86% and 96.09% for the unstratified and stratified cluster estimators, respectively. The corresponding average 95% confidence interval widths were 97 and 85 Mg/ha. The percent reduction in average interval width due to stratification was 13%.

These results mirror the efficiency gains observed earlier in Section 15.7.2 for stratified sampling with single plots and show that effective stratification continues to improve precision when clusters are used as the primary sampling units under an areal sampling frame.

17.8 Advantages and disadvantages of cluster sampling

Cluster sampling is widely used in forest inventory because it can substantially reduce the cost and complexity of field data collection. It's especially useful when measurement is inexpensive but travel between sampling locations is costly—a common situation in large-area surveys. For these reasons, cluster designs are a core feature of many NFI programs.

17.8.1 Advantages

- Operational efficiency: By grouping nearby SSUs into PSUs, cluster sampling reduces travel time and limits the need to move frequently between dispersed sampling locations.
- Simplified field logistics: Field crews can complete several measurements within a single PSU, reducing repositioning effort between measurement locations.
- Alignment with operational structure: Many forest monitoring programs are organized around geographic or administrative PSUs, making cluster sampling a natural and easily implemented choice.

17.8.2 Disadvantages

- Reduced statistical efficiency: SSUs within the same PSU are often spatially autocorrelated and therefore provide less independent information than SSUs sampled across the full population.
- Smaller effective sample size: Because randomization occurs at the PSU level, the effective sample size is the number of PSUs rather than the total number of SSUs, which can reduce precision for a fixed field effort.
- Complex inclusion probabilities under areal frames: When clusters overlap population or stratum boundaries, tree-level inclusion probabilities can be difficult to express analytically. Several principled approaches exist for handling these situations, but each involves tradeoffs among geometric exactness, computational burden, and operational simplicity.

Despite these limitations, cluster sampling remains a practical and effective choice when logistical constraints are high and moderate precision is acceptable. In many large-area inventory settings, it provides a reasonable balance between field efficiency and statistical performance.

17.9 Exercises

17.9.1 Portland park stormwater filtration analysis

Take a moment to reread the description of the Portland park tree inventory in Section 17.7.1. The i-Tree software provides estimates for several tree-level quantities, which appear as columns in `"datasets/PDX/pdx_trees.csv"`:

- `carbon_storage_ton`: total carbon stored (tons; the variable analyzed in Section 17.7.1);
- `carbon_sequestered_ton`: annual carbon sequestered (tons);
- `stormwater_cuft`: annual volume of stormwater filtered (cubic feet);
- `pollution_removal_oz`: annual amount of air pollution removed (ounces);
- `structural_value`: replacement cost for a comparable tree (US dollars).

The data are read in below. In addition to the tree-level i-Tree variables listed above, the dataset includes a unique park and tree ID, inventory date, DBH (in), height (ft), and a species code. Each row represents a single tree in the park census.

```
#> Rows: 25,257
#> Columns: 11
#> $ park_id                <dbl> 1, 1, 1, 1, 1, 1, 1, 1~
#> $ tree_id                <dbl> 1, 2, 3, 4, 5, 6, 7, 8~
#> $ inv_date               <dttm> 2017-09-16, 2017-09-1~
#> $ dbh_in                 <dbl> 15.5, 17.4, 14.0, 5.9,~
#> $ height_ft              <dbl> 24, 65, 20, 21, 24, 36~
#> $ species                <chr> "PRSE", "ACSA3", "PRSR~
#> $ carbon_storage_ton     <dbl> 0.57615, 0.79335, 0.44~
#> $ carbon_sequestered_ton <dbl> 0.01735, 0.02005, 0.01~
#> $ stormwater_cuft        <dbl> 38.1, 49.5, 28.7, 7.2,~
#> $ pollution_removal_oz   <dbl> 10.2, 13.3, 7.7, 1.9, ~
#> $ structural_value       <dbl> 2566.67, 3300.09, 2125~
```

In Section 17.7.1, we used `carbon_storage_ton` (i.e., the y_{ij} values) to estimate total tree carbon across Portland parks using cluster sampling. The repeated sampling analysis in Section 17.7.1.3 showed that the PPS estimator consistently outperformed the equal-probability unbiased and ratio estimators.

In the exercises below, you'll apply the PPS cluster sampling estimator to compute a population total using the `stormwater_cuft` i-Tree variable. The exercises walk through the estimation workflow starting from individual tree-level i-Tree estimates.

Here, `pdx_trees` represents the *entire population* of trees. That gives us the unusual luxury of knowing the truth, so we can directly check how well our estimators perform.

We'll begin by computing the required population-level quantities, then take a PPS sample, explore the sample a bit, apply the estimator, and finally compare the result to the true total.

Exercise 17.1. Using `pdx_trees`, compute a park-level summary (i.e., PSU-level summary) that includes `park_id`, the number of trees (i.e., SSUs) in each park (m_i), and the park-level sum of `stormwater_cuft`, denoted y_i to match the PPS notation in Section 17.4.2.

Call this park-level summary `all_parks`. It should match the output shown below.

Hint, group by `park_id` before calling `summarize()` to compute m_i and y_i.

```
all_parks
```

```
#> # A tibble: 177 x 3
#>     park_id   m_i     y_i
#>       <dbl> <int>   <dbl>
#>  1        1    26   807. 
#>  2        2   253 14890. 
#>  3        3   100  6620. 
#>  4        4    75   982. 
#>  5        5   110  8486. 
#>  6        6    60  4510. 
#>  7        7   318 15365. 
#>  8        8   921 83556. 
#>  9        9     5    32.5
#> 10       10    66  4803. 
#> # i 167 more rows
```

Exercise 17.2. We'll use the number of trees in each park, m_i, to define the selection probability $p_i = m_i/M$, where $M = \sum_{i=1}^{N} m_i$. In this way, parks with more trees—and hence a larger expected contribution to total stormwater filtration—are more likely to be selected.

Add a `p_i` column to `all_parks` via `mutate()`. You can compute $M = \sum_{i=1}^{N} m_i$ directly from `all_parks`. Your modified `all_parks` should match the output below.

Hint, compute M first, then use it inside `mutate()`.

```
all_parks
```

```
#> # A tibble: 177 x 4
#>     park_id   m_i     y_i     p_i
#>       <dbl> <int>   <dbl>   <dbl>
#>  1        1    26   807.  0.00103
#>  2        2   253 14890.  0.0100 
#>  3        3   100  6620.  0.00396
#>  4        4    75   982.  0.00297
#>  5        5   110  8486.  0.00436
```

```
#>  6         6    60  4510.   0.00238
#>  7         7   318 15365.   0.0126
#>  8         8   921 83556.   0.0365
#>  9         9     5    32.5 0.000198
#> 10        10    66  4803.   0.00261
#> # i 167 more rows
```

Exercise 17.3. What is the sum of the `p_i` column in `all_parks`? Is this consistent with the definition of selection probabilities given in Section 17.4.2?

Exercise 17.4. Use the code below to select a PPS sample of $n = 25$ parks *with replacement*. This sample will be used for all subsequent estimation.

Read the help page for `dplyr::slice_sample()`, paying particular attention to the `weight_by` argument. This argument defines each park's selection probability. Because `slice_sample()` internally standardizes the supplied weights to sum to one, we could specify `weight_by = m_i`. Here, we instead pass `weight_by = p_i`; the result is the same.

```
n <- 25

set.seed(1)
pps_parks <- all_parks %>%
  slice_sample(n = n, weight_by = p_i, replace = TRUE)

pps_parks
```

```
#> # A tibble: 25 x 4
#>     park_id   m_i    y_i      p_i
#>       <dbl> <int>  <dbl>    <dbl>
#>  1       87   565 27030. 0.0224
#>  2       94   403 32376. 0.0160
#>  3       58   261 25910. 0.0103
#>  4       10    66  4803. 0.00261
#>  5       70   647 38606. 0.0256
#>  6      109    71  5612. 0.00281
#>  7       92    48  4074. 0.00190
#>  8       11   181 15970. 0.00717
#>  9       36   211 15458. 0.00835
#> 10       79  1016 84078. 0.0402
#> # i 15 more rows
```

Exercise 17.5. Because `pps_parks` was sampled with replacement, some parks may appear multiple times in the sample.

Write code to count the number of occurrences of each `park_id` in the sample; call this count `n_rep`. Also include each park's `m_i` and `p_i` values, and arrange the output in decreasing order of `n_rep`.

Does the output suggest that parks with more trees are more likely to be selected?

Hint, start with `group_by(park_id)` and then use `summarize()`. You can extract `m_i` and `p_i` using `first()`.

```
#> # A tibble: 22 x 4
#>    park_id n_rep   m_i     p_i
#>      <dbl> <int> <int>   <dbl>
#>  1      81     2   633 0.0251
#>  2      87     2   565 0.0224
#>  3      94     2   403 0.0160
#>  4      10     1    66 0.00261
#>  5      11     1   181 0.00717
#>  6      19     1   115 0.00455
#>  7      21     1   338 0.0134
#>  8      27     1   116 0.00459
#>  9      30     1   159 0.00630
#> 10      36     1   211 0.00835
#> # i 12 more rows
```

Exercise 17.6. Finally, implement the PPS estimator for total stormwater filtered by all Portland park trees, along with its standard error and a 95% confidence interval, using (17.12), (17.13), and (17.7), respectively. Units are cubic feet per year.

```
#> # A tibble: 1 x 4
#>      t_hat s_t_hat t_ci_lower t_ci_upper
#>      <dbl>   <dbl>      <dbl>      <dbl>
#> 1 1830639. 114850.   1593600.   2067679.
```

Exercise 17.7. Use `pdx_trees` to compute the true total stormwater filtration across all population trees (this is τ, which you estimated with $\hat{t}$ in Exercise 17.6).

Does the 95% confidence interval from Exercise 17.6, based on this single PPS sample, cover the true total?

18

Multistage sampling

Multistage sampling generalizes cluster sampling by allowing selection to occur in two or more successive stages. Rather than drawing a sample directly from the population, units are selected hierarchically: larger units are sampled first, followed by progressively smaller, nested units. In principle, this process can involve any number of stages, depending on how the population is organized and how data collection is implemented.

While multistage designs can involve more than two stages, two-stage sampling is by far the most common in practice. In a typical two-stage design:

1. a sample of PSUs is selected (e.g., stands, management units, compartments, or ownerships that partition the population of interest);
2. within each sampled PSU, a sample of SSUs is selected.

This structure offers considerable flexibility. It allows sampling intensity to vary across PSUs, accommodates operational constraints, and reduces travel effort by concentrating field measurements within a subset of selected PSUs.

18.1 Connection to cluster sampling

Two-stage sampling is a generalization of cluster sampling. The cluster design introduced in Chapter 17 corresponds to a special case of two-stage sampling in which *all* SSUs within each selected PSU are measured.

Two-stage sampling relaxes this requirement. Instead of measuring every SSU within a selected PSU, only a subset of SSUs is sampled. This distinction is important. In single-stage cluster sampling, sampling variability reflects only differences among PSUs, because SSU-level variation is fully observed within each cluster. In two-stage sampling, variability arises from both stages: between-PSU variation due to first-stage selection and within-PSU variation due to second-stage sampling. As a result, estimators and variance expressions must explicitly account for uncertainty introduced at both stages.

DOI: 10.1201/9781003222699-18

18.2 When multistage sampling is useful

Multistage sampling is widely used in forest inventory because it matches how forests are organized and offers a practical way to balance statistical efficiency, cost, and field logistics. By selecting PSUs first and then sampling SSUs within them, it lets you concentrate effort where it has the biggest impact. In particular, it can:

- use ecological or management units (e.g., stands, compartments, ownership blocks) as convenient PSUs;
- reduce navigation and travel by allowing crews to work intensively within a smaller number of selected PSUs;
- incorporate auxiliary information available at the PSU level (e.g., stand area, prior estimates, remote-sensing predictors) to improve precision;
- align sampling effort with expected variability or importance of different PSUs.

The number of SSUs per PSU can vary, different sampling methods can be used at each stage, and unequal probabilities can be introduced to target PSUs or SSUs expected to contribute more to the population total or mean.

18.3 Strategies for selecting PSUs

The first stage of two-stage sampling involves selecting PSUs. Common strategies include the following.

1. *Simple random sampling (SRS)*: select PSUs with equal probability when no auxiliary information is available.
2. *Stratification*: if PSU-level variables are known (e.g., stand area, stand development stage, forest type), stratify PSUs and sample within strata to improve estimates.
3. *Systematic sampling*: if PSUs can be arranged spatially or sorted by auxiliary information related to the parameter of interest, select PSUs systematically with a random start to ensure broad coverage.
4. *Probability proportional to size (PPS)*: if each PSU has a size measure that's related to the parameter of interest (e.g., area or predicted biomass), select PSUs with probabilities proportional to that measure.

These approaches can be combined. For example, PSUs might be stratified by forest type and then selected using SRS, SYS, or PPS within each stratum. Stratum-specific two-stage estimates for means, totals, and standard errors can then be combined as described in Section 17.6.

Availability of auxiliary information usually guides the choice of first-stage design. Without auxiliary data, equal-probability sampling at the PSU stage is the default. When size- or variability-related variables are available and correlated with the parameter of interest, PPS designs can improve efficiency by allocating sampling effort in proportion to each PSU's contribution.

The appropriate second-stage estimator depends on how SSUs are conceptualized within each PSU. In some applications, SSUs form a finite, list-based set of units. In others, SSUs represent sampling locations placed within a continuous area, as in point or plot sampling under an areal sampling frame. To accommodate these cases, we present estimator forms suited to both interpretations.

We present the estimators in their most general forms, allowing for PSUs of differing sizes and varying numbers of SSUs per PSU. When PSU sizes are equal or the number of SSUs per PSU is constant, the estimators simplify. These special cases are treated in standard survey sampling texts such as Cochran (1977), Thompson (2012), Lohr (2019), and Särndal et al. (1992).

While two-stage estimators can look lengthy at first glance, they follow clear, step-by-step calculations. With the tools in Section 10.2, they're straightforward to implement, and each yields a population mean and total with associated standard errors.

18.4 Estimation using two-stage sampling

We use the following notation:

- N number of PSUs in the population;
- M_i number of SSUs in the i-th PSU;
- $M = \sum_{i=1}^{N} M_i$ total number of SSUs in the population;
- n number of PSUs in the sample;
- m_i number of sampled SSUs in the i-th sampled PSU;
- y_{ij} value of the j-th SSU in the i-th PSU;
- $\bar{y}_i = \frac{1}{m_i} \sum_{j=1}^{m_i} y_{ij}$ sample mean within PSU i;
- $s_i^2 = \frac{1}{m_i - 1} \sum_{j=1}^{m_i} (y_{ij} - \bar{y}_i)^2$ sample variance within PSU i.

This notation and the estimators that follow assume a finite population with a fixed set of N PSUs (and their associated SSUs). Section 18.5 describes how closely related estimators are used when working from an areal sampling frame.

18.4.1 Sampling PSUs with equal probability

18.4.1.1 Unbiased estimators

An unbiased estimator for the population total τ is

$$\hat{t} = \frac{N}{n}\sum_{i=1}^{n} M_i \bar{y}_i. \tag{18.1}$$

The standard error of $\hat{t}$ combines variation among PSU totals and variation among SSUs within PSUs and is estimated as

$$s_{\hat{t}} = \sqrt{\underbrace{\frac{N^2}{n}\left(1-\frac{n}{N}\right)s^2}_{\text{between-PSU variance}} + \underbrace{\frac{N}{n}\sum_{i=1}^{n}\frac{M_i^2}{m_i}\left(1-\frac{m_i}{M_i}\right)s_i^2}_{\text{within-PSU variance}}}, \tag{18.2}$$

where

$$s^2 = \frac{1}{n-1}\sum_{i=1}^{n}\left(M_i\bar{y}_i - \frac{\hat{t}}{N}\right)^2 \tag{18.3}$$

is the sample variance among PSU totals.

Given $\hat{t}$, the population mean μ is estimated as

$$\bar{y} = \frac{\hat{t}}{M}, \tag{18.4}$$

with standard error

$$s_{\bar{y}} = \frac{s_{\hat{t}}}{M}. \tag{18.5}$$

As shown in (18.2), total variance is partitioned into two components.

- The first term (*between-PSU variance*) captures variation among PSU totals and includes a FPC because PSUs are sampled without replacement.
- The second term (*within-PSU variance*) captures variation among SSUs within each PSU. SSUs are also sampled without replacement, so this term includes its own FPC.

This estimator is unbiased for both the population total and mean, as well as their associated variance estimators.

Confidence intervals for the population mean and total follow

$$\bar{y} \pm \mathrm{t}_{\mathrm{df},1-\frac{\alpha}{2}}\, s_{\bar{y}} \tag{18.6}$$

and

$$\hat{t} \pm \mathrm{t}_{\mathrm{df},1-\frac{\alpha}{2}}\, s_{\hat{t}}. \tag{18.7}$$

Here, $\mathrm{df} = n-1$, reflecting that PSUs—not SSUs—are the independent sampling units.

18.4.1.2 Ratio estimators

The ratio estimator provides an alternative to the unbiased estimator and often yields more stable or efficient estimates of the population mean and total, particularly when PSU sizes vary. Although biased, the ratio estimator often represents a reasonable tradeoff for gains in precision. We quantify this bias in Section 18.6.2 using a forest inventory dataset.

The population mean μ is estimated using

$$\bar{y} = \frac{\sum_{i=1}^{n} M_i \bar{y}_i}{\sum_{i=1}^{n} M_i}. \tag{18.8}$$

The standard error of $\bar{y}$ is estimated as

$$s_{\bar{y}} = \frac{1}{M}\sqrt{\underbrace{\frac{N^2}{n}\left(1-\frac{n}{N}\right)\frac{1}{n-1}\sum_{i=1}^{n}(M_i\bar{y}_i - M_i\bar{y})^2}_{\text{between-PSU variance}} + \underbrace{\frac{N}{n}\sum_{i=1}^{n}\frac{M_i^2}{m_i}\left(1-\frac{m_i}{M_i}\right)s_i^2}_{\text{within-PSU variance}}}. \tag{18.9}$$

Given the sample mean $\bar{y}$ and its standard error $s_{\bar{y}}$, the population total and its standard error are estimated as

$$\hat{t} = M\bar{y} \tag{18.10}$$

and

$$s_{\hat{t}} = M s_{\bar{y}}. \tag{18.11}$$

Confidence intervals follow (18.6) and (18.7), with $\text{df} = n - 1$.

The ratio estimator for the population mean in (18.8) doesn't require knowledge of M. However, M is required to compute the standard error of $\bar{y}$ and to estimate the population total and its standard error. When M is unknown, it may be estimated by

$$\hat{M} = \frac{N}{n}\sum_{i=1}^{n} M_i, \tag{18.12}$$

and $\hat{M}$ substituted for M in (18.9), (18.10), and (18.11).

18.4.2 Sampling PSUs with probability proportional to size

The PPS two-stage estimator extends the two-stage framework introduced above by allowing PSUs to be selected with unequal probabilities while incorporating SSU sampling within each selected PSU. When auxiliary information describing PSU size is available prior to sampling, efficiency can often be improved by selecting PSUs with PPS. Here, "size" refers to any nonnegative quantity correlated with the PSU total.

We assume the following design elements.

- Each PSU has a known per-draw selection probability p_i, with $p_i \propto x_i$ for some nonnegative size measure x_i and $\sum_{i=1}^{N} p_i = 1$. A common choice is $x_i = M_i$, giving $p_i = M_i/M$ where $M = \sum_{i=1}^{N} M_i$.
- PSU sampling is conducted with replacement, producing n independent draws.
    - If a PSU is selected more than once, a new SSU sample is drawn each time, and each draw yields its own estimate $\bar{y}_i$.

- Because selection is with replacement, draws are independent and no finite population correction is applied.

Within each selected PSU, SSUs are sampled so that the resulting sample mean $\bar{y}_i$ is an unbiased estimate of the PSU mean, for example using SRS or SYS within the PSU.

The population total τ is estimated using

$$\hat{t} = \frac{1}{n}\sum_{i=1}^{n}\frac{M_i\bar{y}_i}{p_i}, \tag{18.13}$$

where the sum is taken over the n PSUs selected with replacement. A PSU may therefore appear multiple times in the sample, and each appearance contributes its own independently estimated $\bar{y}_i$.

The standard error of $\hat{t}$ is estimated as

$$s_{\hat{t}} = \sqrt{\frac{1}{n(n-1)}\sum_{i=1}^{n}\left(\frac{M_i\bar{y}_i}{p_i} - \hat{t}\right)^2}. \tag{18.14}$$

Given $\hat{t}$, the estimator for the population mean is

$$\bar{y} = \frac{\hat{t}}{M}, \tag{18.15}$$

with standard error

$$s_{\bar{y}} = \frac{s_{\hat{t}}}{M}. \tag{18.16}$$

Confidence intervals for the mean and total follow (18.6) and (18.7).

PPS sampling is particularly useful when PSU sizes vary widely and are correlated with the parameter of interest. Allocating sampling effort proportional to PSU size often yields smaller standard errors than equal-probability sampling.

18.5 Two-stage sampling under an areal sampling frame

Two-stage sampling is commonly used in forest inventory under an areal sampling frame. Here we focus on a setting that arises in many forestry applications: a finite set of PSUs (e.g., stands, management units, or ownerships), with SSUs selected within each sampled PSU using point or plot sampling to characterize local conditions.

The population of interest is defined by the collection of PSUs and their combined spatial extent, with total area A. The PSUs need not form a single continuous region; they may be spatially disjoint. As discussed in Section 12.2, an areal sampling frame implies an effectively infinite set of possible sampling locations. Under two-stage sampling, this interpretation applies at the *SSU stage*: SSUs are treated as points or plots sampled from the areal extent of each PSU. In contrast, the PSUs themselves constitute a finite set of discrete units.

Let A_i be the area of PSU i, with total population area $A = \sum_{i=1}^{N} A_i$. Suppose m_i SSUs are observed in PSU i, and define the SSU sample mean within PSU i as

$$\bar{y}_i = \frac{1}{m_i} \sum_{j=1}^{m_i} y_{ij}. \tag{18.17}$$

Each SSU measurement y_{ij} is expressed on a per-unit-area basis (e.g., tons per hectare), so $\bar{y}_i$ represents a per-unit-area estimate for PSU i. This assumes that SSUs are sampled so that $\bar{y}_i$ is an unbiased estimator of the PSU mean, for example using SRS or SYS within the PSU. Because the SSU population is treated as effectively infinite, the number of SSUs m_i can vary across PSUs without changing the interpretation of $\bar{y}_i$.

The goal is to estimate the population mean per unit area, μ, and the corresponding total $\tau = A\mu$. Estimators in Section 18.4 extend directly to this setting with only a change of notation. We illustrate this by adapting the two-stage ratio and PPS estimators to the areal sampling frame.

18.5.1 Ratio estimator

The two-stage ratio estimator for the population mean per unit area carries over directly from the finite-population setting. Replacing PSU size M_i with PSU area A_i and the population total M with the total area A, the estimator for μ is

$$\bar{y} = \frac{\sum_{i=1}^{n} A_i \bar{y}_i}{\sum_{i=1}^{n} A_i}. \tag{18.18}$$

The corresponding total is $\hat{t} = A\bar{y}$.

The standard error is obtained by modifying the two-stage ratio variance in (18.9) to reflect the areal sampling frame. Because SSUs are sampled from an effectively infinite population within each PSU, the SSU-level FPC is omitted, while the PSU-level FPC is retained. The resulting estimator is

$$s_{\bar{y}} = \frac{1}{A} \sqrt{\underbrace{\frac{N^2}{n}\left(1 - \frac{n}{N}\right) \frac{1}{n-1} \sum_{i=1}^{n} \left(A_i \bar{y}_i - A_i \bar{y}\right)^2}_{\text{between-PSU variance}} + \underbrace{\frac{N}{n} \sum_{i=1}^{n} \frac{A_i^2}{m_i} s_i^2}_{\text{within-PSU variance}}}. \tag{18.19}$$

The first term reflects between-PSU variation, while the second term reflects within-PSU variation arising from second-stage sampling.

The standard error of the total is $s_{\hat{t}} = A s_{\bar{y}}$ and confidence intervals for the mean and total follow (18.6) and (18.7), with $\text{df} = n - 1$.

18.5.2 PPS estimator

The PPS estimators in Section 18.4.2 extend in the same way. We again replace M_i with A_i and M with A in (18.13), (18.15), (18.14), and (18.16), interpreting $A_i \bar{y}_i$ as the contribution of PSU i to the population total.

A natural choice of size measure for PPS is the PSU area, so that $p_i = A_i/A$, assigning per-draw selection probabilities proportional to area under the assumption that larger PSUs tend to contribute more to the total. Other choices of p_i are possible when auxiliary variables correlated with the parameter of interest are available (e.g., predicted biomass or historical volume).

As before, PSUs are selected with replacement, and each time PSU i is selected it yields an independent estimate $\bar{y}_i$ based on a new second-stage sample of SSUs.

18.6 Illustrative analyses

18.6.1 Portland park tree inventory

To illustrate two-stage estimators under a finite first and second stage, we return to the Portland parks dataset introduced in Section 17.7.1, where the goal is to estimate total tree carbon across the city's parks. Because we have a complete census of all parks and all trees, we can directly compare sample-based estimates with the true population total, τ = 26,349 tons.

In the single-stage cluster sampling analysis in Section 17.7.1.3, the PPS estimator outperformed the unstratified and stratified equal-probability estimators in terms of accuracy, precision, and confidence interval coverage. In that setting, stratification provided little additional benefit once PPS was used.

To maintain continuity with those results and highlight how PPS behaves under a two-stage design, we focus this illustration on the PPS estimator. We begin by demonstrating the estimator for a single sample, and then evaluate its performance under repeated sampling so the results can be compared directly with the single-stage cluster sampling analysis.

Recall from Section 17.7.1 that parks serve as PSUs and trees within parks are SSUs. Following the same setup, we assume the number of trees within each park is known (i.e., M_i is known for $i = 1, \ldots, N$), and define the PSU selection probabilities as $p_i = M_i/M$, where $M = \sum_{i=1}^{N} M_i$.

To match the PSU sample size used in Section 17.7.1, we select $n = 25$ parks using PPS with replacement. Then, specific to the two-stage design, for each selected park we use SRS without replacement to select $m = 20$ trees for measurement. If a selected park contains fewer than 20 trees, all trees in that park are measured (i.e., $m_i = \min(20, M_i)$).[1]

18.6.1.1 PPS two-stage estimator

For each sampled park, the m_i measured trees are used to compute the sample mean $\bar{y}_i$. These park-level quantities along with park identifiers, values of M_i, and selection probabil-

[1]We could imagine a potentially even more efficient design that applies 3P sampling (Section 16.6) to estimate mean carbon within each selected park.

ities p_i are provided in `"datasets/PDX/two_stage_pps_park_sample_n25_m20.csv"` and read in below.

```
ppd_parks <-
  read_csv("datasets/PDX/two_stage_pps_park_sample_n25_m20.csv")
ppd_parks %>% glimpse()
```

```
#> Rows: 25
#> Columns: 4
#> $ park_id <dbl> 87, 94, 58, 10, 70, 109, 92, 11, 36, ~
#> $ M_i     <dbl> 565, 403, 261, 66, 647, 71, 48, 181, ~
#> $ p_i     <dbl> 0.0223700, 0.0159560, 0.0103338, 0.00~
#> $ y_bar_i <dbl> 0.46042, 0.90021, 1.05700, 1.89893, 0~
```

Two features of the design are worth emphasizing. First, PSUs are selected with replacement using PPS, so the same park can—and often does—appear multiple times in the sample. Second, each appearance of a park triggers a new, independent second-stage sample of trees, yielding a different realized value of $\bar{y}_i$. Both features are evident below, where parks 81, 87, and 94 each appear twice in the sample but with different $\bar{y}_i$ values due to independent SSU sampling.

```
ppd_parks %>% arrange(desc(p_i))
```

```
#> # A tibble: 25 x 4
#>    park_id   M_i    p_i y_bar_i
#>      <dbl> <dbl>  <dbl>   <dbl>
#>  1      79  1016 0.0402   1.40
#>  2     141   804 0.0318   0.890
#>  3      83   668 0.0264   0.416
#>  4      70   647 0.0256   0.918
#>  5      81   633 0.0251   0.718
#>  6      81   633 0.0251   0.730
#>  7      87   565 0.0224   0.460
#>  8      87   565 0.0224   0.505
#>  9      94   403 0.0160   0.900
#> 10      94   403 0.0160   0.876
#> # i 15 more rows
```

The PPS estimator and its standard error follow (18.13) and (18.14), respectively, and are implemented below.

```
est_pps <- pps_parks %>%
  summarize(
    n = n(),
    t_hat = mean(M_i * y_bar_i / p_i),
```

```
    s_t_hat = sqrt(1 / (n * (n - 1)) *
      sum((M_i * y_bar_i / p_i - t_hat)^2)),
    t = qt(df = n - 1, p = 1 - 0.05 / 2),
    t_ci_lower = t_hat - t * s_t_hat,
    t_ci_upper = t_hat + t * s_t_hat
  )

est_pps %>%
  select(t_hat, t_ci_lower, t_ci_upper)
```

```
#> # A tibble: 1 x 3
#>    t_hat t_ci_lower t_ci_upper
#>    <dbl>      <dbl>      <dbl>
#> 1 25885.     21411.     30360.
```

The census-based total tree carbon across all parks is $\tau = 26{,}349$ tons. For this sample, the two-stage PPS estimate and its 95% confidence interval are 25,885 (21,411, 30,360) tons, which contains the true value.

To assess the estimator's performance more generally—and to compare directly with the single-stage cluster results—we next turn to a repeated-sampling analysis.

18.6.1.2 Repeated-sampling comparison

We evaluate the two-stage PPS estimator under repeated sampling using the same general setup as in Section 17.7.1.3, drawing 500,000 samples from the Portland parks population. As before, each sample yields an estimate of the same fixed population total, allowing us to assess estimator performance across many realizations rather than from a single sample.

Across repeated samples, the two-stage PPS estimator for total tree carbon (tons) is centered very close to the true population total. The mean relative difference across samples is -0.018%, consistent with the expectation that the PPS estimator is centered on the true population total under the sampling design. The estimator shows a mean relative standard error (RSE) of 11.3% and achieves 95% confidence interval coverage of 94.6%.

These results can be compared directly with the single-stage PPS cluster estimator reported in the third row of Table 17.1. Because the single-stage design measures all SSUs within each selected PSU, whereas the two-stage design measures only a subset, some loss of precision is expected under two-stage sampling. In this case, however, the loss is modest: the RSE increases only slightly, from 10.1% under single-stage cluster sampling to 11.3% under the two-stage design.

Many of the sampled PSUs contain hundreds of trees, so restricting measurement to just 20 trees per PSU at the second stage might initially feel like a substantial loss of information. For this population, however, that level of subsampling is sufficient to produce stable estimates of PSU-level mean tree carbon. The resulting gains in operational efficiency therefore come with only a small increase in variance.

Taken together, these results point to a practical takeaway that parallels the single-stage cluster-sampling case. When PSU size is strongly associated with the population total, most efficiency gains come from PPS selection at the PSU stage. Once PSUs are selected efficiently, only a modest number of SSUs are needed within each PSU to obtain stable PSU-level means and reliable estimates of population totals.

18.6.2 Mass timber feedstock analysis

In this example, we apply the two-stage ratio and PPS estimators from Sections 18.4.1.2 and 18.4.2 to an areal-frame setting, with a finite first stage and an areal second stage as described in Section 18.5. The data are simulated, but the structure reflects analyses commonly encountered in applied forestry.

Suppose a mass timber mill and fabrication firm is exploring the feasibility of locating a facility in the region shown in Figure 18.1. The firm expects the state's Department of Natural Resources (DNR) to be its primary supplier and wants to assess how much suitable biomass is available on DNR-managed lands within a practical hauling distance of the proposed site.

This type of pre-operational assessment is often referred to as a *feedstock study*, where *feedstock* refers to the volume of accessible wood available to support sustained operation of a mill or processing facility. Consultants and analysts routinely conduct such studies using limited field data, combined with spatial and stand-level auxiliary information, to estimate both total available biomass and its spatial distribution.

To identify the region of interest, the firm determines that a one-hour travel time is the upper limit for economically feasible wood hauling. We use this threshold to compute an *isochrone*—a polygon showing the area reachable within a fixed travel time using the road network. In this case, the isochrone includes all locations accessible within one hour of the proposed facility.[2]

The fabrication process relies primarily on softwoods, though the firm is experimenting with some hardwood inputs. Our parameters of interest are the mean per-acre and total dry weight of AGB on all DNR conifer and mixed hardwood/softwood stands within the one-hour isochrone.

The population consists of $N = 8{,}927$ stands, and the cruising budget supports a sample of $n = 50$. Stands serve as the PSUs, and the SSUs are inventory plots placed within each stand. We'll use 1/10-acre circular plots laid out on a regular 7-by-7 chain (462 ft) grid—a systematic design commonly used in operational inventory work throughout the region.

Figure 18.1 shows the full PSU population along with 50 PSUs selected using SRS. Figure 18.2 highlights a region containing several selected PSUs, and Figure 18.3 provides a closer view of these PSUs and their SSUs.

---

[2]This type of road-based travel-time polygon can be computed in R using the `osrm` package, which provides an interface to the Open Source Routing Machine. The package supports travel-time and routing calculations using OpenStreetMap's road network data. Other tools and packages are also available for generating isochrones, depending on your needs and data sources (Giraud 2022).

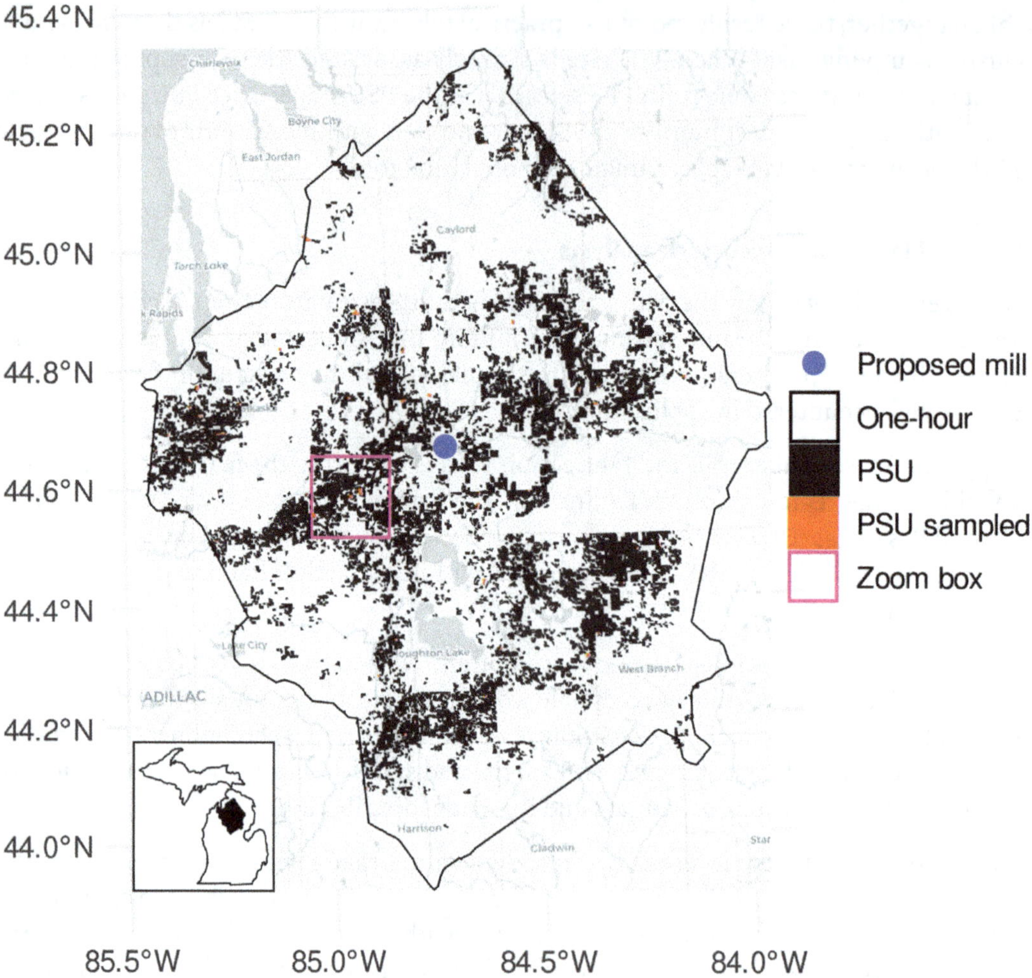

FIGURE 18.1: Conifer and mixed hardwood/softwood stands managed by the Michigan Department of Natural Resources within a one-hour drive of the proposed mass timber mill and fabrication facility. These stands define the population of primary sampling units (PSUs). A simple random sample of 50 PSUs was selected for inventory.

In the sections that follow, we'll apply the two-stage equal-probability ratio and PPS estimators to estimate both mean and total AGB. Because the data are simulated, we know the true population values $\mu = 30.69$ tons/ac and $\tau = 11{,}100{,}216$ tons. This lets us evaluate and compare each estimator's performance against known parameters.

As you work through these sections, you'll see that point estimates (means and totals) are similar across estimators, while the PPS estimator produces a smaller estimated standard error than the equal-probability ratio estimator. Because these results come from a single sample, however, it's unclear whether the observed differences reflect true estimator performance or simply sample-to-sample variability. To explore this question, we turn to a repeated-sampling analysis in Section 18.6.2.3.

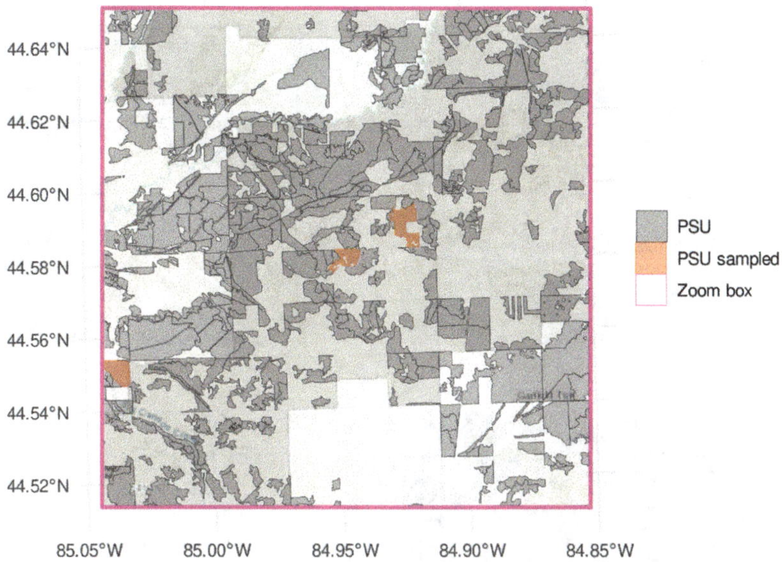

FIGURE 18.2: Zoomed-in view from Figure 18.1, highlighting three selected primary sampling units (PSUs).

18.6.2.1 Estimation using an equal probability sampling

Under this design, each PSU in the population has an equal probability of selection. The following data files are used in this analysis.

- `datasets/mass_timber/all_psu.csv` contains the full set of PSUs in the population. Each row is a PSU, and columns hold a unique PSU index and area in acres.
- `datasets/mass_timber/sampled_equal_prob_psu.csv` contains the sampled PSUs. These were drawn from the full set of PSUs using `dplyr::slice_sample()` with n=50 and all arguments set to their default values.
- `datasets/mass_timber/sampled_ssu_equal_prob_psu.csv` contains the SSUs for each sampled PSU. Each row corresponds to a tree measured on the 1/10-acre circular plot, and columns are:
    - `psu_id` and `ssu_id`, which together uniquely identify the trees measured on a given plot;
    - `tree_count`, which indicates how many trees are associated with each row (used to account for zero-tree plots and trees double-counted via the mirage method);
    - `agb_ton_tree`, the dry weight AGB in tons for each tree, used to compute the y_{ij}'s.

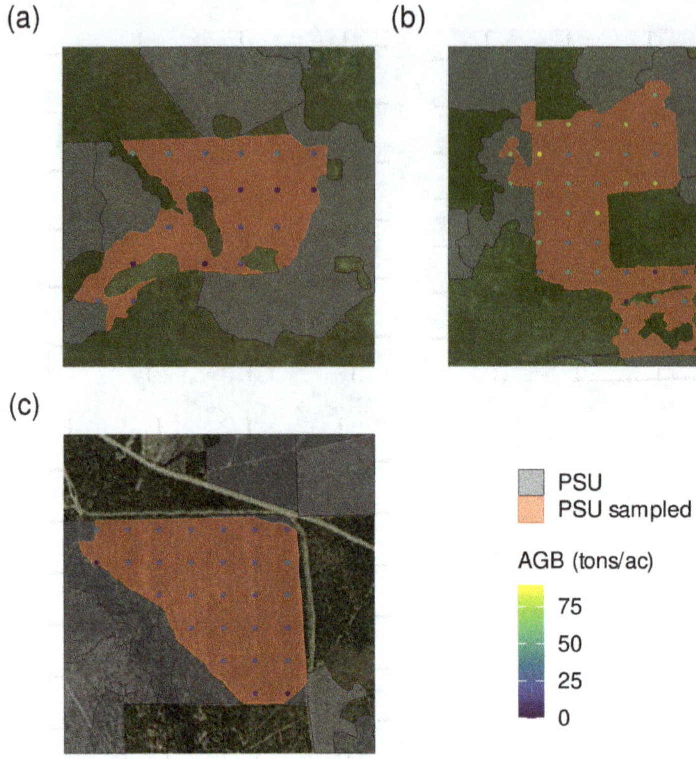

FIGURE 18.3: The three primary sampling units (PSUs) shown in Figure 18.2. PSU IDs are 1895, 1791, and 1948 for (a), (b), and (c), respectively. Secondary sampling units (SSUs) were placed using systematic sampling. Each inventory plot is colored by the observed aboveground biomass (AGB).

The code below reads and takes a quick look at each data file.

```
all_psu <- read_csv("datasets/mass_timber/all_psu.csv")
all_psu %>% glimpse()
```

```
#> Rows: 8,927
#> Columns: 2
#> $ psu_id <dbl> 1, 2, 3, 4, 5, 6, 7, 8, 9, 10, 11, 12,~
#> $ ac     <dbl> 44.565, 21.690, 51.851, 131.325, 75.43~
```

```
sampled_psu <-
  read_csv("datasets/mass_timber/sampled_equal_prob_psu.csv")
sampled_psu %>% glimpse()
```

```
#> Rows: 50
#> Columns: 2
```

```
#> $ psu_id <dbl> 1017, 8004, 4775, 8462, 4050, 8789, 13~
#> $ ac     <dbl> 30.787, 18.785, 11.723, 16.064, 38.443~
```

```
ssu_trees <-
  read_csv("datasets/mass_timber/sampled_ssu_equal_prob_psu.csv")
ssu_trees %>% glimpse()
```

```
#> Rows: 7,750
#> Columns: 4
#> $ psu_id        <dbl> 1017, 1017, 1017, 1017, 1017, 10~
#> $ ssu_id        <dbl> 1, 1, 1, 1, 1, 1, 1, 1, 1, 1, 1,~
#> $ tree_count    <dbl> 1, 1, 1, 1, 1, 1, 1, 1, 1, 1, 1,~
#> $ agb_ton_tree <dbl> 0.16786, 0.16999, 0.14626, 0.151~
```

We'll now rename a few columns to align with our notation and create variables needed in subsequent steps.

```
all_psu <- all_psu %>% rename(A_i = ac)
sampled_psu <- sampled_psu %>% rename(A_i = ac)

N <- nrow(all_psu)
n <- nrow(sampled_psu)

A <- sum(all_psu$A_i)

# Add the tree factor for 1/10-acre plot.
ssu_trees <- ssu_trees %>% mutate(TF = 10)
```

We implement the ratio estimator for the population mean and total, along with their standard errors, as defined in (18.8), (18.10), (18.9), and (18.11). The estimators are applied in two steps, using the areal second-stage setup described in Section 18.5.

First, SSU-level information is summarized within each sampled PSU, including the number of SSUs m_i, the sample mean $\bar{y}_i$, and the sample variance s_i^2. The PSU area A_i is then joined to `psu_summary`. Second, these quantities are aggregated across sampled PSUs to produce the population estimates.

```
psu_summary <- ssu_trees %>%
  group_by(psu_id, ssu_id) %>%
  summarize(
    y_ij = sum(TF * tree_count * agb_ton_tree),
  ) %>%
  group_by(psu_id) %>%
  summarize(
    y_bar_i = mean(y_ij),
```

```
    m_i = n(),
    s_sq_i = 1 / (m_i - 1) * sum((y_ij - y_bar_i)^2) #Or var(y_ij).
  ) %>% # Join PSU areas.
  left_join(sampled_psu, by = join_by(psu_id))
```

```
pop_summary <- psu_summary %>%
  summarize(
    # Ratio estimator for the mean (per unit area).
    y_bar = sum(A_i * y_bar_i) / sum(A_i),

    # Standard error of the mean.
    s_y_bar = (1 / A) * sqrt(
      ## Between-PSU variance (with PSU-level FPC).
      (N^2 / n) * (1 - n / N) *
        (1 / (n - 1)) *
        sum((A_i * y_bar_i - A_i * y_bar)^2) +

      # Within-PSU variance (no SSU-level FPC under areal frame).
      (N / n) *
        sum(A_i^2 / m_i * s_sq_i)
    ),

    # Total and its standard error.
    t_hat    = A * y_bar,
    s_t_hat = A * s_y_bar
  )

pop_summary
```

```
#> # A tibble: 1 x 4
#>   y_bar s_y_bar     t_hat  s_t_hat
#>   <dbl>   <dbl>     <dbl>    <dbl>
#> 1  35.6    3.02 12860648. 1092949.
```

Our estimated mean and total, with 95% confidence intervals, were 35.56 (29.49, 41.63) tons/ac and 12,860,648 (10,664,284, 15,057,013) tons, respectively. Based on this sample, both confidence intervals captured the true values, which are $\mu = 30.69$ tons/ac and $\tau =$ 11,100,216 tons.

18.6.2.2 Estimation using a PPS sample

Here we apply the PPS estimators defined in Section 18.4.2. We again use a sample of 50 PSUs; however, instead of equal-probability selection, PSUs are sampled proportional to their area. In addition to all_psu from the previous section, the following two data files are used in this analysis.

- datasets/mass_timber/sampled_pps_psu.csv contains the sampled PSUs. Each row is a PSU, and columns hold a unique PSU index and area in acres. These PSUs were selected using dplyr::slice_sample() with the weight_by argument set to each PSU's area, applying the desired selection probabilities $p_i = A_i/A$. The argument replace = TRUE allows for selection *with replacement*.

- datasets/mass_timber/sampled_ssu_pps_psu.csv contains the SSUs for each sampled PSU. Each row corresponds to a tree measured on the 1/10-acre circular plot, and columns are:
  - psu_id and ssu_id, which together uniquely identify the trees measured on a given plot and sample;
  - tree_count, which indicates how many trees are associated with each row (used to account for zero-tree plots and trees double-counted via the mirage method);
  - agb_ton_tree, the dry weight AGB in tons for each tree, used to compute the y_{ij}'s.

Because PSU sampling is conducted with replacement under PPS, the same PSU may be selected more than once. Each selection is independent and is associated with its own second-stage sample of SSUs (and, therefore, its own set of tree measurements). As a result, some additional care is needed when organizing the data: we append a sample index to the psu_id values so that tree measurements from different selections of the same PSU are treated as distinct groups in subsequent workflows.[3]

The code below reads and takes a quick look at each data file.

```
sampled_psu <- read_csv("datasets/mass_timber/sampled_pps_psu.csv")
sampled_psu %>% glimpse()
```

```
#> Rows: 50
#> Columns: 2
#> $ psu_id <chr> "316_1", "323_1", "358_1", "420_1", "5~
#> $ ac     <dbl> 19.939, 67.284, 188.677, 128.584, 120.~
```

```
ssu_trees <- read_csv("datasets/mass_timber/sampled_ssu_pps_psu.csv")
ssu_trees %>% glimpse()
```

```
#> Rows: 16,586
#> Columns: 4
#> $ psu_id       <chr> "316_1", "316_1", "316_1", "316_~
#> $ ssu_id       <dbl> 1, 1, 1, 1, 1, 1, 1, 1, 2, 2, 2,~
```

[3] For this particular PSU sample, there happen to be no repeat selections, so all psu_id values have a _1 suffix. This is not true for all samples used in the repeated-sampling analysis in Section 18.6.2.3.

```
#> $ tree_count   <dbl> 1, 1, 1, 1, 1, 1, 1, 1, 1, 1, 1,~
#> $ agb_ton_tree <dbl> 0.00294279, 0.00183660, 0.005673~
```

Rename a few columns to align with our notation and creating variables needed in subsequent steps. The set of all PSUs `all_psu`, `N`, and `A` are the same as the previous section.

```
sampled_psu <- sampled_psu %>%
  rename(A_i = ac)

# Add the tree factor for 1/10-acre plots
ssu_trees <- ssu_trees %>% mutate(TF = 10)
```

Given the PPS sample, we apply the mean and total estimators defined in (18.13) and (18.15), along with their associated standard errors in (18.14) and (18.16).

As with the equal probability estimators, we break the process into two steps. First, we summarize SSU information within each PSU, join to `sampled_psu` to add the area column `A_i`, and compute the selection probabilities `p_i`, both of which are needed for the second step. Second, we compute the required quantities across all sampled PSUs to produce the overall estimates and standard errors.

```
psu_summary <- ssu_trees %>%
  group_by(psu_id, ssu_id) %>%
  summarize(y_ij = sum(TF * tree_count * agb_ton_tree)) %>%
  summarize(
    y_bar_i = mean(y_ij),
    m_i = n(),
    s_sq_i = 1 / (m_i - 1) * sum((y_ij - y_bar_i)^2) #Or var(y_ij).
  ) %>%
  left_join(sampled_psu, by = join_by(psu_id)) %>%
  mutate(p_i = A_i / A)

psu_summary %>% glimpse()
```

```
#> Rows: 50
#> Columns: 6
#> $ psu_id  <chr> "1017_1", "1069_1", "1418_1", "1530_1~
#> $ y_bar_i <dbl> 27.0608, 47.5176, 51.1689, 48.8034, 3~
#> $ m_i     <int> 5, 7, 39, 10, 23, 24, 43, 43, 9, 3, 5~
#> $ s_sq_i  <dbl> 324.535, 139.027, 145.808, 215.842, 5~
#> $ A_i     <dbl> 30.787, 29.630, 216.419, 41.660, 128.~
#> $ p_i     <dbl> 0.000085126, 0.000081929, 0.000598408~
```

```
pop_summary <- psu_summary %>%
  summarize(
```

```
    n = n(),
    t_hat = 1 / n * sum(A_i * y_bar_i / p_i),
    s_t_hat = sqrt(1 / (n * (n - 1)) *
                     sum((A_i * y_bar_i / p_i - t_hat)^2)
                   ),
    y_bar = t_hat / A,
    s_y_bar = s_t_hat / A
  )

pop_summary
```

```
#> # A tibble: 1 x 5
#>       n     t_hat s_t_hat y_bar s_y_bar
#>   <int>     <dbl>   <dbl> <dbl>   <dbl>
#> 1    50 11979201. 672744.  33.1    1.86
```

Our estimated mean and total, with 95% confidence intervals, were 33.12 (29.38, 36.86) tons/ac and 11,979,201 (10,627,272, 13,331,131) tons, respectively. Based on this sample, both confidence intervals captured the true values, which are $\mu = 30.69$ tons/ac and $\tau = 11{,}100{,}216$ tons.

18.6.2.3 Repeated-sampling comparison

Here we assess the performance of the two-stage ratio and PPS estimators using 1,000 samples drawn under the sampling approach described above, where we select $n = 50$ PSUs at the first stage and then sample SSUs within each PSU using a systematic grid of fixed-area plots. For each estimator, we record point estimates, standard errors, and confidence intervals. Performance is evaluated under repeated sampling using the following summaries.

- *Mean difference*—the average difference between the estimate from each sample and the true population mean (see Section 11.2.3 for the formal definition). Under repeated sampling, this quantity is an empirical estimate of bias; values closer to zero indicate that the estimator is centered near the true mean.
- *RMSE*—the root mean squared error, which combines both the mean difference and sampling variability into a single measure of estimator accuracy. Smaller values indicate better overall performance.
- *Confidence interval width*—the average width of the 95% confidence interval, which reflects estimator precision. Narrower intervals are preferred, provided that confidence interval coverage remains close to its nominal level.
- *Confidence interval coverage*—the proportion of confidence intervals that contain the true population mean, expressed as a percentage. This measures how well the variance estimator is calibrated and should be close to the nominal value of 95%.

TABLE 18.1: Comparison of the two-stage equal-probability ratio and PPS estimators for feedstock mean aboveground biomass under repeated sampling from the Michigan feedstock population. Columns show mean difference, root mean squared error (RMSE), and 95% confidence interval width and coverage, based on 1,000 PSU samples each of size n = 50 and a SYS second-stage sample.

				95% Confidence interval	
PSU design	Estimator	Mean difference (tons/ac)	RMSE (tons/ac)	Width (tons/ac)	Coverage (%)
Equal prob.	Ratio	-0.39	2.6	9.2	92
PPS	PPS	-0.35	1.8	7.0	94

Summaries across replicates are given in Table 18.1. The results are consistent with patterns seen earlier in this chapter for PPS designs.

The mean difference is very small for both estimators relative to the population mean. Under equal-probability selection, the mean difference is -0.39 tons/ac, while under PPS it's -0.35 tons/ac. At this scale, these values are negligible, indicating that both estimators are effectively centered on the true mean for this problem. For PPS, any residual departure from zero is expected to diminish as the number of repeated samples increases.

The primary differences between the estimators appear in accuracy and precision. The RMSE decreases from 2.6 tons/ac under equal-probability sampling to 1.8 tons/ac under PPS. This reduction reflects the efficiency gains achieved by selecting PSUs with probability proportional to area, which concentrates sampling effort on stands that contribute more to the population total. The average 95% confidence interval width declines from 9.2 tons/ac under equal-probability sampling to 7 tons/ac under PPS, indicating higher precision under the PPS design.

Confidence interval coverage is closer to the nominal 95% level under PPS as well, although both estimators achieve rates near the target. This suggests that uncertainty is reasonably well characterized in both cases, with PPS performing slightly better.

Overall, while both estimators perform well, PPS provides clear advantages in accuracy, precision, and interval performance. These results reinforce the role of PPS as an effective design choice for two-stage inventories when reliable auxiliary information is available at the PSU level.

18.7 Advantages and disadvantages of two-stage sampling

Two-stage sampling is a flexible design that balances statistical efficiency with operational practicality. It's especially useful when the population is naturally grouped into larger units (PSUs) that can be subdivided into smaller units (SSUs).

18.7.1 Advantages

- Fieldwork can be concentrated within a manageable set of PSUs, reducing travel time and logistical complexity.
- The first-stage design is flexible: PSUs can be selected with equal probability or with PPS when reliable auxiliary size information is available.
- When SSUs within a PSU are relatively homogeneous, measuring only a subset of SSUs can yield results similar to single-stage cluster sampling, while avoiding the cost of measuring all SSUs. This can lead to substantial gains in efficiency with little loss in precision.
- Well suited for inventories over large, heterogeneous regions where direct sampling of all SSUs is impractical.

18.7.2 Disadvantages

- If within-PSU variability is high and only a small number of SSUs are measured within each PSU, precision can be lower than for single-stage designs in which all SSUs are measured.
- When PSU sizes vary widely and the parameter of interest is strongly related to size, equal-probability PSU selection can be inefficient, producing larger standard errors than ratio or PPS designs.

18.8 Exercises

18.8.1 Two-stage sampling with equal-probability PSUs

In this exercise, you'll work through the two-stage unbiased and ratio estimators in Section 18.4.1 using a small simulated dataset. The goal is to practice building a two-stage estimation workflow and to compare the unbiased and ratio estimators in a simple setting.

The two-stage sample is stored in the following files.

- `datasets/two_stage_sim_sampled_psu.csv` contains the PSUs selected in the first stage using SRS. Each row is a sampled PSU, with columns:
  - `psu_id` PSU identifier;
  - `M_i` number of SSUs in that PSU.
- `datasets/two_stage_sim_sampled_ssu.csv` contains the SSUs selected in the second stage from each sampled PSU. Each row is an observed SSU, with columns:
  - `psu_id` PSU identifier;
  - `ssu_id` SSU identifier within PSU;
  - `y_ij` measured SSU values.

For this population, the number of PSUs is $N = 30$ and the total number of SSUs is $M = 1{,}510$.

The code below reads the two-stage sample and shows the file structure.

```
sampled_psu <- read_csv("datasets/two_stage_sim_sampled_psu.csv")
ssu_sample <- read_csv("datasets/two_stage_sim_sampled_ssu.csv")

sampled_psu %>% glimpse()
```

```
#> Rows: 8
#> Columns: 2
#> $ psu_id <dbl> 17, 10, 27, 28, 19, 2, 21, 8
#> $ M_i    <dbl> 73, 37, 65, 56, 29, 23, 28, 62
```

```
ssu_sample %>% glimpse()
```

```
#> Rows: 80
#> Columns: 3
#> $ psu_id <dbl> 2, 2, 2, 2, 2, 2, 2, 2, 2, 2, 8, 8, 8,~
#> $ ssu_id <dbl> 16, 15, 19, 13, 1, 23, 10, 11, 18, 12,~
#> $ y_ij   <dbl> 13.8246, 18.9466, 18.4690, 11.6913, 16~
```

Exercise 18.1. Using `ssu_sample`, summarize SSU measurements within each sampled PSU to compute:

- m_i the number of sampled SSUs in PSU i;
- $\bar{y}_i$ the sample mean within PSU i;
- s_i^2 the sample variance within PSU i.

Call your summarized output `psu_summary`. Then join the PSU sizes `M_i` from `sampled_psu` to `psu_summary`. Your resulting `psu_summary` should match ours below.

```
psu_summary %>% glimpse()
```

```
#> Rows: 8
#> Columns: 5
#> $ psu_id  <dbl> 2, 8, 10, 17, 19, 21, 27, 28
#> $ m_i     <int> 10, 10, 10, 10, 10, 10, 10, 10
#> $ y_bar_i <dbl> 12.967, 13.396, 14.635, 14.696, 10.66~
#> $ s_sq_i  <dbl> 14.3316, 18.4955, 88.2193, 6.7483, 5.~
#> $ M_i     <dbl> 23, 62, 37, 73, 29, 28, 65, 56
```

Exercise 18.2. Using (18.1), (18.2), (18.4), and (18.5), compute the two-stage unbiased estimators for the population total and mean, along with their standard errors and 95% confidence intervals.

Store your results in a tibble called `unb_ests`.

```
unb_ests %>% glimpse()
```

```
#> Rows: 1
#> Columns: 8
#> $ t_hat       <dbl> 18600
#> $ s_t_hat     <dbl> 2631.4
#> $ y_bar       <dbl> 12.318
#> $ s_y_bar     <dbl> 1.7426
#> $ y_bar_lower <dbl> 8.1973
#> $ y_bar_upper <dbl> 16.439
#> $ t_hat_lower <dbl> 12378
#> $ t_hat_upper <dbl> 24822
```

Exercise 18.3. Using (18.8), (18.9), (18.10), and (18.11), compute the two-stage ratio estimators for the population mean and total, along with their standard errors and 95% confidence intervals.

Store your results in a tibble called `ratio_ests`.

```
ratio_ests %>% glimpse()
```

```
#> Rows: 1
#> Columns: 8
#> $ y_bar       <dbl> 13.298
#> $ s_y_bar     <dbl> 0.39009
#> $ t_hat       <dbl> 20079
#> $ s_t_hat     <dbl> 589.03
#> $ y_bar_lower <dbl> 12.375
#> $ y_bar_upper <dbl> 14.22
#> $ t_hat_lower <dbl> 18687
#> $ t_hat_upper <dbl> 21472
```

Exercise 18.4. The data used for this exercise are simulated, so we know the population values:

- $\mu = 13.381$;
- $\tau = 20{,}206$.

Given your estimates from the unbiased and ratio estimators, consider the following questions.

a. Which estimator is closer to the truth for this sample?
b. Which estimator provides the more precise estimate for this sample (i.e., has the smaller standard error or narrower confidence interval)?
c. Why is it not appropriate to infer something about an estimator's overall performance based on a single sample?

18.8.2 Mass timber feedstock reanalysis

In Section 18.6.2.2, we used two-stage PPS to estimate the population mean and total AGB for the feedstock analysis. In this exercise, you'll repeat that analysis using a different PPS sample of PSUs (i.e., different but equally likely). The goal is to reinforce the two-stage PPS estimation workflow and highlight some important features of PPS sampling.

As described in Section 18.6.2, the population is comprised of $N = 8{,}927$ stands with a total population area of $A = 361{,}658$ acres.

The following files are used in this exercise:

- `datasets/mass_timber/all_psu.csv` contains the full PSU population and PSU areas (acres).
- `datasets/mass_timber/sampled_pps_psu_exercise.csv` contains a PPS sample of PSUs drawn with replacement and per-draw selection probability $p_i = A_i/A$.
- `datasets/mass_timber/sampled_ssu_pps_psu_exercise.csv` contains the SSUs (inventory plot) measured within each selected PSU.

To get things started, we read in the data and rename some columns to align with the notation in Section 18.5.

```
all_psu <- read_csv("datasets/mass_timber/all_psu.csv")
sampled_psu <- read_csv("datasets/mass_timber/sampled_pps_psu_exercise.csv")
ssu_trees <- read_csv("datasets/mass_timber/sampled_ssu_pps_psu_exercise.csv")

all_psu <- all_psu %>% rename(A_i = ac)
sampled_psu <- sampled_psu %>% rename(A_i = ac)
```

Exercise 18.5. Because PSUs are sampled with replacement under PPS, the same PSU can be selected more than once. This happens in your current sample—in fact, one PSU is selected three times. Recall from Section 18.6.2.2 that we appended a sample index to the `psu_id` values in `sampled_psu` and `ssu_trees` so that measurements from different selections of the same PSU can be treated as distinct samples.

When you inspect `sampled_psu`, you'll see that PSU ID 3718 was selected three times (e.g., run `print(sampled_psu, n = Inf); sampled_psu %>% filter(str_detect(psu_id, "3718"))`). To better understand why this PSU was selected so often, use its acreage to compute its per-draw selection probability. Then, using `all_psu`, create a histogram (or another plot) showing either the distribution of PSU areas or their per-draw selection probabilities. Where does PSU ID 3718 fall in this distribution (approximately—just describe its position)?

Exercise 18.6. Now start into the estimation workflow presented in Section 18.6.2.2. Begin by adding the tree factor for the 1/10-acre plots to `ssu_trees`, then form `psu_summary`, join the stand areas from `sampled_psu`, and create a new column that holds the selection probabilities $p_i = A_i/A$.

Your `psu_summary` should look like the following.

```
psu_summary %>% glimpse()
```

```
#> Rows: 50
#> Columns: 6
#> $ psu_id  <chr> "1069_1", "148_1", "1670_1", "1712_1"~
#> $ y_bar_i <dbl> 50.95682, 28.77426, 39.54716, 0.48815~
#> $ m_i     <int> 5, 46, 122, 9, 64, 30, 45, 36, 28, 17~
#> $ s_sq_i  <dbl> 34.905752, 190.906975, 202.657507, 0.~
#> $ A_i     <dbl> 29.630, 243.296, 606.799, 54.829, 306~
#> $ p_i     <dbl> 0.000081929, 0.000672723, 0.001677823~
```

Exercise 18.7. Continuing with the workflow in Section 18.6.2.2, use `psu_summary` to form `pop_summary`, implementing (18.13), (18.14), (18.15), and (18.16) to compute:

- the PPS estimators for the population total and mean;
- their standard errors;
- 95% confidence intervals for both quantities.

Store your results in `pop_summary`, which should match ours below.

Compare your estimates and confidence intervals to the true population values $\mu = 30.69$ tons/ac and $\tau = 11{,}100{,}216$ tons. Do the 95% confidence intervals from this PPS sample cover the true mean and total?

```
pop_summary %>% glimpse()
```

```
#> Rows: 1
#> Columns: 9
#> $ n           <int> 50
#> $ t_hat       <dbl> 12151220
#> $ s_t_hat     <dbl> 606451
#> $ y_bar       <dbl> 33.599
```

```
#> $ s_y_bar     <dbl> 1.6769
#> $ y_bar_lower <dbl> 30.229
#> $ y_bar_upper <dbl> 36.968
#> $ t_hat_lower <dbl> 10932512
#> $ t_hat_upper <dbl> 13369928
```

References

Babcock, Chad, Jason Matney, Andrew O. Finley, Aaron Weiskittel, and Bruce D. Cook. 2013. "Multivariate Spatial Regression Models for Predicting Individual Tree Structure Variables Using LiDAR Data." *IEEE Journal of Selected Topics in Applied Earth Observations and Remote Sensing* 6 (1): 6–14.

Beers, Thomas W. 1969. "Slope correction in horizontal point sampling." *Journal of Forestry* 67 (3): 188–92.

Beers, Thomas W. 1977. "Practical correction of boundary overlap." *Southern Journal of Applied Forestry* 1 (1): 16–18.

Bell, John F., and Lucien B. Alexander. 1957. *Application of the Variable Plot Method of Sampling Forest Stands*. Oregon State Board of Forestry.

Bitterlich, Walter. 1952. "Die Winkelzählprobe." *Forstwissenschaftliches Centralblatt* 71 (7): 215–25.

Bitterlich, Walter. 1984. *The Relascope Idea. Relative Measurements in Forestry*. Commonwealth Agricultural Bureaux, Farnham Royal, UK.

Bravo, Adriana, Ana Porzecanski, Eleanor Sterling, et al. 2016. "Teaching for Higher Levels of Thinking: Developing Quantitative and Analytical Skills in Environmental Science Courses." *Ecosphere* 7 (4): e01290.

Bray, Tim, Jean Paoli, C. M. Sperberg-McQueen, Eve Maler, and François Yergeau. 2008. *Extensible Markup Language (XML) 1.0 (5th ed.)*.

Bruce, David. 1955. "A new way to look at trees." *Journal of Forestry* 53 (3): 163–67.

Bruce, Donald. 1961. *Prism Cruising in the Western United States and Volume Tables for Use Therewith*. Mason, Bruce & Girard.

Brunsdon, Chris, and Lex Comber. 2019. *An Introduction to R for Spatial Analysis and Mapping (Spatial Analytics and GIS)*. 2nd ed. SAGE Publications Ltd.

Brus, Dick J. 2022. *Spatial Sampling with R*. Chapman and Hall/CRC the r Series. CRC Press LLC.

Bryan, Mackay B. 1956. "A Simplified Method of Correcting for Slope on Circular Sample Plots." *Journal of Forestry* 54 (7): 442–45.

Bunn, Andrew G. 2010. "Statistical and Visual Crossdating in R Using the dplR Library." *Dendrochronologia* 28 (4): 251–58.

Burkhart, Harold E., Thomas Eugene Avery, and Bronson P. Bullock. 2018. *Forest Measurements*. 6th ed. Waveland Press.

Burrill, Elizabeth A., Andrea M. Wilson, Jeffery A. Turner, et al. 2018. "The Forest Inventory and Analysis Database: Database Description and User Guide Version 8.0 for Phase 2." In *U.S. Department of Agriculture, Forest Service. 946 p.* http://www.fia.fs.fed.us/library/database-documentation.

Chang, Winston. 2018. *R Graphics Cookbook: Practical Recipes for Visualizing Data*. 2nd ed. O'Reilly.

Claes-Magnus, Cassel, Carl-Erik Särndal, and Jan H. Wretman. 1977. *Foundations of Inference in Survey Sampling*. A Wiley-Interscience Publication. Wiley.

Cochran, William G. 1977. *Sampling Techniques*. 3rd ed. John Wiley.

Cornfield, Jerome. 1944. "On Samples from Finite Populations." *Journal of the American Statistical Association* 39 (226): 236–39.

Curtis, Robert O., and David D. Marshall. 2000. "Technical Note: Why Quadratic Mean Diameter?" *Western Journal of Applied Forestry* 15 (3): 137–39.

Davis, Eric C., Brent Sohngen, and David J. Lewis. 2022. "The Effect of Carbon Fertilization on Naturally Regenerated and Planted US Forests." *Nature Communications* 13 (1): 5490.

DiSalvo, Angie, Julie Fukuda, and Jeff Ramsey. 2017. *Street Tree Inventory Report: City of Portland*. Portland Parks & Recreation, Urban Forestry.

Doraisami, Mahendra, Rosalyn Kish, Nicholas J Paroshy, Grant M Domke, Sean C Thomas, and Adam R Martin. 2022. "A Global Database of Woody Tissue Carbon Concentrations." *Scientific Data* 9 (1): 284.

Ducey, Mark J., Jeffrey H. Gove, and Harry T. Valentine. 2004. "A Walkthrough Solution to the Boundary Overlap Problem." *Forest Science* 50: 427–35.

Ducey, Mark J., and John A. Kershaw. 2023. "Alternative Expressions for Stand Diameter in Complex Forests." *Forest Ecosystems* 10: 100114.

Dumelle, Michael, Tom Kincaid, Anthony R. Olsen, and Marc Weber. 2023. "spsurvey: Spatial Sampling Design and Analysis in R." *Journal of Statistical Software* 105 (3): 1–29.

Ellison, Aaron M. 2010. "Repeatability and Transparency in Ecological Research." *Ecology* 91 (9): 2536–39.

Ericson, W. A. 1969. "Subjective Bayesian Models in Sampling Finite Populations." *Journal of the Royal Statistical Society. Series B (Methodological)* 31 (2): 195–233.

ESRI. 1998. *Environmental Systems Research Institute Shapefile Technical Description*. Environmental Systems Research Institute, Inc. https://www.esri.com/library/whitepapers/pdfs/shapefile.pdf.

Fehrmann, Lutz. 2024. *Dendrometer III Instruction Manual*. Forest Inventory; Remote Sensing, Georg-August-Universität Göttingen.

Finley, Andrew O., Hans-Erik Andersen, Chad Babcock, Bruce D. Cook, Douglas C. Morton, and Sudipto Banerjee. 2024. "Models to Support Forest Inventory and Small Area Estimation Using Sparsely Sampled LiDAR: A Case Study Involving G-LiHT LiDAR in Tanana, Alaska." *Journal of Agricultural, Biological and Environmental Statistics* 29 (4): 695–722.

Foster, Jane R., Anthony W. D'Amato, and John B. Bradford. 2014. "Looking for Age-Related Growth Decline in Natural Forests: Unexpected Biomass Patterns from Tree Rings and Simulated Mortality." *Oecologia* 175 (1): 363–74.

Freese, Fank. 1962. *Elementary Forest Sampling*. Agriculture Handbook. U.S. Department of Agriculture, Forest Service.

Fridman, Jonas, Sören Holm, Mats Nilsson, Per Nilsson, Anna Hedström Ringvall, and Göran Ståhl. 2014. "Adapting National Forest Inventories to Changing Requirements – the Case of the Swedish National Forest Inventory at the Turn of the 20th Century." *Silva Fennica* 48 (3).

FVS. 2021. "Northeast (NE) Variant Overview - Forest Vegetation Simulator." In *Internal Rep. Fort Collins, CO: U.S. Department of Agriculture, Forest Service, Forest Management Service Center*. 56p.

Ghosh, Malay., and Glen Meeden. 1997. *Bayesian Methods for Finite Population Sampling*. Chapman & Hall/CRC Monographs on Statistics & Applied Probability. Taylor & Francis.

Giraud, Timothée. 2022. "osrm: Interface Between R and the OpenStreetMap-Based Routing Service OSRM." *Journal of Open Source Software* 7 (78): 4574.

Gollob, Christoph, Tim Ritter, and Arne Nothdurft. 2020. "Forest Inventory with Long Range and High-Speed Personal Laser Scanning (PLS) and Simultaneous Localization and Mapping (SLAM) Technology." *Remote Sensing* 12 (9): 1509. https://doi.org/10.3390/rs12091509.

Gove, Jeffrey H, Timothy G Gregoire, Mark J Ducey, and Thomas B Lynch. 2020. "A Note on the Estimation of Variance for Big BAF Sampling." *Forest Ecosystems* 7 (1): 62.

Grafström, Anton, Niklas L. P. Lundström, and Lina Schelin. 2012. "Spatially Balanced Sampling Through the Pivotal Method." *Biometrics* 68 (2): 514–20.

Grafström, Anton, and Wilmer Prentius. 2024. *BalancedSampling: Balanced and Spatially Balanced Sampling*. https://CRAN.R-project.org/package=BalancedSampling.

Grafström, Anton, Xin Zhao, Martin Nylander, and Hans Petersson. 2017. "A New Sampling Strategy for Forest Inventories Applied to the Temporary Clusters of the Swedish National Forest Inventory." *Canadian Journal of Forest Research* 47 (9): 1161–67.

Green, Edwin .J., Andrew O. Finley, and William E. Strawderman. 2020. *Introduction to Bayesian Methods in Ecology and Natural Resources*. Springer.

Gregoire, Timothy .G., and Harry T. Valentine. 2008. *Sampling Strategies for Natural Resources and the Environment*. Chapman & Hall/CRC Applied Environmental Statistics. Taylor & Francis.

Grosenbaugh, L. R. 1952. "Plotless Timber Estimates–New, Fast, Easy." *Journal of Forestry* 50: 32–37.

Grosenbaugh, L. R. 1958. "Point-Sampling and Line-Sampling Probability Theory, Geometric Implications, Synthesis." In *USDA Forest Service, Southern Forest Experiment Station, Occasional Paper 160*.

Grosenbaugh, L. R. 1965. *Three-Pee Sampling Theory and Program 'THRP' for Computer Generation of Selection Criteria*. U.s. Forest Service Research Paper PSW. Pacific Southwest Forest; Range Experiment Station, Forest Service, U.S. Department of Agriculture.

Grosenbaugh, L. R., and W. S. Stover. 1957. "Point-Sampling Compared with Plot-Sampling in Southeast Texas." *Forest Science* 3 (1): 2–14.

Hanks, Leland F. 1976. "Hardwood Tree Grades for Factory Lumber." In *Res. Pap. NE-333 Upper Darby, PA: U.S. Department of Agriculture, Forest Service, Northeastern Forest Experiment Station. 81 p.*

Heikkinen, Juha. 2006. "Assessment of Uncertainity in Spatially Systematic Sampling." In *Forest Inventory: Methodology and Applications*, edited by Annika Kangas and Matti Maltamo. Springer Netherlands.

Heikkinen, Juha, Helena M. Henttonen, Matti Katila, and Sakari Tuominen. 2025. "Stratified, Spatially Balanced Cluster Sampling for Cost-Efficient Environmental Surveys." *Environmetrics* 36 (5): e70019.

Hendrey, George R, David S Ellsworth, Keith F Lewin, and John Nagy. 1999. "A Free-Air Enrichment System for Exposing Tall Forest Vegetation to Elevated Atmospheric CO2." *Global Change Biology* 5 (3): 293–309.

Hogg, Robert V., Elliot A. Tanis, and Dale L. Zimmerman. 2015. *Probability and Statistical Inference*. 9th ed. Pearson Education.

Honer, T. G. 1967. "Standard Volume Tables and Merchantable Conversion Factors for the Commercial Tree Species of Central and Eastern Canada." In *Can. Dept. Forestry Rural Devel., For. Mgmt. Res. And Serv. Inst. Info. Rep. FMR-X-5*.

IEEE. 2008. "IEEE Standard for Floating-Point Arithmetic." *IEEE Std 754-2008*, 1–70.

Iles, Kim. 2003. *A Sampler of Inventory Topics: A Practical Discussion for Resource Samplers, Concentrating on Forest Inventory Techniques*. Kim Iles & Associates.

Iles, Kim, and Mike Fall. 1988. "Can an Angle Gauge Really Evaluate "borderline trees" Accurately in Variable Plot Sampling?" *Canadian Journal of Forest Research* 18 (6): 776–83.

Iles, Kim, and William H. Wilson. 1988. "Changing Angle Gauges in Variable Plot Sampling: Is There a Bias Under Ordinary Conditions?" *Canadian Journal of Forest Research* 18 (6): 770–75.

Itter, Malcolm S., Andrew O. Finley, Anthony W. D'Amato, Jane R. Foster, and John B. Bradford. 2017. "Variable Effects of Climate on Forest Growth in Relation to Climate Extremes, Disturbance, and Forest Dynamics." *Ecological Applications* 27 (4): 1082–95.

Jenkins, Jennifer C., David C. Chojnacky, Linda S. Heath, and Richard A. Birdsey. 2002. "Comprehensive Database of Diameter-Based Biomass Regressions for North American Tree Species." In *Gen. Tech. Rep. NE-319. Newtown Square, PA: U.S. Department of Agriculture, Forest Service, Northeastern Research Station*.

Jenkins, Jennifer C., David C. Chojnacky, Linda S. Heath, and Richard A. Birdsey. 2003. "National-Scale Biomass Estimators for United States Tree Species." *Forest Science* 49 (1): 12–35.

Johnson, Evert W. 2000. *Forest Sampling Desk Reference*. CRC Press.

Johnson, Richard. 1969. *Grammatical Commentaries, 1706*. English Linguistics 1500-1800. Scolar P.

Jones, Matthew, Margaret O'Brien, Bryce Mecum, et al. 2019. *Ecological Metadata Language Version 2.2.0*. KNB Data Repository.

Kangas, Annika, and Matti Maltamo. 2006. *Forest Inventory: Methodology and Applications*. Vol. 10. Springer Netherlands.

Karnosky, David F., John M. Skelly, Kevin E. Percy, and Art H. Chappelka. 2007. "Perspectives Regarding 50 Years of Research on Effects of Tropospheric Ozone Air Pollution on US Forests." *Environmental Pollution* 147 (3): 489–506.

Kenefic, Laura S., Jeanette A. Allogio, Kathryn M. Gerndt, Rachel A. Knapp, Nicole S. Rogers, and Joshua J. Puhlick. 2025. *Overstory Tree and Regeneration Data from the "Silvicultural Effects on Composition, Structure, and Growth" Study at Penobscot Experimental Forest.* 3rd ed. Fort Collins, CO: Forest Service Research Data Archive.

Kershaw, John A., Mark J. Ducey, Thomas W. Beers, and Bertram Husch. 2016. *Forest Mensuration*. Wiley.

Köhl, Michael, Steen Magnussen, and Marco Marchetti. 2006. *Sampling Methods, Remote Sensing and GIS Multiresource Forest Inventory*. Springer.

Kubiske, M. E., V. S. Quinn, P. E. Marquardt, and D. F. Karnosky. 2007. "Effects of Elevated Atmospheric CO2 and/or O3 on Intra- and Interspecific Competitive Ability of Aspen." *Plant Biology* 9 (2): 342–55.

Kubiske, Mark E. 2013. *Tree Height and Diameter Data from the Aspen FACE Experiment, 1997-2008*. USDA Forest Service, Northern Research Station, Newtown Square, PA.

Laar, A. van, and A. Akça. 2007. *Forest Mensuration*. Managing Forest Ecosystems. Springer Netherlands.

Little, Roderick J. 2004. "To Model or Not to Model? Competing Modes of Inference for Finite Population Sampling." *Journal of the American Statistical Association* 99 (466): 546–56.

Lohr, Sharon L. 2019. *Sampling: Design and Analysis*. Chapman & Hall/CRC Texts in Statistical Science. CRC Press.

Lohr, Steve. 2014. "For Big-Data Scientists, 'Janitor Work' Is Key Hurdle to Insights." In *The New York Times*. Https://www.nytimes.com/2014/08/18/technology/for-big-data-scientists-hurdle-to-insights-is-janitor-work.html.

Luoma, Ville, Ninni Saarinen, Michael A. Wulder, et al. 2017. "Assessing Precision in Conventional Field Measurements of Individual Tree Attributes." *Forests* 8 (2).

Magnussen, Steen, Ronald E. McRoberts, Johannes Breidenbach, et al. 2020. "Comparison of Estimators of Variance for Forest Inventories with Systematic Sampling - Results from Artificial Populations." *Forest Ecosystems* 7 (1): 17.

Mandallaz, D. 2007. *Sampling Techniques for Forest Inventories*. Chapman & Hall/CRC Applied Environmental Statistics. CRC Press.

Marshall, David D, Kim Iles, and John F Bell. 2004. "Using a Large-Angle Gauge to Select Trees for Measurement in Variable Plot Sampling." *Canadian Journal of Forest Research* 34 (4): 840–45.

Matérn, Bertil. 1947. *Methods of Estimating the Accuracy of Line- and Sample-plot Surveys*. Meddelanden från Statens Skogsforskningsinstitut, **36**(1).

McConville, Kelly, and Isabelle Caldwell. 2020. *pdxTrees: Data Package of Portland, Oregon Trees*. https://CRAN.R-project.org/package=pdxTrees.

McLaughlin, S. B., M. Nosal, S. D. Wullschleger, and G. Sun. 2007. "Interactive Effects of Ozone and Climate on Tree Growth and Water Use in a Southern Appalachian Forest in the USA." *New Phytologist* 174 (1): 109–24.

Mehtätalo, Lauri, Adil Yazigi, Kasper Kansanen, et al. 2022. "Estimation of Forest Stand Characteristics Using Individual Tree Detection, Stochastic Geometry and a Sequential Spatial Point Process Model." *International Journal of Applied Earth Observation and Geoinformation* 112: 102920.

Mehtätalo, L., and J. Lappi. 2020. *Biometry for Forestry and Enviornmental Data: With Examples in R*. Chapman & Hall/CRC.

Mesavage, Clement, and James W. Girard. 1956. *Tables for Estimating Board-Foot Volume of Timber*. U.S. Government Printing Office.

Næsset, E. 2011. "Estimating Above-Ground Biomass in Young Forests with Airborne Laser Scanning." *International Journal of Remote Sensing* 32 (2): 473–501.

Neigh, Christopher S. R., Ross F. Nelson, K. Jon Ranson, et al. 2013. "Taking Stock of Circumboreal Forest Carbon with Ground Measurements, Airborne and Spaceborne LiDAR." *Remote Sensing of Environment* 137: 274–87.

Nigh, Gordon D. 2016. *Total and Merchantable Volume Equations for Common Tree Species in British Columbia: By Region and Biogeoclimactic Zone*. Prov. B.C., Victoria, B.C. Tech. Rep. 106.

Nothdurft, Arne, Valentin Sarkleti, Tobias Ofner-Graff, et al. 2025. "Small Area Estimation of Growing Stock Timber Volume, Basal Area, Mean Stem Diameter, and Stem Density for Mountain Forests in Austria." *Canadian Journal of Forest Research* 55: 1–20. https://doi.org/10.1139/cjfr-2024-0302.

Nowak, David J. 2021. *Understanding i-Tree: 2021 Summary of Programs and Methods*. U.S. Department of Agriculture, Forest Service, Northern Research Station.

Oderwald, Richard G. 1981. "Point and Plot Sampling–the Relationship." *Journal of Forestry* 79 (6): 377–78.

OED. 2023. *The Oxford English Dictionary*. Oxford University Press.

Omule, Stephen A. Y. 1980. "Personal Bias in Forest Measurements." *The Forestry Chronicle* 56 (5): 222–24.

Orwig, David A., Jonathan R. Thompson, Nicholas A. Povak, Megan Manner, Donald Niebyl, and David R. Foster. 2012. "A Foundation Tree at the Precipice: Tsuga Canadensis Health After the Arrival of Adelges Tsugae in Central New England." *Ecosphere* 3 (1): art10.

Orwig, David, David Foster, and Aaron Ellison. 2015. *Harvard Forest CTFS-ForestGEO Mapped Forest Plot Since 2014*. Harvard Forest Data Archive: HF253.

Palley, Marshall N., and Leah G. Horwitz. 1961. "Properties of Some Random and Systematic Point Sampling Estimators." *Forest Science* 7 (1): 52–65.

Pebesma, Edzer, and Roger Bivand. 2023. *Spatial Data Science: With Applications in R*. Chapman Hall/CRC.

Pedersen, Thomas Lin. 2024. *patchwork: The Composer of Plots*. https://CRAN.R-project.org/package=patchwork.

Pezoa, Felipe, Juan L Reutter, Fernando Suarez, Martín Ugarte, and Domagoj Vrgoč. 2016. "Foundations of JSON Schema." *Proceedings of the 25th International Conference on World Wide Web*, 263–73.

Posit team. 2025. *RStudio: Integrated Development Environment for r*. Posit Software, PBC. http://www.posit.co/.

Powers, Stephen M., and Stephanie E. Hampton. 2019. "Open Science, Reproducibility, and Transparency in Ecology." *Ecological Applications* 29 (1): e01822. https://doi.org/10.1002/eap.1822.

Python Core Team. 2015. *Python: A Dynamic, Open Source Programming Language*. Python Software Foundation. https://www.python.org/.

R Core Team. 2024. *R: A Language and Environment for Statistical Computing*. R Foundation for Statistical Computing. https://www.R-project.org/.

R Core Team. 2025. *Foreign: Read Data Stored by 'Minitab', 's', 'SAS', 'SPSS', 'Stata', 'Systat', 'Weka', 'dBase', ...* https://CRAN.R-project.org/package=foreign.

Radtke, Philip J., David M. Walker, Aaron R. Weiskittel, Jereme Frank, John W. Coulston, and James A. Westfall. 2015. *Legacy Tree Data: A National Database of Detailed Tree Measurements for Volume, Weight, and Physical Properties*. Gen. Tech. Rep. PNW-GTR-931. Portland, OR: US Department of Agriculture, Forest Service, Pacific Northwest Research Station: 25-30. (Vol. 931, pp. 25-30).

Ranneby, Bo, Torbjörn Cruse, Bengt Hägglund, Hans Jonasson, and Jan Swärd. 1987. "Designing a New National Forest Survey for Sweden." *Studia Forestalia Suecica* 177.

Rao, J. N. K., and Isabel Molina. 2015. *Small Area Estimation*. Wiley Series in Survey Methodology. Wiley.

Räty, Minna, Mikko Kuronen, Mari Myllymäki, Annika Kangas, Kai Mäkisara, and Juha Heikkinen. 2020. "Comparison of the Local Pivotal Method and Systematic Sampling for National Forest Inventories." *Forest Ecosystems* 7 (1): 54.

Robinson, Andrew P., and Jeff D. Hamann. 2011. *Forest Analytics with R: An Introduction*. Springer.

Roesch, Francis A., Edwin J. Green, and Charles T. Scott. 1993. "An Alternative View of Forest Sampling." *Survey Methodology*, 199–204.

Sandve, Geir Kjetil, Anton Nekrutenko, James Taylor, and Eivind Hovig. 2013. "Ten Simple Rules for Reproducible Computational Research." *PLoS Computational Biology* 9 (10): e1003285.

Särndal, C. E., B. Swensson, and J. H. Wretman. 1992. *Model Assisted Survey Sampling*. Springer Series in Statistics. Springer-Verlag.

Schmid, Paul. 1969. "Stichproben Am Waldrand." *Eidgenössische Anstalt für Das Forstliche Versuchswesen* 45 (3): 234–303.

Schreuder, Hans T., Ernst Richard, and Hugo Ramirez-Maldonado. 2004. "Statistical Techniques for Sampling and Monitoring Natural Resources." In *Gen. Tech. Rep. RMRS-GTR-126. Fort Collins, CO: U.S. Department of Agriculture, Forest Service, Rocky Mountain Research Station. 111 p.*

Shiver, Barry D., and Bruce E. Borders. 1995. *Sampling Techniques for Forest Resource Inventory*. Wiley.

Thompson, Steven K. 2012. *Sampling*. 3rd ed. Wiley Series in Probability and Statistics. John Wiley & Sons.

Tomppo, Erkki. 2006. "The Finnish National Forest Inventory." In *Forest Inventory: Methodology and Applications*, edited by Annika Kangas and Matti Maltamo. Springer Netherlands.

Tomppo, Erkki, Thomas Gschwantner, Mark Lawrence, and Ronald E. McRoberts, eds. 2010. *National Forest Inventories: Pathways for Common Reporting*. Springer.

UNFCCC. 2015. *The Paris Agreement*. Https://unfccc.int/process-and-meetings/the-paris-agreement/the-paris-agreement.

UN-REDD. 2008. *Reducing Emissions from Deforestation and Forest Degradation Programme*. Http://www.un-redd.org.

US DOE. 2020. *US Department of Energy Free-Air CO2 Enrichment Experiments: FACE Results, Lessons, and Legacy*. US Department of Energy Office of Science. DOE/SC-0202.

USDA. 2006. *National Forest Log Scaling Handbook*. FSH; 2409.11 U.S. Department of Agriculture, Forest Service.

USDA. 2021. *The PLANTS Database*. U.S. Department of Agriculture, Natural Resources Conservation Service, National Plant Data Team, Greensboro, NC, USA.

USFS. 1996. *Forest Service Handbook, Washington: FSH 2409.12, Timber Cruising Handbook*. Washington, DC: The Forest Service.

Valentine, Harry, Mark Ducey, Jeffrey Gove, Adrian Lanz, and David Affleck. 2006. "Corrections for Cluster-Plot Slop." *Forest Science* 52: 55–66.

Ventura, Jonathan, Camille Pawlak, Milo Honsberger, et al. 2024. "Individual Tree Detection in Large-Scale Urban Environments Using High-Resolution Multispectral Imagery." *International Journal of Applied Earth Observation and Geoinformation* 130: 103848.

Waskiewicz, Justin D., Laura S. Kenefic, Nicole S. Rogers, Joshua J. Puhlick, John C. Brissette, and Richard J. Dionne. 2015. "Sampling and Measurement Protocols for Long-Term Silvicultural Studies on the Penobscot Experimental Forest." In *Gen. Tech. Rep. NRS-147. Newtown Square, PA: U.S. Department of Agriculture, Forest Service, Northern Research Station*. 32 p.

Weiskittel, Aaron R., David W. Hann, John A. Kershaw, and Jerome K. Vanclay. 2011. *Forest Growth and Yield Modeling*. Wiley.

Welch, B. L. 1947. "The Generalization of Student's Problem When Several Different Population Variances Are Involved." *Biometrika* 34 (1-2): 28–35.

Wensel, Lee C., Jack Levitan, and Klaus Barber. 1980. "Selection of Basal Area Factor in Point Sampling." *Journal of Forestry* 78 (2): 83–84.

West, Philip. 2011. "Potential for Wider Application of 3P Sampling in Forest Inventory." *Canadian Journal of Forest Research* 41 (July): 1500–1508.

Westfall, James A., John W. Coulston, Andrew N. Gray, et al. 2024. *A National-Scale Tree Volume, Biomass, and Carbon Modeling System for the United States*. General Technical Report WO-104. U.S. Department of Agriculture, Forest Service, Northern Research Station.

Wiant, Harry V. 1986. "Computer Corner: Formulas for Mesavage and Girard's Volume Tables." *Northern Journal of Applied Forestry* 3 (3): 124. https://doi.org/10.1093/njaf/3.3.124.

Wickham, Hadley. 2010. "A Layered Grammar of Graphics." *Journal of Computational and Graphical Statistics* 19 (1): 3–28.

Wickham, Hadley. 2019. *Advanced R*. CRC Press.

Wickham, Hadley, Mine Çetinkaya-Rundel, and Garrett Grolemund. 2023. *R for Data Science: Import, Tidy, Transform, Visualize, and Model Data*. 2nd ed. O'Reilly Media, Inc.

Wickham, Hadley, Evan Miller, Danny Smith, and PBC Posit Software. 2025. *haven: Import and Export 'SPSS', 'Stata' and 'SAS' Files*. https://CRAN.R-project.org/package=haven.

Wickham, Hadley, Danielle Navarro, and Thomas Lin Pedersen. 2023. *ggplot2: Elegant Graphics for Data Analysis*. 3rd ed. Springer New York.

Wilke, Claus O. 2024. *cowplot: Streamlined Plot Theme and Plot Annotations for 'Ggplot2'*. https://CRAN.R-project.org/package=cowplot.

Wilkinson, Leland. 2005. *The Grammar of Graphics*. 2nd ed. Statistics and Computing. Springer New York.

Wilson, Greg, Jennifer Bryan, Karen Cranston, Justin Kitzes, Lex Nederbragt, and Tracy K Teal. 2017. "Good Enough Practices in Scientific Computing." *PLoS Computational Biology* 13 (6): e1005510.

Wood, Frederica, James N. Kochenderfer, and Mary B. Adams. 2016. *Felled Tree Biomass for Four Hardwood Species of the Central Appalachians, West Virginia*. Fort Collins, CO: Forest Service Research Data Archive.

Young, H. E. 1957. *Additional Volume Tables for Maine*. Maine Agricultural Experiment Station Miscellaneous Publication 627.

Index

R Code Index

For Product Safety Concerns and Information please contact our EU
representative GPSR@taylorandfrancis.com
Taylor & Francis Verlag GmbH, Kaufingerstraße 24, 80331 München, Germany

www.ingramcontent.com/pod-product-compliance
Lightning Source LLC
LaVergne TN
LVHW081311190926
848399LV00036B/366

9781032120270